T0329854

METEOROLOGICAL MONOGRAPHS

VOLUME 33   NOVEMBER 2008   NUMBER 55

SYNOPTIC–DYNAMIC METEOROLOGY AND WEATHER ANALYSIS AND FORECASTING: A TRIBUTE TO FRED SANDERS

Edited by

Lance F. Bosart
Howard B. Bluestein

American Meteorological Society
45 Beacon Street, Boston, Massachusetts 02108

ISBN 978-1-878220-84-4
ISSN 0065-9401

Published by the American Meteorological Society
45 Beacon St., Boston, MA 02108

For a catalog of AMS Books, see www.ametsoc.org/pubs/books.
To order, call (617) 227-2426, extension 686, or
e-mail amsorder@ametsoc.org.

Printed in the United States of America
by Allen Press, Inc., Lawrence, KS

DEDICATION

There was once an MIT professor named Fred
Who in his leisure time sailed out of Marblehead
He expounded on the weather
To a bevy of those who on the 16th floor of the Green Building he kept on a
    tether
Wondering about the next front, the next storm, and the Logan ob the next
    morning at 12 "zed"

—Howie "Cb" Bluestein

This monograph is dedicated to Fred Sanders (AKA "Olde Dad"), shown here in full foul weather gear, aboard his 39-foot sailboat, the *Stillwater* (not aptly named, for Fred and his sailboat cruised in water that was far from still), summer 1973. Many of Fred's students spent memorable afternoons sailing with him out of Marblehead, MA. Courtesy of Howie Bluestein.

TABLE OF CONTENTS

FOREWORD

The Fred Sanders Legacy in Synoptic–Dynamic Meteorology and Weather Analysis and Forecasting

---

The field of synoptic meteorology, which seeks to understand weather systems such as fronts and cyclones by careful analysis and interpretation of weather observations, was much influenced by the late Frederick Sanders, emeritus professor of meteorology at the Massachusetts Institute of Technology (MIT). Professor Sanders, universally known as "Fred," made important contributions to the analysis, understanding, and prediction of fronts, extratropical cyclones, hurricanes, squall lines and other warm-season convective weather systems, and flood-producing storms. He coined the term "bomb" to describe explosively intensifying winter storms. His classic oft-cited 1955 paper on an investigation of the structure and dynamics of an intense frontal zone established the critical role of low-level horizontal confluence and convergence in leading to "frontal collapse" in intense fronts. He also invented the field of oceanic mesometeorology, applying careful analysis of meteorological conditions experienced by a fleet of sailboats participating in races from Newport, Rhode Island, to Bermuda.

Born in Detroit on 17 May 1923, Fred was the eldest of the three children of Frederick William and Dorothy Martin Sanders. After spending much of his childhood in Bloomfield Hills, Michigan, Fred attended Amherst College, where he studied mathematics, economics, and music. This was during World War II, and the Army Air Corps was determined to train 10,000 weather forecasters, canvassing colleges and universities for students who were studying math and physics. Fred signed up around Christmas 1941 and was sent for infantry basic training at Jefferson Barracks outside St. Louis, thence to MIT for six months of intensive study in math and physics, followed by nine months of meteorology.

Fred graduated as a second lieutenant shortly after D-Day in Normandy, and requested and was granted assignment to Greenland, where he discovered that flight crews keenly valued his forecasting skills. On return to the United States at the end of the war, he worked briefly as an air inspector at Headquarters Eighth Weather Group at Grenier Air Force Base in New Hampshire.

Just before separating from the military, Fred met Nancy Brown, whom he married in 1946. In the same year, he decided to become a professional weather forecaster rather than to join his father's candy manufac-

turing business. He became a transatlantic aviation forecaster for the U.S. Weather Bureau at La Guardia Field, but after two years returned to MIT as a graduate student under the G.I. Bill, intending to get a master's degree and return to the Weather Bureau as a researcher. But he was persuaded by his MIT mentors to enter the doctoral program, and he earned an Sc.D. degree in 1954 under the guidance of Thomas Malone, after which he joined the faculty of the MIT Department of Meteorology, where he remained until his retirement in 1984.

In the 1960s, Fred began to think about the importance of forecast verification studies as one key to unlocking some of the scientific mysteries of the atmosphere. His 1963 paper on subjective probability forecasting demonstrated the scientific insights that could be obtained by rigorously and quantitatively evaluating the skill of daily weather forecasts and applying the knowledge gained to improve weather forecasts. His 1975 paper (with Pike and Gaertner) reported on one of the first successful computer models based on the barotropic vorticity equation (SANBAR) that was used for operational hurricane track forecasting. In 1980, Fred published two seminal papers. The first paper (with Miller) analyzed the mesoscale conditions associated with the jumbo tornado outbreak of 3–4 April 1974, from which it was deduced that the many tornadoes in that event tended to cluster in three main bands that possessed distinctive atmospheric structure. The second paper (with Gyakum) shed light for the first time on the systematic distribution and structure of explosively deepening oceanic cyclones (aka bombs) in the Northern Hemisphere and resulted in an avalanche of research papers on this topic over the next 10–20 years.

In the 1980s, Fred turned his attention to individual case studies of warm-season convective weather systems and cold-season winter storms with a particular emphasis on mesoscale structures embedded within these weather systems; many publications resulted from this effort. In 1988, Fred published a seminal paper on the life history of mobile troughs in the upper westerlies. In this paper, Fred established that preferred regions for 500-hPa trough genesis events in the Northern Hemisphere occurred over and downstream of major north–south-oriented mountain barriers such as the Rockies, while 500-hPa trough lysis events occurred preferentially over the eastern-ocean basins. A critical aspect of this paper was Fred's demonstration that upper-level trough genesis typically occurred well upstream and prior to surface cyclogenesis in the oceanic storm-track entrance regions immediately adjacent to the east coasts of North America and Asia rather than simultaneously with surface cyclone development, as postulated in theories of classical baroclinic instability.

In the 1990s, Fred became interested in assessing the skill of operational dynamical models in predicting oceanic cyclogenesis as typified by his 1992 paper that showed that skills were improving, and he returned to his roots with the publication of a paper in 1995 (with Doswell) on the case for detailed surface analyses. The latter paper, and several others on the same topic, were motivated by what Fred saw as a need to arrest a perceived decline in the quality of operational surface frontal analyses by calling attention to the importance of these analyses to understanding observed weather systems, and by new science opportunities that could be uncovered from careful mesoscale analyses of the life cycles of surface fronts. Fred argued that real surface fronts should be defined on the basis of the magnitude of the observed surface potential temperature gradient. In his last paper, published in 2005, he applied a potential temperature gradient criterion to distinguish between what he called "real fronts" (surface boundaries characterized by significant potential temperature differences) and "baroclinic troughs" (surface boundaries marked by wind shifts but little or no potential temperature contrast).

Fred's strong interest in weather analysis and forecasting enabled him to pioneer methods for evaluating the skill of both human and computer weather forecasts, stressing the need for quantifying the uncertainty of the forecasts; this work also led to improvements in numerical weather prediction models and his demonstration that a consensus forecast made up of a group of equally skilled forecasters would usually beat individual forecasters in the group over the long haul. Fred watched the weather every day and impressed his students by what he could "see" on the many weather maps that were posted in the 16th-floor hallway of the Green Building at MIT. There was always a "story to be told" by the weather maps. The story was different every day, but the overall theme never changed. The story was always about physics, dynamics, thermodynamics, and new scientific insights that could be gleaned from synthesizing these processes and applying them in real-time weather analysis and forecasting. In addition to the scientific story, there was also some psychological drama when he sauntered past the weather maps, behind students preparing their forecasts, and muttered under his breath, "You don't suppose that . . . " The bulk of the forecasters, reaching for their erasers, would hurriedly amend portions of their forecasts. With Fred, learning was made fun.

But the story didn't end there. Fred also taught his students the importance of transferring the scientific knowledge gained from studying the weather to operations, to the benefit of weather forecasters—and ultimately the general public. Fred used the daily weather forecasting contest at MIT to teach his students about how the atmosphere worked. At stake were a prize cigar and the potential for prospective thesis topics. Although his students worked long and hard to try to beat him during the semester-length forecasting contests, when the bell rang at the end of the semester more often than not Fred was at the top of the heap. New scientific ideas and insights about the workings of the atmosphere were continually put on the table during these map discussions, which proved to be the highlight of the day for many students. Together with his colleague, Richard Reed, Fred elevated the field of synoptic meteorology to the status of a respected science, to the benefit of the field and to generations of students. He was the recipient of many awards, and was a fellow of the AMS as well as the American Association for the Advancement of Science.

While mixing very well with students and maintaining an air of informality, Fred held them to very high standards. He constantly challenged them and made them think very deeply about scientific issues. Fred was much beloved and esteemed not only by the many students he mentored, but also by the colleagues he worked with, so much so that, in 2004, the AMS held a scientific colloquium in his honor. Most of those who today teach and do research in synoptic meteorology have profited directly from Fred's guidance. In Fred's words, "My career was heavily weighted toward teaching, in which I enjoyed sharing the enthusiasm I felt for weather analysis and forecasting." That Fred's publication rate in the refereed scientific literature went up after he retired from MIT in 1984 is testimony to the considerable amount of his time that he had invested in teaching. Steve Mullen, a scientific collaborator in recent years, has noted that "There are few people in the history of the field who have trained and mentored as many outstanding meteorologists as Fred; his legacy in terms of his offspring is just legendary."

That many of the authors and coauthors of articles appearing in this monograph are Fred's academic "children" and "grandchildren," and that other authors and coauthors are close Sanders "family" members, attests to his enduring legacy as a mentor, colleague, and educator.

Fred's first Ph.D. student, Lance Bosart, and Lance's former Ph.D. students Alicia Wasula, Eric Hoffman, and David Schultz (co-advised with Daniel Keyser); his present Ph.D. student, Tom Galarneau; and his former M.S. students Greg Hakim (Dan Keyser was Greg's Ph.D. advisor) and Keith Meier are represented in this monograph. Fred's Ph.D. student Howie Bluestein and Howie's former Ph.D. student Christopher Weiss are also represented, as are Fred's former Ph.D. students Bob Burpee, Randy Dole, John Gyakum, and Steve

Tracton, and Fred's former M.S. student Paul Roebber. Finally, Ryan Torn, who was Greg Hakim's Ph.D. student, making him Fred's academic "great grandchild," is also represented in this monograph.

At MIT, Kerry Emanuel interacted with and was influenced by Fred Sanders; Kerry's Ph.D. student, John Nielsen-Gammon, and John's Ph.D. student Dave Gold have also made contributions to this monograph. Ed Kessler, the first director of the National Severe Storms Laboratory, interacted with Fred at MIT beginning in the 1950s and continuing up until Fred's passing in October 2006.

Fred joined the MIT faculty at a time when federal funding of scientific research was ramping up rapidly in response to the challenge posed by the U.S.S.R.'s Sputnik satellite. To his great credit, Fred resisted the sea change toward research and away from teaching that affected most premier institutions of higher education, preferring to spend most of his time preparing lectures and interacting with students. Consequently, he became a greatly beloved professor and mentor, and has had a large influence on his field not just through his own research, but through the carefully nurtured talent of his students. Fred could often be found tutoring lagging students over lunch, or taking entire classes for an outing on his sailing yacht, Stillwater, bringing joy as well as knowledge to the study of weather. Fred was also beloved by staff members at MIT, including Ann Corrigan, Ed Nelson (who took care of the map room and assisted students with their data needs), and Isabel Kole (who drafted all kinds of figures for Fred and his students). A common sight on the 16th floor of the Green Building at MIT was Fred huddling with Ann, Ed, Isabel, and various students surrounded by maps, teletype paper, and figures being drafted.

In later years, Fred maintained his friendship and scientific collaboration with many of his students. He was a frequent scientific visitor at the University of Arizona, and visited Norman, Oklahoma, almost yearly to storm chase. While he never did see a tornado, he completed numerous collaborative studies with colleagues at the National Severe Storms Laboratory and the Cooperative Institute for Mesoscale Meteorological Studies while there.

Fred was a passionate sailor and participated in many ocean races, including the Newport–Bermuda and Marblehead–Halifax races. He also loved to cruise the coast of Maine and the Canadian Maritimes with his family and friends, to whom he brought much pleasure. An accomplished tenor, he sang with the MIT Choral Society and more recently with the choir of the Old North Church in Marblehead.

The spirit of Fred Sanders is well captured in this remembrance by his friend and colleague, Ed Zipser: "I don't think we will ever see his equal—not just for his scientific insight, but his outgoing nature, his helpfulness, his sometimes acerbic wit, and without fail remaining the consummate gentleman at all times."

—Adapted from Lance Bosart, Howie Bluestein, and Kerry Emanuel, obituary of Frederick Sanders, *Bull. Amer. Meteor. Soc.*, **88**, 425–427

Fred Sanders' life was commemorated in the *Boston Globe* on 27 October 2006 (the article may be found online at www.boston.com) and by the MIT News Office on the same date (the article maybe found online at web.mit.edu/newsoffice/2006/obit-sanders.html).

ACKNOWLEDGMENTS

The publication of this monograph was made possible by the collective efforts of many people. The idea to create a monograph to honor Fred Sanders' many scientific, educational, and operational contributions to the atmospheric sciences originated during the planning for the *AMS Fred Sanders Symposium* by the "Gang of Four" (Howie Bluestein, Lance Bosart, Brad Colman, and Todd Glickman). The well-attended Sanders Symposium was held in Seattle, Washington, in January 2004 in conjunction with the *84th AMS Annual Meeting*.

We thank Peter Ray, AMS meteorological monographs series editor, for giving us the green light to proceed with this monograph. We would also like to express our great appreciation to AMS Executive Director, Keith Seitter, and AMS Publications Director, Ken Heideman, for their continuing support and encouragement as the monograph moved forward from a concept to reality. Special thanks also go to Sarah Jane Shangraw at the AMS and Celeste Iovinella at the University at Albany, SUNY, for expertly managing all the details and the technical issues with a firm hand on the tiller that otherwise would have overwhelmed us.

We also thank the reviewers of the articles: Robert Black, Warren Blier, Fred Carr, Brian Colle, N. Andrew Crook, Chris Davis, David Dowell, Tom Hamill, Richard H. Johnson, Daniel Keyser, Steve Koch, Paul Kocin, T. N. Krishnamurti, Gary Lackmann, Chris Landsea, Tony Lupo, Jonathan E. Martin, Frank Marks, Clifford F. Mass, Steve Mudrick, Steve Mullen, John Nielsen-Gammon, Fred Sanders, David Schultz, Jim Steenburgh, and Roger Wakimoto.

Special thanks also go to the generations of Fred's students (and their students) as well as his many friends and colleagues who enthusiastically supported the *Sanders Symposium* and the idea to produce a Sanders monograph. We also greatly appreciate the continuing support and interest of Fred's beloved wife, Nancy Sanders, and his children, grandchildren, and other family members as the Sanders monograph moved forward from a concept to reality.

—Lance F. Bosart and Howard B. Bluestein

CONTRIBUTORS

ANANTHA R. AIYYER
Department of Marine, Earth, and Atmospheric
    Sciences
North Carolina State University
2800 Faucette Dr.
Raleigh, NC 27695
E-mail: afractal@gmail.com

MICHAEL BAKER
NOAA/NWS/MDL
W/OST22
1325 East West Hwy.
Sta. 11316
Silver Springs, MD 20910
E-mail: michael.n.baker@noaa.gov

HOWARD B. BLUESTEIN
School of Meteorology
University of Oklahoma
120 David L. Boren Blvd., Suite 5900
Norman, OK 73072
E-mail: hblue@ou.edu

LANCE F. BOSART
Department of Earth and Atmospheric Sciences
University of Albany, State University of New York
1400 Washington Ave.
Albany, NY 12222
E-mail: bosart@atmos.albany.edu

ROBERT W. BURPEE[1]
Cooperative Institute for Marine and Atmospheric
    Studies
4600 Rickenbacker Causeway
Miami, FL 33149

RANDALL M. DOLE
NOAA/Earth System Research Laboratory
325 Broadway
Boulder, CO 80305
E-mail: randall.m.dole@noaa.gov

WALTER H. DRAG
NOAA/NWS
445 Myles Standish Blvd.
Taunton, MA 02780-1041
E-mail: wdrag111@comcast.net

KERRY EMANUEL
Program in Atmospheres, Oceans, and Climate
    (PAOC)
Rm. 54-1620
Massachusetts Institute of Technology
77 Massachusetts Avenue
Cambridge, MA 02139
E-mail: emanuel@texmex.mit.edu

THOMAS J. GALARNEAU JR.
Department of Earth and Atmospheric Sciences
University at Albany, State University of New York
ES-234, 1400 Washington Ave.
Albany, NY 12222
E-mail: tomjr@atmos.albany.edu

BART GEERTS
Department of Atmospheric Sciences
University of Wyoming
P. O. Box 3038
Laramie, WY 82071-3038
E-mail: geerts@uwyo.edu

DAVID A. GOLD
Department of Atmospheric Sciences
Texas A&M University
College Station, TX 77843-3150
E-mail: dr_david_gold@earthlink.net

JOHN R. GYAKUM
Department of Atmospheric and Oceanic Sciences
McGill University
805 Sherbrooke Street West
Montreal, QC H3A 2K6, Canada
E-mail: john.gyakum@mcgill.ca

GREGORY J. HAKIM
Department of Atmospheric Sciences
Box 351640
University of Washington
Seattle, WA 98195-1640
E-mail: hakim@atmos.washington.edu

ERIC G. HOFFMAN
Department of Chemical, Earth, Atmospheric, and
    Physical Sciences
Plymouth State University
313 Boyd Hall
Plymouth, NH 03264
E-mail: ehoffman@plymouth.edu

[1] Deceased. See Lance F. Bosart for correspondence.

EDWIN KESSLER
Department of Geography, and School of Meteorology
University of Oklahoma
Norman, OK 73072-6337
E-mail: kess3@swbell.net

ROBERT KISTLER
NOAA/NWS/NCEP
Ocean Prediction Center
5200 Auth Road
Camp Springs, MD 20746
E-mail: rek067@gmail.com

PAUL J. KOCIN
NOAA/NWS/NCEP
Hydrometeorological Prediction Center
5200 Auth Road
Camp Springs, MD 20746
E-mail: pkocin@aol.com

KEITH W. MEIER
National Weather Service
2170 Overland Ave.
Billings, MT 59102-6455
E-mail: keith.meier@noaa.gov

JOHN W. NIELSEN-GAMMON
Department of Atmospheric Sciences
Texas A&M University
3150 TAMUS
College Station, TX 77843-3150
E-mail: n-g@tamu.edu

ANDREW L. PAZMANY
ProSensing, Inc.
107 Sunderland Rd.
Amherst, MA 01002-1117
E-mail: apazmany@yahoo.com

PAUL J. ROEBBER
Atmospheric Science Group
Department of Mathematical Sciences
University of Wisconsin—Milwaukee
P. O. Box 413
Milwaukee, WI 53201
E-mail: roebber@uwm.edu

DAVID M. SCHULTZ
Finnish Meteorological Institute
Erik Palmenin Aukio 1
P. O. Box 503
FI-00101 Helsinki, Finland
E-mail: david.schultz@fmi.fi.

JOSEPH SIENKIEWICZ
NOAA/NWS/NCEP
Ocean Prediction Center
5200 Auth Road
Camp Springs, MD 20746
E-mail: joseph.sienkiewicz@noaa.gov

RYAN D. TORN
National Center for Atmospheric Research
Box 3000
Boulder, CO 80307
E-mail: torn@atmos.washington.edu

M. STEVEN TRACTON
Office of Naval Research
One Liberty Center
875 Randolph St., Suite 1425
Arlington, VA 22203
E-mail: s.tracton@hotmail.com

LOUIS W. UCCELLINI
NOAA/NWS/NCEP
5200 Auth Road
Camp Springs, MD 20726
E-mail: louis.uccellini@noaa.gov

ALICIA C. WASULA
Department of Earth and Atmospheric Sciences
University at Albany, State University of New York
1400 Washington Ave.
Albany, NY 12222
E-mail: alicia@atmos.albany.edu

CHRISTOPHER C. WEISS
Department of Geoscience
Texas Tech University
Box 42101
Lubbock, TX 79409
E-mail: chris.weiss@ttu.edu

INTRODUCTION TO PARTS I AND II

By Howard B. Bluestein

Fred Sanders published research in many areas of meteorology. In this volume contributions from current studies and from reviews are included on the observations and theory of surface fronts and other surface boundaries, the techniques used for the analysis and diagnosis of observational and model data, weather forecasting, and climatology-related issues. The contributions in the volume by no means represent all of Fred's interests, but perhaps best showcase how his mentorship has inspired several generations of students to continue to make progress in areas in which he has made contributions. The papers presented herein can be used to augment graduate courses in synoptic and mesoscale meteorology.

In chapter 1, Howard Bluestein reviews the characteristics of surface boundaries in the Southern Plains and discusses their role in the initiation of convective storms. This paper updates the seminal work Fred did in the 1950s and continued on with applications to convective systems. Fred visited Oklahoma many times to chase storms and try to forecast them. He collaborated with scientists at the National Severe Storms Laboratory on convection-related issues. Although Fred's early interests were on fronts in the central United States, he later considered fronts along the east coast. Lance Bosart, Alicia Wasula, Walt Drag, and Keith Meier discuss the characteristics and dynamics of strong surface fronts over sloping terrain and the coastal plains. Fred had a keen interest in surface fronts that interacted with the mountains in New England and the Appalachians in the Carolinas. Kerry Emanuel has contributed a provocative essay on the strengths of the Norwegian School of cyclones and fronts. It is interesting to note how certain modes of thought are in favor, then fall out favor, and then return again. Ed Kessler discusses small-scale observations of frontal passages on his farm in Oklahoma and how they conform and don't conform to conven-

tional ideas. Dave Schultz presents a review of Fred's work on surface cold fronts and updates his work using contemporary observations. Dave Schultz and Paul Roebber then discuss a numerical simulation using a state-of-the art mesoscale model of the front analyzed by Fred back in the mid-1950s; Fred's work has since been accepted as a classic study. It is interesting to see how significant aspects of his seminal work bear up under the scrutiny of modern model simulations.

In doing so many observational studies, Fred tackled the problems of how to assess data quality, analyze the data, and make carefully thought out inferences that are tested against alternative explanations. In Part II there are papers on analysis and diagnostic techniques. Using ensemble model forecasts to produce analyses on the synoptic scale is the subject of a paper by Greg Hakim and Ryan Torn. In this paper, a technique for improving analyses of data by using a suite of numerical simulations to produce dynamically and statistically consistent analyses on the synoptic scale is detailed. Fred long advocated using analyses of surface potential temperature rather than analyses of temperature, especially when stations are not at the same elevation. Eric Hoffman discusses the implementation of Fred's ideas. Fred Sanders taught generations of students the intricacies of quasisgeostrophic theory and carefully applied it to the analysis of many types of cyclones and anticyclones. John Nielsen-Gammon and Dave Gold propose that the analysis of Ertel's potential vorticity has significant advantages to simple quasigeostrophic diagnosis. Fred was always interested in the analysis of new types of data. Chris Weiss, Howard Bluestein, and Andrew Pazmany describe the development of a new technique for analyzing the vertical circulation across drylines using data from a mobile Doppler radar. The authors focus on data collected during IHOP (International H$_2$O Project) in 2002.

Part I
Fronts and Surface Boundaries

Chapter 1

Surface Boundaries of the Southern Plains: Their Role in the Initiation of Convective Storms

HOWARD B. BLUESTEIN

School of Meteorology, University of Oklahoma, Norman, Oklahoma

(Manuscript received 28 January 2004, in final form 16 June 2006)

Bluestein

Our approach is to resist the traditional appeal to "indications" or "ingredients" as described in a historical account by Schaefer (1986), however successful it has been in development of the present modest predictive skill. We rely instead on a simple physical consideration: intense convection will occur provided that large convective available potential energy (CAPE) is present in the air column and provided that the typical negative area (CIN, or convective inhibition) below the level of free convection for surface air is somehow removed or reduced to a small value that can be overcome by random cloud-scale pulses at the top of the surface boundary layer.

— Sanders and Blanchard (1993)

ABSTRACT

The nature of the different types of surface boundaries that appear in the southern plains of the United States during the convectively active season is reviewed. The following boundaries are discussed: fronts, the dryline, troughs, and outflow boundaries. The boundaries are related to their environment and to local topography. The role these boundaries might play in the initiation of convective storms is emphasized. The various types of boundary-related vertical circulations and their dynamics are discussed. In particular, quasigeostrophic and semigeostrophic dynamics, and the dynamics of solenoidal circulations, density currents, boundary layers, and gravity waves are considered.

Miscellaneous topics pertinent to convective storms and their relationship to surface boundaries such as along-the-boundary variability, boundary collisions, and the role of vertical shear are also discussed. Although some cases of storm initiation along surface boundaries have been well documented using research datasets collected during comprehensive field experiments, much of what we know is based only on empirical forecasting and nowcasting experience. It is suggested that many problems relating to convective-storm formation need to be explored in detail using real datasets with new observing systems and techniques, in conjunction with numerical simulation studies, and through climatological studies.

Corresponding author address: Prof. Howard B. Bluestein, School of Meterology, University of Oklahoma, 100 E. Boyd, Rm. 1310, Norman, OK 73019.
E-mail: hblue@ou.edu

1. Introduction

Fred Sanders has taught his students the importance of plotting surface-observation data and analyzing surface weather maps, especially using standard data from the operational network. He has shown that much can be learned about weather forecasting and the physical mechanisms driving weather systems by examining surface weather maps and by describing what actually happens in nature. The purposes of analyzing the maps are to critically assess how well existing conceptual models are valid and to formulate newer and more accurate conceptual models of the features analyzed on the maps. In recent years, numerical forecast models that have assimilated data on both large and small scales have attempted to predict the onset of convective storms (e.g., http://www.caps.ou.edu/forecasts.htm; http://wrf-model.ogr/index.php).

I have been inspired by watching Fred Sanders analyze sequences of hourly radar summaries and other data spread out on the 16th floor of the Green Building at the Massachusetts Institute of Technology (MIT), in an attempt to find order in the evolution of a convective system. I recall the tremendous excitement he instilled in us students on 3 April 1974 as he anxiously monitored one of the largest outbreaks of severe weather ever. In addition, I remember him engaging Ed Kessler, who was at the time at MIT and on leave from the National Severe Storms Laboratory (NSSL), in conversations on severe convection. Finally, I recall how he regaled me with stories of waterspouts seen by moonlight, while he was sailing.

For more than 25 yr, this student of Fred's has spent countless hours plotting and analyzing surface weather maps and, in the past several years, output from numerical forecast models, with the goal of being able to forecast the initiation of severe convective storms in the southern plains. The ultimate goal has been even more practical: being able to position storm-intercept vehicles near tornadoes and their parent convective storms for observation and probing by various types of instruments (Bluestein 1999). Fred Sanders joined our team on many occasions. He shocked some of the student participants by meticulously plotting and analyzing surface equivalent potential temperature, while the rest of us were fixated on the sky.

Much has been learned about the conditions under which convective storms form. It is generally acknowledged that storms that are rooted in the boundary layer (i.e., those whose main, buoyant updraft contains air that originated in the boundary layer) often are initiated along or near surface "boundaries" (e.g., Byers and Braham 1948; Rhea 1966; Gaza and Bosart 1985; Shapiro et al. 1985; Schaefer 1986; Wilson and Schreiber 1986; Dorian et al. 1988; Lee et al. 1991; Galway 1992), especially in the southern plains during the spring season. [In other instances, convection is not triggered along surface boundaries and not rooted in the boundary layer (e.g., Martin et al. 1995).] In this paper a boundary is defined as a line of discontinuity or a narrow zone separating air having distinctly different meteorological conditions. A boundary may separate two different air masses or might mark a shift in wind direction and/or speed across which there may or may not be a change

in air mass. Careful and detailed surface analyses of data are essential for recognizing and describing the boundaries on a weather map (Sanders and Doswell 1995).

The purpose of this chapter is to review what we have learned about the initiation of convective storms in the southern plains of the United States, during the convectively active season, and how initiation is related to surface boundaries. Because convective storms are observed most frequently during the spring months, from April to June, we will focus our attention only on these months. In the next section we review briefly the basic physics of storm formation. In section 3 the characteristics of the primary surface boundaries found in the southern plains are discussed. The nature of vertical circulations along the edges of these surface boundaries is discussed in section 4. The chapter ends with a list of suggested research topics.

2. Storm initiation

How is storm initiation defined? It could be when the first convective cloud becomes visible, which is a sign that the level of free convection (LFC) has been reached. The appearance of cumulus congestus on satellite images or reports of the same from spotters might be enough to convince forecasters to issue statements concerning the possibility of convective activity. However, sometimes cumulus congestus do not develop into convective storms because they grow in an environment of too much vertical shear or because they are too narrow, and the entrainment of dry, environmental air destroys their buoyancy before they can develop any precipitation. The beginning of a storm might be when precipitation aloft is first detected. In radar studies, such a definition is common (e.g., Rhea 1966; Bluestein and Jain 1985; Bluestein et al. 1987; Bluestein and Parker 1993). However, in some instances the convective cell dissipates immediately when a downdraft develops and there is no further development. In any case, the process of storm initiation is highly nonlinear: there is only a very brief intermediate stage between that of no convective clouds and that of deep convective clouds bearing precipitation (Crook 1996), just as there is no intermediate stage at all between air that is saturated and air that has condensation; air is not partially saturated and storms do not exist in a partially formed state, except perhaps for a very short length of time.

Convective storms can be initiated when air parcels are heated at the surface to their convective temperature, lifted to their LFC, or are both heated and lifted (Bluestein 1993). It is generally thought that synoptic-scale vertical motions (~ 1 cm s^{-1}) are not directly responsible for initiating convective clouds (because the vertical motions are so weak), but rather for preconditioning the environment (e.g., Rockwood and Maddox 1988). There are many cases in which sinking motion on the synoptic scale suppresses convective development (e.g., Richter and Bosart 2002), while rising motion can make the difference between the initiation of and the suppression of convection (Roebber et al. 2002). The reader is referred to Doswell and Bosart (2001) for further discussions on the relationship between synoptic-scale processes and convection. Mesoscale vertical motions (approximately 10 cm s^{-1}–1 m s^{-1}), on the other hand, are thought to be capable of both initiating convective clouds and modifying the local environment so that convection becomes possible or so that the type of possible convection is modified (Doswell 1987; Johnson and Mapes 2001). Mesoscale lift can decrease the convective inhibition, increase the convective available potential energy (CAPE), and moisten the air column.

Since surface boundaries are often the locations of mesoscale regions of upward motion, it is not surprising that convective storms often begin along boundaries. The precise nature of storm initiation is not understood, mainly because it is very difficult to observe clouds in the act of growing into storms, while at the same time collecting observational data on scales small enough to resolve the features associated with storm formation. While many field programs have addressed the problem of convective-storm behavior, not many have specifically addressed storm initiation. The International H$_2$O Project (IHOP) was conducted in the late spring and early summer of 2002 in the southern plains (Weckwerth et al. 2004). One of the goals of this project was to study storm initiation, with particular focus on the role of moisture variability. Because the results from this field experiment are still forthcoming, analyses of data from IHOP and significant findings are not available at the time of this writing for inclusion into this review paper. However, it is expected that there will be significant new findings, especially from the airborne (e.g., Murphey et al. 2003) and ground-based mobile multiple-Doppler radar analyses (e.g., Richardson et al. 2003; Ziegler et al. 2003) of the wind field, and analyses of the moisture field from dropsonde data and lidar data around boundaries, as convective storms were forming.

Because it is so difficult to use observational instruments to document all the meteorological variables during storm formation along boundaries, it is useful to employ numerical simulations. Numerical simulation studies of convective-storm evolution, however, often use unrealistic methods for triggering the storms. For example, in many idealized studies, unrealistically wide and highly buoyant thermal bubbles are introduced to initiate the storms and the environmental soundings must be moistened, especially at midlevels (e.g., Weisman and Klemp 1982). In models in which the nature of the heating and lift are explicitly modeled, the nature of the subgrid-scale turbulence parameterization must be called into question when the horizontal resolution is reduced below 1 km (Bryan et al. 2003).

3. Surface boundaries in the southern plains

The primary surface boundaries in the southern plains are fronts (cold, warm, and stationary), the dryline, troughs, and outflow boundaries. While frontal zones are usually located on the cold side of the axis of troughs, frontal zones also have a significant temperature gradient across them (Sanders 1999). Of these boundaries, the dryline is unique to the plains region. However, fronts and troughs also have characteristics unique to the plains, owing to their link to local topography. Also present, but perhaps underappreciated by the community, are discontinuities associated with surface sensible heat-flux gradients (Segal and Arritt 1992) and bores (e.g., Doviak and Ge 1984).

The characteristics of fronts and prefrontal troughs are discussed in detail elsewhere in this volume (Schultz 2008). Because convective storms often form along and just behind surface cold fronts (e.g., Crook 1987), it is important to understand how they behave. Cold fronts that frequent the southern plains are affected by the terrain, which slopes relatively gently upward to the west and then more steeply to the far west, along the Rocky Mountains. In particular, it has been hypothesized that many fronts that are zonally oriented propagate rapidly southward as either Kelvin waves or trapped density currents in the western regions of the southern plains (Bluestein 1993). Colle and Mass (1995) concluded, however, that they do not propagate as a result of rotationally trapped waves. If they propagate too rapidly, then convective cells that grow are removed from their boundary layer roots before they have had a chance to mature. On the other hand, for fronts that are meridionally oriented, extend southward from surface cyclones, and propagate toward the east, storm formation may, under some circumstances, be more likely.

The dryline marks the boundary between relatively cool, moist air of maritime origin and relatively warm, dry air of continental origin (Bluestein 1993). It may also be collocated with a trough axis (Fig. 1) (e.g., Bluestein 1993; Bluestein and Crawford 1997; Martin et al. 1995). Since convective storms often form near and just to the east of them (Rhea 1966; Ziegler and Rasmussen 1998; Ziegler et al. 1997), it is important to understand where they are located and how they move. A "quiescent" dryline is one not embedded in an environment of strong, synoptic-scale forcing. Such a dryline advances eastward during the day into the eastern Texas panhandle or western Oklahoma and retreats westward at night into the western Texas panhandle and eastern New Mexico (e.g., Ziegler et al. 1995).

When a dryline is influenced by strong synoptic-scale forcing (e.g., associated with a mobile, short-wave trough aloft), a surface cyclone can form and the dryline that extends to its south may then advance far to the east as the surface cyclone propagates away from the Rocky Mountain region (Carr and Millard 1985; Hane et al. 2001) (Fig. 2). Although the main mechanism through which a dryline propagates eastward during the day is vertical mixing, when there is strong synoptic-scale forcing, horizontal advection of dry air aloft is the

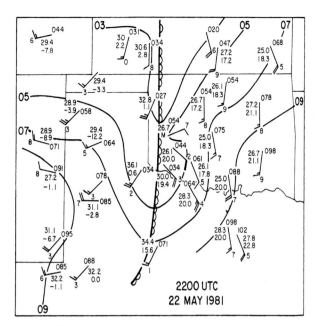

FIG. 1. The dryline, not under the influence of strong synoptic-scale forcing. The dryline, late in the afternoon, remains in far western OK. Analysis of surface pressure reduced to sea level as altimeter setting (solid lines, hPa, without the leading "10") around a dryline (scalloped line) at 2200 UTC 22 May 1981. Temperature and dewpoint plotted in °C; whole (half) wind barbs = 5 (2.5) m s^{-1}. Tornadic supercells formed along the dryline and propagated eastward. From Bluestein (1993).

major reason why the dryline advances through vertical mixing so much farther to the east (Hane et al. 2001).

When there is a surface cyclone in the lee of the Rockies, formed in part from compressional warming in air flowing downslope and in part from synoptic-scale forcing associated with an upper-level trough, the dryline often intersects fronts at the center of the cyclone (Fig. 3). However, the dryline often intersects fronts even in the absence of strong synoptic-scale forcing; in this case, the surface cyclone is orographically forced (Bluestein and Parks 1983) (Fig. 3a). Sometimes the dryline intersects a front and there is no cyclone at the intersection point (e.g., Ziegler and Hane 1993). In other situations when there is no surface cyclone, but only a meridionally oriented trough associated with the dryline, an outflow boundary from earlier convection may intersect the dryline and behave like the intersection of a front with the dryline (e.g., Bluestein and MacGorman 1998; Bluestein and Gaddy 2001; Weiss and Bluestein 2002) (Figs. 3b and 4).

Maddox et al. (1980) have shown how the low-level vertical wind shear is enhanced just behind an outflow boundary, near its intersection with the dryline (Fig. 3d). This increase in vertical shear could be responsible for providing an environment more conducive for supercells (Weisman and Klemp 1982) and possibly for tornadoes. Even in the absence of a preexisting outflow boundary, an isolated convective storm that forms along the dryline and leaves behind an outflow boundary (as

it propagates away from the dryline) could intersect the dryline to the west and set the stage for new supercells; in the absence of the outflow boundary from the preceding storm, it is possible that subsequent convection might not be as severe because the vertical shear would be weaker. In the case of low-precipitation (LP) supercells (Bluestein and Parks 1983), there would not be strong outflow because the potential for evaporative cooling is less.

Troughs not associated with fronts or the dryline ["baroclinic troughs" (Sanders 1999)] may be classified as lee troughs when they have propagated away from the lee slopes of mountains (e.g., Karyampudi et al. 1995a) or "prefrontal troughs" when they are located just ahead of frontal zones (Fig. 5) (Hutchinson and Bluestein 1998), and "inverted troughs" (Keshishian et al. 1994), which are troughs in easterly flow, poleward of surface cyclones (Fig. 6).

Inverted troughs form in the lee of the Rocky Mountains and separate a relatively cold air mass dammed up along the eastern foothills of the Rocky Mountains from a modified (not as cold, because of a longer residence time away from its source) cold-air mass over the plains. Not much has appeared in the literature about convective-storm formation along these inverted troughs, probably because they tend to occur most frequently during the cold season and because the air to their east is not often susceptible to boundary-layer-based convection. Weisman et al. (2002) have shown how during the cold season, precipitation in general can be found on either side of the trough, but is much more common east of the trough. There are some instances in which "inverted troughs" are found north of the intersection of the dryline and an outflow boundary (Fig. 3a); these troughs might represent an extension of the trough found along the dryline, above the cool pool of air behind the outflow boundary.

Outflow boundaries separate evaporatively cooled air produced in convective-storm downdrafts (in which precipitation has fallen into unsaturated air) and ambient, warm air (e.g., Byers and Braham 1948; Young and Fritsch 1989; Stensrud and Fritsch 1993; Fritsch and Vislocky 1996). Outflow boundaries sometimes propagate as density currents, especially during their mature stages when they mark the leading edge of a deep cold pool. When they behave like density currents they are called gust fronts. Usually troughs are not found along outflow boundaries (Fig. 7) because air parcels do not reside along them long enough for earth's vorticity to be amplified significantly. Since the location and intensity of outflow boundaries are determined by the details of the spatial extent and intensity of prior convective storms, the location of outflow boundaries, unlike the location of fronts and the dryline, are not as easily well forecast. It is a significant challenge to predict the initiation of storms along an outflow boundary, especially before the convection producing the outflow boundary has broken out (Carbone et al. 1990). In many instances,

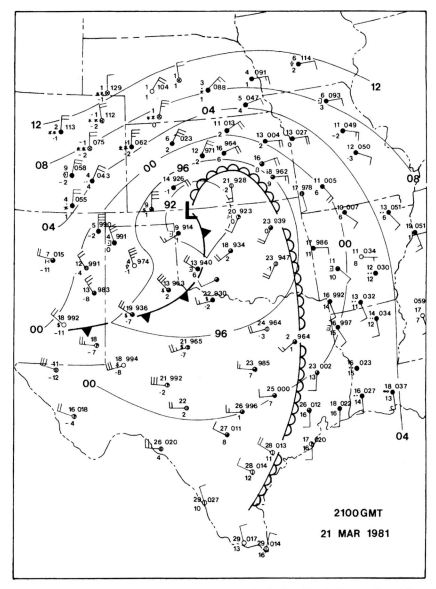

FIG. 2. The dryline, under the influence of strong synoptic-scale forcing. The dryline (scalloped line), during the afternoon, has pushed into eastern OK and TX. Analysis as in Fig. 1, but at 2100 UTC 21 Mar 1981; the reduced sea level pressure is plotted without the leading "9" or "10." Severe convection developed along the dryline and moved eastward. From Carr and Millard (1985).

convective storms have been initiated along outflow boundaries and they have gone on to produce tornadoes and other severe weather. In the absence of an outflow boundary, it is unlikely that any convection would have been triggered at all.

When outflow boundaries form in response to convective activity triggered along a front, the front may jump discontinuously ahead as a result of the cold surface air (Bryan and Fritsch 2000). The old surface front dissipates and a new one forms at the edge of the outflow boundary, as a midlevel front passes over the pool of cold air and reaches the leading edge of the cold pool.

Bores are boundaries across which the wind shifts but the temperature may or may not change. Bores are produced when a relatively dense fluid impinges on a low-level stable layer. In a bore, changes in temperature are produced through adiabatic vertical motions; no mass is transported and it is a type of a gravity wave. The passage of a bore is characterized by a wind shift that is accompanied by no drop in temperature or even an increase in temperature, while the passage of a gust front/density current is characterized by a wind shift that is accompanied by a drop in temperature. It is not, however, always easy to distinguish a bore from a density current (Simpson 1997) because there is in nature a continuum between flows that are pure bores and flows

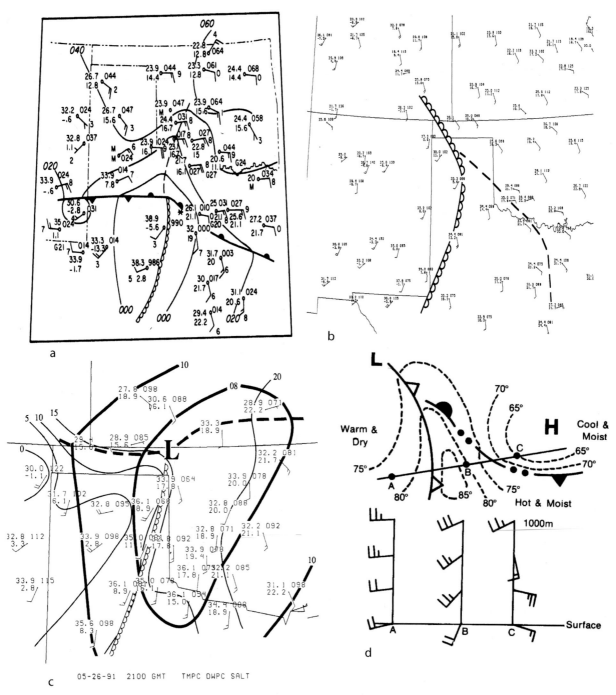

FIG. 3. Three examples of the intersection of the dryline with a front or outflow boundary, often at a surface cyclone, and an idealized depiction. In all cases, the only or the most significant convective storm was initiated near the dryline–front/outflow boundary intersection. (a) With little synoptic-scale forcing. An isolated supercell formed near the dryline–front intersection. Analysis of altimeter settings at 0000 UTC 17 May 1978. Temperature and dewpoint plotted in °C. From Bluestein and Parks (1983). (b) With little synoptic-scale forcing at 2300 UTC 31 May 1990. Series of isolated tornadic supercells formed near the dryline–outflow boundary intersection. From Bluestein and MacGorman (1998). (c) With significant synoptic-scale forcing at 2100 UTC 26 May 1991. A tornadic supercell formed near the intersection of the dryline and outflow boundary. From Hane et al. (1997). Analyses similar to those in Figs. 1 and 2. (d) Idealized depiction, including variation of vertical wind profile across the dryline (shown as a cold front symbol with open triangular barbs) and outflow boundaries. From Maddox et al. (1980).

2106-2112 UTC

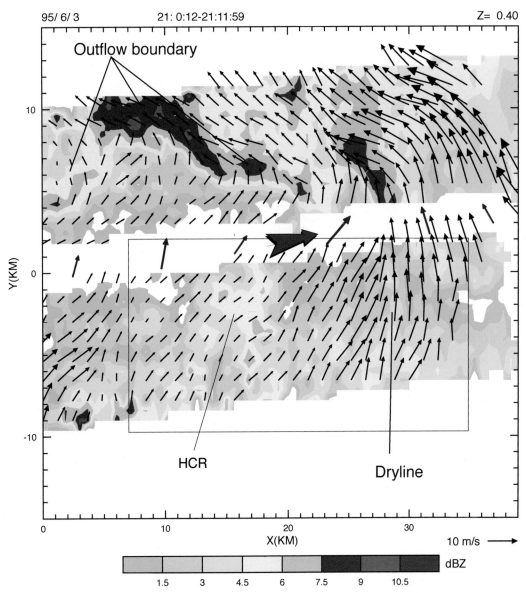

FIG. 4. Intersection of an outflow boundary with the dryline at 400 m AGL on the small scale in northwest TX, as depicted by wind (vectors) and radar reflectivity (color coded) based on data from the Electra Doppler Radar (ELDORA) from 2106 to 2122 UTC 3 Jun 1995. The red arrow depicts the direction of the Electra flight track, which coincides approximately with the data-free swath. The blue arrows along the track indicate in situ flight-level wind measurements. From Weiss and Bluestein (2002).

that are pure density currents. Haertel et al. (2001) argued that that density currents and gravity waves lie at the opposite ends of a spectrum of phenomena: bores, which are in the middle of the spectrum, are due partly to the advection of cold air that accompanies the pressure difference across a density current and partly to propagation that is associated with buoyancy as a restoring force.

4. Vertical circulations associated with surface boundaries

a. Frontal circulations

Vertical circulations are induced along fronts in response to changes in the across-front temperature gradient by deformation, convergence, and cross-frontal gradients in diabatic heating. [The reader is referred to

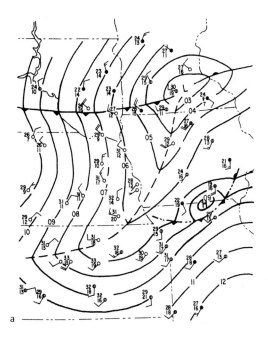

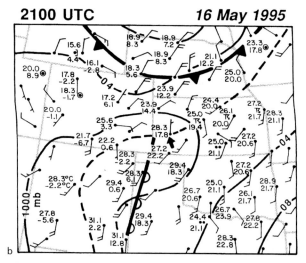

FIG. 5. Prefrontal surface troughs not collocated with a front or the dryline. (a) Surface isobars (solid lines) depicted every 1 hPa at 1900 UTC 4 Jun 1979. Temperature and dewpoint in °C. Severe convective storms formed along the trough. From Gaza and Bosart (1985). (b) Analysis of surface pressure (solid lines) at 2100 UTC 16 May 1995; temperature and dewpoint in °C. A tornadic supercell formed along the trough. From Wakimoto et al. (1998).

studies by Petterssen, Bergeron, Eliassen, and Hoskins and Bretherton, which are summarized in Bluestein (1993).] When the forcing is frontogenetical (fronto-lytical), a thermally direct (indirect) circulation is forced. The thermally direct circulation favors convective development because the upward branch of the circulation originates on the warm side of the front. A thermally direct vertical circulation forced adiabatically may be enhanced during the day (night) by diabatic heating when the air behind (ahead of) the front is

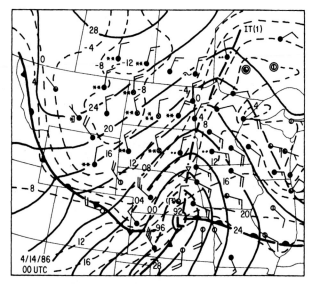

FIG. 6. Inverted trough extending northward from a surface cyclone. Analysis of sea level pressure at 4-hPa intervals at 0000 UTC 14 Apr 1986. Temperature and dewpoint plotted in °C. From Keshishian et al. (1994).

cloudy (clear) (Koch 1984; Koch et al. 1995). In some instances, however, there is more surface moisture on the cold side of the front, especially when the surface air on the warm side of the front is of continental, not maritime origin, and the boundary layer is heated so that it is deep and surface moisture is diluted via mixing

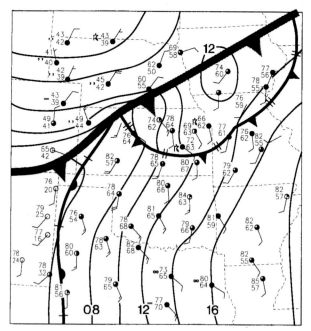

FIG. 7. Outflow boundary ahead of a cold front. Analysis of sea level pressure (hPa, solid lines) at 1800 UTC 11 May 1982. Temperature and dewpoint plotted in °F for emphasis. From Stensrud and Fritsch (1993).

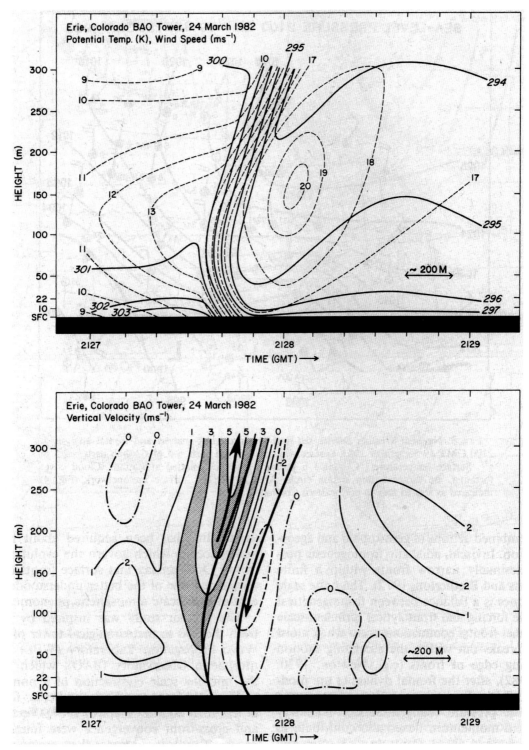

FIG. 8. Vertical cross section (via time-to-space conversion in the cross-front direction) through a front that behaved like a density current of (a) potential temperature and wind speed normal to the front and (b) vertical velocity. Data from the Boulder Atmospheric Observatory 300-m tower, 24 Mar 1982. From Shapiro et al. (1985).

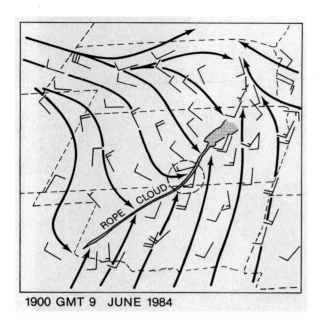

1900 GMT 9 JUNE 1984

FIG. 9. Surface streamline analysis associated with a cold front and depiction of a rope cloud along it (cf. Fig. 10) where convection is being initiated at 1900 UTC 9 Jun 1984. Full (half) wind barb = 5 (2.5) m s^{-1}. From Shapiro et al. (1985).

through a deep layer. In these cases, convection along the front may be suppressed.

When fronts behave like density currents (Fig. 8), the earth's rotation plays little or no role in forcing the vertical circulation at the leading edge of the front; the major forcing mechanism is the across-the-front (hydrostatic) pressure-gradient force. It is thought that fronts that are not associated with precipitation and that behave like density currents were formed through the collapse of the frontal temperature gradient through the convergent action of the ageostrophic vertical circulation (Shapiro et al. 1985). When precipitation falls on the cold side of fronts and evaporates into unsaturated air below, the density-current-like character of a front can be enhanced through differential diabatic heating (cooling on the cold side, no diabatic temperature changes on the warm side) (Browning and Pardoe 1973).

Fronts sometimes have properties of both semigeostrophic phenomena (i.e., those affected by earth's rotation such that the advection of momentum and heat by the ageostrophic components of the wind is significant and those for which their variations along the boundary can be ignored) and density currents, depending on how wide, deep, and intense the frontal zone is. Frontal zones on the order of 100 km behave semigeostrophically, while frontal zones on the order of 1–10 km behave like density currents if they are intense and deep enough. The magnitude of the lift along a density current (1–10 m s^{-1}) is greater than the lift associated with semigeostrophic processes (10 cm s^{-1}–1 m s^{-1}), and is therefore more efficient at initiating convection (Figs. 9 and 10).

Fronts propagating into a strong low-level inversion, such as that produced at night, can trigger a bore. For example, Karyampudi et al. (1995b) have discussed the role a prefrontal bore (e.g., Crook and Miller 1985) can play in triggering a squall line.

The difference in the ability of cold, warm, and stationary fronts to trigger convection is the difference in the trajectories of air parcels that are lifted along them. The front-normal relative surface wind speed and vertical motion are of most importance because they determine how long and how far air parcels are lifted. In the southern plains, it is an empirical observation (but not as yet rigorously proven through climatological analysis) that rapidly southward moving, zonally oriented cold fronts are not very efficient at initiating convection. On the other hand, zonally oriented stationary fronts are efficient at initiating convection. In the case of the former, air trajectories originating south of the front may not be lifted very much or very long; in the case of the latter, air parcels may be lifted substantially for a longer period of time.

b. Outflow boundary circulations

As air approaches an outflow boundary, it is lifted over it (e.g., Wilhelmson and Chen 1982; Bluestein 1993). The susceptibility of the air to reaching its LFC depends to a large extent on the depth of the outflow boundary and its intensity (i.e., temperature deficit). Rotunno et al. (1988), in what has come to be known as Rotunno, Klemp, and Weisman (RKW) theory, have shown how the magnitude of the vertical wind shear in the direction normal to the outflow boundary, and extending over the depth of the outflow boundary, plays a crucial role in determining how far air is lifted along a steady-state, frictionless density current of constant depth. The farther air is lifted, the more likely it will reach its LFC. When the rate of import of horizontal vorticity (associated with vertical shear) from the region ahead of the outflow boundary is counterbalanced by the baroclinic generation of horizontal vorticity at the leading edge of the outflow boundary, then it is most likely that the vertical circulation will remain erect and air parcels can be lifted to their LFC (Fig. 11a).

Xu (1992) also showed that the amount of lifting of air along an outflow boundary depends to a great extent on the magnitude of the vertical shear over the depth of the boundary layer (Fig. 11b). He demonstrated that when the shear is relatively weak, the depth of the density current head increases with shear (density-current relative wind speeds decreasing with height) and the flow is supercritical; the depth of the vertical excursion of air lifted over the boundary depends on the depth of the head. As the shear is increased, the flow may eventually become subcritical, and its depth decreases. In both Rotunno et al.'s (1988) and Xu's (1992) analyses, there is an optimal value of low-level vertical shear for which the chances for storm initiation are maximized.

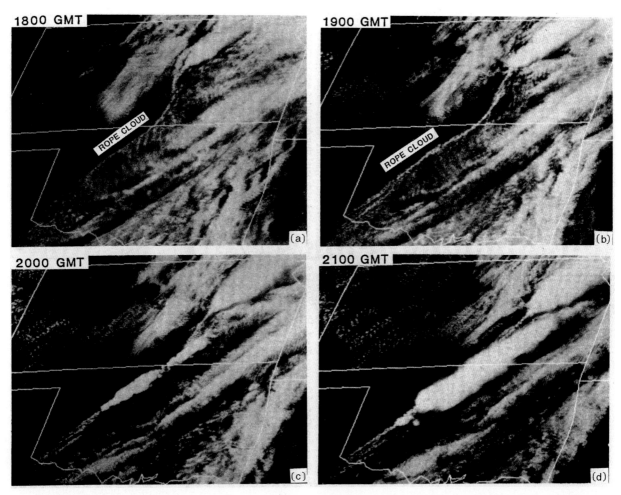

FIG. 10. Deep convection being initiated along a cold front on 9 Jun 1984 as viewed by National Oceanic and Atmospheric Administration (NOAA)/*Geostationary Operational Earth Satellite-5* (*GOES-5*) visible satellite images. From Shapiro et al. (1985).

When outflow boundaries impinge on an environment having a stable layer, a bore may be triggered (Doviak and Ge 1984; Fulton et al. 1990), especially at night when there is a nocturnal inversion. Rising motion associated with the bore may or not be able to initiate new convection. Such a process is similar to that described by Karyampudi et al. (1995b) for a front propagating into a nocturnal inversion.

c. Dryline circulations

1) INLAND SEA BREEZE

It has been proposed that the vertical circulation across the dryline is forced solenoidally by the horizontal gradient in diabatic heating across it, in the same way that a sea-breeze circulation is forced by the diabatic heating difference across a land–water interface (Sun and Ogura 1979; Sun 1987; Sun and Wu 1992; Bluestein and Crawford 1997). Such a circulation has therefore been given the oxymoronic name, the "inland sea-breeze." The variation in the across-the-dryline

heating may be caused by the nature of the surface vegetation and soil moisture (Grasso 2000). As convergence develops at the surface under the rising branch of the vertical circulation in response to the solenoidal forcing, both the temperature and moisture gradients will increase.

2) DENSITY-CURRENT-LIKE BEHAVIOR

The dryline may behave like a weak density current late in the day, when the difference in virtual temperature across the dryline is the greatest, which is when the virtual temperature is the highest on its west side (Parsons et al. 1991) (Fig. 12). It is possible that the inland-sea-breeze circulation, which had been acting much of the day frontogenetically to increase the surface density gradient, is the trigger that makes the density contrast at the dryline strong enough that it behaves like a density current.

The analysis of data from a scanning Doppler lidar during the Texas Frontal Experiment (TEXEX) in 1985

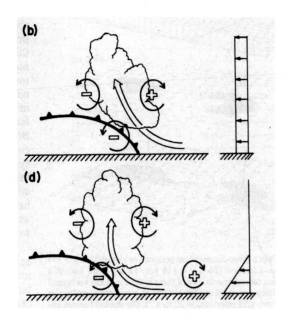

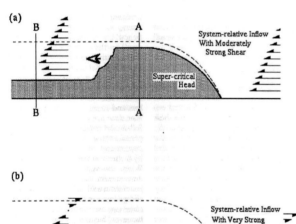

FIG. 11. Illustration of the effects of low-level vertical shear on the vertical circulation across a density current and on the shape of a density current. (top) RKW theory: density-current-relative horizontal wind depicted at the right. (top, b) When the shear is zero air is lifted at a relatively low angle with respect to the leading edge of the density current (cold-front symbol). (top, d) When there is shear directed in the same direction as the motion of the density current the air is lifted at a steeper angle. The sense of horizontal vorticity associated baroclinically with the leading edge of the density current and the edges of the buoyant cloud/updraft, and associated with the environmental low-level shear, are shown by curved arrows and plus signs (into the figure) and minus signs (out from the figure) according to the right-hand rule. From Rotunno et al. (1988). (bottom) Shapes of head of a density current when the shear is (a) moderately strong and in supercritical flow, and (b) very strong and in subcritical flow. (bottom, a) The head is deep and (bottom, b) the head is shallow. From Xu (1992).

shows air being lifted up and over the cooler, westward-retreating air on the east side of the dryline during the early evening (Fig. 13a). The lifting was as intense as 5 m s^{-1}. Atkins et al. (1998), using airborne data from

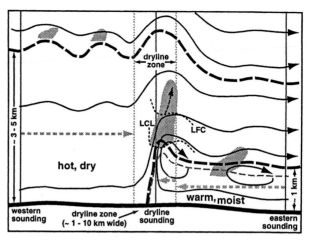

FIG. 12. Conceptual model of convection initiation along the dryline. Vertical cross section depicting streamlines and clouds; lower heavy dashed line represents the extent of the moist, convective boundary layer; the upper heavy dashed line represents the top of the deep, dry convective boundary layer west of the dryline and the top of the elevated "residual" layer east of the dryline (above the moist layer). The heavy dashed streamline represents a buoyantly accelerated cloudy air parcel trajectory. From Ziegler and Rasmussen (1998).

the Verification of the Origin of Rotation in Tornadoes Experiment (VORTEX) in 1995, also showed how the vertical circulation along the dryline (in Texas) can behave like that associated with a density current (Fig. 13b). More recently, the analysis of a scanning, mobile, W-band Doppler radar during IHOP in 2002 shows lift on even finer scales (Fig. 14). It is not thought that the slope of the terrain over the high plains is high enough to slow down significantly the upslope retreat of the cool, moist air mass (Parsons et al. 1991). A significant tornado, which struck Lubbock, Texas, on 11 May 1970, was spawned from a storm initiated along a retreating dryline (Fujita 1970).

When a density current impinges on a stable layer, such as a nocturnal inversion, a bore may be triggered. Wakimoto and Kingsmill (1995) described a case in which a sea-breeze front collided with a gust front and generated a bore. The sea-breeze current undercut the gust front and the bore propagated against the ambient flow in the stable layer associated with the cold pool from the sea-breeze air mass. It is therefore possible that a dryline, which behaves like a density current, could trigger a bore if it collided with a gust front.

3) GRAVITY WAVE MOMENTUM MIXED DOWNWARD BEHIND THE DRYLINE

Koch and McCarthy (1982) have presented evidence of waves along the dryline and shown how they might be associated with convective development. Sanders and Blanchard (1993) found periodic fluctuations in the dryline (Fig. 15a) during Oklahoma–Kansas Preliminary Regional Experiment for Storm-scale Operational and

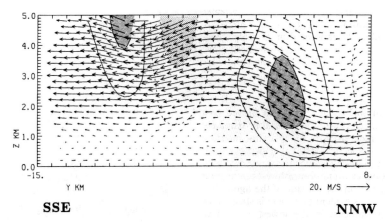

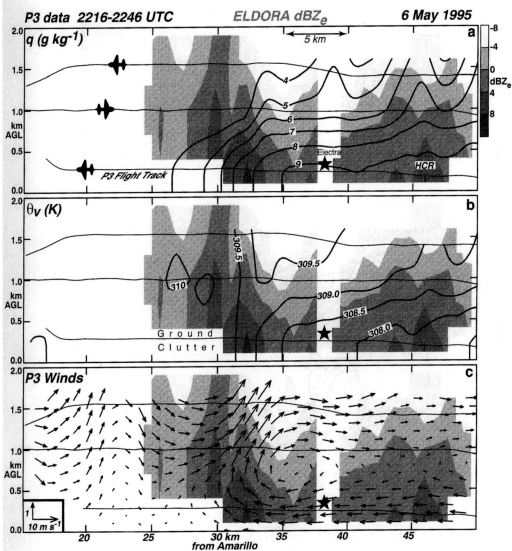

FIG. 13. Density-current-like vertical circulations at the edge of the dryline as shown in vertical cross sections of wind. (top) Dryline-relative wind vectors and contours of vertical motion based on data from a scanning Doppler lidar. The solid and dashed lines indicate vertical motions (upward and downward, respectively) greater than 1 m s^{-1}. The shading indicates vertical motions in excess of 4 m s^{-1} and the stippling indicates vertical motions less than -4 m s^{-1}. Analysis for 0100 UTC 22 Apr 1985, at Midland, TX. From Parsons et al. (1991). (bottom panels) Analysis of (top) water vapor mixing ratio, (middle) virtual potential temperature, and (bottom) winds from a NOAA P-3 aircraft from 2216 to 2246 UTC 6 May 1995 in west TX. Shaded regions represent radar reflectivity field from ELDORA. The thin black line marks the P-3 flight track and the star indicates the position of the Electra aircraft. From Atkins et al. (1998).

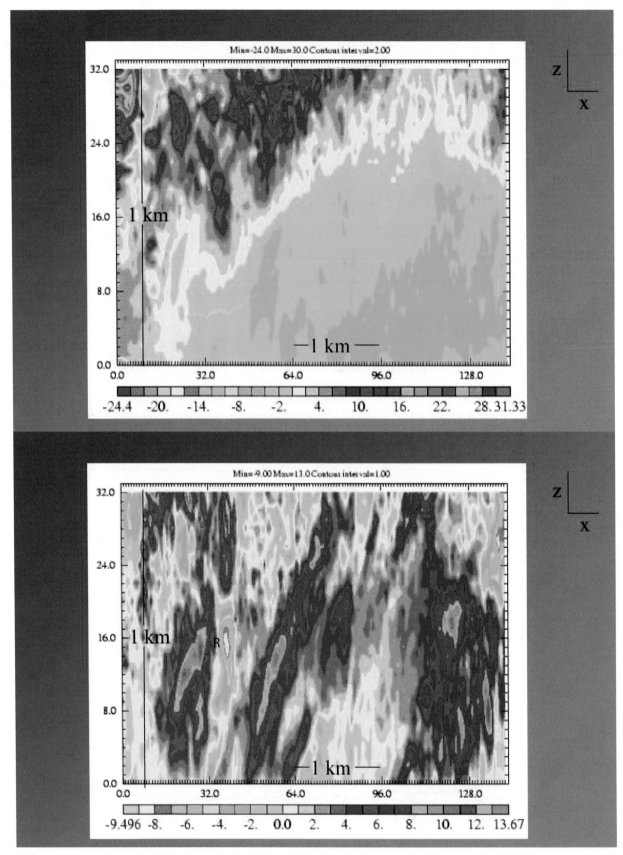

FIG. 14. Analysis of vertical cross section of the (top) ground-relative wind component normal to the dryline [m s⁻¹; positive (negative) speeds denote a westerly (easterly) wind component] and (bottom) vertical velocity (m s⁻¹) at 0007–0036 UTC 23 May 2002 in the Oklahoma Panhandle. The wedged-shaped dryline boundary is located approximately along the yellow–green interface. The spacing between each numbered tick mark along the abscissa and ordinate represents 30 m. The "R" denotes the center of a rotor circulation. Based on data collected by a truck-mounted, W-band Doppler radar (Weiss et al. 2003). Courtesy of C. Weiss.

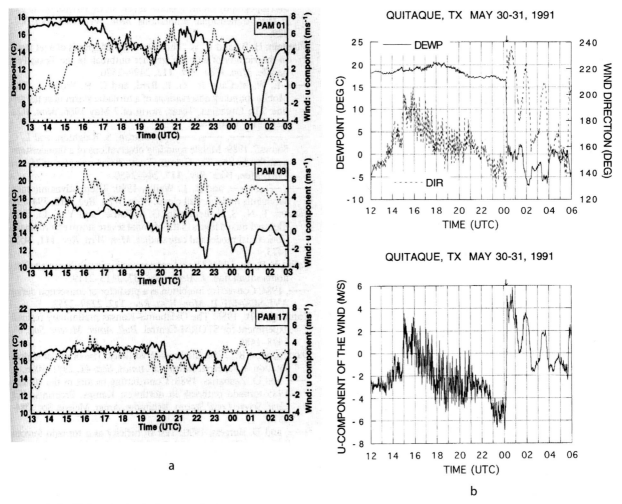

FIG. 15. Oscillations in the dryline-normal wind component and the dewpoint near the dryline. (a) Zonal wind component (dashed line) and dewpoint (solid line) at surface mesonet stations as a function of time on 10–11 May 1985. Note how the dewpoint in general rises (drops) when the zonal wind component decreases (increases) with a periodicity of around 2–3 h. From Sanders and Blanchard (1993). (b) Wind direction (dashed line) and dewpoint (solid line) are shown in the top panel and the zonal component of the wind is shown in the bottom panel. Note also how zonal wind speed and changes in wind direction are correlated with changes in the dewpoint. From Crawford and Bluestein (1997).

Research Meteorology (O–K PRESTORM) in 1985, which were associated with convective development. The fluctuations were associated with mesoscale waves having a wavelength of about 200 km, and were thought to be forced in the exit region of a jet streak. Crawford and Bluestein (1997), during Cooperative Oklahoma Profiler Studies (COPS-91), also documented similar periodic fluctuations along the dryline, though they were not necessarily associated with convective development. The wind periodically backed and the dewpoint rose, while later the wind veered and the dewpoint dropped, and so on, with a period of a few hours (Fig. 15b). It was hypothesized the waves were caused by gravity waves generated in the upper troposphere and their momentum mixed down to the surface in the deep dry-adiabatic layer to the west of the dryline.

4) "SIX-O'CLOCK MAGIC"

Storm chasers have noted that the initiation of convection along the dryline in the southern plains of the United States is often delayed until about 6 P.M. local time. (There is anecdotal evidence that such a delay is observed elsewhere, also.) When storms have not been initiated by then, it is not likely that they will form later that day or early evening. This empirical behavior is colloquially known as "6-o'clock magic." It has been shown, using a mixed-layer model, that behavior like the six-o'clock magic phenomenon may be associated with a local maximum in the height of the inversion that caps the moist air east of the dryline around dusk (Jones and Bannon 2002), when the daytime eastward movement of the dryline has ceased and its westward movement begins. A "spike" in inversion height (Fig.

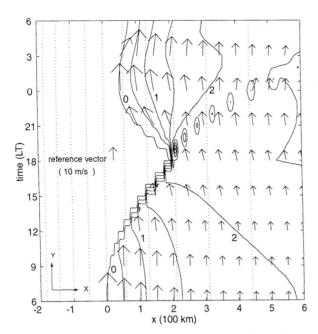

FIG. 16. Location of the dryline (mixed-layer depth of zero) east of its position at 0600 LT as a function of time from a numerical simulation. Also shown are wind vectors east of the dryline and the mixed-layer depth (km). The surface slopes exponentially upward to the west; $x = 0$ corresponds to a terrain height of 2 km. From Jones and Bannon (2002).

16) indicates that air has been lifted substantially; the spike is caused both by entrainment at the dryline where the inversion height is small and by the westward-directed pressure gradient associated with the warmer air to the west. Using a shallow-water model of the dryline, Miller et al. (2001) showed that the dryline behavior at dusk is similar to that of the dam-break problem for a rotating fluid.

5) THE RESIDUAL DRYLINE SECONDARY CIRCULATION

The vertical circulation across the dryline can also be maintained as it is advected northward, up and over a zonally oriented outflow boundary that intersects it (Weiss and Bluestein 2002). The remnants of the vertical circulation, ostensibly forced by the inland-sea-breeze mechanism, is called the "residual dryline secondary circulation" (RDSC) when it appears north of an outflow boundary (Fig. 17). The magnitude of the vertical velocities of the RDSC is several meters per second. It is thought that the RDSC may be responsible for initiating convection near the intersection of the dryline and an outflow boundary either because the upward branch of the dryline circulation is augmented by the lift air gets when it passes over the outflow boundary, or because air is lifted to its lifting condensation level (LCL) west of the dryline and then to its LFC as it ingests more moisture. Storms often, but certainly not always, tend to occur along the intersection of the dry-

line and an outflow boundary or front (sometimes called the "triple point") even when there is no convection elsewhere along the dryline or along the outflow boundary or front (e.g., Bluestein and Parks 1983). The "inverted" trough sometimes found north of the dryline–outflow boundary intersection might sometimes be associated with the RDSC.

d. Trough circulations

Gaza and Bosart (1985) and Wakimoto et al. (1998) have documented storm formation along surface-pressure troughs. The nature of the vertical circulation along the axis of a surface trough has not been studied as much as that along fronts, outflow boundaries, and the dryline.

There are several adiabatic, quasigeostrophic explanations for producing surface troughs not associated with significant baroclinicity. Sanders (1999) argued that the wind shift line associated with a surface cold front will separate from the zone of strong temperature gradient on the cold side because the latter is advected eastward by the front-normal wind component, while the former propagates eastward because of convergence and divergence at the surface to the east and west, respectively, of the trough axis. The wind shift line and trough axis therefore propagate into a region of much less baroclinicity. The rising motion along the wind shift line may be thought of as part of a vestige of the old frontal circulation.

It is also possible that the surface trough is simply the surface reflection of a trough aloft. In this case, the vertical circulation might be diagnosed quasigeostrophically; the rising branch is simply the region where vorticity advection becomes more cyclonic with height, that is, where the intensity of the trough increases with height.

It is yet also possible that the trough is orographically forced, in which case it has a warm core (Bluestein 1993). Rising motion downstream from the trough would be associated with warm advection; however, in a warm core trough, the trough weakens with height, so that vorticity advection becomes less cyclonic with height. It is not clear how any strong upward motion could be generated quasigeostrophically in the absence of diabatic heating.

Another possible role a meridionally oriented surface trough could play in the formation of convective storms is that resulting from the wind shift across it, the air on the east side of the trough would be moister than the air on the west side. If the terrain slopes upward to the west, storms might tend to form first at the western edge of the moist region, which is near the trough axis.

e. Bore circulations

Air oscillates vertically in a bore as a gravity wave. Rising motion in a bore can weaken a nocturnal inver-

Averaged Dryline Extension Cross Section

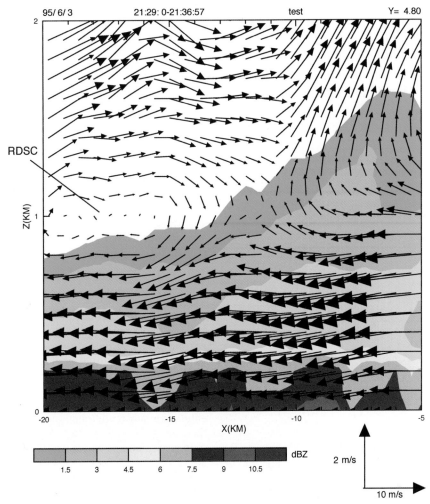

FIG. 17. Vertical cross section in the east–west (right–left) direction highlighting the RDSC behind (north of) an outflow boundary. Cross section averaged in the north–south direction from 2129 to 2137 UTC 3 Jun 1995 in northwest TX. Wind vectors represent average zonal and vertical wind components; colors represent radar reflectivity. RDSC lies to the north of where the dryline would be if it extended northward across the outflow boundary. Winds (vectors) and radar reflectivity (color coded) are from ELDORA. From Weiss and Bluestein (2002).

sion and deepen a low-level moist layer (Weckwerth et al. 2004).

f. The nature of along-boundary variability

While boundaries are often the sites of the initiation of convection, convection does not necessarily break out everywhere along the boundaries (e.g., Bluestein et al. 1988, 1989, 1990) because moisture, temperature, and the depth of the boundary layer may vary significantly along the boundaries. Even if the thermodynamic environment were uniform along boundaries, it is possible that the magnitude of the lift experienced by air parcels could vary along the boundaries. Some mechanisms responsible for effecting along-the-boundary variability are now considered.

1) BOUNDARY LAYER ROLLS

It has been shown from data collected during the Convection Initiation and Downburst Experiment (CINDE) in eastern Colorado in 1987 how thunderstorms that form along a line of convergence in the boundary layer can be triggered preferentially at the intersection of boundary layer rolls [also called horizontal convective rolls (HCRs) (LeMone 1973; Weckwerth et al. 1997, 1999)] and the convergence line (Wilson et al. 1992). Wakimoto and Atkins (1994) found similar results for cloud development along the sea-breeze front in Florida during the Convection and Precipitation/Electrification (CaPE) experiment in 1991. It is thought that the boundary layer convergence is enhanced in the upward branch of boundary layer rolls.

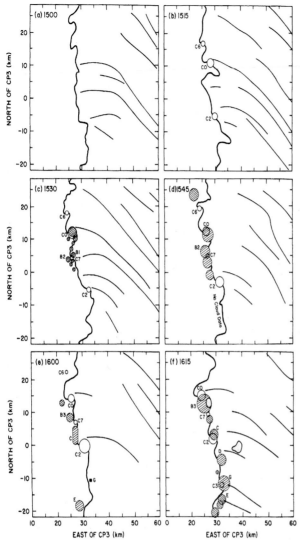

FIG. 18. The relationship between locations of boundary layer rolls (light solid lines) and their intersections with a surface convergence zone (heavy solid line), and locations of convective-storm formation (clouds denoted by cross-hatched features). Between 2100 and 2215 UTC 17 Jul 1987 in eastern CO. From Wilson et al. (1992).

Atkins et al. (1998), using airborne Doppler radar data from VORTEX, found that the rising branch of boundary layer rolls that formed west of the dryline and intersected the dryline were preferential sites for the formation of clouds (the dryline is a proxy for the convergence zone in Fig. 18).

2) KELVIN–HELMHOLTZ AND INTERNAL GRAVITY WAVES

It was found, from an analysis of data from the Microburst and Severe Thunderstorm (MIST) project in Alabama in 1986, that when the low-level boundary layer shear vector is parallel to an outflow boundary, Kelvin–Helmholtz (K–H) waves and internal gravity

waves may be generated, which effect along-the-boundary variability; the lines of constant phase of the waves are then aligned normal to the boundary (Weckwerth and Wakimoto 1992) so that convective cells are initiated preferentially at the intersection of the outflow boundary with the upward crests of K–H waves (Fig. 19).

3) DRYLINE BULGES

Tegtmeier (1974), McCarthy and Koch (1982), and Hane et al. (1997, 2002) have documented cases in which the dryline was deformed such that it bulged eastward locally (Fig. 20). The likelihood of storm formation seemed to be enhanced near the bulge. The bulges could represent the crest of a dryline wave or a region where vertical mixing is more effective, because of warmer surface temperatures or stronger dryline-normal winds aloft.

4) SYMMETRIC INSTABILITY

McGinley and Sasaki (1975) proposed that symmetric instability may play a role in convective initiation along the dryline. The mechanism for triggering convection along the dryline might be the enhancement of upward motion in the upward branch of vertical circulations associated with symmetrically unstable vertical circulations. The thermal wind vector would have to be oriented at a substantial angle to the dryline in order that there be along-the-dryline variability.

In environments where the stratification is nearly dry-adiabatic, such as west of the dryline or above the moist layer east of the dryline, gravitational instability is more likely, since symmetric instability is associated with much slower growth rates. A review of issues concerning the coexistence of gravitational and symmetric instabilities is found in Schultz and Schumacher (1999). Xu and Clark (1985) have suggested that gravitational and symmetric instabilities lie at the ends of a continuous spectrum of instabilities. Xu (1986) has described two types of convective development: During "upscale development," gravitational instability is released first in relatively small convective clouds and mesoscale bands develop later when the atmosphere has become stabilized with respect to gravitational instability. During "downscale development," mesoscale bands are initiated first, while the latent heat released by them destabilizes the midtroposphere with respect to gravitational instability.

5) DRYLINE–CLOUD BAND INTERSECTION

Hane et al. (1997) described a case during COPS-91 (Hane et al. 1993), in which a tornadic supercell, the main storm of the day, formed at the intersection of the dryline and a "secondary cloud line," which was com-

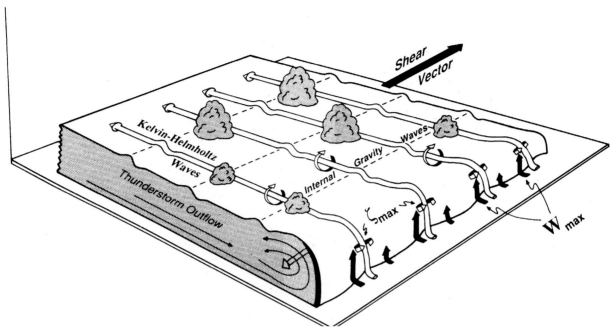

FIG. 19. Conceptual model of along-the-outflow-boundary variations in wind due to Kelvin–Helmholtz waves and internal gravity waves. From Weckwerth and Wakimoto (1992).

posed of high-based convective clouds, west of the dryline (Fig. 21). It was suggested that the secondary cloud line could have been triggered by a "nonclassical mesoscale circulation" (NCMC) forced diabatically by gradients in the vegetative and radiative properties of the surface (Segal and Arritt 1992). An NCMC is forced solenoidally, but unlike the dryline circulation, is not necessarily oriented parallel to the dryline.

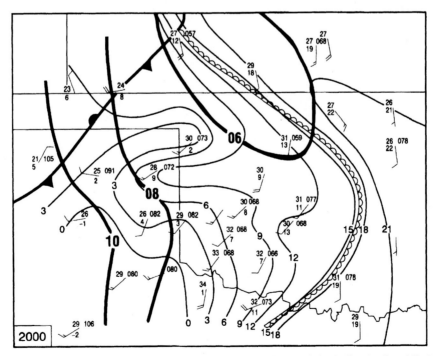

FIG. 20. A dryline bulge case. Analysis of sea level pressure (solid lines) and the dryline (scalloped line) at 2000 UTC 16 May 1991. Temperature and dewpoint plotted in °C. From Hane et al. (2001).

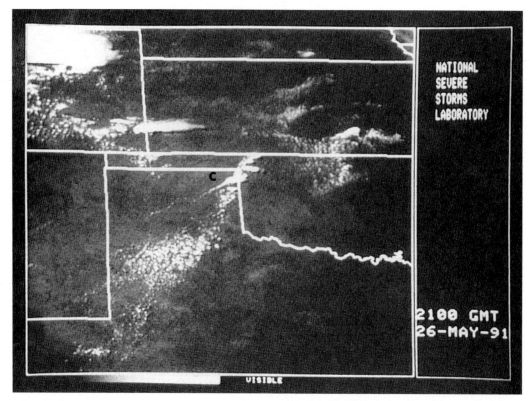

FIG. 21. Example of a cloud band possibly forced by a NCMC; the band intersects the dryline at the location where a tornadic supercell formed. The cloud band is oriented from southwest to northeast and in the visible *GOES-7* satellite image at 2100 UTC 26 May 1991 is just below the "c." From Hane et al. (1997).

6) BAROCLINICITY ASSOCIATED WITH CIRRUS
   SHADOWS

Markowski et al. (1998b) have suggested that during the daytime, surface radiative heating gradients created by the juxtaposition of regions of strong insolation in clear air next to regions of reduced insolation under anvils from convective storms might be the source of solenoidally generated horizontal vorticity. It is possible that lifting could be enhanced at the intersection of the dryline and the upward branch of the solenoidally generated vertical circulation.

It is also possible that there is a solenoidally generated vertical circulation at the northern edge of the thick cirrus band associated with the subtropical jet (Whitney 1977; Durran and Weber 1988) when it intersects the dryline. The author has noted that convection is frequently suppressed beneath the cirrus band itself (see also Roebber et al. 2002) and that storm formation pref-

erentially occurs just to the north of the edge of the band. There are some cases in which deep convection developed only under holes in the cirrus shield. These informal findings need to be confirmed quantitatively through further climatological investigation.

7) DRYLINE–FRONT MERGER

Sometimes a cold front that trails a dryline catches up to the dryline (Fig. 22) (Shapiro 1982; Burgess and Curran 1985; Koch and Clark 1999; Neiman and Wakimoto 1999; Parsons et al. 2000). Often, when convection is not initiated along the dryline because of high values of convective inhibition (CIN) associated with a capping inversion, the CIN is overcome by lift along the front when it reaches the moist air east of the dryline so that the LFC is reached and storms do form. On other occasions, when storms do form along the dryline, there

→

FIG. 22. Examples of a cold front catching up to and overtaking the dryline. (top) Analysis of altimeter setting (solid lines) (inches of mercury × 100, without the leading "29") at 2100, 0000, and 0300 UTC 26/27 Apr 1984. Dryline is denoted by scalloped line. Temperature and dewpoint plotted in °F; pressure plotted is altimeter setting in inches of mercury × 100, without the leading "2." Whole (half) wind barbs denote 5 (2.5) m s⁻¹. From Burgess and Curran (1985). (lower left) Analyses of surface isotherms every 2°F at 1500, 2100, and 0300 UTC 26/27 Apr 1991. Temperature and dewpoint plotted in °C. From Koch and Clark (1999). (lower right) Conceptual representation of a vertical cross section while a Pacific cold front merges with the dryline. Δt ~3 h. (a) Distinct separation between the advancing front and

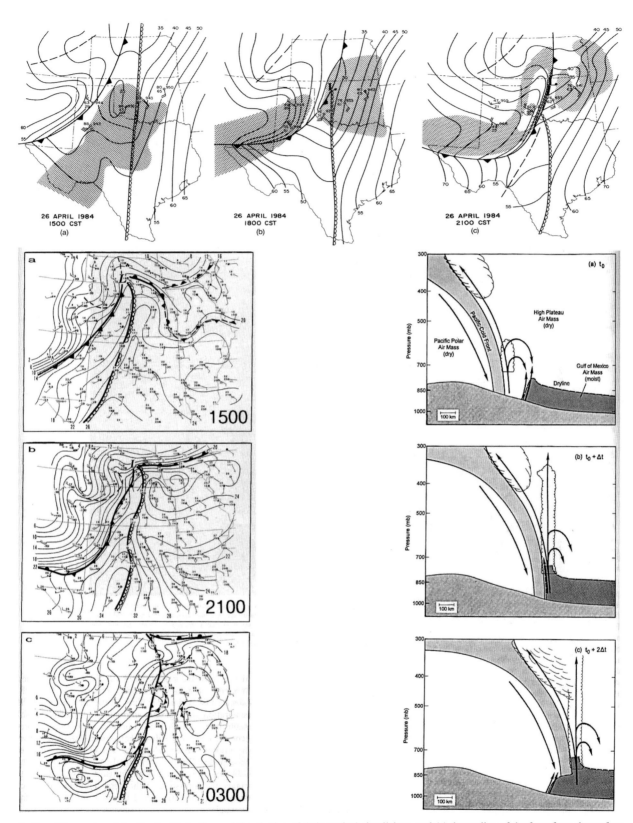

dryline, (b) merging of the front with dryline and the phasing of their vertical circulations, and (c) decoupling of the front from the surface by the shallow Gulf of Mexico air mass. The Pacific cold frontal zone and Gulf of Mexico air mass are shaded light and dark, respectively. The dashed lines mark the dryline. The gray-shaded arrows portray the vertical motions associated with the Pacific cold front and dryline. Cloud schematics are also shown. From Neiman and Wakimoto (1999).

can be a second round of convection later in the day when the front catches up to the dryline.

Koch and Clark (1999) found evidence from data collected during COPS-91 that the initiation of convection can be a result of the combined lifting by a bore triggered along a front and a density current along the front, as the bore and density current propagate into stable, moist air during the evening, after a nocturnal inversion has developed.

8) OUTFLOW–BOUNDARY MERGER

The merger of outflow boundaries can promote new cell growth even before the outflow boundaries actually reach each other (Droegemeier and Wilhelmson 1985a). In a stably stratified environment of no vertical shear, the greatest surface convergence and lift is found along the flanks of the region where outflows first make contact (Fig. 23, left panel). The behavior of new cells when outflows collide in the presence of vertical shear is described, using numerical simulation experiments, for a limited set of shears, in Droegemeier and Wilhelmson (1985b).

Colliding outflow boundaries can also affect other aspects of convective-storm behavior. For example, Weaver and Nelson (1982) showed how tornado formation may be favored when outflow boundaries from two storms collide (Fig. 23, right panel).

It is also noted that semicircular outflow boundaries from isolated regions of convection can combine and form a large outflow boundary, whose leading edge assumes a more linear shape. When the convective storms are triggered along a front, it can become difficult to distinguish between the front and the outflow boundary. As noted earlier, the front may effectively propagate forward into the warm air mass. Such behavior is of great importance to forecasters who must consider where subsequent convection may redevelop.

9) LOW-LEVEL JET INTERSECTING A SURFACE BOUNDARY

Where the southerly, nocturnal, low-level jet (LLJ) of the Great Plains intersects an approximately zonally oriented outflow boundary or front, storm initiation north of the boundary or front is most likely (Augustine and Caracena 1994; Trier and Parsons 1993). It is thought that upward motion associated with frontogenesis is enhanced quasigeostrophically by a local maximum in warm advection above the low-level cold air (Fig. 24). It is also possible that the process is not related to frontal (semigeostrophic) or large-scale quasigeostrophic processes at all, but rather to the enhanced lift of the LLJ itself over the shallow pool of cool air.

g. The influence of surface boundaries on storms that cross them

Because the nature of convective storms depends to a large degree on the environmental vertical shear profile and the CAPE (Weisman and Klemp 1982), it is expected that a storm's character will change as it crosses a surface boundary. In particular, it would be expected that as a storm crosses from the warm side to the cold side of a surface boundary its CAPE will decrease and the severity of the storm will also decrease. In addition, the shear profile may also change, especially at low levels (Fig. 3d). Numerical experiments have shown how convective storms can behave as they propagate through a nonhomogeneous environment (Richardson et al. 2000).

There have been cases in which storms form along the intersection of a dryline and an outflow boundary, but eventually cross the outflow boundary and decay. In some of these cases, convection is suppressed at most locations, except right near the intersection of the dryline and the outflow boundary, so that there can be periodic development of the same type of storm at the same location (Bluestein and MacGorman 1998).

It has been the author's experience that supercells rarely produce tornadoes after they have crossed an outflow boundary or front and moved a substantial distance into the colder air. Because the supercells usually retain a mesocyclone aloft, tornado warnings based on a mesocyclone signature are frequently issued by the National Weather Service. However, this empirical finding needs to be quantified through an intensive climatological study.

On the other hand, the nearness to an outflow boundary may promote tornadogenesis, as suggested by Maddox et al. (1980), Weaver and Nelson (1982), Markowski et al. (1998a), and Rasmussen et al. (2000). Atkins et al. (1999) have shown, using numerical experiments, how the presence of a boundary can hasten surface mesocyclogenesis, but not increase its final intensity. The role of the outflow boundary is to provide a region in which horizontal vorticity can be produced baroclinically; the horizontal vorticity may subsequently be tilted onto the vertical. Bluestein and Pazmany (2000) estimated the vertical shear profile in the clear air, using a mobile Doppler radar, just upwind from a supercell about to produce a tornado near an outflow boundary, and found very strong shear at low levels.

When a storm crosses a surface boundary the electrical characteristics of the storm may also change (Gilmore and Wicker 2002). Bluestein and MacGorman (1998) also showed how the nature of cloud-to-ground lightning strikes changes, perhaps in response to a change in environment.

h. Relationship of the orientation of the hodograph (the vertical profile of environmental wind) to the orientation of surface boundaries, and convective storm initiation and behavior

The relationship between the component of low-level vertical wind shear normal to density currents and storm initiation has been noted earlier (Rotunno et al. 1988;

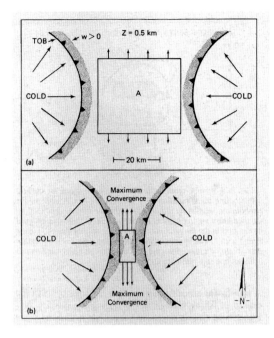

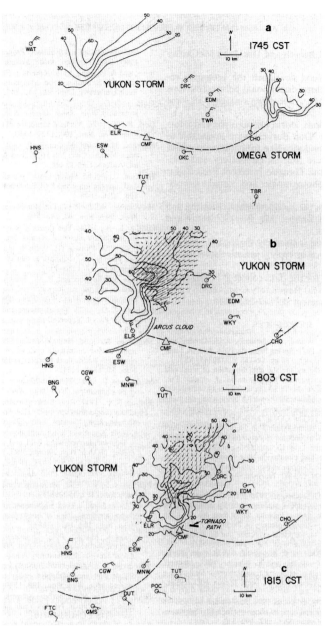

FIG. 23. Collision of outflow boundaries (left) head on (idealized depiction) and (right) at an angle (from observations). (left) Schematic diagram for the case of no vertical shear for (a) two outflows ~40 km apart and moving toward each other. The gust fronts are indicated by the boldface solid lines with barbs, and the regions of upward motion along the gust fronts are stippled. An arbitrary area A is shown by the box, and the small arrows indicate horizontal flow out of A as the outflows approach each other. (b) The outflows are now ~10 km apart. The regions of maximum horizontal convergence due to the rapid flow out of A from mass continuity are indicated. From Droegemeier and Wilhelmson (1985a). (right) Composite pictures of Doppler radar and surface data showing the relationship between outflow boundaries from two storms, the "Omega" and "Yukon" storms. Reflectivity data (dBZ) in the 1745 CST figure are from the NSSL WSR-57 radar (westernmost storm; Yukon storm). The Cimarron radar site is shown by a triangle and surface sites by circles. Full (half) wind barbs represent 5 (2.5) m s⁻¹. Outflow boundaries are indicated by the heavy dashed line. From Weaver and Nelson (1982).

Xu 1992). Peckham and Wicker (2000) have numerically investigated the relationship between the magnitude of both the vertical shear component and the reference wind normal to the dryline (and directed zonally, from the dry to the moist side). They found that the initiation of convective storms is most favored by weak cross-dryline reference flow (Fig. 25). When the ref-

erence flow becomes too strong, warm, subsiding air is advected eastward, producing surface-pressure falls east of the dryline; thus the westward-directed hydrostatic pressure-gradient force weakens, along with surface convergence. In addition, the capping inversion is strengthened and is lower.

When convective storms are initiated along a bound-

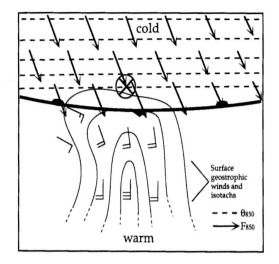

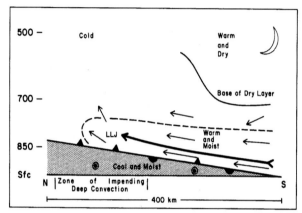

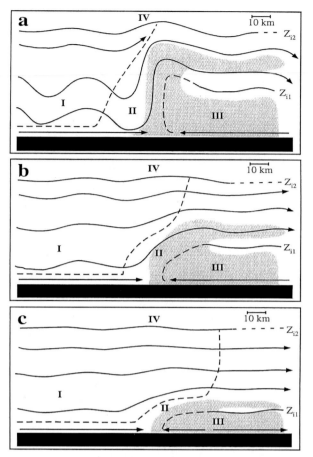

Fig. 24. Idealized depictions of the intersection of the LLJ with an outflow boundary or surface front. (top) Plan view; probable location of convection initiation at circle with an inscribed "X." From Augustine and Caracena (1994). (bottom) North–south vertical cross section. Deep convection develops above the wedge-shaped cold air mass (lightly shaded). The vectors represent the flow in the x–z plane (vertical component is greatly exaggerated). The dot inside the circle represents easterly flow (out of the page) within the cool, moist air mass below the frontal surface. The dashed line represents the boundary of the warm, moist air mass transported northward above the frontal surface by the southerly LLJ, whose axis is denoted by the boldface streamline. From Trier and Parsons (1993).

Fig. 25. Variation of dryline behavior with respect to strength of reference flow across the dryline from west to east, as determined from numerical simulation experiments. Conceptual model of the dryline environment at midafternoon for three zonal flow situations. Solid curves with arrows are streamlines representing the flow normal to the dryline. Dashed curves represent airmass discontinuities, and solid curves represent the upper (Z_{i2}) and lower (Z_{i1}) inversions. Roman numerals refer to air masses that are (I) hot, dry, and well mixed; (II) the mixing zone, containing mixtures of air from regions I and III; (III) warm and moist; and (IV) the overlying free atmosphere. The shaded regions represent regions containing relatively high water vapor content. (a) Conceptual model for weak zonal flow, (b) conceptual model for moderate zonal flow, and (c) conceptual model for strong zonal flow. From Peckham and Wicker (2000).

ary in an environment of ample CAPE, and the vertical shear is strong enough (approximately >20 m s^{-1} over the lowest 6 km), supercells form (Weisman and Klemp 1982). Because most supercells undergo some degree of splitting during their early life, each of the split pair of cells moves in different directions. Because the motion of the storms depends primarily on the orientation of mean vertical shear in the lowest 6 km or so, it is expected that neighboring storms also triggered along the boundary may interact with each other and/or cross the boundaries according to the relative orientation of the boundaries with the mean vertical wind shear vector (Lilly 1979). Bluestein and Weisman (2000) found, using numerical simulation experiments, that for storms

triggered along a line every 30 km, shear oblique to (45° from) the line is most favorable for discrete cyclonic supercells within the line (Fig. 26). Shear normal to the line supports a squall line with isolated supercells at both ends of the line, and shear parallel to the line supports isolated supercells only on the downshear end of the line. A major question remains, however: What controls the spacing of the storms that are triggered in nature?

5. Suggested research topics

Our knowledge about the role of surface boundaries in the southern plains of the United States in initiating

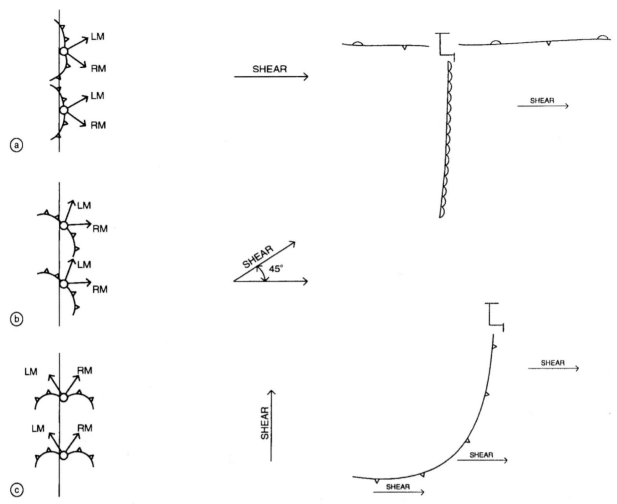

FIG. 26. How the relative orientation of a boundary with respect to the orientation of the vertical shear vector affects the behavior of convective storms triggered along the boundary. (left) Idealized illustration of the relationship between the orientation of the boundary along which convection is triggered (initial thermal bubbles indicated by circles along a vertical line) and the vertical shear vector. Outflow boundaries denoted by cold front symbols; RM and LM denote storm motion (vectors) of subsequent right- and left-moving cells. (right) Typical synoptic features at the surface. Vectors represent vertical shear; L denotes low pressure area associated with the cyclone. The dryline is represented by a scalloped line. From Bluestein and Weisman (2000).

convection is rather limited, even though this area is one the most intensely observed and studied regions on earth. While at the time of this writing the results from IHOP are forthcoming, and should advance our knowledge considerably, much still remains to be investigated. Studies of both an observational nature, using new observing systems such as mobile Doppler radars, lidars, radiometers, automated mesonets, and profilers need to be employed. Controlled numerical simulation experiments using higher-resolution models having more sophisticated rendering of boundary layer processes are also necessary.

In the spirit of Fred Sanders' careful analysis of observational data and continual questioning of conventional wisdom, the following fundamental questions are suggested as the basis for further study:

1) What controls the spacing of convective elements along a boundary?
2) What controls the size of mature convective elements along boundaries?
3) Under what conditions will isolated cells develop along a boundary? Under what conditions will convective lines develop along a boundary?
4) How can the location and depth of outflow boundaries be forecast more accurately with lead times of one day or more?
5) How can the location of the dryline be forecast more accurately?
6) How can one measure the depth of the cold pool behind an outflow boundary operationally?
7) How can one measure vertical shear at boundaries operationally?

8) What is the role of cirrus overcast in modulating convective initiation?

9) How are surface troughs formed and maintained? Can Sanders' (1999) model of the separation between a frontal zone and a trough be verified observationally and numerically?

10) What is the thermal structure of surface boundaries on the small scale?

While mobile-multiple-Doppler-radar observations can document the small-scale aspects of the three-dimensional wind field (e.g., Richardson et al. 2003), and in situ mobile instruments (Straka et al. 1996) can map out the small-scale temperature field in one dimension, few techniques exist for mapping remotely the three-dimensional temperature field on the small scale. Thermodynamic retrieval techniques based on the wind field (e.g., Gal-Chen and Kropfli 1984) and techniques based on radar phase measurements of ground targets (Fabry et al. 1997) are yet to be put to their full potential.

In addition to answering the aforementioned questions, the available archive of Weather Surveillance Radar-1988 Doppler (WSR-88D) data needs to be studied climatologically to quantify the many different modes of convective development along surface boundaries and to relate the modes to the synoptic pattern and to the environmental shear and CAPE. Corresponding numerical simulations also need to be done to further elucidate the physical processes involved.

Acknowledgments. This work was supported by NSF Grant ATM-0241037. Fred Sanders' mentorship in graduate school was the inspiration for much of the author's interest and research in the topics described in this paper. The author is also grateful to his students and other participants in annual spring severe weather field programs based at the University of Oklahoma and NSSL. Andrew Crook and an anonymous reviewer made helpful suggestions on how to improve this manuscript. Dave Schultz also provided very useful comments.

REFERENCES

Atkins, N. T., R. M. Wakimoto, and C. L. Ziegler, 1998: Observations of the finescale structure of a dryline during VORTEX 95. *Mon. Wea. Rev.,* **126,** 525–550.

——, M. L. Weisman, and L. J. Wicker, 1999: The influence of preexisting boundaries on supercell evolution. *Mon. Wea. Rev.,* **127,** 2910–2927.

Augustine, J. A., and F. Caracena, 1994: Lower-tropospheric precursors to nocturnal MCS development over the central United States. *Wea. Forecasting,* **9,** 116–135.

Bluestein, H. B., 1993: *Synoptic-Dynamic Meteorology in Midlatitudes.* Vol II. *Observations and Theory of Weather Systems,* Oxford University Press, 594 pp.

——, 1999: A history of storm-intercept field programs. *Wea. Forecasting,* **14,** 558–577.

——, and C. R. Parks, 1983: A synoptic and photographic climatology of low-precipitation severe thunderstorms in the southern plains. *Mon. Wea. Rev.,* **111,** 2034–2046.

——, and M. H. Jain, 1985: Formation of mesoscale lines of precipitation: Severe squall lines in Oklahoma during the spring. *J. Atmos. Sci.,* **42,** 1711–1732.

——, and S. S. Parker, 1993: Modes of isolated, severe convective storm formation along the dryline. *Mon. Wea. Rev.,* **121,** 1354–1372.

——, and T. M. Crawford, 1997: Mesoscale dynamics of the near-dryline environment: Analysis of data from COPS-91. *Mon. Wea. Rev.,* **125,** 2161–2175.

——, and D. R. MacGorman, 1998: Evolution of cloud-to-ground lightning characteristics and storm structure in the Spearman, Texas, tornadic supercells of 31 May 1990. *Mon. Wea. Rev.,* **126,** 1451–1467.

——, and A. L. Pazmany, 2000: Observations of tornadoes and other convective phenomena with a mobile, 3-mm wavelength, Doppler radar: The spring 1999 field experiment. *Bull. Amer. Meteor. Soc.,* **81,** 2939–2951.

——, and M. L. Weisman, 2000: The interaction of numerically simulated supercells initiated along lines. *Mon. Wea. Rev.,* **128,** 3128–3149.

——, and S. G. Gaddy, 2001: Airborne pseudo-dual-Doppler analysis of a rear-inflow jet and deep convergence zone within a supercell. *Mon. Wea. Rev.,* **129,** 2270–2289.

——, G. T. Marx, and M. H. Jain, 1987: Formation of mesoscale lines of precipitation: Nonsevere squall lines in Oklahoma during the spring. *Mon. Wea. Rev.,* **115,** 2719–2727.

——, E. W. McCaul, G. P. Byrd, and G. R. Woodall, 1988: Mobile sounding observations of a tornadic storm near the dryline: The Canadian, Texas storm of 7 May 1986. *Mon. Wea. Rev.,* **116,** 1790–1804.

——, ——, ——, G. R. Martin, S. Keighton, and L. C. Showell, 1989: Mobile sounding observations of a thunderstorm near the dryline: The Gruver, Texas storm complex of 25 May 1987. *Mon. Wea. Rev.,* **117,** 244–250.

——, ——, ——, R. L. Walko, and R. Davies-Jones, 1990: An observational study of splitting convective clouds. *Mon. Wea. Rev.,* **118,** 1359–1370.

Browning, K. A., and C. W. Pardoe, 1973: Structure of low-level jet streams ahead of mid-latitude cold fronts. *Quart. J. Roy. Meteor. Soc.,* **99,** 619–638.

Bryan, G. H., and J. M. Fritsch, 2000: Discrete propagation of surface fronts in a convective environment: Observations and theory. *J. Atmos. Sci.,* **57,** 2041–2060.

——, J. C. Wyngaard, and J. M. Fritsch, 2003: Resolution requirements for the simulation of deep moist convection. *Mon. Wea. Rev.,* **130,** 2917–2928.

Burgess, D. W., and E. B. Curran, 1985: The relationship of storm type to environment in Oklahoma on 26 April 1984. Preprints, *14th Conf. on Severe Local Storms,* Indianapolis, IN, Amer. Meteor. Soc., 208–211.

Byers, H. R., and R. R. Braham Jr., 1948: Thunderstorm structure and circulation. *J. Meteor.,* **5,** 71–86.

Carbone, R. E., J. W. Conway, N. A. Crook, and M. W. Moncrieff, 1990: The generation and propagation of a nocturnal squall line. Part I: Observations and implications for mesoscale predictability. *Mon. Wea. Rev.,* **118,** 26–49.

Carr, F. H., and J. P. Millard, 1985: A composite study of comma clouds and their association with severe weather over the Great Plains. *Mon. Wea. Rev.,* **113,** 370–387.

Colle, B. A., and C. F. Mass, 1995: The structure and evolution of cold surges east of the Rocky Mountains. *Mon. Wea. Rev.,* **123,** 2577–2610.

Crawford, T. M., and H. B. Bluestein, 1997: Characteristics of dryline passage during COPS-91. *Mon. Wea. Rev.,* **125,** 463–477.

Crook, N. A., 1987: Moist convection at a surface cold front. *J. Atmos. Sci.,* **44,** 3469–3494.

——, 1996: Sensitivity of moist convection forced by boundary layer processes to low-level thermodynamic fields. *Mon. Wea. Rev.,* **124,** 1767–1785.

——, and M. J. Miller, 1985: A numerical and analytical study of atmospheric undular bores. *Quart. J. Roy. Meteor. Soc.,* **111,** 225–242.

Dorian, P. B., S. E. Koch, and W. C. Skillman, 1988: The relationship between satellite-inferred frontogenesis and squall line formation. *Wea. Forecasting,* **3,** 319–342.

Doswell, C. A., III, 1987: The distinction between large-scale and mesoscale contribution to severe convection: A case study example. *Wea. Forecasting,* **2,** 3–16.

——, and L. F. Bosart, 2001: Extratropical synoptic-scale processes and severe convection. *Severe Convective Storms, Meteor. Monogr.,* No. 50, Amer. Meteor. Soc., 27–69.

Doviak, R. J., and R. Ge, 1984: An atmospheric solitary gust observed with a radar, a tall tower and a surface network. *J. Atmos. Sci.,* **41,** 2559–2573.

Droegemeier, K. K., and R. B. Wilhelmson, 1985a: Three-dimensional numerical modeling of convection produced by thunderstorm outflows: Part I: Control simulation and low-level moisture variations. *J. Atmos. Sci.,* **42,** 2381–2403.

——, and ——, 1985b: Three-dimensional numerical modeling of convection produced by interacting thunderstorm outflows: Part II: Variations in vertical wind shear. *J. Atmos. Sci.,* **42,** 2404–2414.

Durran, D. R., and D. B. Weber, 1988: An investigation of the poleward edges of cirrus clouds associated with midlatitude jet streams. *Mon. Wea. Rev.,* **116,** 702–714.

Fabry, F., C. Frush, I. Zawadzki, and A. Kilambi, 1997: On the extraction of near-surface index of refraction using radar phase measurements from ground targets. *J. Atmos. Oceanic Technol.,* **14,** 978–987.

Fritsch, J. M., and R. L. Vislocky, 1996: Enhanced depiction of surface weather features. *Bull. Amer. Meteor. Soc.,* **77,** 491–506.

Fujita, T. T., 1970: The Lubbock tornadoes: A study of suction spots. *Weatherwise,* **23,** 160–173.

Fulton, R., D. S. Zrnic, and R. J. Doviak, 1990: Initiation of a solitary wave family in the demise of a nocturnal thunderstorm density current. *J. Atmos. Sci.,* **47,** 319–337.

Gal-Chen, T., and R. A. Kropfli, 1984: Buoyancy and pressure perturbations derived from dual-Doppler radar observations of the planetary boundary layer: Application for matching models with observations. *J. Atmos. Sci.,* **41,** 3007–3020.

Galway, J. G., 1992: Early severe thunderstorm forecasting and research by the United States Weather Bureau. *Wea. Forecasting,* **7,** 564–587.

Gaza, R. S., and L. F. Bosart, 1985: The Kansas City severe weather event of 4 June 1979. *Mon. Wea. Rev.,* **113,** 1300–1320.

Gilmore, M. S., and L. J. Wicker, 2002: Influences of the local environment on supercell cloud-to-ground lightning, radar characteristics, and severe weather on 2 June 1995. *Mon. Wea. Rev.,* **130,** 2349–2372.

Grasso, L. D., 2000: A numerical simulation of dryline sensitivity to soil moisture. *Mon. Wea. Rev.,* **128,** 2816–2834.

Haertel, P. T., R. H. Johnson, and S. N. Tulich, 2001: Some simple simulations of thunderstorm outflows. *J. Atmos. Sci.,* **58,** 504–516.

Hane, C. E., C. L. Ziegler, and H. B. Bluestein, 1993: Investigation of the dryline and convective storms initiated along the dryline: Field experiments during COPS-91. *Bull. Amer. Meteor. Soc.,* **74,** 2133–2145.

——, H. B. Bluestein, T. M. Crawford, M. E. Baldwin, and R. M. Rabin, 1997: Severe thunderstorm development in relation to along-dryline variability: A case study. *Mon. Wea. Rev.,* **125,** 231–251.

——, M. E. Baldwin, H. B. Bluestein, T. M. Crawford, and R. M. Rabin, 2001: A case study of severe storm development along a dryline within asynoptically active environment. Part I: Dryline motion and an Eta Model forecast. *Mon. Wea. Rev.,* **129,** 2183–2204.

——, R. M. Rabin, T. M. Crawford, H. B. Bluestein, and M. E. Baldwin, 2002: A case study of severe storm development along a dryline within a synoptically active environment. Part II: Mul-

tiple boundaries and convective initiation. *Mon. Wea. Rev.,* **130,** 900–920.

Hutchinson, T. A., and H. B. Bluestein, 1998: Prefrontal wind-shift lines in the plains of the United States. *Mon. Wea. Rev.,* **126,** 141–166.

Johnson, R. H., and B. E. Mapes, 2001: Mesoscale processes and severe convective weather. *Severe Convective Storms, Meteor. Monogr.,* No. 50, Amer. Meteor. Soc., 71–122.

Jones, P. A., and P. R. Bannon, 2002: A mixed-layer model of the diurnal dryline. *J. Atmos. Sci.,* **59,** 2582–2593.

Karyampudi, V. M., M. L. Kaplan, S. E. Koch, and R. J. Zamora, 1995a: The influence of the Rocky Mountains on the 13–14 April 1986 severe weather outbreak. Part I: Mesoscale lee cyclogenesis and its relationship to severe weather and dust storms. *Mon. Wea. Rev.,* **123,** 1394–1422.

——, S. E. Koch, C. Chen, J. W. Rottman, and M. L. Kaplan, 1995b: The influence of the Rocky Mountains on the 13–14 April 1986 severe weather outbreak. Part II: Evolution of a prefrontal bore and its role in triggering a squall line. *Mon. Wea. Rev.,* **123,** 1423–1446.

Keshishian, L. G., L. F. Bosart, and E. W. Bracken, 1994: Inverted troughs and cyclogenesis over interior North America: A limited regional climatology and case studies. *Mon. Wea. Rev.,* **122,** 565–607.

Koch, S. E., 1984: The role of an apparent mesoscale frontogenetic circulation in squall line initiation. *Mon. Wea. Rev.,* **112,** 2090–2111.

——, and J. McCarthy, 1982: The evolution of an Oklahoma dryline. Part II: Boundary-layer forcing of mesoconvective systems. *J. Atmos. Sci.,* **39,** 237–257.

——, and W. L. Clark, 1999: A nonclassical cold front observed during COPS-91: Frontal structure and the process of severe storm initiation. *J. Atmos. Sci.,* **56,** 2862–2890.

——, J. T. McQueen, and V. M. Karyampudi, 1995: A numerical study of the effects of differential cloud cover on cold frontal structure and dynamics. *J. Atmos. Sci.,* **52,** 937–964.

Lee, B. D., R. D. Farley, and M. R. Hjelmfelt, 1991: A numerical case study of convection initiation along colliding convergence boundaries in northeast Colorado. *J. Atmos. Sci.,* **48,** 2350–2366.

LeMone, M. A., 1973: The structure and dynamics of horizontal roll vortices in the planetary boundary layer. *J. Atmos. Sci.,* **30,** 1077–1091.

Lilly, D. K., 1979: The dynamical structure and evolution of thunderstorms and squall lines. *Annu. Rev. Earth Planet. Sci.,* **7,** 117–161.

Maddox, R. A., L. R. Hoxit, and C. F. Chappell, 1980: A study of tornadic thunderstorm interactions with thermal boundaries. *Mon. Wea. Rev.,* **108,** 322–336.

Markowski, P. M., E. N. Rasmussen, and J. M. Straka, 1998a: The occurrence of tornadoes in supercells interacting with boundaries during VORTEX-95. *Wea. Forecasting,* **13,** 852–859.

——, ——, ——, and D. C. Dowell, 1998b: Observations of low-level baroclinity generated by anvil shadows. *Mon. Wea. Rev.,* **126,** 2942–2958.

Martin, J. E., J. D. Locatelli, P. V. Hobbs, P.-Y. Wang, and J. A. Castle, 1995: Structure and evolution of winter cyclones in the central United States and their effects on the distribution of precipitation. Part I: A synoptic-scale rainband associated with a dryline and lee trough. *Mon. Wea. Rev.,* **123,** 241–264.

McCarthy, J., and S. E. Koch, 1982: The evolution of an Oklahoma dryline. Part I: A meso- and subsynoptic-scale analysis. *J. Atmos. Sci.,* **39,** 225–236.

McGinley, J. A., and Y. K. Sasaki, 1975: The role of symmetric instabilities in thunderstorm development along drylines. Preprints, *Ninth Conf. on Severe Local Storms,* Norman, OK, Amer. Meteor. Soc., 173–180.

Miller, J. A., T. A. Kovacs, and P. R. Bannon, 2001: A shallow-water model of the diurnal dryline. *J. Atmos. Sci.,* **58,** 3508–3524.

Murphey, H. V., R. M. Wakimoto, E. V. Browell, D. E. Kingsmill, and C. N. Flamant, 2003: Initiation of deep convection on May

24 and June 19, 2002 during IHOP. Preprints, *31st Conf. on Radar Meteor.*, Vol. II, Seattle, WA, Amer. Meteor. Soc., 775–777.

Neiman, P. J., and R. M. Wakimoto, 1999: The interaction of a Pacific cold front with shallow air masses east of the Rocky Mountains. *Mon. Wea. Rev.*, **127**, 2102–2127.

Parsons, D. B., M. A. Shapiro, M. Hardesty, R. J. Zamora, and J. M. Intrieri, 1991: The finescale structure of a west Texas dryline. *Mon. Wea. Rev.*, **119**, 1242–1258.

——,——, and E. Miller, 2000: The mesoscale structure of a nocturnal dryline and of a frontal–dryline merger. *Mon. Wea. Rev.*, **128**, 3824–3838.

Peckham, S. E., and L. J. Wicker, 2000: The influence of topography and lower-tropospheric winds on dryline morphology. *Mon. Wea. Rev.*, **128**, 2165–2189.

Rasmussen, E. N., S. Richardson, J. M. Straka, P. M. Markowski, and D. O. Blanchard, 2000: The association of significant tornadoes with a baroclinic boundary on 2 June 1995. *Mon. Wea. Rev.*, **128**, 174–191.

Rhea, J. O., 1966: A study of thunderstorm formation along drylines. *J. Appl. Meteor.*, **5**, 58–63.

Richardson, Y. P., K. K. Droegemeier, and R. P. Davies-Jones, 2000: The influence of horizontal variations in vertical shear and low-level moisture on numerically simulated convective storms. Preprints, *20th Conf. on Severe Local Storms*, Orlando, FL, Amer. Meteor. Soc., 11–15.

——, J. M. Wurman, and C. Hartman, 2003: Multi-Doppler analysis of convective initiation on 19 June 2002 during IHOP. Preprints, *31st Conf. on Radar Meteor.*, Vol. II, Seattle, WA, Amer. Meteor. Soc., 793–795.

Richter, H., and L. F. Bosart, 2002: The suppression of deep moist convection near the Southern Great Plains dryline. *Mon. Wea. Rev.*, **130**, 1665–1691.

Rockwood, A. A., and R. A. Maddox, 1988: Mesoscale and synoptic scale interactions leading to intense convection: The case of 7 June 1982. *Wea. Forecasting*, **3**, 51–68.

Roebber, P. J., D. M. Schultz, and R. Romero, 2002: Synoptic regulation of the 3 May 1999 tornado outbreak. *Wea. Forecasting*, **17**, 399–429.

Rotunno, R., J. B. Klemp, and M. L. Weisman, 1988: A theory for strong, long-lived squall lines. *J. Atmos. Sci.*, **45**, 463–485.

Sanders, F., 1999: A proposed method of surface map analysis. *Mon. Wea. Rev.*, **127**, 945–955.

——, and D. O. Blanchard, 1993: The origin of a severe thunderstorm in Kansas on 10 May 1985. *Mon. Wea. Rev.*, **121**, 133–149.

——, and C. A. Doswell, 1995: A case for detailed surface analysis. *Bull. Amer. Meteor. Soc.*, **76**, 505–522.

Schaefer, J. T., 1986: Severe thunderstorm forecasting: A historical perspective. *Wea. Forecasting*, **1**, 164–189.

Schultz, D. M., 2008: A review of cold fronts, including prefrontal troughs and wind shifts. *Synoptic–Dynamic Meteorology and Weather Analysis and Forecasting: A Tribute to Fred Sanders, Meteor. Monogr.*, No. 55, Amer. Meteor. Soc.

——, and P. N. Schumacher, 1999: The use and misuse of conditional symmetric instability. *Mon. Wea. Rev.*, **127**, 2709–2732.

Segal, M., and R. W. Arritt, 1992: Nonclassical mesoscale circulations caused by surface sensible heat-flux gradients. *Bull. Amer. Meteor. Soc.*, **73**, 1593–1604.

Shapiro, M. A., 1982: Mesoscale weather systems of the central United States. CIRES, 78 pp. [Available from the Cooperative Institute for Research in Environmental Sciences, 216 University of Colorado, Boulder, CO 80309.]

——, T. Hampel, D. Rotzoll, and F. Mosher, 1985: The frontal hydraulic head: A micro-α scale (~1 km) triggering mechanism for mesoconvective weather sytems. *Mon. Wea. Rev.*, **113**, 1166–1183.

Simpson, J. E., 1997: *Gravity Currents in the Environment and the Laboratory.* Cambridge University Press, 244 pp.

Stensrud, D. J., and J. M. Fritsch, 1993: Mesoscale convective systems in weakly forced large-scale environments. Part I: Observations. *Mon. Wea. Rev.*, **121**, 3326–3344.

Straka, J. M., E. N. Rasmussen, and S. E. Fredrickson, 1996: A mobile mesonet for finescale meteorological observations. *J. Atmos. Oceanic Technol.*, **13**, 921–936.

Sun, W. Y., 1987: Mesoscale convection along the dryline. *J. Atmos. Sci.*, **44**, 1394–1403.

——, and Y. Ogura, 1979: Boundary layer forcing as a possible trigger to a squall line formation. *J. Atmos. Sci.*, **36**, 235–254.

——, and C. C. Wu, 1992: Formation and diurnal variation of the dryline. *J. Atmos. Sci.*, **49**, 1606–1619.

Tegtmeier, S. A., 1974: The role of the surface, sub-synoptic, low pressure system in severe weather forecasting. M.S. thesis, School of Meteorology, University of Oklahoma, 66 pp. [Available from School of Meteorology, University of Oklahoma, 100 E. Boyd, Rm. 1310, Norman, OK 73019.]

Trier, S. B., and D. B. Parsons, 1993: Evolution of environmental conditions preceding the development of a nocturnal mesoscale convective complex. *Mon. Wea. Rev.*, **121**, 1078–1098.

Wakimoto, R. M., and N. T. Atkins, 1994: Observations of the seabreeze front during CaPE. Part I: Single-Doppler, satellite, and cloud photogrammetry analysis. *Mon. Wea. Rev.*, **122**, 1092–1114.

——, and D. E. Kingsmill, 1995: Structure of an atmospheric undular bore generated from colliding boundaries during CaPE. *Mon. Wea. Rev.*, **123**, 1374–1393.

——, C. Liu, and H. Cai, 1998: The Garden City, Kansas, storm during VORTEX 95. Part I: Overview of the storm's life cycle and mesocyclogenesis. *Mon. Wea. Rev.*, **126**, 372–392.

Weaver, J. F., and S. P. Nelson, 1982: Multiscale aspects of thunderstorm gust fronts and their effects on subsequent storm development. *Mon. Wea. Rev.*, **110**, 707–718.

Weckwerth, T. M., and R. M. Wakimoto, 1992: The initiation and organization of convective cells atop a cold-air outflow boundary. *Mon. Wea. Rev.*, **120**, 2169–2187.

——, J. W. Wilson, R. M. Wakimoto, and N. A. Crook, 1997: Horizontal convective rolls: Determining the environmental conditions supporting their existence and characteristics. *Mon. Wea. Rev.*, **125**, 505–526.

——, T. W. Horst, and J. W. Wilson, 1999: An observational study of the evolution of horizontal convective rolls. *Mon. Wea. Rev.*, **127**, 2160–2179.

——, and Coauthors, 2004: An overview of the International H₂O Project (IHOP 2002) and some preliminary highlights. *Bull. Amer. Meteor. Soc.*, **85**, 253–277.

Weisman, M. L., and J. B. Klemp, 1982: The dependence of numerically simulated convective storms on vertical wind shear and buoyancy. *Mon. Wea. Rev.*, **110**, 504–520.

Weisman, R. A., K. G. McGregor, D. R. Novak, J. L. Selzler, M. L. Spinar, B. C. Thomas, and P. N. Schumacher, 2002: Precipitation regimes during cold-season central U.S. inverted trough cases. Part I: Synoptic climatology and composite study. *Wea. Forecasting*, **17**, 1173–1193.

Weiss, C. C., and H. B. Bluestein, 2002: Airborne pseudo-dual Doppler analysis of a dryline-outflow boundary intersection. *Mon. Wea. Rev.*, **130**, 1207–1226.

——, ——, and A. L. Pazmany, 2003: Fine-scale radar observations of a dryline during the International H₂O Project. Preprints, *31st Conf. on Radar Meteorology*, Vol. II, Seattle, WA, Amer. Meteor. Soc., 846–849.

Whitney, L. F., Jr., 1977: Relationship of the subtropical jet stream to severe local storms. *Mon. Wea. Rev.*, **105**, 398–412.

Wilhelmson, R. B., and C.-S. Chen, 1982: A simulation of the development of successive cells along a cold outflow boundary. *J. Atmos. Sci.*, **39**, 1466–1483.

Wilson, J. W., and W. E. Schreiber, 1986: Initiation of convective storms at radar-observed boundary-layer convergence lines. *Mon. Wea. Rev.*, **114**, 2516–2536.

——, G. B. Foote, N. A. Crook, J. C. Fankhauser, C. G. Wade, J. D. Tuttle, C. K. Mueller, and S. K. Krueger, 1992: The role of boundary-layer convergence zones and horizontal rolls in the initiation of thunderstorms: A case study. *Mon. Wea. Rev.*, **120**, 1785–1815.

Xu, Q., 1986: Conditional symmetric instability and mesoscale rainbands. *Quart. J. Roy. Meteor. Soc., 112,* 315–334.

——, 1992: Density currents in shear flows—A two-fluid model. *J. Atmos. Sci., 49,* 511–524.

——, and J. H. E. Clark, 1985: The nature of symmetric instability and its similarity to convective and inertial instability. *J. Atmos. Sci., 42,* 2880–2883.

Young, G. S., and J. M. Fritsch, 1989: A proposal for general conventions in analyses of mesoscale boundaries. *Bull. Amer. Meteor. Soc., 70,* 1412–1421.

Ziegler, C. L., and C. E. Hane, 1993: An observational study of the dryline. *Mon. Wea. Rev., 121,* 1134–1151.

——, and E. N. Rasmussen, 1998: The initiation of moist convection at the dryline: Forecasting issues from a case study perspective. *Wea. Forecasting, 13,* 1106–1131.

——, W. J. Martin, R. A. Pielke, and R. L. Walko, 1995: A modeling study of the dryline. *J. Atmos. Sci., 52,* 263–285.

——, J. T. Lee, and R. A. Pielke, 1997: Convective initiation at the dryline: A modeling study. *Mon. Wea. Rev., 125,* 1001–1026.

——, E. N. Rasmussen, Y. P. Richardson, R. M. Rabin, and M. S. Buban, 2003: Relation of radar-derived kinematic features and in-situ moisture to cumulus development on 24 May 2002 during IHOP. Preprints, *31st Conf. on Radar Meteorology,* Vol. II, Seattle, WA, Amer. Meteor. Soc., 796–798.

Chapter 2

Strong Surface Fronts over Sloping Terrain and Coastal Plains

LANCE F. BOSART AND ALICIA C. WASULA

Department of Earth and Atmospheric Sciences, University at Albany, State University of New York, Albany, New York

WALTER H. DRAG

National Weather Service, Taunton, Massachusetts

KEITH W. MEIER

National Weather Service, Billings, Montana

(Manuscript received 4 November 2004, in final form 7 September 2005)

Bosart   Wasula   Drag   Meier

ABSTRACT

This paper begins with a review of basic surface frontogenesis concepts with an emphasis on fronts located over sloping terrain adjacent to mountain barriers and fronts located in large-scale baroclinic zones close to coastlines. The impact of cold-air damming and differential diabatic heating and cooling on frontogenesis is considered through two detailed case studies of intense surface fronts. The first case, from 17 to 18 April 2002, featured the westward passage of a cold (side-door) front across coastal eastern New England in which 15°–20°C temperature decreases were observed in less than one hour. The second case, from 28 February to 4 March 1972, featured a long-lived front that affected most of the United States from the Rockies to the Atlantic coast and was noteworthy for a 50°C temperature contrast between Kansas and southern Manitoba, Canada.

In the April 2002 case most of New England was initially covered by an unusually warm, dry air mass. Dynamical anticyclogenesis over eastern Canada set the stage for a favorable pressure gradient to allow chilly marine air to approach coastal New England from the east. Diabatic cooling over the chilly (5°–8°C) waters of the Gulf of Maine allowed surface pressures to remain relatively high offshore while diabatic heating over the land (31°–33°C temperatures) enabled surface pressures to fall relative to over the ocean. The resulting higher pressures offshore resulted in an onshore cold push. Frontal intensity was likely enhanced prior to leaf out and grass green-up as virtually all of the available insolation went into sensible heating.

The large-scale environment in the February–March 1972 case favored the accumulation of bitterly cold arctic air in Canada. Frontal formation occurred over northern Montana and North Dakota as the arctic air moved slowly southward in conjunction with surface pressure rises east of the Canadian Rockies. The arctic air accelerated southward subsequent to lee cyclogenesis–induced pressure falls ahead of an upstream trough that crossed the Rockies. The southward acceleration of the arctic air was also facilitated by dynamic anticyclogenesis in southern Canada beneath a poleward jet-entrance region. Frontal intensity varied diurnally in response to

Corresponding author address: Lance F. Bosart, University of Albany, State University of New York, 1400 Washington Ave., Albany, NY 12222.
E-mail: bosart@atmos.albany.edu

differential diabatic heating. Three types of cyclogenesis events were observed over the lifetime of the event: 1) low-amplitude frontal waves with no upper-level support, 2) low-amplitude frontal waves that formed in a jet-entrance region, and 3) cyclones that formed ahead of advancing upper-level troughs. All cyclones were either nondeveloping or weak developments despite extreme baroclinicity, likely the result of large atmospheric static stability in the arctic frontal zone and unfavorable alongfront stretching deformation. Significant frontal–mountain interactions were observed over the Rockies and the Appalachians.

---

1. Introduction

This paper is motivated by Fred Sanders' many pioneering contributions to the study of atmospheric fronts and frontogenetical processes [see Schultz (2008) elsewhere in this monograph volume for a discussion of his contributions in this area]. Fred taught us how to analyze and extract the maximum possible information out of the observations and how to understand and interpret the extracted information in terms of fundamental dynamical principles. The focus of this paper will be on intense surface fronts in which diurnally varying differential diabatic heating plays a significant role in frontogenesis. This task will be accomplished through two detailed case studies after a brief review of relevant frontogenetical processes. The first case study involves a short-lived but very intense "side-door" cold front that affected eastern New England on 17–18 April 2002, 49 yr to the day since the now-famous 17–18 April 1953 intense plains cold front studied by Sanders (1955). The second case study involves a long-lived, intense front that stretched from the Rockies to New England between 28 February and 3 March 1972. This case also had some similarities to the intense plains front studied by Sanders (1955).

Bjerknes (1919) conducted one of the first comprehensive studies of atmospheric weather systems through use of a dense network of surface stations over northwest Europe. These observations enabled Bjerknes and Solberg (1922) to first establish the structure and evolution of what are now called cold, warm, stationary, and occluded fronts. This now-classic paper increased the awareness of the meteorological community to the importance of surface boundaries to sensible weather in association with mobile cyclones in middle latitudes. This awareness culminated in the development of the Norwegian cyclone model [NCM; see Volkert (1999) for a review of the ideas behind the NCM] and contributed to an avalanche of published papers on fronts and frontogenesis. The invention of the radiosonde in the 1930s enabled researchers to begin exploring the vertical structure of cyclones and fronts (e.g., Bjerknes and Palmén 1937; Palmén 1949) and culminated in the discovery of upper-level fronts by Reed and Sanders (1953), Newton (1954), and Reed (1955) [see Bosart (2003) for a recent review of upper-level fronts]. These early pioneering studies showed that upper-level fronts were characterized by baroclinic zones of characteristic width (depth) of 100 km (1 km), horizontal temperature gradients of 5°–10°C (100 km)$^{-1}$, and potential tem-

perature increases of 10°–20°C through the frontal stable layer. The concept of a front as a three-dimensional entity motivated Miller (1948) to write a general frontogenesis equation and represented an advance over the early "two density" fluid frontal model used by Margules (1906). [Detailed reviews of surface and upper-level fronts and frontogenetical processes can be found in Eliassen (1959, 1990); Bluestein (1986); Keyser (1986, 1999); (Shapiro and Keyser (1990); Davies (1999); and Bosart (2003).]

Sanders (1955) used detailed surface and upper-air observations to conduct a comprehensive study of the intense surface cold front of 17–18 April 1953 over the high plains of the United States. This now-classic paper was the first in which radiosonde observations were used systematically to derive the vertical structure of an intense surface front. Sanders (1955) showed that the surface front, as measured by the horizontal temperature gradient, horizontal wind shear, and horizontal divergence, was strongest near the surface and weakened upward. He used the Miller (1948) frontogenesis equation to show that frontogenesis was maximized near the surface because of horizontal confluence along the frontal wind shift at the leading edge of the temperature break, accounting for the maximum in frontal intensity near the ground. Strong low-level horizontal convergence in conjunction with the frontal wind shift resulted in a narrow plume of intense ascent above the leading edge of the sloping frontal zone. The frontogenesis weakened upward in association with a relaxation of the contributions of horizontal confluence and twisting to frontogenesis in conjunction with a cross-front thermally direct circulation. Sanders (1955) also showed that the equatorward movement of the cold front was determined by the strength of the front-normal surface winds in the cold air just behind the frontal wind shift. Important unresolved questions from the Sanders (1955) study include how the cold front was initially organized near the lee (eastern) slopes of the Rockies and how diurnally varying differential diabatic heating contributed to frontal strength. These questions provide a part of the focus for the two case studies that follow.

It is important to recall that sharp horizontal temperature gradients on small scales (1–10 km) can be found, for example, across thunderstorm-driven outflow boundaries (e.g., Charba 1974; Goff 1976; Matthews 1981; Wakimoto 1982), near sea-breeze front convergence zones (e.g., Simpson et al. 1977), along cold fronts adjacent to sloping terrain (e.g., Carbone 1982; Hobbs and Persson 1982), near the heads of atmospheric density currents adjacent to higher terrain (e.g., Koch and Clark 1999; Zhang and Koch 2000), at the leading edge of "backdoor" and "side-door" cold fronts (e.g., Carr 1951; Bosart et al. 1973; Hakim 1992), and in conjunction with the passage of coastal fronts (e.g., Bosart et al. 1972, 1992; Bosart 1975, 1981, 1984; Nielsen 1989; Nielsen and Neilley 1990; Atallah and Bosart 2003; Colle 2003). The strength of baroclinic zones on these scales (1–10 km) can vary diurnally as a function of inland diabatic heating or cooling relative to coastal waters, diabatic heating and cooling over sloping and elevated terrain, and evaporative cooling associated with deep convection.

Sharp horizontal temperature gradients on somewhat larger scales (10–100 km) can be found in conjunction with terrain-influenced cool surges. Examples include the Australian and New Zealand "southerly (cool) changes" (e.g., Colquhoun et al. 1985; Garratt et al. 1985, 1989; Smith et al. 1982, 1987; Steiner et al. 1987; Garratt 1988; McInnes and McBride 1993; McBride and McInnes 1993); cold-air damming and cold surges east of the Rockies (e.g., Shapiro 1984; Shapiro et al. 1985; Dunn 1987, 1988, 1992; Hartjenstein and Bleck 1991; Keshishian et al. 1994; Colle and Mass 1995; Schultz et al. 1997), east of the Appalachians (e.g., Forbes et al. 1987; Stauffer and Warner 1987; Bell and Bosart 1988; Doyle and Warner 1990; Lapenta and Seaman 1990; Fritsch et al. 1992; Doyle and Warner 1993a,b,c,d), east of the Tibetan Plateau (e.g., Boyle 1986; Chen et al. 2002), and east of the Andes (e.g., Marengo et al. 1997; Garreaud 1999, 2000; Vera and Vigliarolo 2000; Lupo et al. 2001; Seluchi et al. 2006); topographic distortion of cold fronts over the Snake River plain (e.g., Steenburgh and Blazek 2001); orographic channeling of cold fronts along the Pyrenees (e.g., Hoinka and Heimann 1988); cold frontogenesis near the Alps (e.g., Hoinka and Volkert 1987; Volkert et al. 1991); cold surges east of the Sierra Madre and ensuing gap winds (Tehuantepecers) through the Chivela Pass into the Gulf of Tehuantepec (e.g., Schultz et al. 1997; Schultz and Steenburgh 1999; Steenburgh et al. 1998); and cold surges east of the Andes and Brazilian highlands (e.g., Marengo et al. 1997; Garreaud and Wallace 1998; Garreaud 1999, 2000; Vera and Vigliaroldo 2000; Lupo et al. 2001).

Surface temperature contrasts across frontal zones on these somewhat larger scales (10–100 km) can also vary diurnally in conjunction with the timing of lee cyclogenesis over sloping terrain (e.g., Keshishian et al. 1994; Colle and Mass 1995), prefrontal lee troughs that strengthen during afternoon heating (e.g., Hutchinson and Bluestein 1998; Schultz 2008), differential diabatic heating across frontal boundaries in association with sensible heating under clear skies in the warm air (e.g., Sanders 1955; Shapiro 1982; Koch 1984; Koch et al. 1995), and evaporatively assisted cold-air damming east of the Appalachians (e.g., Fritsch et al. 1992; Langmaid and Riordan 1998). Recent modeling studies have also

shed additional light on the impacts of complex terrain and differential diabatic heating on frontogenesis on these 10–100-km scales. Brennan et al. (2003) conducted a numerical investigation of a progressive rainband on Appalachian cold-air damming. They found that an area of inland pressure falls associated with the rainband triggered an isallobaric response that resulted in the reformation of the associated Carolina coastal front inland. Doyle (1997) studied the impact of mesoscale orography on a coastal jet and rainband. He found that flow blocking in the lowest 500 m and flow over the coastal barrier above this level contributed to mesoscale pressure perturbations and the development of a mesoscale windward ridge that helped to concentrate the coastal jet. A modeling study of Hurricane Floyd (1999) by Colle (2003) showed that a coastal front beneath a strong sloping baroclinic zone that extended to the middle troposphere played a crucial role in the formation of a narrow band of intense ascent and heavy precipitation just inland of the position of the coastal front.

The observational and numerical studies described above motivated us to conduct case studies of two intense surface fronts that illustrate the importance of diurnally varying differential diabatic heating over sloping terrain and across coastlines on frontogenesis. This paper is organized as follows. Section 2 contains a brief review of important frontal concepts and cold-air damming processes. A case study of an intense eastern New England coastal (side-door) front of 17–18 April 2002 is presented in section 3. It is followed in section 4 by a case study of a long-lived, intense front that extended from the northern Rockies to the Atlantic coast (28 February–3 March 1972). The discussion and conclusions follow in section 5.

2. Some frontogenesis concepts: A review

This section will briefly review basic frontogenesis concepts as applicable to coastal- and mountain-influenced fronts and associated cold-air damming. Much more comprehensive reviews of fronts and frontogenetical processes can be found in Bluestein (1986), Keyser (1986), and Schultz (2008).

Miller (1948) derived a general equation for two- and three-dimensional frontogenesis. A simplified version of this equation, applicable at the surface, and with the x (y) axis oriented parallel to the isentropes (perpendicular and directed toward colder air), can be written as

$$F = \frac{d}{dt}\left(-\frac{\partial \theta}{\partial y}\right)_p$$

$$= \left(\frac{\partial v}{\partial y}\right)_p \left(\frac{\partial \theta}{\partial y}\right)_p + \left(\frac{\partial \omega}{\partial y}\right)_p \left(\frac{\partial \theta}{\partial p}\right) - \frac{\partial}{\partial y}\left(\frac{d\theta}{dt}\right). \quad (1)$$

Here F is the frontogenesis function that expresses the rate of change of the potential temperature (θ) gradient

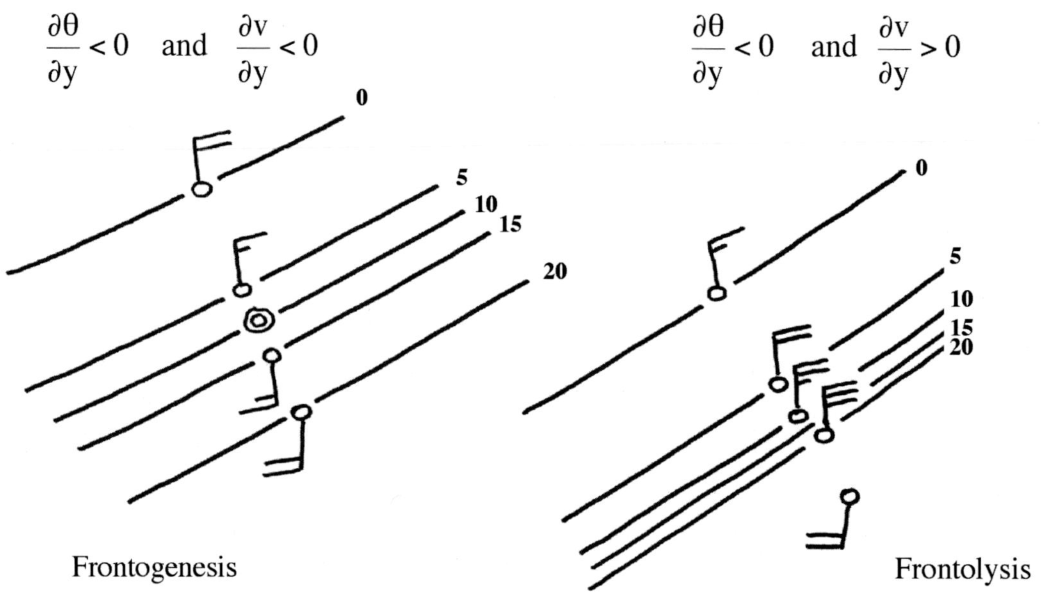

$$\frac{\partial\theta}{\partial y} < 0 \quad \text{and} \quad \frac{\partial v}{\partial y} < 0 \qquad\qquad \frac{\partial\theta}{\partial y} < 0 \quad \text{and} \quad \frac{\partial v}{\partial y} > 0$$

Frontogenesis

Frontolysis

$$F = \frac{d}{dt}\left(-\frac{\partial\theta}{\partial y}\right)_p = \left(\frac{\partial v}{\partial y}\right)_p\left(\frac{\partial\theta}{\partial y}\right)_p + \left(\frac{\partial\omega}{\partial y}\right)_p\left(\frac{\partial\theta}{\partial p}\right) - \frac{\partial}{\partial y}\left(\frac{d\theta}{dt}\right)$$

Fig. 1. Schematic of confluent frontogenesis and frontolysis and simplified Miller (1948) frontogenesis equation. Solid lines are isentropes. The x axis is taken parallel to the isentropes with the y axis directed toward colder air.

following a parcel, u and v are the wind components parallel to the x and y axes, respectively, and ω is the vertical motion (vanishes if the surface is level). The three terms on the right-hand side of (1) represent contributions to frontogenesis from horizontal confluence, twisting, and differential diabatic heating, respectively. Petterssen (1936, 1956) derived an alternative equation to (1) that could be expressed in terms of the contributions of deformation and divergence to frontogenesis (not shown). In his equation surface convergence is always frontogenetical while horizontal deformation contributes to frontogenesis only when the axis of dilatation is oriented at an angle of less than 45° to the isentropes. Although the Miller (1948) and Petterssen (1936) surface frontogenesis equations are equivalent computationally, here we opt to use (1) for all further discussions.

A schematic diagram illustrating the terms on the right-hand side of (1) is shown in Fig. 1. Surface frontogenesis occurs in regions where there is an overlap between the wind shift and the surface baroclinic zone and when wind speeds are relatively light in the vicinity of the wind shift and are stronger deeper into the cold and warm air, respectively. Under these conditions horizontal confluence acts to tighten the horizontal temperature gradient. When the wind shift and strongest winds coincide with the leading edge of the temperature

break, the isentropes at the leading edge of the colder air will "outrun" the isentropes deeper into the colder air and horizontal confluence contributes to frontolysis. A second schematic diagram shown in Fig. 2 extends the concept illustrated in Fig. 1 to frontogenesis over sloping terrain and across a coastline. Horizontal confluence contributes to frontogenesis when the warmest air lies closest to the mountains and the winds are downslope as compared with upslope winds farther east, a common situation over the high plains during much of the year. The twisting term also contributes to frontogenesis over sloping terrain when descent (ascent) maximizes in the warmer (cooler) air under the assumption that the atmosphere is statically stable ($\partial\theta/\partial p < 0$) as shown in Fig. 2.

Differential diabatic heating over sloping terrain and across coastlines can also contribute to frontogenesis. When surface sensible heating is maximized in the warmest air adjacent to the mountains [$-\partial/\partial y(d\theta/dt) > 0$] frontogenesis results, as shown in Fig. 2. Similarly, strong surface fronts can form over sloping terrain adjacent to higher mountains in conjunction with strong diabatic cooling immediately adjacent to the mountains relative to farther east on the plains. Likewise, differential diabatic heating across a coastline can contribute to frontogenesis when diabatic cooling (warming) is maximized over colder land (warmer ocean) even when

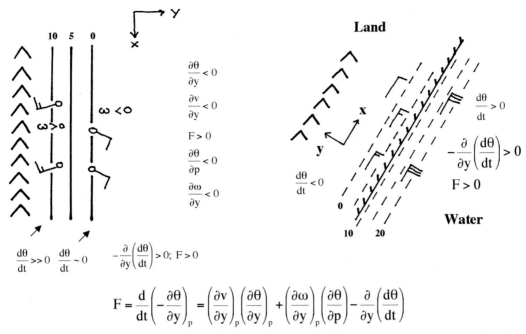

$$F = \frac{d}{dt}\left(-\frac{\partial \theta}{\partial y}\right)_p = \left(\frac{\partial v}{\partial y}\right)_p\left(\frac{\partial \theta}{\partial y}\right)_p + \left(\frac{\partial \omega}{\partial y}\right)_p\left(\frac{\partial \theta}{\partial p}\right) - \frac{\partial}{\partial y}\left(\frac{d\theta}{dt}\right)$$

FIG. 2. Schematic of frontogenesis over sloping terrain and across a coastline. Orientation of x and y axes as shown. Isentropes shown (left) solid and (right) dashed.

the frontogenesis contributions from horizontal confluence and twisting are small (Fig. 2).

An example of the latter process, taken from a case along the Atlantic coast of the southeastern United States during 9–10 November 2003, is shown in Fig. 3. The strong diurnal variation in the horizontal surface θ gradient is due to large (small) diurnal potential temperature variations over land (sea). The horizontal θ gradient strengthens overnight and is strongest at 1200 UTC 10 November in response to clear-sky nocturnal radiative cooling over land together with little temperature change over the much warmer ocean waters. By 1800 UTC 10 November the surface θ gradient has weakened considerably in response to the rapid warming of the air over land relative to the air over the ocean.

Equation (1) has also been applied to coastal frontogenesis situations (e.g., Bosart et al. 1972; Bosart 1975, 1984). A schematic diagram of cool-season coastal frontogenesis along the Atlantic coast is given in Fig. 4. Surface θ boundaries and associated cyclonic wind shifts are indicated in three regions: 1) along the coast, 2) where the cold coastal water (Labrador Current) encounters the warmer shelf waters, and 3) where the warmer shelf water meets the even warmer water marking the Gulf Stream. Coastal frontogenesis can occur across all three boundaries in response to horizontal confluence and differential diabatic heating. Differential roughness, not specifically indicated in (1), can also contribute to frontogenesis across the land–sea boundary in conjunction with greater (lesser) cross-contour flow angles over land (water). Additionally, differential roughness may contribute to frontogenesis across at-

mospheric thermal boundaries where horizontal gradients of atmospheric stability and surface wind speed may occur, such as near the landward edge of the shelf and Gulf Stream waters, respectively.

Cold-air damming, defined as the blocking of low-level flow by mountain barriers, can occur in conjunction with coastal frontogenesis where mountain ranges lie near coastlines such as east of the Appalachians of North America (e.g., Forbes et al. 1987; Bell and Bosart 1988), the Alps of southeastern Australia (e.g., Colquhoun et al. 1985), and the Brazilian highlands of northeastern Brazil (e.g., Lupo et al. 2001). A common way for cold-air damming to occur in the Northern Hemisphere is when a cold anticyclone noses equatorward east of a north–south-oriented mountain barrier. Under these conditions there will be a geostrophic easterly wind component directed toward the mountains. The strength of the winds, the depth of the cold air, and the height of the mountain barrier will all act to determine how much of the mountain-directed low-level flow is blocked and slowed by the mountains (e.g., O'Handley and Bosart 1996; Schumacher et al. 1996). Once the mountain-directed low-level flow is slowed and blocked by the mountains it is forced equatorward (downgradient) in response to insufficient Coriolis acceleration to balance the horizontal pressure gradient force (e.g., Forbes et al. 1987; Bell and Bosart 1988).

This blocking process is illustrated schematically for the Appalachians in Fig. 5. Cold-air damming will persist so long as a component of the low-level flow is directed toward the mountains. Coastal frontogenesis, by lowering sea level pressures along the coast relative

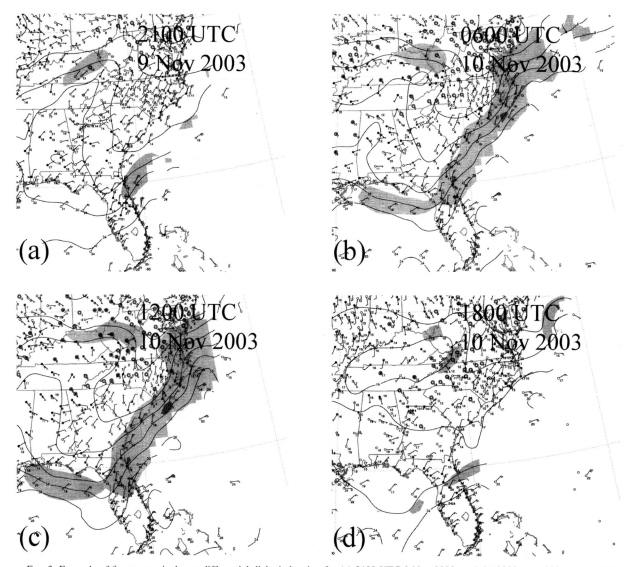

FIG. 3. Example of frontogenesis due to differential diabatic heating for (a) 2100 UTC 9 Nov 2003, and (b) 0000, (c) 1200, and (d) 1800 UTC 10 Nov 2003. Surface potential temperature contoured (solid) every 4°C. Surface potential temperature gradient (shaded) with light, moderate, and dark shading connoting a potential temperature gradient of 2.5°, 5.0°, and 7.5°C (100 km)$^{-1}$, respectively. Conventional plotting for surface winds (kt) and potential temperature (°C).

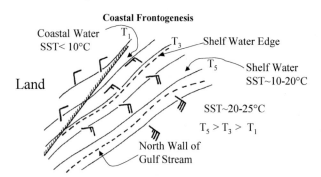

FIG. 4. Schematic of coastal frontogenesis across the land–ocean interface and thermal boundaries in the ocean.

to farther inland, can contribute to the maintenance of cold-air damming as a (wedge) ridge of higher pressure is maintained along the axis of coldest air between the mountains and the coast. Coastal frontogenesis and cold-air damming are often precursors to significant near-shore cyclogenesis (e.g., Bosart 1981). The impact of coastal cyclogenesis through its associated sea level pressure falls is to gradually allow a geostrophic westerly wind component to develop in the rear of the cyclone to the east of the mountains. The resulting offshore-directed low-level flow allows any cold air dammed up east of the Appalachians to be drained away. An example of cold-air damming is shown by cross sections of θ and winds in Fig. 6. The well-defined dome of stable, cold air wedged between the Appalachians

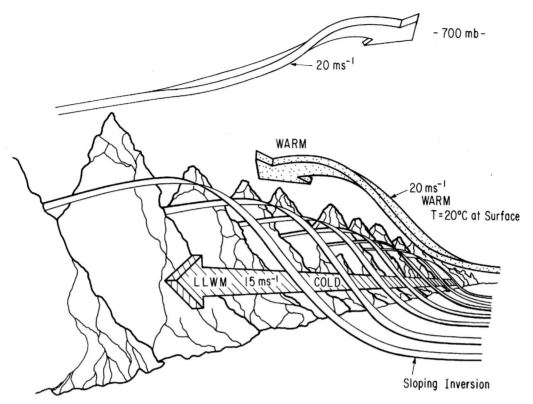

FIG. 5. Schematic of cold-air damming. The low-level wind maximum (LLWM) is indicated. From Bell and Bosart (1988, their Fig. 22).

and the coast is a persistent feature since there is no offshore component to the surface flow in the cold-air-damming region immediately to the west of the coastal front located just west of Cape Hatteras (HAT), North Carolina.

Cold-air damming and frontogenesis can also occur in conjunction with convective weather systems (e.g., Branick et al. 1988) and landfalling and transitioning tropical cyclones. Hurricane Agnes (1972) was associated with exceptionally heavy 20–40-cm rains in much of the Susquehanna River valley (e.g., Bosart and Carr 1978; DiMego and Bosart 1982a,b). The heaviest rains were focused along a weak surface thermal boundary that formed in situ along a line parallel to the coast and approximately 100 km inland (Bosart and Dean 1991). This thermal boundary, which acted as an inland coastal front, is shown in a time–distance (~50 km) cross section through the Washington, D.C., area in Fig. 7. The heaviest rains fell after the wind shift from easterly to northwesterly as typified by the 116 mm (128 mm) that fell at Dulles Airport (IAD) [Washington Reagan National Airport (DCA)] in the 6-h period ending 0600 UTC 22 June 1972. Although the thermal contrast across the boundary was small, ~3°C, there was sufficient convergence and deep ascent in association with the boundary to produce copious precipitation (Bosart and Dean 1991).

The most common way to produce cold-air damming

in conjunction with coastal frontogenesis to the east of the Appalachians is with a cold surface anticyclone centered over New England and adjacent Canada. The sea level pressure must be higher to the north so that a geostrophic easterly wind component can help to advect air toward the Appalachians. Less appreciated is that lower and decreasing sea level pressure to the south and west relative to the north and east can also create an easterly geostrophic wind component directed toward the Appalachians and induce cold-air damming and coastal frontogenesis. The landfall of Tropical Storm Marco in Florida on 11 October 1990 is an example of this process as sea level pressures fall to the south relative to the north (Fig. 8). Very heavy (15–25 cm) rains were observed along and to the west of the inland coastal front over Georgia and the Carolinas well north and well prior to the arrival of Marco (1990).

Cold-air damming also occurs east of the Rockies in conjunction with a cold anticyclone pushing equatorward (e.g., Dunn 1987, 1988, 1992). It is not unusual for cold air to make it all the way from eastern Montana to west Texas in 24 h in such situations, even in the absence of any northerly wind components above 850 hPa. The classic study of an intense low-level front by Sanders (1955) involved a strong cold surge east of the Rockies. A common characteristic of a cold-air surge east of the Rockies is that an inverted trough may be present on the plains to the east. Inverted troughs are

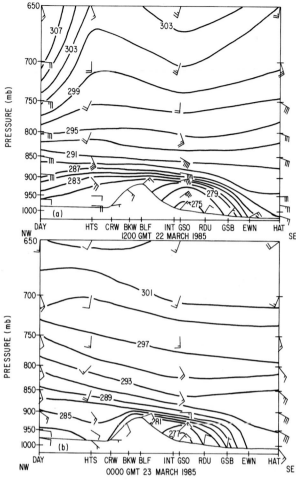

FIG. 6. Cross section of potential temperature (solid, every 2 K) and winds (m s⁻¹, with one full barb and half barb denoting 5 and 2.5 m s⁻¹, respectively). Cross section runs northwest–southeast from Dayton (DAY), OH, to HAT. Station locations are shown in Fig. A1. From Bell and Bosart (1988, their Fig. 14).

baroclinic features that are typically situated beneath the region of maximum low-level warm-air advection (Keshishian et al. 1994). Inverted troughs, which have many of the attributes of coastal fronts, tend to mark the eastern edge of a heavy precipitation shield. They have also been shown to be important in the distribution of precipitation associated with midwest cyclones (Weisman et al. 2002).

3. An eastern New England side-door cold front: 17–18 April 2002

This section presents the first of two case studies of intense surface fronts. Emphasis will be placed on the frontogenetical contribution of diurnally varying differential diabatic heating over sloping terrain and across the land–sea boundary. Cold fronts that cross the northeastern part of the United States, especially New England, from the north or northeast are known as back-

door fronts (e.g., Bosart et al. 1973). Cold fronts that reach the Middle Atlantic and New England states from the east are known as side-door fronts (e.g., Hakim 1992). Distinguishing features of many backdoor and side-door cold fronts include 1) shallow (<1 km deep) cold air, 2) an abrupt wind shift and temperature decrease, 3) minimal cloudiness and an absence of precipitation (the previous continental air mass is warm and dry), and 4) a large diurnal variation in frontal intensity in response to diurnally varying surface differential diabatic heating.

Late in the afternoon of 17 April 2002 a side-door cold front crossed coastal eastern New England, abruptly ending a brief early-season heat wave. The front originated over the cold waters of the Gulf of Maine. Rising (falling) sea level pressures over Maine and Nova Scotia (interior southern New England) helped to accelerate the front westward. At Milton (Blue Hill, MQE), Massachusetts, the prefrontal temperature of 32°–33°C decreased sharply (~19°C) to 13°C and the relative humidity increased from less than 30% to more than 90% between 2245 and 2315 UTC (Fig. 9). The prefrontal west-northwest wind of 10–15 kt abruptly shifted to a postfrontal northeast wind at 35 kt, a vector wind shift of more than 45 kt. By any standard measure this was an impressive cold front passage.

The impact of the side-door frontal passage on the local population was significant. The unexpected intensity of the afternoon heat (temperatures above 32°C in New England are *very* rare in April) induced a massive bout of severe spring fever in the populace. Hordes of people left work early (and students cut school), went home, ransacked their closets and drawers for shorts and T-shirts, and headed outside. Crowds descended on the beaches, the riverbanks, and the parks. Many people opted to celebrate a surprise hot summer day in April that was devoid of bugs by staging impromptu barbecues. Still other people turned on their air conditioners. The dinner hour throughout coastal eastern New England began on a very blissful note. Virtually the entire populace, intoxicated by the record warmth, forgot a vital lesson of life: Wise indeed is the New Englander who knows which way the wind is blowing. In the spring, a delightfully warm, sunny day with a nice offshore breeze is seldom "improved" when an onshore wind arrives, thanks to the cold waters offshore. As attested by the abrupt local weather change to chilly, windy, and damp conditions seen in Fig. 9, the end of the dinner hour for those caught outside (and that included most of the populace) was anything but blissful. One can also well imagine that most of the populace was irate because the local forecasts were mostly irrelevant as to advising the public of the timing and magnitude of the observed sensible weather changes.

To illustrate one aspect of the forecast challenge posed by strong side-door frontal passages a brief story is in order. A similar side-door frontal passage occurred on 25 April 1962 while the lead author, an undergraduate

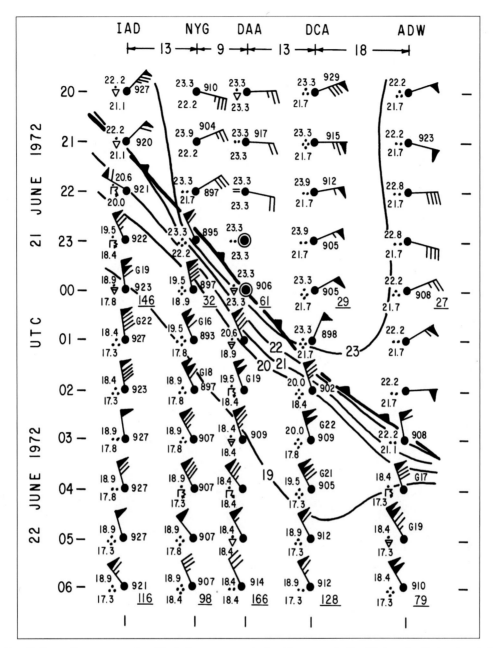

FIG. 7. Space–time cross section illustrating a mesoscale frontal passage in the Chesapeake Bay region during Hurricane Agnes (1972). Cross section extends from Sterling (Dulles, IAD), VA, to Andrews Air Force Base (ADW), MD. Time increases downward from 2000 UTC 21 Jun to 0600 UTC 22 Jun 1972. Isotherms (solid) are contoured every 1°C. Conventional surface observations except winds are plotted in knots with one pennant, full barb, and half barb denoting 10, 2, and 1 kt, respectively. Numbers between stations, denoted by 3-letter identifiers at the top, represent distances (km). Station locations are shown in Fig. A1. From Bosart and Dean (1991, their Fig. 7).

student at the Massachusetts Institute of Technology (MIT), was skippering a Tech Dingy on the Charles River as a part of his intramural sailing class. He was unprepared for the sudden wind shift (from west at ~25 kt to north-northeast at 20 kt) and capsized. Dozens of other sailboats shared a similar fate. To add insult to injury, the temperature drop from 25° to 9°C (most of which occurred in 10–15 min) in the hour ending 2100 UTC and accompanying strong northeast wind thoroughly chilled everyone adrift in the cold water. There we were, all dressed in shorts and T-shirts, soaked, wretched, miserable, and wondering what happened. Fast forward 40 yr to 17 April 2002 when once again a whole slew of sailboats capsized on the Charles River. What is wrong with this picture? Despite demonstrable rapid progress in numerical weather prediction and

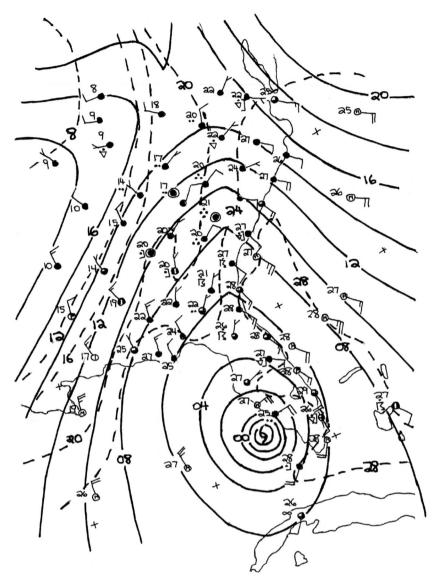

FIG. 8. Surface map for 0000 UTC 11 Oct 1990. Sea level isobars (solid) and isotherms (dashed) are drawn every 2 hPa and 4°C, respectively. Surface observations are plotted conventionally. Open hurricane symbol denotes the position of Tropical Storm Marco.

weather forecasting over the last 40 yr, the sailboats still fell like dominoes on the Charles River with the passage of a strong side-door cold front (another frontal event in April 2003 produced a similar result; K. Emanuel 2004, personal communication).

Events like the 17–18 April 2002 case create a credibility problem for forecasters and scientists alike. Given the rapid gains in scientific knowledge over the last 40 yr, it is difficult to understand why the forecast weather variability continues to be so much less than the observed weather variability for a select class of mesoscale weather phenomena that are readily noticed by the general public. There is clearly a gap between what scientists and forecasters know and understand about selected mesoscale weather phenomena and what they tell

the general public about these same phenomena. Mindful of his unplanned swim in the chilly waters of the Charles River more than 40 yr ago, motivated by a similar event in April 2002 that sent another generation of young sailors for an unexpected swim in the still-chilly waters of the Charles River, and driven by an appreciation of Fred Sanders' pioneering the field of oceanic mesometeorology (Sanders 1972) as the result of a storm-wracked Newport, Rhode Island, to Bermuda yacht race, the lead author decided to include the April 2002 case study in this article.

Although surface cooling with the side-door frontal passage near the coast was substantial as seen at MQE in Fig. 9, the flow aloft was quite ordinary. Anticyclonic conditions prevailed in the upper troposphere over much

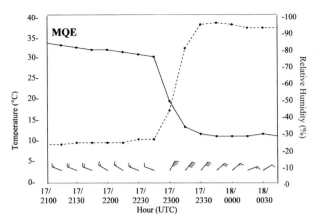

FIG. 9. Meteogram of temperature (solid, °C) and relative humidity (dashed, %) for MQE from 2100 UTC 17 through 0030 UTC 18 Apr 2002. Winds (kt) are plotted conventionally. See Fig. A1 for the location of MQE.

of eastern North America as seen by the distribution of θ on the dynamic tropopause (DT), and the 300–200-hPa mean-layer potential vorticity (PV)/winds at 0000 UTC 18 April (Figs. 10a,b). At 850 hPa a westerly flow of warm air covered most of the Northeast (Fig. 10c). Warm-air advection and ascent (700 hPa) was observed over parts of New England along the southwestern edge of the primary baroclinic zone that ran from Hudson Bay to Nova Scotia (Fig. 10d). A surface anticyclone was pushing southward across Labrador in the wake of a deep trough over extreme northeast Canada (Fig. 10d) that helped to reinforce this baroclinic zone. Surface pressure rises (falls) over Nova Scotia (New England) in response to this upper-level flow pattern (strong surface heating over New England) helped to drive the side-door cold front onshore over eastern New England as is also evident in a time series of selected surface observations for Boston (BOS), Massachusetts, and Yarmouth (YQI), Nova Scotia (Fig. 11). At YQI, in the cool air throughout, the sea level pressure rose ~5 hPa in the 6 h ending 0200 UTC 18 April in conjunction with anticyclogenesis across Nova Scotia. At BOS, situated in the strongly heated offshore flow, the sea level pressure fell slowly through 2200 UTC 17 April and the YQI–BOS pressure difference reached a maximum of 6.2 hPa just prior to the abrupt frontal wind shift to the northeast and the nearly 20°C temperature decrease.

The 1800 UTC surface analysis showed that the Northeast, with the exception of most of Maine, was covered by a hot and dry westerly offshore flow (Fig. 12a). A boundary separating the hot, dry westerly offshore flow from the cooler, moist easterly flow (note wind shift and associated isotherm packing in Fig. 12a) extended southeastward from north of New York to southwestern Maine. Within the associated sea level pressure trough a very weak low pressure center was situated near extreme southern coastal Maine. Thunderstorms observed at two stations, one in northern New Hampshire and the other in southern Maine, were in-

dicative of scattered convection along the boundary (Fig. 12a). An area of 700-hPa ascent that moved southeastward in the warm-air advection region along the warm flank of the baroclinic zone supported these thunderstorms (Fig. 10d). By 2200 UTC sea level pressures had lowered still further over interior southern New England in response to continued surface heating (note the expansion of the closed 1012-hPa isobar in Fig. 12b) and strengthening of the surface baroclinic zone. To the northeast, sea level pressures continued to rise through 2200 UTC, setting the stage for the onshore cool surge. At the Isle of Shoals (IOSN3) buoy located off the New Hampshire coast a sustained east-northeast wind of 19.1 m s^{-1} (peak gust ~25 m s^{-1} at 2054 UTC) at 2200 UTC (Fig. 12b) attested to the surge intensity (see also MQE winds in Fig. 9).

Based on the data shown in Figs. 9 and 12a,b, the surface temperature gradient across the frontal boundary was ~15°C (30 km)$^{-1}$. This large temperature gradient, combined with an estimated frontal speed of ~45 km h^{-1} (~12.5 m s^{-1}), produced 15°C (and more) temperature decreases in ~30 min (or less) along the coast of northeastern Massachusetts. As evidenced by the sea surface temperature (SST) analysis for the 2.7-day period ending 2145 UTC 17 April shown in Fig. 13, over much of the Gulf of Maine—the source for the chilly marine air behind the front—the SSTs were in the 5°–8°C range.

The Chatham (CHH), Massachusetts, sounding from 0000 UTC 18 April was representative of the hot, dry air mass (characterized by a deep mixed layer with a θ of ~33°C) with the exception of a very stable, shallow surface-based inversion (Fig. 14). Winds were offshore at all heights and veered upward through 700 hPa, consistent with the previously inferred low-level warm-air advection. The shallow cooler, moister marine air at Gray (GYX), Maine, was confined below 850 hPa (Fig. 14). No mixed layer was present between 850 and 700 hPa at GYX as this station was located just to the northeast of the poleward edge of the hot air. Above 700 hPa the CHH and GYX temperature soundings were virtually identical. The west-southwesterly surface airflow at CHH was cooled by passage over the still-chilly ocean waters south of New England (SSTs < 10°C; Fig. 13) and accounted for the very shallow surface-based inversion seen in the CHH sounding. The presence of this surface-based inversion made it very difficult for the strong surface thermal boundary evident at 1800 and 2200 UTC along the coast of eastern New England to be seen in the chilly coastal waters around Cape Cod, Martha's Vineyard, and Nantucket Island (Figs. 12a,b). Accordingly, sea level pressures were higher there relative to interior southern New England, leading to a brisk west-southwesterly flow along the southern New England coast while a center of minimum sea level pressure remained anchored over the strongly heated landmass (Figs. 12a,b).

Gridded initialized data, available at 20-km resolu-

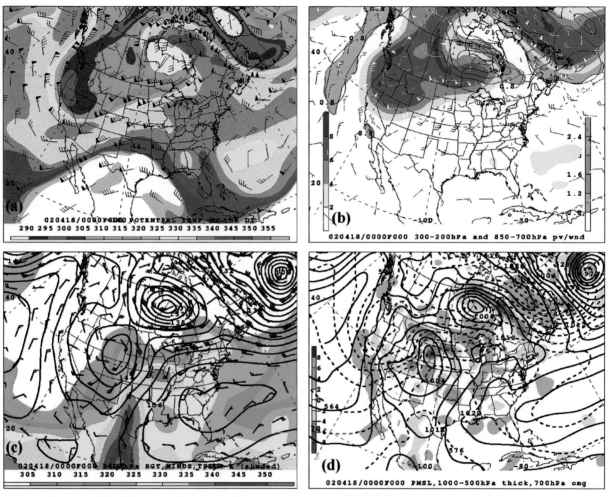

FIG. 10. Maps for 0000 UTC 18 Apr 2002 as follows: (a) potential temperature (shaded according to the color bar, every 5 K), and winds (m s^{-1}) on the dynamic tropopause [defined by the 1.5 PV unit (PVU) surface, where 1 PVU is 1×10^{-6} K m^2 kg^{-1} s^{-1}]; (b) 300–200-hPa (850–700-hPa) layer-averaged PV indicated by warm (cool) colors shaded according to the color bar at the left and right with 300–200-hPa (850–700-hPa) layer-averaged winds plotted in white (black; m s^{-1}) with 1 pennant, full barb, and half barb denoting 25, 5, and 2.5 m s^{-1}, respectively; (c) 850-hPa heights (solid, every 3 dam), winds (m s^{-1}), and equivalent potential temperature (shaded according to the color bar, every 5 K); and (d) sea level isobars (solid, every 4 hPa), 1000–500-hPa thickness (dashed, every 6 dam), and 700-hPa vertical motion with warm (cool) colors denoting ascent (descent) and shaded according to the color bar (10^{-3} hPa s^{-1}).

tion, were obtained from the National Centers for Environmental Prediction (NCEP) Rapid Update Cycle (RUC) model to use as a "surrogate" for observations between the standard 0000 and 1200 UTC upper-air data times. Shown in Fig. 15 is a cross section of θ, vertical motion, winds, and frontogenesis [horizontal confluence term in Eq. (1) only] through the side-door front for 2100 UTC 17 April as the front was approaching the coast of northeastern Massachusetts (note shallow, cool easterly flow offshore). Frontogenesis, concentrated at the leading edge of the thermal gradient and mostly confined below 900 hPa, was maximized at the surface [>10°C (100 km)$^{-1}$ (3 h)$^{-1}$] in the convergence region below the updraft maximum ($<-21 \times 10^{-3}$ hPa s^{-1}) centered between 850 and 800 hPa. Although the frontally induced ascent was quite vigorous, only scattered clouds were reported with the frontal passage because

of the dryness of the warm continental air mass ahead of the frontal boundary.

The thunderstorms over New Hampshire, Maine, and offshore noted previously were part of a broken area of convection that was moving southeast along the thermal boundary evident in Figs. 12a,b. This patchy convection can be seen in the 1932 UTC 17 April Weather Surveillance Radar-1988 Doppler (WSR-88D) base reflectivity image from GYX shown in Fig. 16a. As this area of convection moved offshore, embedded convective cells were still present at 2203 and 2301 UTC as seen in the base reflectivity radar images from Taunton (KBOX), Massachusetts (Figs. 16b,d). Given the very cold water over the Gulf of Maine (Fig. 13) and the 0000 UTC GYX sounding (Fig. 14), it is likely that the scattered convective cells had their roots above the cold, stable planetary boundary layer. The extent to which

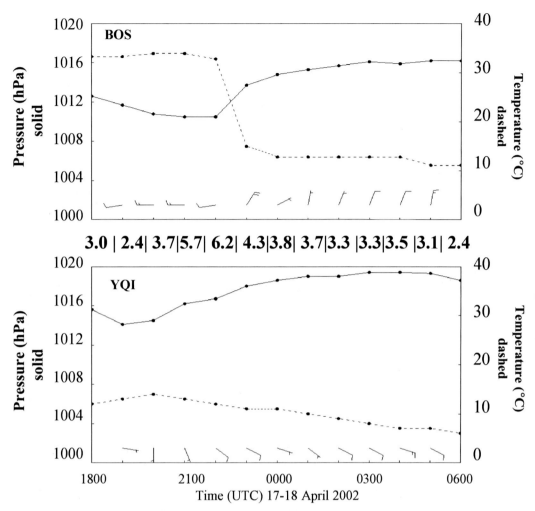

FIG. 11. Meteogram of sea level pressure (solid, hPa) and temperature (dashed, °C) for (top) BOS and (bottom) YQI for 1800 UTC 17 Apr–0600 UTC 18 Apr 2002. The YQI–BOS pressure difference (hPa) is given between the two panels. Station locations are shown in Fig. A1.

rain-cooled air may have reinforced the cold marine air behind the cold front is unknown. A fine line on the KBOX base reflectivity image for 2301 UTC (Fig. 16d) that marked the arrival of the frontal wind shift was well ahead of the scattered convective cells farther to the northeast. The KBOX base velocity image for 2255 UTC showed the abrupt transition from 20–26 kt (approximately 10–13 m s^{-1}) outbound to 20–36 kt (approximately 10–18.5 m s^{-1}) inbound flow across the leading edge of the front (Fig. 16c). The isolated pixels of greater than 36 kt (18.5 m s^{-1}) inbound flow were consistent with the reported 19 m s^{-1} surface wind at IOSN3 shown in Fig. 12b. Several weak wavelike perturbations along the leading edge of the frontal wind shift could be detected in the base velocity and reflectivity images as the frontal fine line approached and passed the radar site at 2203 and 2301 UTC (Figs. 16c,d).

The observations suggest that the side-door front may have had some characteristics of an atmospheric density

current. This possibility is explored by estimating the velocity of the front and the magnitude of the (hydrostatic) pressure rise following the frontal from an expression given by Seitter (1986) and used by Hakim (1992) in his side-door frontal case as follows:

$$V = k\left(\frac{\Delta P}{\rho_w}\right)^{1/2} + 0.61U. \qquad (2)$$

Here V is the speed of the front, ΔP is the hydrostatic surface pressure difference across the front, ρ_w is the density of the warm air, and U is the front-relative speed in the warm air. The Froude constant k is taken as 0.79. With estimates of U, ΔP, and ρ_w of 5 m s^{-1}, 3.5 hPa, and 1.14 kg m^{-3} (for a temperature of 32°C at 1012 hPa), respectively, V is calculated to be 16.8 m s^{-1}. This value of V is slightly more than 4 m s^{-1} larger than the estimated 12.5 m s^{-1} 2-h average frontal propagation speed (the front slowed down from nearly 15 m s^{-1} to almost 10 m s^{-1} during this period as it moved inland).

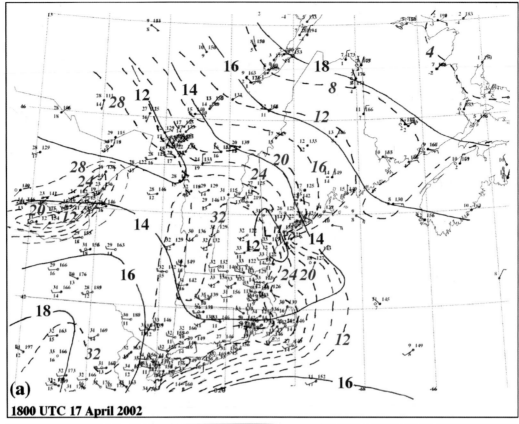

1800 UTC 17 April 2002

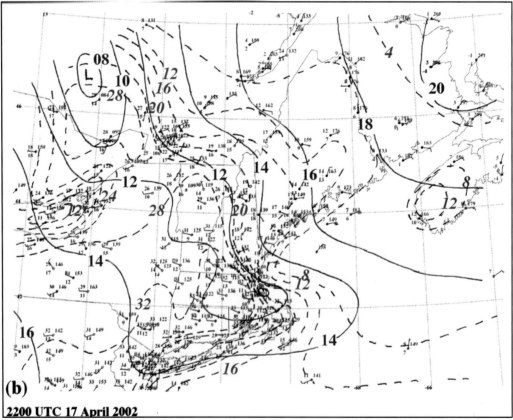

2200 UTC 17 April 2002

FIG. 12. Manually analyzed surface map for (a) 1800 and (b) 2200 UTC 17 Apr 2002. Sea level isobars (solid, every 2 hPa) and isotherms (dashed, every 4°C). Conventional station plotting.

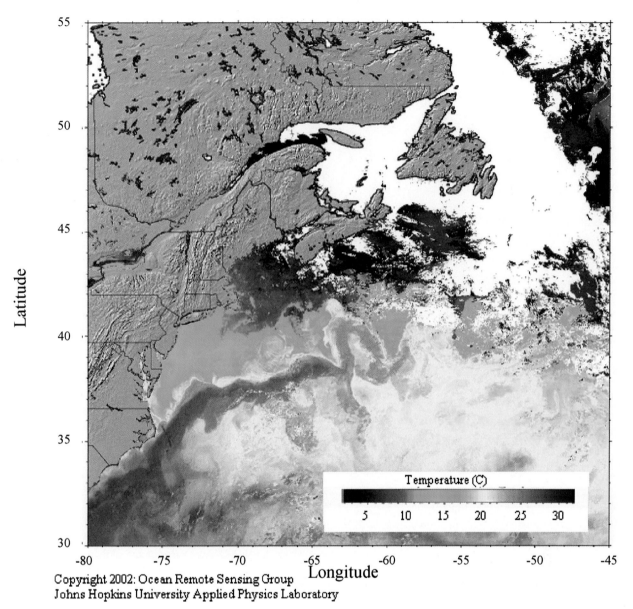

FIG. 13. Mean sea surface temperature (°C) for the 2.66-day period ending 2145 UTC 17 Apr 2002. Source: Ocean Remote Sensing Group, Johns Hopkins Applied Physics Laboratory.

The uncertainty of the computed value of V, given the frontal intensity and the observed strength of the cold air mass, is ~2 m s^{-1} [ambiguities in the choice for a value for k could increase the uncertainty further (Smith and Reeder 1988)].

A similar estimate of the hydrostatic pressure rise was made from the expression given by Schoenberger (1984) and Hakim (1992) as follows:

$$\Delta P = \frac{Pg\Delta Z\Delta T_v}{RT_{vw}T_{vc}}. \tag{3}$$

Here ΔP is the hydrostatic pressure change for a virtual temperature change ΔT_v across the frontal boundary, ΔZ is the depth of the cold air, R is the gas constant, g is gravity, P is the pressure, and T_{vw} and T_{vc} are the mean virtual temperatures in the warm and cold air, respectively. Based upon the observed sounding data shown in Fig. 13 and aircraft-derived soundings out of BOS [Aircraft Communication, Addressing, and Reporting System (ACARS); not shown], the depth of the cold air was estimated as ~750 m. With P taken to be 1012 hPa, $\Delta T_v = 12.5°C$, $T_{vw} = 305$ K, and $T_{vc} = 284$ K, ΔP was calculated as 3.7 hPa with an uncertainty of ~0.5 hPa, in reasonable agreement with the observed 2-h average change of ~3.5 hPa. The estimates from (2) and (3) and reported peak gusts >20 m s^{-1} at coastal locations suggest that the frontal surge had characteristics of an atmospheric density current as it moved onshore

74494 CHH   LI = 2   CAPE = 0   74389 GYX   LI = 3   CAPE = 0

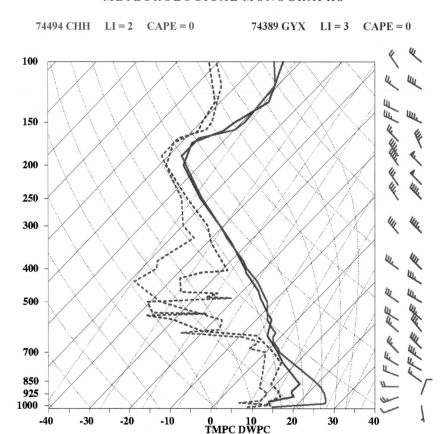

FIG. 14. Soundings (skew T–log$p$ format) from CHH (red) and GYX (blue) for 0000 UTC 18 Apr 2002. Winds (m s^{-1}) as in Fig. 10. Station locations are shown in Fig. A1.

and across eastern Massachusetts in the 2-h period ending 0000 UTC 18 April. Although the inference of an atmospheric density current must be viewed within the context of the uncertainties involved in evaluating the various parameters and constants in (2) as noted by Smith and Reeder (1988), the presence of a prefrontal deep mixed layer, the absence of prefrontal precipitation, and the indication of a fairly broad area of positive relative flow behind the front argues in favor of an atmospheric density current. That said, geostrophic adjustment likely contributes to the initiation and maintenance of the atmospheric density current in response to a diurnally driven increase in the cross-coast mesoscale surface pressure gradient from differential diabatic heating.

Meteograms of the observed and model output statistics (MOS) forecasts of temperature and winds for BOS for the period 1800 UTC 17 April to 0600 UTC 18 April are shown in Fig. 17. The MOS forecasts were derived from the NCEP Nested Grid Model (NGM), the Eta Model, and the Aviation (AVN) model. Collectively, the assorted model MOS temperature and wind forecasts showed no indication of the passage of a significant side-door cold front. Even the MOS forecast from the 1800 UTC AVN, initialized < 5 h before the arrival of the frontal wind shift at BOS, was unable to forecast the

abrupt temperature decrease and the change from an offshore to an onshore wind component at BOS. Clearly, the objective forecasts of the strong frontal passage across eastern New England in the late-afternoon hours of 17 April 2002 were irrelevant. This automated forecast failure of a very significant nonconvective sensible weather event suggests to us that opportunities exist to extract more mesoscale clues in the real observations on days when severe weather is unlikely and that doing so on these occasions might be more valuable to the local taxpayers than overly worrying about the week-2 temperature forecast. We suspect that much of the local populace might agree with this suggestion. Many years ago Napoleon was quoted as saying that nothing good comes from the east (he was referring both to cold weather and foreign armies). It would appear that Napoleon's admonition should be extended to the arrival of early-spring side-door cold fronts in eastern New England.

4. A long-lived intense front: 28 February to 4 March 1972

a. Overview of event

The long-lived intense cold front of 28 February–4 March 1972 was noteworthy for having a temperature

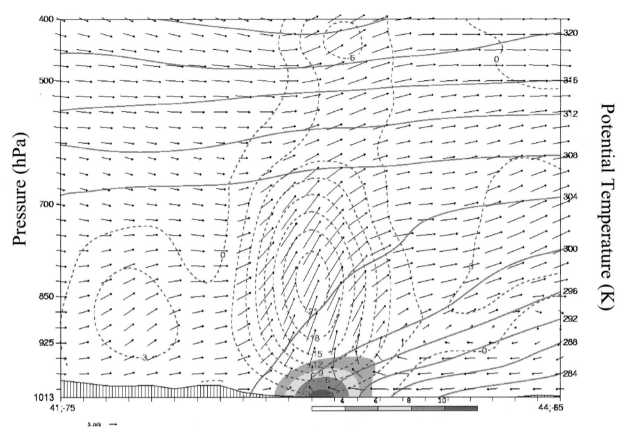

FIG. 15. Cross section of potential temperature (blue, every 4 K), vertical motion (dashed red with zero and ascent contours only every -3×10^{-3} hPa s^{-1}), winds in the plane of the cross section (scale at lower left), and frontogenesis due to horizontal confluence [shaded according to the color bar at the bottom in °C (100 km)$^{-1}$ (3 h)$^{-1}$] for 2100 UTC 17 Apr 2002. Orientation of the cross section is from 41.0°N, 75.0°W to 44.0°N, 65.0°W (Q–R) and is shown in Fig. A1.

contrast of more than 50°C between southern Manitoba, Canada, and Kansas (if that doesn't stir the synoptic blood then nothing ever will). This case is also an excellent example of a shallow, intense cold front of the type studied by Sanders (1955) and modeled 50 yr later by Schultz and Roebber (2008). Fred Sanders and the lead author also used this frontal case in our graduate-level synoptic laboratory classes for a number of years. Meier (1993) wrote his master's thesis on this case and some of his results will be integrated into this section.

The 1972 frontal event is also of interest because it featured 1) well-defined front–mountain interactions over the Rockies and Appalachians, 2) an abrupt equatorward movement east of the Rockies subsequent to 0000 UTC 1 March, 3) the absence of significant cyclogenesis despite (arguably) the strongest surface temperature gradient the first author has ever seen (going back to 1964), and 4) a number of relatively short-lived (~24 h) weak frontal waves that developed and propagated along the frontal boundary. These attributes of this case motivated us to study frontal behavior in the vicinity of complex terrain, to assess constraints on frontal movement, to document the role of differential diabatic heating in the observed diurnal variation of frontal

strength, and to determine the relationships between surface disturbances that propagated along the front and upper-level disturbances embedded in a strong jet stream.

A frontal isochrone analysis shows that through 0000 UTC 1 March the front remained quasi stationary along the lee (eastern) slopes of the northern Rockies eastward to southern Nebraska and eastward to New England (Fig. 18). Small (~100 km) north–south frontal oscillations were observed and the signature of cold-air damming was evident in the position of the frontal isochrones east of the northern Appalachians. Subsequent to 0000 UTC 1 March the front plunged equatorward, reaching south-central Texas by 0000 UTC 2 March, the central Gulf of Mexico by 0000 UTC 3 March, and the Atlantic Ocean by 1200 UTC 3 March. Several locations, notably near Omaha (OMA), Nebraska; Chicago O'Hare (ORD), Illinois; Cleveland (CLE), Ohio; Philadelphia (PHL), Pennsylvania; and New York LaGuardia (LGA), New York, experienced multiple frontal passages during the 6-day period.

A distinctive strong zonal flow pattern at 300 hPa was evident across North America with low-amplitude troughs (ridges) located over east-central North America

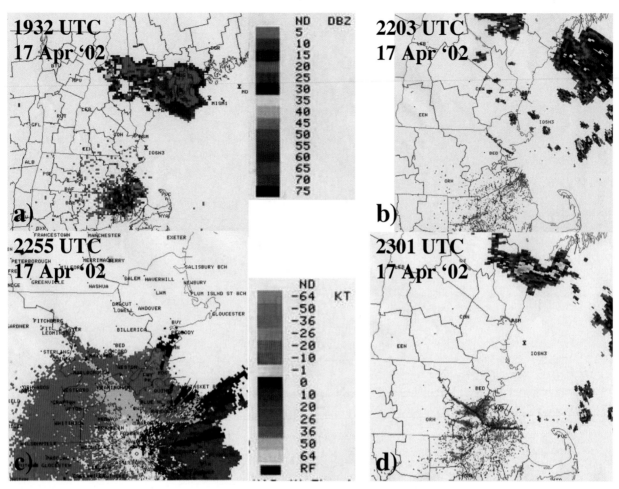

FIG. 16. Base reflectivity image from (a) Gray (KGYX), ME, at 1932 UTC; (b) as in (a) but for KBOX at 2203 UTC; (c) base velocity image from KBOX for 2255 UTC; and (d) base reflectivity image from KBOX for 2301 UTC. All radar beam elevation angles are 0.5°. All imagery is from 17 Apr 2002. Station locations are shown in Fig. A1.

and east of the date line (the eastern Pacific and the central and eastern Atlantic Ocean basins) in conjunction with the intense front (Fig. 19). Given the anomalously high heights located just west of California and just east of the mid-Atlantic coast, and the anomalously low (high) heights situated over the extreme northern United States and southern Canada (the southern United States), an anomalously strong westerly jet was present across the northern third of the United States (Figs. 19a,b). The large-scale flow pattern shown in Fig. 19, together with extensive snow cover over Canada and parts of the northern plains (Fig. 20), favored the accumulation of very cold air in these regions and set the stage for the formation of the intense front discussed in this section.

b. Data analysis procedures

To conduct a synoptic diagnostic analysis of this event a gridded upper-air dataset was generated using objective analysis procedures (Gandin 1963; Eddy 1967) de-

scribed in DiMego and Bosart (1982a,b). Objective analyses of height, temperature, winds, and relative humidity were made on a 51×31 latitude–longitude grid with a horizontal resolution of 1.0°. For the period 0000 UTC 28 February through 0000 UTC 2 March (1200 UTC 2–3 March) the analysis grid was bounded by 25°–55°N, 120°–70°W (25°–55°N, 115°–65°W). Each field was analyzed at 10 equally spaced pressure levels from 1000 to 100 hPa for all 10 synoptic observation periods. The input data region extended 20° of latitude–longitude out from each side of the analysis grid. The first-guess field was obtained from the then National Meteorological Center (NMC) gridded operational analyses obtained from the National Center for Atmospheric Research (NCAR). All available upper-air soundings, aircraft data, and surface observations within 3, 2, and 1 h of the 0000 and 1200 UTC observation times were used to generate the objective analyses. Once the objective analyses were completed they were interpolated to the intermediate 50-hPa levels by applying a linear-log p interpolation scheme. The General Meteorological

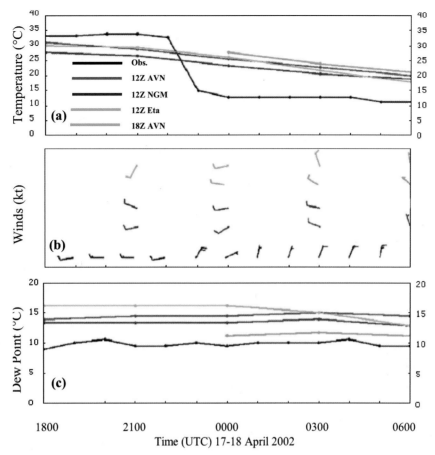

FIG. 17. Meteogram of BOS observed surface temperature (°C) and winds (kt), plotted in black, for the period 1800 UTC 17 Apr–0600 UTC 18 Apr 2002. MOS temperature and wind forecasts for BOS from the NCEP AVN, NGM, and Eta Model for selected times are shown in the indicated colors. Station location is shown in Fig. A1.

Package (GEMPAK; e.g., Koch et al. 1983) was used to compute and display all diagnostic results. Additional computational details can be found in Meier (1993).

An alternative diagnostic package developed by Loughe (1992) was also used for the core diagnostic calculations. This diagnostic package was initialized with the 1.0° gridded objective analyses described in the previous paragraph. These grids were then modified on a Lambert conic conformal map projection centered at 40°N and 90°W with a grid point spacing of 100 km. This second dataset was generated so that the psi-vector technique (Keyser et al. 1989) could be used to calculate alongfront and cross-front vertical circulations. These calculations were compared with those derived directly from GEMPAK and were found to produce results in agreement with the findings of Loughe (1992) who compared results obtained from this alternative diagnostic package with other published diagnostic analyses (e.g., Bosart 1981; Bosart and Lin 1984; Uccellini et al. 1984, 1985; Sanders and Bosart 1985a,b; Whitaker et al. 1988).

Detailed surface analyses were also prepared using all available first- and second-order National Weather Service (NWS) observations, Federal Administration Aviation (FAA) reports, and military observations. These raw observations were obtained from the original observations on microfiche from the National Climatic Data Center (NCDC). Ship observations were obtained from the NCAR Navy Spot dataset and the Comprehensive Ocean–Atmosphere Data Set (COADS). Canadian surface observations were obtained from NCAR and the Atmospheric Environment Service. The gridded dataset was generated using GEMPAK directly from the assembled raw observations that were manually digitized. The objectively analyzed surface data were prepared on a 0.5° × 0.5° grid from 25°–55°N, 120°–70°W through 0000 UTC 2 March and from 25°–55°N, 115°–65°W from 1200 UTC 2 March through 1200 UTC 3 March. Additional computation details can be found in Meier (1993).

c. *Synoptic-scale features*

At 0000 UTC 28 February a strong west-northwest flow at 300 hPa (jet core > 80 m s^{-1}) was found east of the Canadian Rockies and poleward of a broad ridge

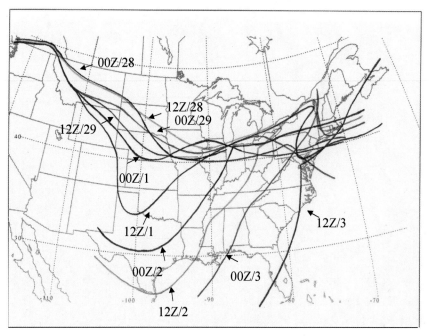

FIG. 18. Frontal continuity map for the period 0000 UTC 28 Feb–1200 UTC 3 Mar 1972. Isochrones are shown every 12 h.

300 hPa Z (dam)   300 hPa Z'(m)

28 Feb - 3 Mar 1972

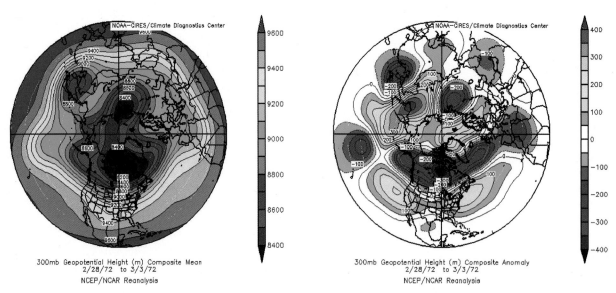

FIG. 19. (left) Time-mean and (right) anomalous 300-hPa heights shaded according to the color bars at the right for the period 28 Feb–3 Mar 1972. Height anomalies are computed relative to climatology (1968–96). These plots were constructed using an interactive data analysis package available from the National Oceanic and Atmospheric Administration (NOAA)–Cooperative Institute for Research in the Environmental Sciences (CIRES) Climate Diagnostic Center's (CDC) Web site at www.cdc.noaa.gov/Composites/Day.

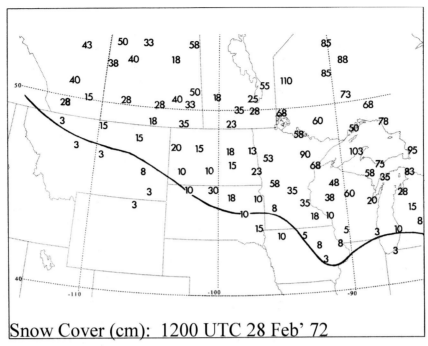

FIG. 20. Observed snow cover (cm) for 1200 UTC 28 Feb 1972.

located over western North America (Fig. 21b). A lee trough east of the Rockies at the surface and 850 hPa coincided with a subsidence region and a 1000–500-hPa-thickness ridge (Figs. 21a,c,d). The surface front, located on the warm side of a strong 1000–500-hPa-thickness gradient, lay along the lee slopes of the northern Rockies in an area of frontogenesis at 850 hPa from eastern British Columbia to western North Dakota (Figs. 18 and 21a,d). A cyclone over the Great Lakes was moving eastward and was allowing cold air to filter southward into the northern United States in its wake (Fig. 21a).

By 0000 UTC 1 March a strong lee cyclone had developed over Kansas ahead of a trough crossing the intermountain region (Fig. 22). As the downstream ridge aloft moved eastward to the western Great Lakes, a very strong west-northwesterly flow (>80 m s⁻¹) developed in the confluent jet-entrance region between this ridge and a deep, cold cutoff cyclone (1000–500-hPa-thickness minimum < 462 dam and 850-hPa temperatures <−40°C) situated over northern Hudson Bay at 300 hPa (Figs. 22a,b,d). Surface ridging over western Canada expanded eastward toward James Bay in conjunction with cold-air advection on the poleward side of the confluent jet-entrance region at 300 hPa (Figs. 22a,b). A vertical motion dipole at 700 hPa was located in the confluent jet-entrance region at 300 hPa (Fig. 22c). Descent coincided with the eastward ridging over Canada while ascent occurred with 850-hPa warm-air advection in the cold, easterly upslope flow north of the cyclone over western Kansas (Fig. 22c). With sea level pressures

over northern Manitoba and Saskatchewan more than 40 hPa higher than over western Kansas, the stage was set for a strong cold-air push equatorward along the eastern slopes of the Rockies subsequent to 0000 UTC 1 March (Fig. 22a). The cold push began once the confluent jet-entrance region at 300 hPa over Canada moved eastward onto the plains (Fig. 18).

By 0000 UTC 3 March the synoptic-scale flow pattern amplified somewhat and the strong surface front became more meridionally oriented, extending from north of New England to the northeastern Gulf of Mexico (Figs. 23a,b). Frontogenesis at 850 hPa was concentrated in the confluent 300-hPa equatorward jet-entrance region from the Ohio Valley to extreme southern Quebec beneath an elongated area of ascent at 700 hPa (Figs. 23b–d). Farther south, an area of strong convection had developed over the southeastern United States ahead of a trailing secondary trough in the southern branch of the westerlies (not shown). The cold air trapped from the Rockies to the northern plains at 0000 UTC 1 March plunged into the Gulf of Mexico (Figs. 23a,d).

d. Front along the lee slopes of the northern Rockies and over the northern high plains

Initially, the arctic front was most intense along the eastern slopes of the northern Rockies in Montana as seen in an objective surface θ analyses (Fig. 24). Through 1800 UTC 29 February the frontal boundary slowly slipped southward from Montana and North Dakota into Wyoming and South Dakota. The arctic air

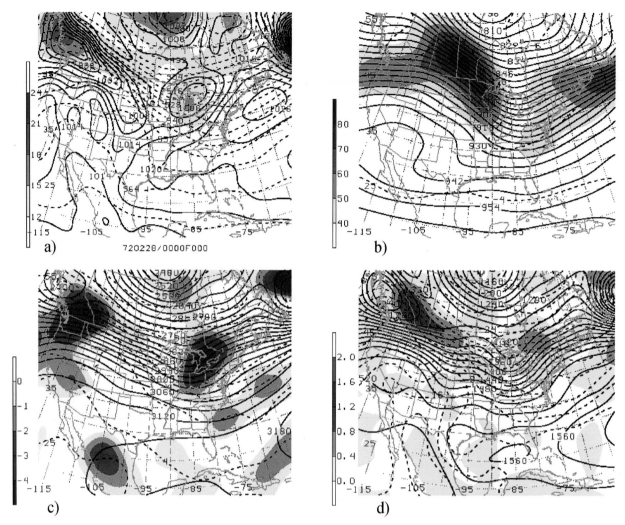

FIG. 21. Selected maps for 0000 UTC 28 Feb 1972. (a) Sea level isobars (solid, every 3 hPa), 1000–500-hPa thickness (dashed, every 6 dam), and 700-hPa absolute vorticity (shaded according to the color bar beginning at 12×10^{-5} s^{-1}); (b) 300-hPa heights (solid, every 6 dam), temperatures (dashed, every 4°C), and isotachs (m s^{-1}, shaded according to the color bar); (c) 700-hPa heights (solid, every 3 dam), temperatures (dashed, every 4°C), and vertical motion ($\times10^{-3}$ hPa s^{-1}, shaded according to the color bar for ascent only); and (d) 850-hPa heights (solid, every 3 dam), temperatures (dashed, every 4°C), and frontogenesis [°C (100 km)$^{-1}$ (3 h)$^{-1}$, shaded according to the color bar].

behind the front was very shallow as shown by the soundings from Great Falls (GTF), Montana, for 0000 UTC 28–29 February (Fig. 25). At 0000 UTC 28 and 29 February the strength of the near surface-based inversion was ~20°C. The shallow arctic air mass was characterized by easterly (upslope) wind components at both times (Fig. 25). Very strong west-southwesterly flow was present just above the surface where wind speeds exceeded 25 m s^{-1}. This strong west-southwesterly (Chinook) flow mixed down to the surface at 1200 UTC 28 February with resultant warming as seen by a temperature close to 10°C (Fig. 25).

Surface meteograms for Cut Bank (CTB), GTF, and Malmstrom Air Force Base (GFA) in Great Falls, all in Montana, attest to the frontal intensity (Fig. 26). Temperature rises of approximately 15°–20°C occurred in

one hour at both GFA and CTB as shallow arctic air was replaced by a warm southwest Chinook flow after 0200 UTC 28 February. At GTF, to the southwest of GFA, the Chinook-driven warm air arrived earlier (by 0000 UTC), was replaced temporarily by arctic air again at 0100 UTC, and returned by 0200 UTC 29 February. The intensity of the warm, dry Chinook winds was indicated by the reported gust of 69 kt (~35 m s^{-1}) at GTF at 0500 UTC.

The arctic air returned with equally abrupt (and spectacular) temperature falls at CTB, GFA, and GTF after 0900, 1500, and 1600 UTC 29 February, respectively, and was accompanied by a wind shift to northeast, 6–8-hPa pressure rises, and the onset of a steady light snow in the upslope northeasterly flow (Fig. 26). The resulting southward surge of the arctic air (Figs. 24a–e) is con-

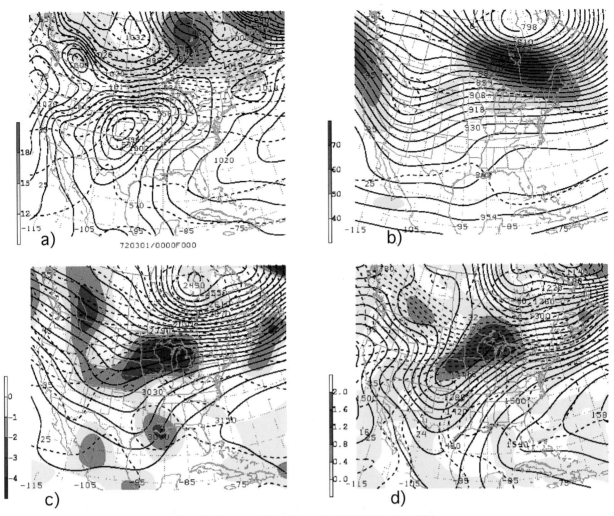

720301/0000F000

FIG. 22. Same as in Fig. 21 but for 0000 UTC 1 Mar 1972.

firmed by the elongated area of surface frontogenesis values greater than 20×10^{-10}°C m^{-1} s^{-1} located across Montana with a weaker maximum over extreme eastern Wyoming at 0000 UTC 28 February (Fig. 27a). Maximum frontogenesis values increased to greater than 100×10^{-10}°C m^{-1} s^{-1} in central Montana by 0600 and 1200 UTC 28 February (Figs. 27b,c). This substantial increase in frontogenesis was probably indicative of the previously inferred frontal strengthening as warm southwesterly Chinook winds developed (Fig. 27) with an attendant increase in horizontal convergence across the strengthening surface baroclinic zone. As the front began to move southward subsequent to 1200 UTC 28 February the maximum frontogenesis values decreased to less than 100×10^{-10}°C m^{-1} s^{-1} (except in isolated pockets) (Figs. 27d–h).

A PV cross section along 98°W (111°W) for 1200 (0000) UTC 28 (29) February shows that the warm air mass was characterized by near-zero values of PV (Figs. 28a,b) and steep lapse rates in the troposphere (not shown). The large values of static stability in the frontal

zone north of the surface boundary were associated with a distinct low-level PV maximum in the frontal zone. This frontal inversion PV anomaly developed in the absence of significant precipitation as only light snow was reported at the leading edge of the arctic air (Figs. 26 and 28a,b).

e. Equatorward frontal acceleration

A 2100 UTC 29 February surface analysis showed that temperatures ranged from 30°C over southwestern Kansas to −23°C in southern Manitoba (Fig. 29). The associated well-defined thermal boundary extended from southwestern South Dakota east-southeastward to eastern Nebraska, where a weak low-amplitude wave was present. The thermal boundary continued eastward across Iowa, extreme northern Illinois, and southern Michigan. A hint of a prefrontal trough was seen in the warm air over northwest Kansas and southwest Nebraska.

Subsequent to 0000 UTC 1 March the cold front ac-

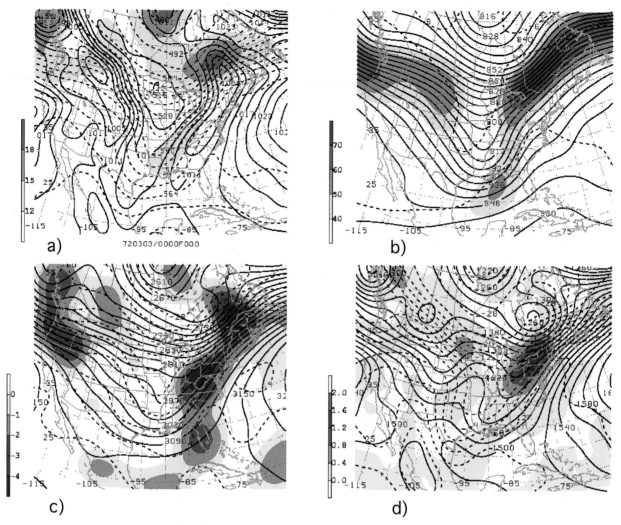

a)

720303/0000F000

b)

c)

d)

FIG. 23. Same as in Fig. 21 but for 0000 UTC 3 Mar 1972.

celerated southward and reached northern Texas by 1800 UTC 1 March (Fig. 30). The frontal θ gradient, strongest at 0000 UTC 1 March, weakened overnight across Kansas and Oklahoma in response to enhanced cooling in the warm, dry air (Fig. 29), as the front accelerated southward (Figs. 30a–c). By 1800 UTC 1 March, the θ gradient increased over southeastern Oklahoma and central Missouri as the warm air heated up again (Fig. 30d). Frontal passage across Dodge City (DDC), Kansas, is illustrated by soundings for 0000 and 1200 UTC 1 March in Fig. 31. The cooling was concentrated below 700 hPa ($\sim$33°C at the surface) and was driven by 15–25 m s^{-1} northerly winds. Above 700 hPa the cooling was minimal (3°–4°C) as winds remained west-southwesterly.

The arctic plunge toward the Gulf of Mexico began just before 0000 UTC 1 March. This plunge was driven by 3-h pressure rises that originated in the arctic air over eastern Montana at 1800 UTC 29 February (Fig. 32a) and expanded eastward into the western Dakotas by

2100 UTC (Fig. 32b). Three-hour pressure falls increased to 3–5 hPa from central Nebraska to central Minnesota by 2100 UTC 29 February because of strong insolation in the warm air and troughing in the colder air northeast of the weak wave development in eastern Iowa (Figs. 29 and 32d). An east–west pressure tendency gradient was located across Nebraska at 2100 UTC 29 February with the largest falls concentrated to the east of the developing prefrontal trough. The onset of the southward plunge of the arctic air just before 0000 UTC 1 March was manifest by a doubling of the 3-h pressure rises to 7–8 hPa over extreme northern Nebraska and extreme southern South Dakota (Fig. 32c). This enhanced pressure rise region expanded southward across Nebraska by 0300 UTC as the arctic air rushed southward (Figs. 30 and 32d).

Arctic surge onset is shown in the surface analyses for 0000 and 0600 UTC 1 March (Figs. 33a,b). A prefrontal trough noted earlier (Fig. 29) extended west-southwestward from a weak wave in southwest Iowa to

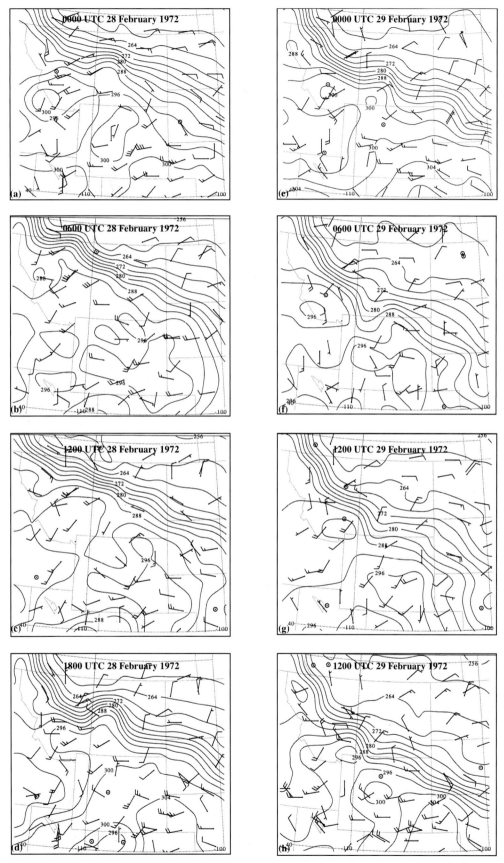

FIG. 24. Surface potential temperature (solid, every 4°C) and winds (kt) at (a) 0000, (b) 0600, (c) 1200, (d) 1800 UTC 28 Feb 1972, and (e) 0000, (f) 0600, (g) 1200, and (h) 1800 UTC 29 Feb 1972.

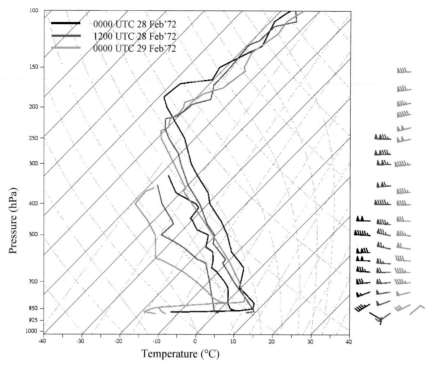

FIG. 25. As in Fig. 14 but for GTF at 0000 (black) and 1200 UTC (red) 28 Feb 1972, and 0000 UTC 29 Feb 1972 (green).

south-central Colorado (Fig. 33a). Three-hour pressure falls were concentrated ahead of the prefrontal trough (Fig. 32c). The arctic front extended from northeastern Wyoming to northern Nebraska and was marked by strong pressure rises (Figs. 32d and 33a). By 0600 UTC 1 March, the leading weak wave had raced eastward along the quasi-stationary portion of the frontal boundary to extreme southern Lake Michigan while a new weak wave that developed over extreme southeastern Nebraska at 0300 UTC 29 February (not shown) had moved to east-central southern Iowa (Fig. 33b). The southward-surging arctic frontal boundary intersected the prefrontal trough over west-central Kansas at 0600 UTC 1 March (Fig. 33b). Nocturnal cooling in the warm, dry air reduced the frontal temperature difference between Oklahoma and southern Canada to a still-amazing 41°C at this time.

A PV cross section at 0000 UTC 1 March through the weak Iowa frontal wave showed that PV was concentrated in the arctic frontal inversion while near-zero PV values prevailed in the warm air above (Fig. 28c). A θ, ω, and ageostrophic motion cross section farther east showed that a southward-sloping zone of ascent extended from the leading edge of the arctic air to above 500 hPa in the warm air near the prefrontal trough position over eastern Colorado (Fig. 34a). Comparison of Figs. 33a and 34a with Fig. 22 suggests that this deep ascent along 102°W occurred ahead of a 300-hPa trough that was crossing the Rockies. The 3-h pressure falls ahead of the prefrontal trough at 0000 UTC 1 March may also reflect this advancing trough

(Fig. 32c). The northerly ageostrophic wind components in the arctic air are small (3–5 m s⁻¹) and are consistent with surface winds blowing at right angles to the isobars (Fig. 33a).

By 1200 UTC 1 March, arctic air was deeply entrenched over the high plains northward of the Texas and Oklahoma panhandles (Figs. 18 and 30). High PV values prevailed at this time in the arctic frontal inversion that was now situated over northern Kansas and southern Nebraska (Fig. 28d) where only scattered light snow was reported (not shown). A cross section of θ, ω, and ageostrophic motion farther east (88°W) showed that this deep ascent was fed by strong low-level northerlies (10–12 m s⁻¹) at the leading edge of the arctic front (Fig. 34b).

f. Weak frontal waves

Meteograms for Des Moines (DSM), Cedar Rapids (CID), and Ottumwa (OTM), Iowa, were constructed to illustrate the passage of the weak frontal wave seen over Iowa at 0600 UTC 1 March (Fig. 33b) between these stations (Fig. 35). At DSM, wave passage was just before 0200 UTC 1 March when winds backed from southeasterly to northerly. The wind shift was accompanied by a temperature decrease from ~14° to ~3°C. The wave had a minimum sea level pressure near 996 hPa and was associated with a 1–2-hPa pressure perturbation. Further cooling occurred in the stronger northerlies behind the wave as pressures rose. At CID, the winds backed from southeasterly to northerly between 0300

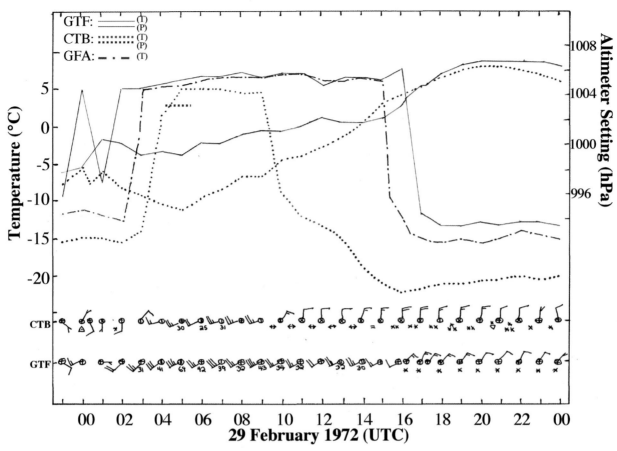

FIG. 26. Time series of temperature (°C), altimeter setting (hPa) and winds (kt), present weather, and sky cover, plotted conventionally for GFA, GTF, and CTB from 2300 UTC 27 Feb through 0000 UTC 1 Mar 1972. Station locations are shown in Fig. A1.

and 0400 UTC 1 March, and a similar 1–2-hPa pressure perturbation was observed. At OTM, to the south of DSM and CID, wave passage was reflected by veering winds from south-southeasterly to gusty south-south-westerly before 0400 UTC 1 March. The pressure per-turbation at OTM was also 1–2 hPa, but was broader, and cooling was delayed until the wind shift to north-westerly near 0900 UTC 1 March.

g. Diurnal variation in frontal intensity

A noteworthy aspect of the frontal boundary over the plains was the significant diurnal variation in frontal intensity as measured by $|\nabla \theta|$ and is illustrated by a time series of the maximum computed $|\nabla \theta|$ (based on hourly analyses) along 100°W in Fig. 36. The arctic front was most intense near 0000 UTC 29 February and again near 0000 UTC 1 March with maximum values of $|\nabla \theta|$ of ~17.5° and 16.2°C, respectively. The mini-mum values of $|\nabla \theta|$ were between 5°–7°C and occurred just before 1200 UTC on 28–29 February and 1 March. The diurnal variation in $|\nabla \theta|$ was a factor of 3 and was mostly attributable to abundant insolation in the warm, dry air ahead of the arctic frontal boundary. Small (1°–

3°C) perturbations in $|\nabla \theta|$ during the 60-h period were attributed mostly to analysis uncertainties.

Observed hourly profiles of $|\nabla \theta|$ and frontogenesis along 98°W were constricted to illustrate frontal evo-lution over western plains (Fig. 37). At 1200 UTC 29 February the primary frontal boundary (largest $|\nabla \theta|$) was situated between Huron (HON), South Dakota, and Grand Island (GRI), Nebraska (Fig. 37a). A secondary boundary, located between Hutchinson (HUT), Kansas, and Enid (END), Oklahoma, mostly disappeared after sunrise as differential nocturnal cooling weakened. Af-ter 1500 UTC 29 February the quasi-stationary primary frontal boundary just north of GRI strengthened as dia-batic heating began in the warm air. The frontal $|\nabla \theta|$ weakened as the boundary shifted southward between 0000 and 0200 UTC 1 March and then restrengthened as it slowed down (0200–0500 UTC) (Fig. 37b). Sub-sequent to 0500 UTC, $|\nabla \theta|$ lessened again as the frontal boundary accelerated southward. A discontinuous southward frontal movement was also suggested by the observed frontogenesis strengthening (weakening) when the frontal boundary decelerated (accelerated) (Fig. 37b).

As the arctic air advanced eastward and southward

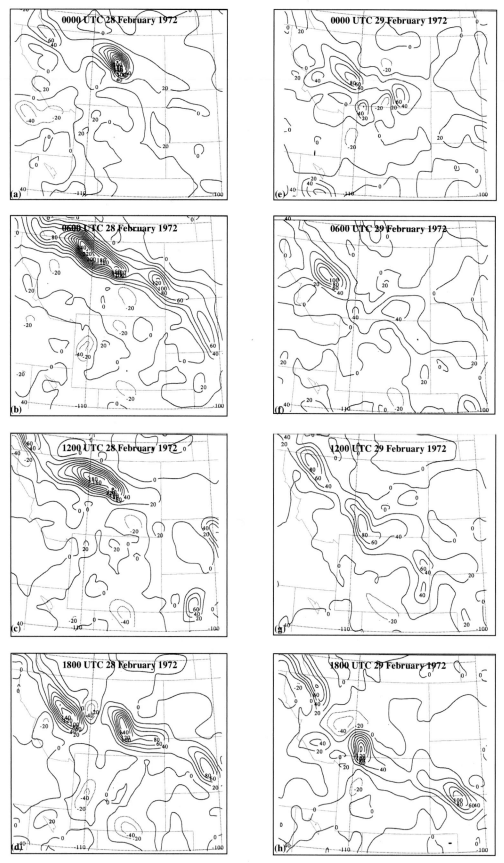

FIG. 27. Same as in Fig. 24 but for surface frontogenesis contoured every 20×10^{-10} °C m^{-1} s^{-1} with positive (negative) values shown by solid (dashed) contours.

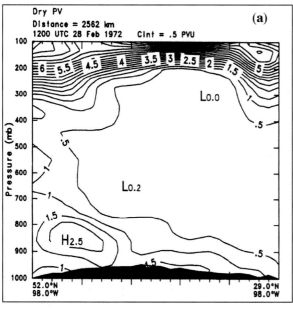

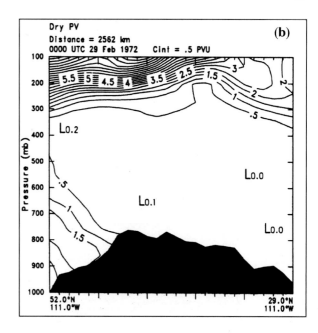

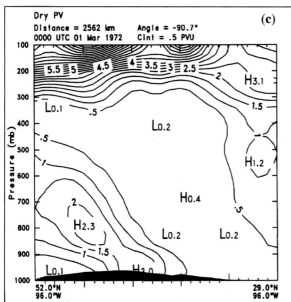

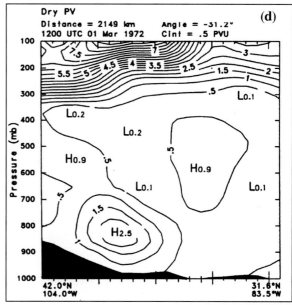

FIG. 28. Cross sections of PV, contoured every 0.5 PVU, at (a) 1200 UTC 28 Feb, (b) 0000 UTC 29 Feb, and (c) 0000 and (d) 1200 UTC 1 Mar 1972 along the orientations given below each panel. Cross section lines (A–B, C–D, I–J, and K–L) are shown in Fig. A1.

through 1200 UTC 2 March the Northeast and the Southeast became the primary frontal battleground (Fig. 18) while over the lower Mississippi River valley a convective outbreak developed ahead of the advancing cold air (not shown). At 1200 UTC a weak frontal wave (central pressure ~1000 hPa) was located over extreme southwestern Lake Ontario just the north of Buffalo (BUF), New York (Fig. 38a). This wave originated along the prefrontal trough over southern Nebraska and northern Kansas between 1800 UTC 29 February and 0000 UTC 1 March (Fig. 33a). It redeveloped southeastward to central Oklahoma by 1200 UTC 1 March

(central pressure ~996 hPa) as the arctic front accelerated southward. Subsequently, it turned northeastward and raced along the front toward the eastern Great Lakes as it weakened slightly.

h. Frontal interaction with the northern Appalachians

By 1200 UTC 2 March, a strong frontal boundary extended eastward to New England as judged by surface temperatures that ranged from −26°C across Ontario to 17°C over central and northern New Jersey (Fig. 38a). The 43°C temperature contrast was remarkable because

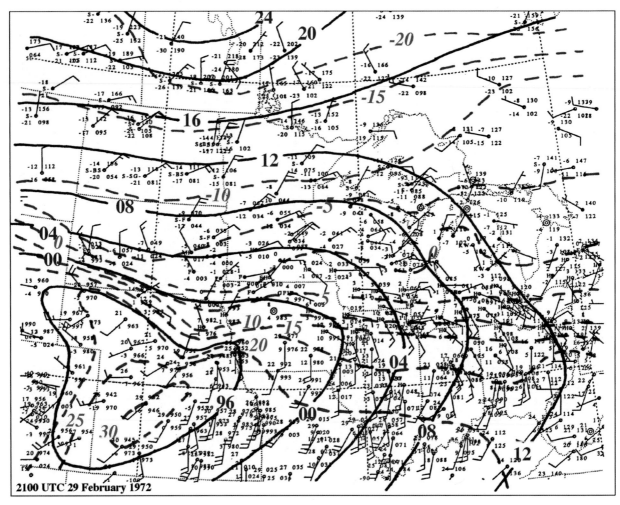

FIG. 29. Manually analyzed surface map for 2100 UTC 29 Feb 1972. Sea level isobars (solid, every 4 hPa) and surface temperature (dashed, every 5°C). Surface observations are plotted conventionally.

of the high minimum temperatures in the warm air and the low minimum temperatures in the cold air despite the reported widespread clouds and light snow (recall the area of sub −40°C air at 850 hPa over eastern Canada at 0000 UTC 1 March in Fig. 22d). Terrain influences on surface boundaries were quite apparent. Cold-air damming across Maine southwestward to northeastern Massachusetts was manifest by a wedge of high pressure and cold northeasterlies. A frontal boundary extended eastward from the weak wave near BUF to north of Syracuse (SYR) and Utica (UCA), New York, before turning southeastward toward southern New York and southwestern Connecticut. Along the south coast of New England the strength of the frontal boundary was tempered by a southwesterly flow off the chilly ocean waters (Fig. 38a).

A separate frontal boundary extended northeastward from eastern Lake Ontario down the St. Lawrence River Valley at 1200 UTC 2 March (Fig. 38a). The associated surface lee trough likely reflected downslope southerly and southeasterly flow across the mountain ranges just

to the south (windward ridging is apparent over Maine in the cold-air damming region). Southerly flow over much of Vermont and extreme southeastern Quebec was a manifestation of this lee trough. Surface temperatures in the upper Hudson and Champlain Valleys and Vermont were relatively homogeneous (−3° to 3°C). This air mass was clearly different than the warmer air mass to the southwest and the arctic air mass to the north (Fig. 38a).

By 2100 UTC 2 March, the frontal wave that was near BUF at 1200 UTC was located along the Quebec–Vermont border (Fig. 38b). This nondeveloping wave (central pressure ~1000 hPa) advanced northeastward along the separate frontal boundary that comprised the lee trough in the St. Lawrence Valley. A narrow tongue of warm air (temperatures > 10°C) was pulled northward through the upper Hudson and Champlain Valleys and western Vermont into extreme southeastern Quebec ahead of the advancing frontal wave (Fig. 38b). The eastern edge of this tongue of warm air was bounded by the Green Mountains of Vermont. Cold-air damming

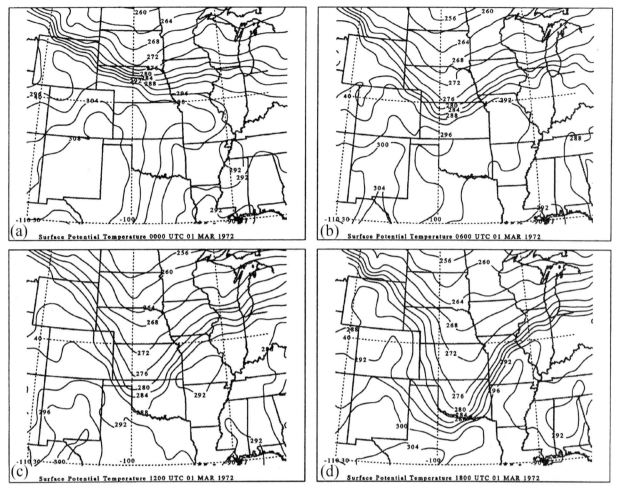

FIG. 30. Same as in Fig. 24 but at (a) 0000, (b) 0600, (c) 1200, and (d) 1800 UTC 1 Mar 1972. Winds are not shown.

from extreme northeastern Connecticut northeastward to Maine ensured that cold air east of the Green Mountains would remain entrenched across much of central and northern New England (Fig. 38b). A cold front advancing eastward across New York and Pennsylvania (it had passed UCA by 2100 UTC) was encroaching on the tongue of warm air from the west (Fig. 38b).

Meteograms for Albany (ALB) and Glens Falls (GFL), New York, are used to show terrain influences on the cold front passage across the Hudson Valley (Fig. 39). At ALB the tongue of warm air arrived near 0900 UTC 2 March when winds shifted briefly to southerly (freezing rain ended; not shown) and temperatures increased to 1°–2°C. Although the warm air retreated briefly near 1200 UTC as the winds shifted to light northwesterly and the surface temperature rise was arrested, by 1300 UTC it had arrived in earnest as southerly winds strengthened and temperatures soared above 10°C. An abrupt wind shift to west-northwest and rising (decreasing) pressures (temperatures) at ALB just before 0000 UTC 3 March heralded the arrival of the cold air (Fig. 39) and was typical of postfrontal conditions at ALB as

cold air advanced down the west-northwest- to east-southeast-oriented Mohawk Valley (near ALB the Mohawk Valley opens into the north–south-oriented Hudson Valley).

At GFL, located in the Hudson Valley ~80 km to north of ALB, the initial arrival of the warm air and southerlies was delayed until just before 1900 UTC (Fig. 39). The return of the cold air was delayed until just after 0000 UTC 3 March, roughly an hour later than the cold air reached ALB. The pressure rise at GFL beginning near 0000 UTC 3 March was less abrupt than at ALB, indicative that the arrival of the cold air at GFL was more gradual, and the northwesterly wind shift was short-lived as the winds returned to southerly near 0300 UTC before shifting to northeast by 0600 UTC. The initial temperature break at GFL was delayed by ~1 h relative to ALB and further cooling was delayed until the final wind shift to north-northeast near 0600 UTC (Fig. 39). Our interpretation of the ALB/GFL meteograms is that the advancing shallow cold air was partially blocked by the Adirondack Mountains to the west of GFL because in the absence of mountains and with

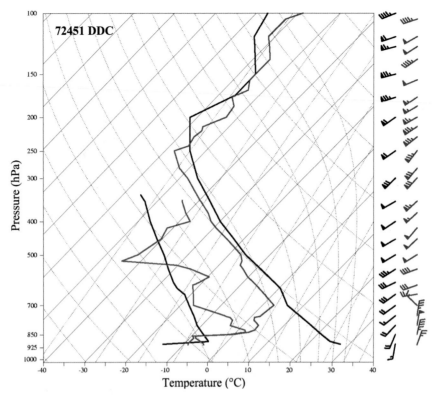

FIG. 31. Same as in Fig. 14 but for DDC at 0000 (black) and 1200 UTC (red) 1 Mar 1972. Winds are not shown.

a northeast–southwest-oriented cold front (Fig. 38b), the cold air should have reached GFL before ALB. A speculation is offered that the orientation of the Mohawk Valley favored the funneling of cold air into ALB before GFL and that the cold air reached GFL both from the south (up the Hudson Valley) and from the west (over the Adirondacks).

Soundings from Portland (PWM) and Caribou (CAR), Maine, for 0000 UTC 2 and 3 March illustrate the arctic push into New England (Fig. 40). At 0000 UTC 2 March the wedge of cold air at PWM was below 850 hPa as low-level northeasterly flow lay under strong west-southwesterly flow above. Cold advection at CAR, based on the observed backing winds between the surface and 750 hPa, supported a strong southward surge of much colder low-level arctic air across eastern New England (Figs. 38a,b). By 0000 UTC 3 March very strong low-level inversions at PWM and CAR were capped by a very strong warm southwesterly flow that had reached speeds of near 35 and 30 m s^{-1} near 850 hPa at CAR and PWM, respectively (Fig. 40). At CAR the temperature increased 28°C from near −20°C at 925 hPa to near 8°C at 850 hPa. Ice pellets were reported with a surface temperature of −21°C shortly before 0000 UTC 3 March, indicative of the strong low-level temperature inversion. The wet-bulb temperature above the arctic frontal inversion was near 15°C and was typical of surface conditions in the warm air across parts

of Pennsylvania, New Jersey, and southern New York (Figs. 38a,b).

The shallow arctic air at PWM and CAR was associated with a strong cold-air damming and a persistent north-northeasterly surface flow (Figs. 38a,b and 40) that extended southward into central New England as seen in meteograms for PWM and South Weymouth (NZW), Massachusetts (Fig. 41). During a prolonged period of freezing rain with occasional ice pellets at PWM the temperature warmed between 0600 and 1200 UTC 3 March to near 0°C as surface winds veered to more northeasterly, indicative of the modified marine air from the Gulf of Maine that was reaching coastal southwestern Maine (Fig. 41). By 1400 UTC the winds backed to northerly and by 1800 UTC the temperature decreased again with a more continental air trajectory. At NZW, almost 200 km to the south-southwest of PWM and very close to the frontal boundary, the warm air arrived just before 0200 UTC 3 March with an abrupt ~10°C temperature rise and a wind shift to moderate southerly. The warm air's toehold in NZW was briefly interrupted at 0600 UTC by a short-lived wind shift to light northerly and ~6°C cooling and was eliminated after 0800 UTC as the cold air returned with moderate northerly winds and ~14°C of cooling by 1000 UTC (Fig. 41).

At both NZW and PWM there was also an indication of the passage of small-amplitude inertia–gravity waves.

3h Pressure Change (hPa)

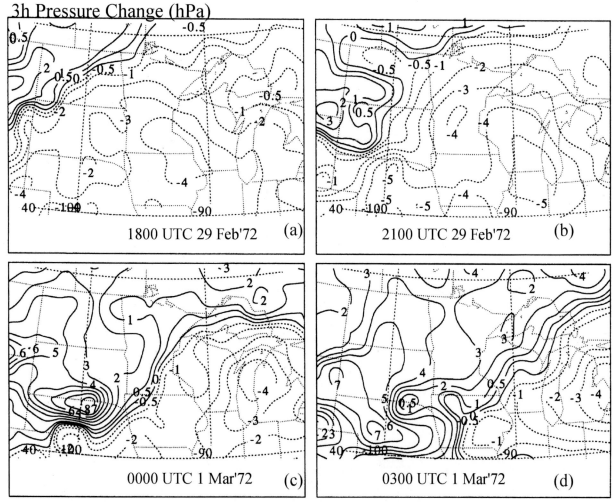

FIG. 32. Three-hour surface pressure change at (a) 1800 and (b) 2100 UTC 29 Feb, and (c) 0000 and (d) 0300 UTC 1 Mar 1972. Isallobars contoured every 1 hPa (3 h)$^{-1}$, except for values of 0.5 hPa (3 h)$^{-1}$, with positive (negative) values shown solid (dashed).

At NZW the pressure increased ~2 hPa (wave of elevation) in less than one hour ending at 1400 UTC then decreased slowly thereafter (Fig. 41). This ~2-hPa pressure perturbation was accompanied by slightly backed winds and a brief period of moderate and heavy rain. At PWM, deeper in the arctic air, a similar but sharper 2–3-hPa pressure perturbation was observed near 0400 UTC 3 March and was followed by minor (~1 hPa) pressure perturbations thereafter (Fig. 41). The first pressure perturbation at PWM was also accompanied by briefly backed winds. Light freezing rain at PWM temporarily ended as the pressure returned to ambient values by 0600 UTC. The reported thunderstorms at PWM and NZW suggested that the surface pressure perturbations may have also been linked to changes in the depth of the frontal inversion in conjunction with elevated base convection with roots in the warm air above the inversion (Colman 1990a,b). Whatever role convection played in the observed inertia–gravity waves, the presence of a very strong duct (stable layer) in the

lower troposphere at PWM and CAR was favorable for the propagation of long-lived inertia-gravity wave disturbances (Fig. 40).

i. Frontal interaction with the southern Appalachians

Significant frontal interactions with the southern Appalachians were also observed (Figs. 42a,b). At 1200 UTC 3 March cold air was approaching northern Maryland and New Jersey and was also moving eastward around the southern end of the Appalachians. However, from southern Virginia to northeastern Georgia most of the cold air was blocked from reaching the coastal plain by the mountains. In the blocked flow region a weak boundary (lee trough) extended southward from a weak cyclone located over southern Virginia. This boundary marked the western edge of an area of broken showers and thunderstorms. Behind the boundary the surface winds veered 20°–40° and temperatures decreased 2°–4°C, suggestive of a weak cold front passage. How-

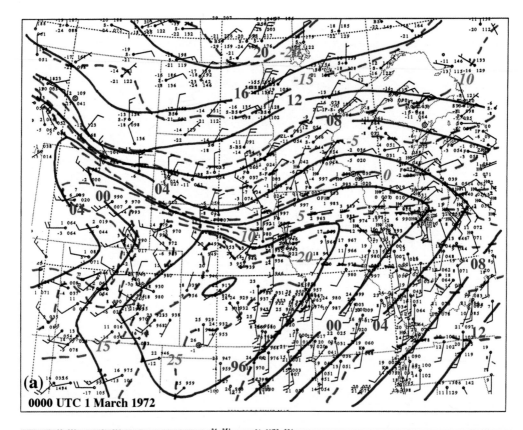

0000 UTC 1 March 1972

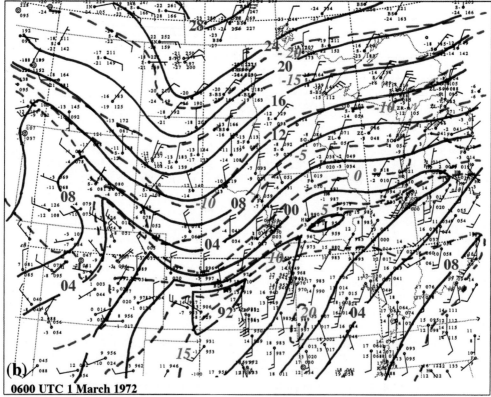

0600 UTC 1 March 1972

FIG. 33. Same as in Fig. 29 but for (a) 0000 and (b) 0600 UTC 1 Mar 1972.

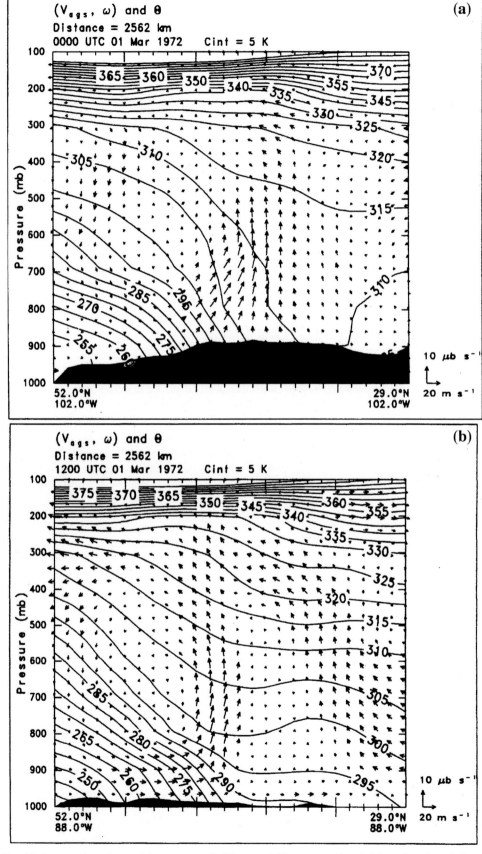

FIG. 34. Cross section of potential temperature (solid, every 5 K), ageostrophic wind (m s⁻¹), and vertical motion (×10⁻³ hPa s⁻¹) in the plane of the cross section with magnitudes indicated by the plotted reference scales for (a) 0000 and (b) 1200 UTC 1 Mar 1972. Cross-section lines (E–F and M–N) are shown in Fig. A1.

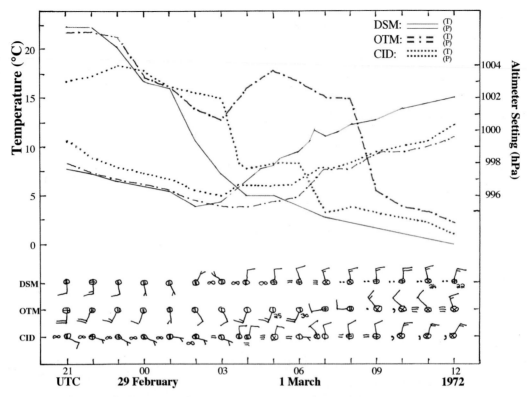

FIG. 35. Same as in Fig. 26 but for DSM, CID, and OTM from 2100 UTC 29 Feb through 1200 UTC 1 Mar 1972. Station locations are shown in Fig. A1.

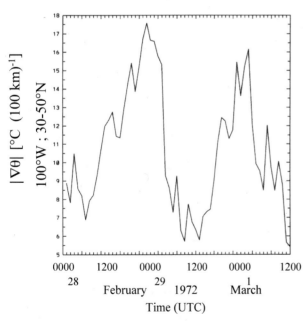

FIG. 36. Diurnal variation of the magnitude of the maximum surface potential temperature gradient [°C (100 km)⁻¹ (3 h)⁻¹], between 30° and 50°N along 100°W for the period 0000 UTC 28 Feb–1200 UTC 1 Mar 1972. Cross section line (G–H) is shown in Fig. A1.

ever, most of the cold air lay along the western slopes of the southern Appalachians to the west of the lee trough that was situated over western North Carolina. Much of the cold air was still blocked by the southern Appalachians at 1800 UTC despite coastal winds having turned westerly almost everywhere behind the weak surface cyclone located east of Virginia (Fig. 42b). The surface observations indicated that colder air would arrive in western South Carolina from the southwest and northern North Carolina from the north (Fig. 42b).

The delayed arrival of the air in the western Carolinas is illustrated by meteograms for Bristol (TRI), Tennessee; Hickory (HKY), South Carolina; and Charlotte (CLT), North Carolina (Fig. 43). A line of showers and thunderstorms, associated with the upper-level disturbance, crossed the southern Appalachians near 0000 UTC 3 March and ushered in a cooling of 6°–7°C at all three stations (Fig. 43). However, a wind shift to northwest only occurred at TRI while winds veered slightly to southwest at HKY and CLT (Fig. 43). Little additional cooling was noted at HKY and CLT while the temperature decreased overnight at TRI (Fig. 43). Between 1200 and 1800 UTC the temperature difference between HKY/CLT and TRI exceeded 10°C and the pressure difference was more than 4 hPa as colder air was mostly blocked from crossing the mountains (Fig. 43). As the lee trough east of the mountains began to move eastward and winds shifted to northwest at HKY and CLT by

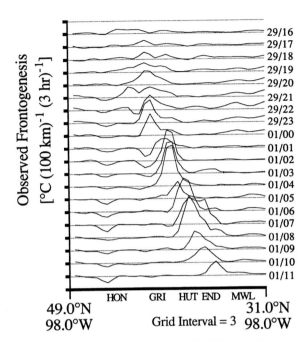

FIG. 37. (left) Cross section of the magnitude of the surface potential temperature gradient [vertical scale every 4°C (100 km)⁻¹] between 31° and 49°N along 98°W for 1200 UTC 29 Feb–0700 UTC 1 Mar 1972 and (right) cross section of observed surface frontogenesis [vertical scale every 4°C (100 km)⁻¹ (3 h)⁻¹] along 98°W for 1600 UTC 29 Feb–1100 UTC 1 Mar 1972. Cross section line (I–J) shown in Fig. A1.

1500 and 2000 UTC 3 March, respectively, delayed cooling occurred as the remainder of the cold air finally arrived (Figs. 42b and 43).

j. Surface and upper-air linkages

In this section surface mesoscale features discussed in the previous sections are linked with specific upper-level features from manually prepared continuity maps of surface cyclone position and intensity (central pressure) and 500-hPa absolute vorticity maxima for the period 0000 UTC 28 February through 4 March (Figs. 44 and 45). The NCEP–NCAR gridded sea level pressure analyses (Kalnay et al. 1996; Kistler et al. 2001) were used as an independent check on our manual analyses. The surface cyclones shown in Fig. 44 were organized into three broad categories: 1) weak cyclones without obvious upper-level support, 2) weak cyclones near downstream jet-entrance regions, and 3) cyclones ahead of mobile upstream troughs.

Five weak cyclones fell into the first category. The first cyclone of interest was a quasi-stationary surface (lee) cyclone (filled square) along the extreme western U.S.–Canada border. It was important because persistent easterly and northeasterly flow to its north enabled very cold air to accumulate along the lee slopes of the Canadian Rockies from where it poured southward into Montana and North Dakota (Fig. 44). A very short-lived

lee cyclone (half-filled square) that formed and died in Wyoming on 28 February was of little consequence. Weak cyclones were also noted over Nebraska–Kansas (asterisk) and Iowa–Illinois (open circle) on 29 February and 1 March, respectively; the latter cyclone was associated with the weak frontal wave discussed previously (Figs. 32a,b). A fifth short-lived cyclone (filled wedge) was seen near Nova Scotia on 2 March (Fig. 44).

Two other weak cyclones (cross and plus sign) that were situated in jet-entrance regions over Michigan and western New York at 0000 and 1200 UTC 1 March, respectively, were included in the second category of weak cyclones (Fig. 44). These cyclones remained weak or weakened as they moved east-northeastward down the St. Lawrence River valley and eventually dissipated (Fig. 44).

An additional seven cyclones were in the upstream trough category (Fig. 44). Cyclone (filled triangle) formed along the arctic frontal boundary in Montana at 0000 UTC 28 February, moved eastward across the Great Lakes, and dissipated near Nova Scotia after 0000 UTC 1 March. It was associated with two 500-hPa vorticity maxima (open square and filled wedge) (Fig. 45). Arctic air moved eastward beyond the Great Lakes in its wake. An earlier 500-hPa vorticity maximum (filled circle[1]), seen over the upper peninsula of Michigan at 0000 UTC 28 February in Fig. 45, was associated with

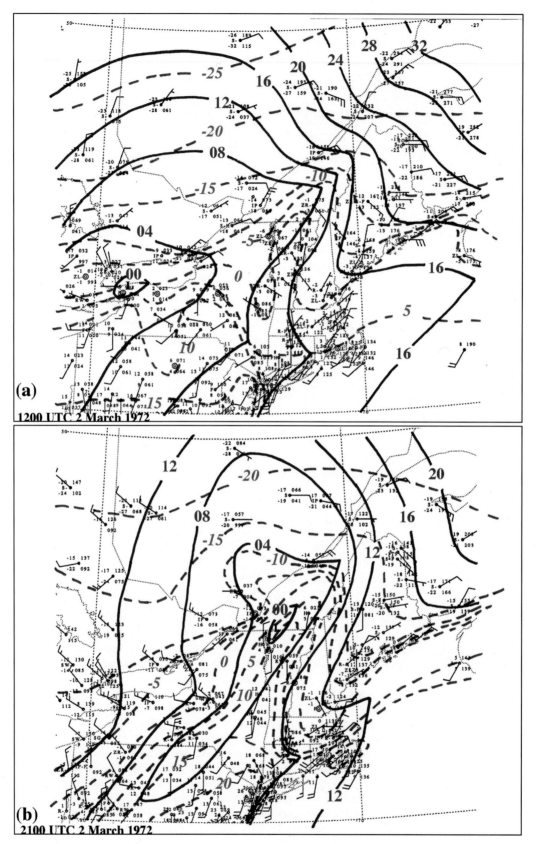

FIG. 38. Same as in Fig. 29 but at (a) 1200 and (b) 2100 UTC 2 Mar 1972.

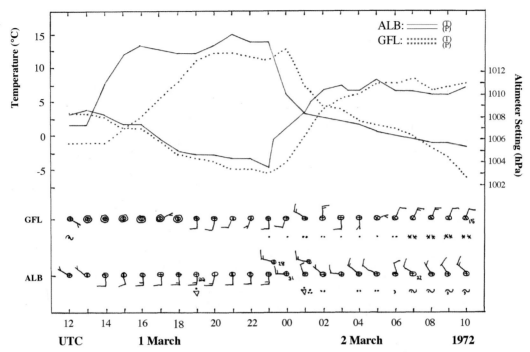

FIG. 39. Same as in Fig. 26 but for GFL and ALB for 1200 UTC 2 Mar–1000 UTC 3 Mar 1972. Station locations are shown in Fig. A1.

a moderately deepening cyclone (crossed circle) that moved from the upper Great Lakes to east of Labrador by 1200 UTC 29 February (Fig. 44). This disturbance triggered the initial arctic air surge into eastern Canada. The arctic air was further reinforced behind the second surface disturbance (filled triangle).

One mobile 500-hPa vorticity maximum (filled tri-

angle in Fig. 45) was associated with three surface disturbances (circle with filled-bottom half, circle with filled-right half, and open wedge; Fig. 44). This vorticity maximum originated over the eastern Pacific prior to 29 February, climbed the low-amplitude ridge over the West Coast, dropped south-southeastward across the southern Rockies, turned eastward across Texas and

02 Mar 1972 0000 UTC   03 Mar 1972 0000 UTC

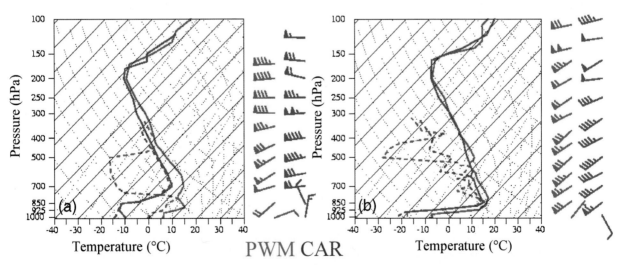

FIG. 40. Same as in Fig. 14 but for CAR (blue) and PWM (red) at (a) 0000 UTC 2 Mar and (b) 0000 UTC 3 Mar 1972. Station locations are shown in Fig. A1.

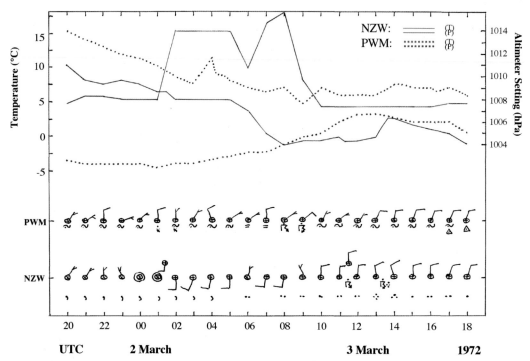

FIG. 41. Same as in Fig. 26 but for PWM and NZW for 2000 UTC 2 Mar–1800 UTC 3 Mar 1972. Station locations are shown in Fig. A1.

northern Louisiana, and accelerated east-northeastward to Nova Scotia by 0000 UTC 4 March. The first cyclone (circle with filled-bottom half) over Utah lasted 12 h through 1800 UTC 29 February and then dissipated as it lost its upper-level support as the 500-hPa vorticity center moved southeastward into New Mexico.

The second cyclone (circle with filled-right half) associated with 500-hPa vorticity center (filled triangle) formed along the Nebraska–Kansas border after 1200 UTC 29 February, and at 0000 UTC 1 March was over extreme northern Kansas embedded in a broad region of low pressure in Fig. 32a (Figs. 44 and 45). This broad area of low pressure over western Kansas (Fig. 22) was indicative of lee cyclone development in the southwesterly flow ahead of 500-hPa vorticity center (filled triangle; Figs. 22 and 45). As is typical of lee cyclone developments east of the Rockies (e.g., Colle and Mass 1995), the Nebraska–Kansas cyclone settled southward without developing to Oklahoma by 1200 UTC 1 March whereupon it turned northeastward toward the lower Great Lakes before dissipating after 0000 UTC 3 March near northern Maine (this cyclone was just north of Vermont at 2100 UTC 2 March; Fig. 38b). The north–south pressure gradient over the high plains strengthened between this second (lee) cyclonic development and the arctic anticyclone that ridged eastward toward the U.S.–Canadian border in a poleward jet-entrance region (Fig. 43), setting the stage for the observed southward cold surge beginning near 0000 UTC 1 March (Figs. 32a,b). Meanwhile, 500-hPa vorticity center

(filled triangle) initiated a third surface cyclone (open wedge) over Louisiana at 0600 UTC 2 March (Figs. 44 and 45). This third cyclone moved northeastward without intensifying along the trailing cold front to a position south of Nova Scotia at 0000 UTC 4 March.

5. Discussion and conclusions

This paper consisted of three parts: 1) a review of surface frontogenesis associated with differential diabatic heating over sloping terrain and across coastlines, 2) a case study (17–18 April 2002) of an intense side-door cold front passage in eastern New England in which coastal temperature decreases of 15°–20°C in less than 60 min were common, and 3) a case study (28 February–4 March 1972) of a long-lived intense front that developed along the lee slopes of the northern Rockies, expanded eastward to New England, and featured surface temperature differences of greater than 50°C over 500 km. The analysis of both case studies was motivated by the pioneering and classical study of an intense plains front by Sanders (1955).

a. 17–18 April 2002 case study

A critical issue for this case was the production of a surface pressure gradient capable of driving and sustaining an onshore flow of chilly marine air. The large-scale flow pattern favored anomalously warm conditions over much of New England and surface pressure falls

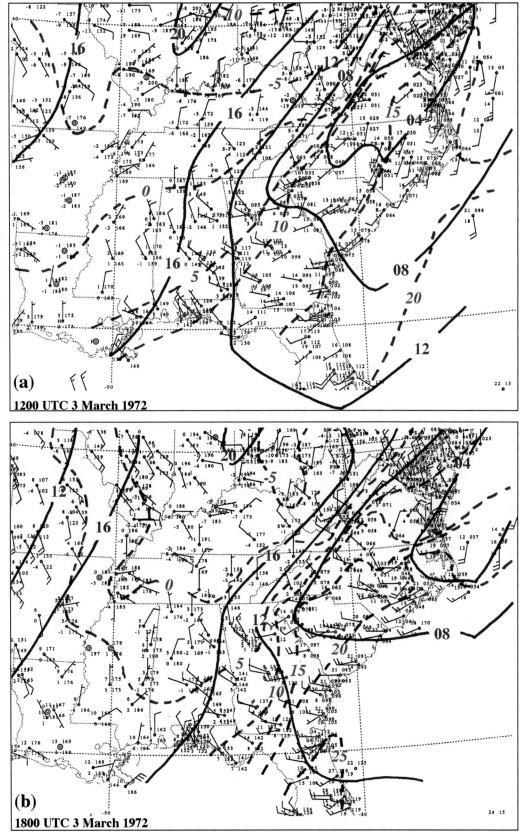

(a)
1200 UTC 3 March 1972

(b)
1800 UTC 3 March 1972

FIG. 42. Same as in Fig. 29 but for (a) 1200 and (b) 1800 UTC 3 Mar 1972.

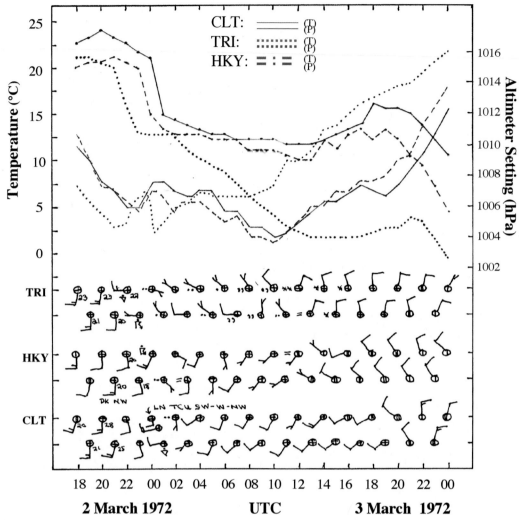

FIG. 43. Same as in Fig. 26 but for TRI, HKY, and CLT for 1800 UTC 2 Mar–0000 UTC 4 Mar 1972. Station locations are shown in Fig. A1.

in conjunction with ridging aloft (Fig. 10). Dynamic anticyclogenesis over eastern Canada on the poleward side of a confluent jet-entrance region on the western side of this deep trough set the stage for surface pressure rises over Atlantic Canada as a surface anticyclone ridged southward toward Nova Scotia. Surface pressure rises (falls) over Atlantic Canada (New England) established a sufficiently strong pressure gradient necessary to drive chilly marine air onshore. We speculate that geostrophic adjustment associated with this process played an important role in the initiation of the resulting gravity (density) current.

Bosart et al. (1973) showed that a surface anticyclone somewhere north and/or northeast of New England was necessary to sustain a southward push of cooler air across New England. A side-door cold front differs from a backdoor cold front in that the necessary anticyclone is weaker and situated farther east than is the case for a backdoor cold front. This anticyclone position differ-

ence means that in general side-door cold frontal passages in New England should be more of a glancing blow, affecting primarily coastal regions, than backdoor cold frontal passages that can affect the entire region. We hypothesize that there is a window of opportunity in early spring for the formation of intense side-door cold fronts after the snow has melted and prior to leaf out and green-up. Our hypothesis rests on the idea that during these periods virtually all of the insolation is available for sensible heating because transpiration processes have yet to begin. Support for this hypothesis rests on studies of the evapotranspiration process that have shown that transpiration from moist forests and mixed forests and grasslands is the dominant contributor to overall evapotranspiration (e.g., Moreira et al. 1997; Fitzjarrald et al. 2001; Freedman et al. 2001; Czikowsky and Fitzjarrald 2004).

When a period of strong sensible heating and little transpiration coincides with a large-scale weather pat-

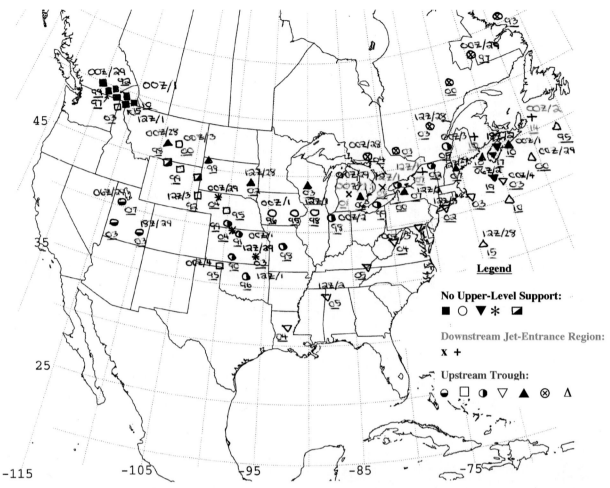

FIG. 44. Tracks of individual surface cyclone centers every 6 h for the period 0000 UTC 28 Feb–4 Mar 1972. Symbols denoting individual cyclone centers are given in the legend at the right. Underlined two-digit numbers indicate cyclone central pressure (hPa, leading 10 or 9 omitted). UTC time (Z) and day shown by Z/day numbers adjacent to the individual symbols.

tern that favors well above normal temperatures in New England, a side-door cold frontal passage is possible if the pressure difference between New England and Atlantic Canada exceeds a critical threshold. The YQI–BOS pressure difference reached 6 hPa at the outset of the side-door frontal cold push across coastal eastern Massachusetts (Fig. 11) in response to the aforementioned dynamic anticyclogenesis over Atlantic Canada and sensible heating-induced pressure falls over New England. Inland diabatic heating under conditions of strong insolation and little transpiration leads to surface pressure falls that maximize in the late afternoon, setting the stage for a late-day onshore push of chilly marine air. Similar onshore pushes of marine air occur along the west coast of North America during the warm season when surface pressures over the heated continent lower sufficiently relative to the pressure in the immediate coastal waters (e.g., Mass and Albright 1987; Mass et al. 1987; Bond et al. 1996; Mass and Bond 1996; Chien et al. 1997).

An important extra ingredient that favors the passage

of intense side-door cold fronts across coastal eastern New England is the Gulf of Maine. The SSTs in the Gulf of Maine in mid-April 2002 averaged between 4° and 7°C (Fig. 13). We hypothesize that these chilly waters produce a marine polar air mass through diabatic cooling analogous to the production of continental polar air masses over cold land surfaces. The associated diabatic cooling will help to keep surface pressures relatively high over the Gulf of Maine relative to the heated interior. Furthermore, the near-uniform SSTs in the Gulf of Maine help to ensure that there will be little surface modification of air parcels that are driven southwestward and then onshore over coastal eastern New England. It is also possible that evaporative cooling associated with scattered convection over the Gulf of Maine may have contributed to additional surface cooling (Fig. 16).

Frontogenesis due to horizontal confluence at the leading edge of the marine push over the Gulf of Maine was mostly confined below 900 hPa and maximized in the late afternoon (Fig. 15). Peak late-afternoon front-

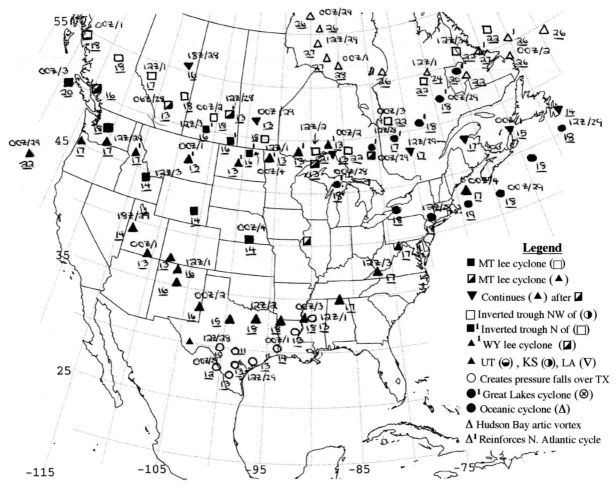

FIG. 45. Tracks of individual 500-hPa absolute vorticity centers every 6 h for the period 0000 UTC 28 Feb–4 Mar 1972. Symbols denoting individual vorticity centers are given in the legend at the right. Symbols in parentheses in the legend refer to surface cyclones given in Fig. 44. Underlined two-digit numbers give absolute vorticity ($\times 10^{-5}$ s^{-1}). UTC time (Z) and day shown by Z/day numbers adjacent to the individual symbols.

ogenesis rates > 10°C (100 km)$^{-1}$ (3 h)$^{-1}$ and an increasing cross-coast surface pressure gradient likely helped the side-door front take on characteristics of an atmospheric density current. Coastal and offshore winds > 20 m s^{-1} (e.g., MQE, Fig. 9, and IOSN3) exceeded the observed speed of the front. These observed wind speeds provide some support for viewing the side-door cold front as an atmospheric density current (e.g., Hakim 1992). Also, alongfront wave structures seen by KBOX (Fig. 16) were similar to alongfront lobe and cleft structures seen at the leading edge of density currents in laboratory studies by Simpson (1987, his Figs. 2.4 and 11.10; Simpson 1994, his Figs. 3.1, 3.2, 11.10, 11.11). However, mindful of the caveats expressed by Smith and Reeder (1988), our inferences of atmospheric density current behavior should be tested by direct numerical experimentation.

This case also raised important forecast issues as the available objective MOS guidance failed to indicate that any significant wind shifts or temperature changes

would occur during the late-afternoon hours of 17 April 2002 at BOS (Fig. 17). This forecast failure, repeated often in these situations, points toward difficulties simulating shallow mesoscale frontogenesis that is driven predominantly by diurnally varying differential diabatic heating. That sailboats are still capsizing en masse on the Charles River with the passage of intense springtime side-door cold fronts indicates that there is a problem communicating expected important nonconvective mesoscale weather changes to the public (or that potential users are disregarding useful forecast guidance of impending abrupt weather changes). Improved means need to be fashioned to communicate the likelihood of highly significant mesoscale weather changes that are outside the traditional convection-based severe weather occurrences to the general public. Efforts should be made to make the forecast weather variability more closely match the observed weather variability to help increase the public perception that weather forecasting is improving, especially for mesoscale weather systems.

b. 28 February–4 March 1972 case study

The large-scale flow pattern during this period favored the accumulation of frigid air in southern Canada. Below-normal 500-hPa heights over most of Canada and above-normal heights in the flanking flat ridges at lower latitudes along the west and east coasts of the United States and parts of the southern United States favored fast westerlies across the northern United States and southern Canada and a strong meridional temperature gradient (Fig. 19). Sub 462-dam 1000–500-hPa thicknesses and sub −40°C 850-hPa temperatures over central and eastern Canada attested to the severity of the arctic chill. A persistent trough over northern British Columbia and a deep, cold vortex over northern Hudson Bay created a strong confluent jet-entrance region over western Canada by 0000 UTC 28 February (Fig. 21). An east–west-oriented surface trough formed to the east of the extreme southern Canadian Rockies ahead of the British Columbia trough. A building high pressure in the poleward-entrance region of the strengthening northwesterly jet between the Hudson Bay vortex and British Columbia trough enabled arctic air to edge southward in the easterly geostrophic flow to the north of a surface trough near the U.S.–Canadian border subsequent to 0000 UTC 27 February. By 0000 UTC 28 February, the arctic air was moving southward into northern Montana, where it accumulated along the northward-facing mountain slopes, and North Dakota, where it spilled eastward across the plains (Figs. 20–21 and 24).

Arctic air accumulated over Montana and gradually spilled into snow-covered regions of northern and eastern Wyoming and South Dakota (Figs. 20, 24, and 27) in conjunction with a strengthening arctic anticyclone and a developing cyclone to the lee of the central Rockies between 28 and 29 February (Figs. 21 and 22). At 300 hPa a jet-entrance region located over the Alberta–Saskatchewan region on 28–29 February (Fig. 21) shifted eastward into Manitoba by 0000 UTC 1 March (Fig. 22) and into Ontario and northern Minnesota by 1200 UTC (0000 UTC) 1 (2) March (Figs. 22 and 23). As the arctic anticyclone moved toward the U.S. border the nose of highest pressure remained in the vicinity of the poleward jet-entrance region. The eastward advance of the jet-entrance region to well east of the Rockies by 1200 UTC 1 March enabled the arctic anticyclone to build southward and helped set the stage for the southward plunge of arctic air toward the Gulf of Mexico that began subsequent to 0000 UTC 1 March (Fig. 18).

Although this case exhibited the strongest baroclinicity over North America ever seen by the lead author, no significant cyclogenesis occurred (Fig. 44). Bomb fans (Sanders and Gyakum 1980) were no doubt frustrated that so much baroclinicity yielded so few closed isobars, so little wind, and so little snow. Those cyclones that did form were mostly nondeveloping and reflected the influence of terrain, jet-entrance regions, and unimpressive mobile troughs (Figs. 44 and 45). Past theoretical work and idealized modeling studies (e.g., Joly and Thorpe 1990, 1991; Schär and Davies 1990; Thorncroft and Hoskins 1990; Bishop and Thorpe 1994a,b; Renfrew et al. 1997; Mallet et al. 1999b) suggests that alongfront stretching deformation above a certain threshold inhibits cyclogenesis. As stated by Renfrew et al. (1997), "The role of environmental deformation appears to be crucial: as part of a baroclinic life cycle, stretching deformation acts to build up a front, but suppresses along-front waves." A relaxation of the stretching deformation near the front as could occur if a significant upper-level disturbance approached and interacted with the front could allow new frontal cyclogenesis to proceed. More recently, Patoux et al. (2005), in a study of three Southern Ocean fronts, found that a relaxation of the stretching deformation did not guarantee cyclogenesis if the large-scale environment was frontolytic.

Mallet et al. (1999a) examined a Fronts and Atlantic Storm-Track Experiment (FASTEX, 1997) case and suggested that the approach of a cyclone toward an elongated frontal boundary was sufficient to weaken the contribution of alongfront stretching deformation to frontogenesis to permit frontal wave growth. Arbogast (2004) studied another FASTEX case and showed that if a disturbance containing sufficient vorticity approached to within 1000–1500 km of the primary frontal boundary along a line approximately normal to the frontal axis of dilatation it could trigger the growth of a new frontal wave. Schwierz et al. (2004), in an idealized theoretical study of forced waves associated with a zonally oriented jet, showed that a narrow corridor of high meridional PV gradient on an isentropic surface that crossed the jet axis could be sufficient to trap waves along the PV gradient. In cases where a significant upper-level disturbance approached the jet axis and the associated enhanced meridional PV gradient region the resulting interaction of the disturbance with the jet could be sufficient to permit the trapped waves to grow. Both the Arbogast (2004) and Schwierz et al. (2004) studies suggest that the key to allowing cyclogenesis to proceed along a strong frontal boundary defined by large values of alongfront stretching deformation is the local disruption and lessening of the magnitude of this deformation that can occur when a strong enough disturbance approaches the frontal boundary.

Although our long-lived frontal case from 1972 has some similarities to the oceanic FASTEX cases (e.g., strong jets, strong baroclinicity, and implied large values of stretching deformation as estimated from Figs. 21–23, it is possible that large values of atmospheric stability in the continental arctic frontal zone (e.g., Figs. 25, 31, and 40) contributed significantly to limiting cyclogenesis over the plains by crushing dynamically forced vertical motions. The existence of deep ascent in a 250-km-wide band above the leading edge of the front at 1200 UTC 1 March (Fig. 34b) suggests that the ascent was dynamically driven by the trough to the west and/or the jet-entrance region well to the north. This deep ascent was immediately to the west of a nondeveloping frontal wave that at 0600 UTC 1 March was located over southeastern Iowa (Fig. 33b). While it is possible that the residence times of individual air parcels in this narrow ascent band were insufficient to allow significant low-level vorticity growth by vertical stretching to sustain significant cyclogenesis, the likely competing effects of alongfront stretching deformation and dynamically driven ascent, modified by frontal zone atmospheric static stability, on cyclogenesis need to be considered more carefully.

Prior to 0000 UTC 1 March, when the arctic front was strongest near the mountains and situated beneath west-northwesterly flow there was very little deep ascent associated with the leading edge of the arctic front (cf. Figs. 28 and 33a). Sinking motion was situated below 500 hPa at the leading edge of the front while very weak ascent was located deeper into the shallow cold air (Fig. 28). The defining character of the arctic front was the existence of a well-defined band of PV (~1.5 PV units) that coincided nicely with the arctic frontal inversion. The presence of this PV strip likely attests to the importance of diabatic cooling at the base of the frontal inversion in contributing to frontal zone PV generation above the level of maximum cooling, analogous to the more common case where widespread latent heat release contributes to the generation of PV below the level of maximum diabatic heating in conjunction with warm-air, advection-driven precipitation (embedded convective and stratiform) near the frontal zone.

Our analyses showed that there were numerous examples of mountain–front interactions, ranging from cold-air damming, windward ridging, lee troughing, and flow blocking. Lee troughing east of the Rockies by 0000 UTC 1 March ahead of a 500-hPa trough (Figs. 22 and 45) and west of a 300 hPa jet-entrance region paved the way for a cold-air, damming-driven cold surge east of the Rockies. We hypothesize that 1) in the absence of the Rockies the strong frontal boundary would have remained much more latitudinally confined, given the quasi-zonal nature of the flow aloft, and 2) the greatly enhanced static stability in the shallow arctic air mass behind the frontal boundary over the plains limited the opportunity for robust cyclogenesis along the eastern margins of the frontal boundary over the lower Mississippi Valley (frontal forcing was sufficient to produce a significant convective outbreak, however; not shown).

Strong cold-air damming was also observed across the northern Appalachians and New England as very warm air was thrust poleward across Pennsylvania, New Jersey, and the Hudson Valley and arctic air remained entrenched east of the mountains across western New England (Figs. 38a,b). Lee troughing and associated frontogenesis was observed in the St. Lawrence Valley in conjunction with a very strong (30–35 m s^{-1}) low-level southerly downslope flow across the mountains of northern New England (Figs. 38a,b and 40). As a result,

a very complex distribution of terrain-influenced thermal boundaries were observed over the northeastern United States on 2–3 March that defied description by any classical frontal cyclone models.

The timing of frontal passages across the complex terrain region in the Northeast was also unconventional. The final cold frontal passage through eastern New York occurred ~1 h earlier at ALB than at GFL even though the broad northeast–southwest frontal orientation suggested that frontal passage should be expected first at GFL (~80 km north of ALB). The earlier frontal passage at ALB was likely a result of cold air being funneled down the Mohawk Valley while the advancing cold air was partially blocked from reaching GFL by the Adirondacks to the west (Fig. 39). Additionally, some cold air may have reached GFL from the south by flowing up the Hudson Valley, given that the Mohawk Valley opens up into the Hudson Valley near ALB (Fig. 39). Finally, windward ridging, flow blocking, and lee troughing were observed with frontal passage across the southern Appalachians (Figs. 42a,b), a common characteristic of frontal passages across Georgia and the Carolinas in the lee of the Appalachians (e.g., O'Handley and Bosart 1996; Schumacher et al. 1996). Windward ridging and flow blocking delayed the arrival of the cold air in the western Carolinas and northeastern Georgia by 6–12 h (Figs. 42a,b). Instead, the cold air had to work around the southern end of the Appalachians and approach northeastern Georgia and western South Carolina from the southwest, while deeper cold air, able to cross the mountains farther north, entered North Carolina from the north (Figs. 42a,b).

c. Forecast issues

On the basis of Fig. 18 it is likely that had the 1972 long-lived frontal case occurred today there would have been huge forecast errors as to the timing of the frontal passages, large temperature changes, and the associated weather variations with resultant high impact on daily lives and commerce (especially the transportation industry and in particular airline hubs). Evidence for this assertion comes from the complete failure of MOS to predict a significant mesoscale frontal passage over eastern New England in the 17–18 April 2002 case (Fig. 17). Bosart (2003) noted that the unavailability of real-time, high-quality, high-resolution regional and national mesoscale analyses from the numerous automated weather observations routinely available could be a detriment to forecaster ability to recognize subtle but important mesoscale features of interest to the general public. Without the availability of high-quality mesoscale analyses on a routine basis it is unreasonable to expect that the forecast weather variability will come close to matching the observed weather variability, given that existing mesoscale models often have difficulty simulating the location and timing of high-impact weather events.

Roebber and Bosart (1998) showed that uncertainties in the precipitation distribution in midlatitude cyclones were very sensitive to small, diurnally varying uncertainties in the mass field, rendering pattern recognition questionable. Roebber and Bosart (1998) arrived at their conclusion by showing that the best regional mass field analogs showed little corresponding agreement in the distribution of precipitation and vice versa. As Fred Sanders reminded us numerous times over the last few decades, the observations matter, frontal boundaries need to be defined carefully on the basis of observed potential temperature gradients, and lots of interesting and not necessarily well-forecast mesoscale weather events, often precipitation-related, occur along these frontal boundaries (e.g., Sanders 1963, 1979, 1999, 2000; Sanders and Doswell 1995; Sanders and Kessler 1999; Sanders and Hoffman 2002).

Failure to properly replicate important thermal boundaries of the type shown in the two case studies contained in this paper is a recipe for forecast trouble and, given the numerous surface observations widely available but insufficiently used throughout the country, invites the paying taxpayers, who do not live at the 500-hPa level, to dismiss the demonstrable increase in forecast skill over the last several decades as mostly irrelevant to their routine day-to-day lives. Gains in forecasting the mass field do not necessarily translate into improvements in forecasting sensible weather elements.

Any serious forecast verification study needs to address skill levels associated with *mesoscale* weather elements of interest to the general public. It is speculated that by doing so all sorts of problems ranging from analysis uncertainties to communications difficulties will be uncovered. These forecast problems need to be uncovered in order to be better appreciated before they can be fixed.

Acknowledgments. The lead author thanks Fred Sanders for giving him a terrific education, for teaching him that the data always matters and that the answers to any scientific questions lie within the data, for always being supportive, and for inspiring generations of students and colleagues. The manuscript was improved because of the constructive comments of the reviewers and David Schultz. Celeste Iovinella is thanked for assistance in putting the manuscript and figures in final form. Tom Galarneau Jr., assisted with the preparation of Fig. 16. This research was supported by NSF Grants ATM-8311106, ATM-9120331, ATM-9413012, ATM-9912075, and CSTAR Grants NA07WA0458 and NA04NWS4680005.

APPENDIX

Figure A1 shows a simplified terrain map for the domain included by both case studies. The locations of all key stations mentioned in the text are indicated, along with the orientations of all cross sections.

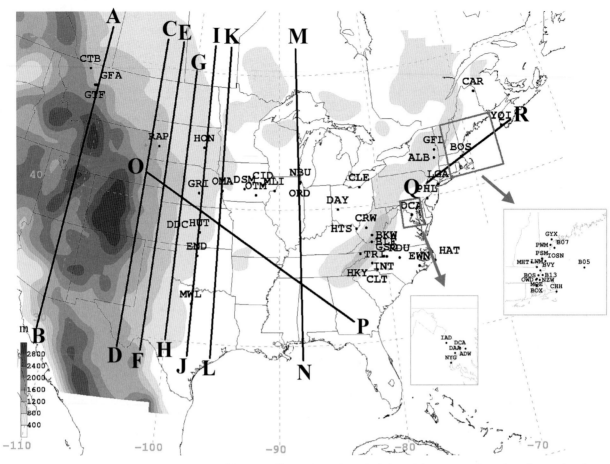

FIG. A1. Map showing terrain (shaded every 500 m according to the scale at the left) and the cross section orientations and station locations as discussed in the text.

REFERENCES

Arbogast, P., 2004: Frontal-weave development by interaction between a front and a cyclone: Application to FASTEX IOP 17. *Quart. J. Roy. Meteor. Soc.,* **130,** 1675–1696.

Atallah, E. H., and L. F. Bosart, 2003: Extratropical transition and precipitation distribution of Floyd '99. *Mon. Wea. Rev.,* **131,** 1063–1081.

Bell, G. D., and L. F. Bosart, 1988: Appalachian cold-air damming. *Mon. Wea. Rev.,* **116,** 137–161.

Bishop, C. H., and A. J. Thorpe, 1994a: Frontal wave stability during moist deformation frontogenesis. Part I: Linear wave dynamics. *J. Atmos. Sci.,* **51,** 852–873.

——, and ——, 1994b: Frontal wave stability during moist deformation frontogenesis. Part II: The suppression of nonlinear wave development. *J. Atmos. Sci.,* **51,** 874–888.

Bjerknes, J., 1919: On the structure of moving cyclones. *Geofys. Publ.,* **1,** 1–8

——, and H. Solberg, 1922: Life cycle of cyclones and the polar front theory of atmospheric circulation. *Geofys. Publ.,* **3,** 1–18.

——, and E. Palmén, 1937: Investigations of selected European cyclones by means of serial ascents. *Geofys. Publ. Norske Videnskaps-Akad. Oslo,* **12,** 1–62.

Bluestein, H. B., 1986: Fronts and jet streaks: A theoretical perspective. *Mesoscale Meteorology and Forecasting,* P. S. Ray, Ed., Amer. Meteor. Soc., 173–215.

Bond, N. A., C. F. Mass, and J. E. Overland, 1996: Coastally trapped wind reversals along the United States West Coast during the warm season. Part I: Climatology and temporal evolution. *Mon. Wea. Rev.,* **124,** 430–445.

Bosart, L. F., 1975: New England coastal frontogenesis. *Quart. J. Roy. Meteor. Soc.,* **101,** 957–978.

——, 1981: The Presidents' Day snowstorm of February 1979: A sub-synoptic scale event. *Mon. Wea. Rev.,* **109,** 1542–1566.

——, 1984: The Texas coastal rainstorm of 17–21 September 1979: An example of synoptic-mesoscale interaction. *Mon. Wea. Rev.,* **112,** 1108–1133.

——, 2003: Tropopause folding: Upper-level frontogenesis, and beyond. *A Half Century of Progress in Meteorology: A Tribute to Richard J. Reed, Meteor. Monogr.,* No. 31, Amer. Meteor. Soc., 13–47.

——, and F. H. Carr, 1978: A case study of excessive rainfall centered around Wellsville, New York, 20–21 June 1972. *Mon. Wea. Rev.,* **106,** 348–362.

——, and S. C. Lin, 1984: A diagnostic analysis of the Presidents' Day storm of February 1979. *Mon. Wea. Rev.,* **112,** 2148–2177.

——, and D. B. Dean, 1991: The Agnes Rainstorm of June 1972: Surface feature evolution culminating in inland storm redevelopment. *Wea. Forecasting,* **6,** 515–537.

——, C. J. Vaudo, and J. H. Helsdon Jr., 1972: Coastal frontogenesis. *J. Appl. Meteor.,* **11,** 1236–1258.

——, V. Pagnotti, and B. Lettau, 1973: Climatological aspects of eastern United States back-door cold frontal passages. *Mon. Wea. Rev.,* **101,** 627–635.

——, C.-C. Lai, and R. A. Weisman, 1992: A case study of heavy rainfall associated with weak cyclogenesis in the northwest Gulf of Mexico. *Mon. Wea. Rev.,* **120,** 2469–2500.

Boyle, J. S., 1986: Comparison of the synoptic conditions in midlatitudes accompanying cold surges over eastern Asia for the months of December 1974 and 1978. Part I: Monthly mean fields and individual events. *Mon. Wea. Rev.,* **114,** 903–918.

Branick, M. L., F. Vitale, C.-C. Lai, and L. F. Bosart, 1988: The synoptic and subsynoptic structure of a long-lived severe convective system. *Mon. Wea. Rev.,* **116,** 1335–1370.

Brennan, M. J., G. M. Lackmann, and S. E. Koch, 2003: An analysis of the impact of a split-front rainband on Appalachian cold-air damming. *Wea. Forecasting,* **18,** 712–731.

Carbone, R. E., 1982: A severe frontal rainband. Part I: Stormwide hydrodynamic structure. *J. Atmos. Sci.,* **39,** 258–279.

Carr, J. A., 1951: The east coast back-door cold front of 16-20 May 1951. *Mon. Wea. Rev.,* **79,** 100–105.

Charba, J., 1974: Application of gravity current model to analysis of squall-line gust front. *Mon. Wea. Rev.,* **102,** 140–156.

Chen, T.-C., M.-C. Yen, W.-R. Huang, and W. A. Gallus Jr., 2002: An east Asian cold surge: Case study. *Mon. Wea. Rev.,* **130,** 2271–2290.

Chien, F.-C., C. F. Mass, and Y.-H. Kuo, 1997: Interaction of a warm-season frontal system with the coastal mountains of the western United States. Part I: Prefrontal onshore push, coastal ridging, and along shore southerlies. *Mon. Wea. Rev.,* **125,** 1705–1729.

Colle, B. A., 2003: Numerical simulations of the extratropical transition of Floyd (1999): Structural evolution and responsible mechanisms for the heavy rainfall over the northeast United States. *Mon. Wea. Rev.,* **131,** 2905–2926.

——, and C. F. Mass, 1995: The structure and evolution of cold surges east of the Rocky Mountains. *Mon. Wea. Rev.,* **123,** 2577–2610

Colman, B. R., 1990a: Thunderstorms above frontal surfaces in environments without positive CAPE. Part I: A climatology. *Mon. Wea. Rev.,* **118,** 1103–1121.

——, 1990b: Thunderstorms above frontal surfaces in environments without positive CAPE. Part II: Organization and instability mechanisms. *Mon. Wea. Rev.,* **118,** 1123–1144.

Colquhoun, J. R., D. J. Shepherd, C. E. Coulman, R. K. Smith, and K. McInnes, 1985: The southerly buster of southeastern Australia: An orographically forced cold front. *Mon. Wea. Rev.,* **113,** 2090–2107.

Czikowsky, M. J., and D. R. Fitzjarrald, 2004: Evidence of seasonal changes in evapotranspiration in eastern U.S. hydrological records. *J. Hydrometeor.,* **5,** 974–988.

Davies, H. C., 1999: Theories of frontogenesis. *The Life Cycles of Extratropical Cycles: Bergen Symposium Book,* C. W. Newton and S. Grønås, Eds., Amer. Meteor. Soc., 215–238.

DiMego, G. J., and L. F. Bosart, 1982a: The transformation of Tropical Storm Agnes into an extratropical cyclone. Part I: The observed fields and vertical motion computations. *Mon. Wea. Rev.,* **110,** 385–411.

——, and ——, 1982b: The transformation of Tropical Storm Agnes into an extratropical cyclone. Part II: Moisture, vorticity and kinetic energy budgets. *Mon. Wea. Rev.,* **110,** 412–433.

Doyle, J. D., 1997: The influence of mesoscale orography on a coastal jet and rainband. *Mon. Wea. Rev.,* **125,** 1465–1488.

——, and T. T. Warner, 1990: Mesoscale coastal processes during GALE OP 2. *Mon. Wea. Rev.,* **118,** 283–308.

——, and ——, 1993a: The impact of the sea surface temperature resolution on mesoscale coastal processes during GLE IOP 2. *Mon. Wea. Rev.,* **121,** 313–334.

——, and ——, 1993b: A three-dimensional numerical investigation of a Carolina coastal low-level jet during GALE IOP 2. *Mon. Wea. Rev.,* **121,** 1030–1047

——, and ——, 1993c: A numerical investigation of coastal frontogenesis and mesoscale cyclogenesis during GALE IOP 2. *Mon. Wea. Rev.,* **121,** 1048–1077.

——, and ——, 1993d: Nonhydrostatic simulations of coastal mesobeta-scale vortices and frontogenesis. *Mon. Wea. Rev.,* **121,** 3371–3392.

Dunn, L. B., 1987: Cold air damming by the front range of the Colorado Rockies and its relationship to locally heavy snows. *Wea. Forecasting,* **2,** 177–189.

——, 1988: Vertical motion evaluation of a Colorado snowstorm from a synoptician's perspective. *Wea. Forecasting,* **3,** 261–272.

——, 1992: Evidence of ascent in a sloped barrier jet and an associated heavy-snow band. *Mon. Wea. Rev.,* **120,** 914–924.

Eddy, A., 1967: Statistical objective analysis of scalar data fields. *J. Appl. Meteor.,* **6,** 597–609.

Eliassen, A., 1959: On the formation of fronts in the atmosphere. *The Atmosphere and the Sea in Motion,* B. Bolin, Ed., Rockefeller Institute Press, 277–287.

——, 1990: Transverse circulations in frontal zones. *Extratropical Cyclones: The Erik Palmén Memorial Volume,* C. W. Newton and E. O. Holopainen, Eds., Amer. Meteor. Soc., 155–164.

Fitzjarrald, D. R., O. C. Acevedo, and K. E. Moore, 2001: Climatic consequences of leaf presence in the eastern United States. *J. Climate,* **14,** 598–614

Forbes, G. S., R. A. Anthes, and D. W. Thomson, 1987: Synoptic and mesoscale aspects of an Appalachian ice storm associated with cold air damming. *Mon. Wea. Rev.,* **115,** 564–591.

Freedman, J. M., D. R. Fitzjarrald, K. E. Moore, and R. K. Sakai, 2001: Boundary layer clouds and vegetation–atmosphere feedbacks. *J. Climate,* **14,** 180–197.

Fritsch, J. M., J. Kapolka, and P. A. Hirschberg, 1992: The effects of subcloud-layer diabatic processes on cold air damming. *J. Atmos. Sci.,* **49,** 49–70.

Gandin, L. S., 1963: *Objective Analysis of Meteorological Fields.* Israel Program for Scientific Translation, 242 pp.

Garratt, J. R., 1988: Summertime cold fronts in southeast Australia—Behavior and low-level structure of main frontal types. *Mon. Wea. Rev.,* **116,** 636–649.

——, W. L. Physick, R. K. Smith, and A. J. Troup, 1985: The Australian summertime cool change. Part II: Mesoscale aspects. *Mon. Wea. Rev.,* **113,** 202–223.

——, P. A. C. Howells, and E. Kowalczyk, 1989: The behavior of dry cold fronts traveling along a coastline. *Mon. Wea. Rev.,* **117,** 1208–1220.

Garreaud, R. D., 1999: Cold air incursions over subtropical and tropical South America: A numerical case study. *Mon. Wea. Rev.,* **127,** 2823–2853.

——, 2000: Cold air incursions over subtropical South America: Mean structure and dynamics. *Mon. Wea. Rev.,* **128,** 2544–2559.

——, and J. M. Wallace, 1998: Summertime incursions of midlatitude air into subtropical and tropical South America. *Mon. Wea. Rev.,* **126,** 2713–2733.

Goff, R. C., 1976: Vertical structure of thunderstorm outflows. *Mon. Wea. Rev.,* **104,** 1429–1440.

Hakim, G. J., 1992: The eastern United States side-door cold front of 22 April 1987: A case study of an intense atmospheric density current. *Mon. Wea. Rev.,* **120,** 2738–2762.

Hartjenstein, G., and R. Bleck, 1991: Factors affecting cold-air outbreaks east of the Rocky Mountains. *Mon. Wea. Rev.,* **119,** 2280–2292.

Hobbs, P. V., and P. O. G. Persson, 1982: The mesoscale and microscale structure and organization of clods and precipitation in midlatitude cyclones. Part V: The substructure of narrow cold-frontal rainbands. *J. Atmos. Sci.,* **39,** 280–295.

Hoinka, K. P., and H. Volkert, 1987: The German front experiment 1987. *Bull. Amer. Meteor. Soc.,* **68,** 1424–1427.

——, and D. Heimann, 1988: Orographic channeling of a cold front by the Pyrenees. *Mon. Wea. Rev.,* **116,** 1817–1823.

Hutchinson, T. A., and H. B. Bluestein, 1998: Prefrontal wind-shift lines in the plains of the United States. *Mon. Wea. Rev.,* **126,** 141–166.

Joly, A., and J. A. Thorpe, 1990: The stability of a steady horizontal shear front with uniform potential vorticity. *J. Atmos. Sci.,* **47,** 2612–2623.

——, and ——, 1991: The stability of time-dependent flows: An application to fronts in developing baroclinic waves. *J. Atmos. Sci.,* **48,** 163–182.

Kalnay, E., and Coauthors, 1996: The NCEP/NCAR 40-Year Reanalysis Project. *Bull. Amer. Meteor. Soc.,* **77,** 437–471

Keshishian, L. G., L. F. Bosart, and W. E. Bracken, 1994: Inverted troughs and cyclogenesis over interior North America: A limited regional climatology and case studies. *Mon. Wea. Rev.,* **122,** 565–607.

Keyser, D., 1986: Atmospheric fronts: An observational perspective. *Mesoscale Meteorology and Forecasting*, P. S. Ray, Ed., Amer. Meteor. Soc., 216–258.

——, 1999: On the representation and diagnosis of frontal circulations in two and three dimensions. *The Life Cycles of Extratropical Cyclones: Bergen Symposium Book,* C. W. Newton and S. Grønås, Eds., Amer. Meteor. Soc., 239–264.

——, B. D. Schmidt, and D. G. Duffy, 1989: A technique for representing three-dimensional vertical circulations in baroclinic disturbances. *Mon. Wea. Rev.,* **117,** 2463–2494.

Kistler, R., and Coauthors, 2001: The NCEP-NCAR 50-Year Reanalysis: Monthly means CD-ROM and documentation. *Bull. Amer. Meteor. Soc.,* **82,** 247–268.

Koch, S. E., 1984: The role of an apparent mesoscale frontogenetic circulation in squall line initiation. *Mon. Wea. Rev.,* **112,** 2090–2111.

——, and W. L. Clark, 1999: A nonclassical cold front observed during COPS-91: Frontal structure and the process of severe storm initiation. *J. Atmos. Sci.,* **56,** 2862–2890.

——, M. desJardins, and P. J. Kocin, 1983: An interactive Barnes objective map analysis scheme for use with satellite and conventional data. *J. Climate Appl. Meteor.,* **22,** 1487–1503.

——, J. T. McQueen, and V. M. Karyampudi, 1995: A numerical study of the effects of differential cloud cover on cold frontal structure and dynamics. *J. Atmos. Sci.,* **52,** 937–964.

Langmaid, A. H., and A. J. Riordan, 1998: Surface mesoscale processes during the 1994 Palm Sunday tornado outbreak. *Mon. Wea. Rev.,* **126,** 2117–2132.

Lapenta, W. M., and N. L. Seaman, 1990: A numerical investigation of East Coast cyclogenesis during the cold-air damming event of 27–28 February 1982. Part I: Dynamic and thermodynamic structure. *Mon. Wea. Rev.,* **118,** 2668–2695.

Loughe, A. F., 1992: Real-data diagnosis of partitioned ageostrophic vertical circulations. M.S. thesis, Department of Earth and Atmospheric Sciences, University at Albany, State University of New York, 133 pp.

Lupo, A. R., J. J. Nocera, L. F. Bosart, E. G. Hoffman, and D. J. Knight, 2001: South American cold surges: Types, composites, and case studies. *Mon. Wea. Rev.,* **129,** 1021–1041.

Mallet, I., P. Arbogast, C. Baehr, J.-P. Cammas, and P. Mascart, 1999a: Effects of cloud diabatic heating on the early development of the FASTEX IOP17 cyclone. *Quart. J. Roy. Meteor. Soc.,* **125,** 3439–3467

——, J.-P. Cammas, and P. Mascart, 1999b: Dynamical characterization of the FASTEX cyclogenesis cases. *Quart. J. Roy. Meteor. Soc.,* **125,** 3469–3494.

Marengo, J., A. Cornejo, P. Satyamurty, C. Nobre, and W. Sea, 1997: Cold surges in tropical and extratropical South America: The strong event in June 1994. *Mon. Wea. Rev.,* **125,** 2759–2786.

Margules, M., 1906: Über temperaturschichtung in stationär bewegter und ruhender luft. *Hann-Band. Meteor. Z.,* **2,** 245–254.

Mass, C. F., and M. D. Albright, 1987: Coastal southerlies and alongshore surges of the West Coast of North America: Evidence of mesoscale topographically trapped response to synoptic forcing. *Mon. Wea. Rev.,* **115,** 1707–1738.

——, and N. A. Bond, 1996: Coastally trapped wind reversals along the United States West Coast during the warm season. Part II: Synoptic evolution. *Mon. Wea. Rev.,* **124,** 446–461.

——, H. J. Edmon, H. J. Friedman, N. R. Cheney, and E. E. Recker, 1987: The use of compact discs for the storage of large meteorological and oceanographic data sets. *Bull. Amer. Meteor. Soc.,* **68,** 1556–1558.

Matthews, D. A., 1981: Observation of a cloud arc triggered by thunderstorm outflow. *Mon. Wea. Rev.,* **109,** 2140–2157.

McBride, J. L., and K. L. McInnes, 1993: Australian southerly busters. Part II: The dynamical structure of the orographically modified front. *Mon. Wea. Rev.,* **121,** 1921–1935.

McInnes, K. L., and J. L. McBride, 1993: Australian southerly busters. Part I. Analysis of a numerically simulated case study. *Mon. Wea. Rev.,* **121,** 1904–1920.

Meier, K. W., 1993: Analysis of a long-lived intense low-level front: A case study of 28 February through 3 March 1972. M.S. thesis, Dept. of Earth and Atmospheric Sciences, University at Albany, State University of New York, 274 pp.

Miller, J. E., 1948: On the concept of frontogenesis. *J. Meteor.,* **5,** 169–171.

Moreira, M., L. Sternburg, L. Martinelli, R. Victoria, E. Barbosa, L. Bonates, and D. Nepstad, 1997: Contribution of transpiration to forest ambient vapour based on isotopic measurements. *Global Change Biol.,* **3,** 439–450.

Newton, C. W., 1954: Frontogenesis and frontolysis as a three-dimensional process. *J. Meteor.,* **11,** 449–461.

Nielsen, J. W., 1989: The formation of New England coastal fronts. *Mon. Wea. Rev.,* **117,** 1380–1401.

——, and P. P. Neilley, 1990: The vertical structure of New England coastal fronts. *Mon. Wea. Rev.,* **118,** 1793–1807

O'Handley, C., and L. F. Bosart, 1996: The impact of the Appalachian Mountains on cyclonic weather systems. Part I: A climatology. *Mon. Wea. Rev.,* **124,** 1353–1373

Palmén, E., 1949: On the origin and structure of high-level cyclones south of the maximum westerlies. *Tellus,* **1,** 22–31

Patoux, J., G. J. Hakim, and R. A. Brown, 2005: Diagnosis of frontal instabilities over the Southern Ocean. *Mon. Wea. Rev.,* **133,** 863–875.

Petterssen, S., 1936: Contribution to the theory of frontogenesis. *Geofys. Publ.,* **11,** 1–27.

——, 1956: *Weather Analysis and Forecasting.* Vol. 1, McGraw-Hill, 428 pp.

Reed, R. J., 1955: A study of a characteristic type of upper-level frontogenesis. *J. Meteor.,* **12,** 226–237.

——, and F. Sanders, 1953: An investigation of the development of a mid-tropospheric frontal zone and its associated vorticity field. *J. Meteor.,* **10,** 338–349.

Renfrew, I. A., A. J. Thorpe, and C. H. Bishop, 1997: The role of environmental flow in the development of secondary frontal cyclones. *Quart. J. Roy. Meteor. Soc.,* **123,** 1653–1675.

Roebber, P. J., and L. F. Bosart, 1998: The sensitivity of precipitation to circulation details. Part I: An analysis of regional analogues. *Mon. Wea. Rev.,* **126,** 437–455.

Sanders, F., 1955: An investigation of the structure and dynamics of an intense surface frontal zone. *J. Meteor.,* **12,** 542–552

——, 1963: On subjective probability forecasting. *J. Appl. Meteor.,* **2,** 191–201.

——, 1972: Meteorological and oceanographic conditions during the 1970 Bermuda Yacht Race. *Mon. Wea. Rev.,* **100,** 597–606

——, 1979: Trends in skill of daily forecasts of temperature and precipitation, 1966–78. *Bull. Amer. Meteor. Soc.,* **60,** 763–769.

——, 1999: A proposed method of surface map analysis. *Mon. Wea. Rev.,* **127,** 945–955.

——, 2000: Frontal focusing of a flooding rainstorm. *Mon. Wea. Rev.,* **128,** 4155–4159.

——, and J. R. Gyakum, 1980: Synoptic-dynamic climatology of the "bomb." *Mon. Wea. Rev.,* **108,** 1589–1606.

——, and L. F. Bosart, 1985a: Mesoscale structure in the megalopolitan snowstorm of 11–12 February 1983. Part I: Frontogenetical forcing and symmetric instability. *J. Atmos. Sci.,* **42,** 1050–1061.

——, and ——, 1985b: Mesoscale structure in the megalopolitan snowstorm, 11–12 February 1983. Part II: Doppler radar study of the New England snowband. *J. Atmos. Sci.,* **42,** 1398–1407.

——, and C. A. Doswell III, 1995: A case for detailed surface analysis. *Bull. Amer. Meteor. Soc.,* **76,** 505–521.

——, and E. Kessler, 1999: Frontal analysis in the light of abrupt temperature changes in a shallow valley. *Mon. Wea. Rev.,* **127,** 1125–1133.

——, and E. G. Hoffman, 2002: A climatology of surface baroclinic zones. *Wea. Forecasting,* **17,** 774–782.

Schär, C., and H. C. Davies, 1990: An instability of mature cold fronts. *J. Atmos. Sci.,* **47,** 929–950.

Schoenberger, L. M., 1984: Doppler radar observation of a land breeze cold front. *Mon. Wea. Rev.,* **112,** 2455–2464.

Schultz, D. M., 2004: Cold fronts with and without prefrontal wind shifts in the central United States. *Mon. Wea. Rev.,* **132,** 2040–2053.

——, 2008: A review of cold fronts, including prefrontal troughs and wind shifts. *Synoptic–Dynamic Meteorology and Weather Analysis and Forecasting: A Tribute to Fred Sanders, Meteor. Monogr.,* No. 55, Amer. Meteor. Soc.

——, and W. J. Steenburgh, 1999: The formation of a forward-tilting cold front with multiple cloud bands during Superstorm 1993. *Mon. Wea. Rev.,* **127,** 1108–1124.

——, and P. J. Roebber, 2008: The fiftieth anniversary of Sanders (1955): A mesoscale model simulation of the cold front of 17–18 April 1953. *Synoptic–Dynamic Meteorology and Weather Analysis and Forecasting: A Tribute to Fred Sanders, Meteor. Monogr.,* No. 55, Amer. Meteor. Soc.

——, W. E. Bracken, L. F. Bosart, G. J. Hakim, M. A. Bedrick, M. J. Dickinson, and K. R. Tyle, 1997: The 1993 Superstorm cold surge: Frontal structure, gap flow, and tropical impact. *Mon. Wea. Rev.,* **125,** 5–39.

Schumacher, P. N., D. J. Knight, and L. F. Bosart, 1996: Frontal interaction with the Appalachian Mountains. Part I: A climatology. *Mon. Wea. Rev.,* **124,** 2453–2468.

Schwierz, C., S. Dirren, and H. C. Davies, 2004: Forced waves on a zonally aligned jet stream. *J. Atmos. Sci.,* **61,** 73–87

Seitter, K. L., 1986: A numerical study of atmospheric density current motion including the effects of condensation. *J. Atmos. Sci.,* **43,** 3068–3076.

Seluchi, M. E., R. D. Garreaud, F. A. Norte, and A. C. Saulo, 2006: Influence of the subtropical Andes on baroclinic disturbances: A cold front case study. *Mon. Wea. Rev.,* **134,** 3317–3335.

Shapiro, M. A., 1982: Mesoscale weather systems of the central United States. NOAA–CIRES Tech. Rep., University of Colorado, 78 pp.

——, 1984: Meteorological tower measurements of a surface cold front. *Mon. Wea. Rev.,* **112,** 1634–1639.

——, and D. Keyser, 1990: Fronts, jet streams and the tropopause. *Extratropical Cyclones: The Erik Palmén Memorial Volume,* C. W. Newton and E. O. Holopainen, Eds., Amer. Meteor. Soc., 167–191.

——, T. Hampel, D. Rotzoll, and F. Mosher, 1985: The frontal hydraulic head: A micro-scale (~1 km) triggering mechanism for mesoconvective weather systems. *Mon. Wea. Rev.,* **113,** 1166–1183.

Simpson, J. E., 1987: *Gravity Currents: In the Environment and the Laboratory.* Halstead Press, 244 pp.

——, D. A. Mansfield, and J. R. Milford, 1977: Inland penetration of sea-breeze fronts. *Quart. J. Roy. Meteor. Soc.,* **103,** 47–76

Smith, R. B., and M. J. Reeder, 1988: On the movement and low-level structure of cold fronts. *Mon. Wea. Rev.,* **116,** 1927–1944.

Smith, R. K., B. F. Ryan, A. J. Troup, and K. J. Wilson, 1982: Cold fronts research: The Australian summertime "cool change." *Bull. Amer. Meteor. Soc.,* **63,** 1028–1034.

Smith, S. A., D. C. Fritts, and T. E. Vanzandt, 1987: Evidence for a saturated spectrum of atmospheric gravity waves. *J. Atmos. Sci.,* **44,** 1404–1410

Stauffer, D. R., and T. T. Warner, 1987: A numerical study of Appalachian cold-air damming and coastal frontogenesis. *Mon. Wea. Rev.,* **115,** 799–821.

Steenburgh, W. J., and T. R. Blazek, 2001: Topographic distortion of a cold front over the Snake River Plain and central Idaho mountains. *Wea. Forecasting,* **16,** 301–314.

——, D. M. Schultz, and B. A. Colle, 1998: The structure and evolution of gap outflow over the Gulf of Tehuantepec, Mexico. *Mon. Wea. Rev.,* **126,** 2673–2691

Steiner, J. T., C. G. Revell, R. N. Ridley, R. K. Smith, M. A. Page, K. L. McInnes, and A. P. Sturman, 1987: The New Zealand southerly change experiment. *Bull. Amer. Meteor. Soc.,* **68,** 1226–1229.

Thorncroft, C. D., and B. J. Hoskins, 1990: Frontal cyclogenesis. *J. Atmos. Sci.,* **47,** 2317–2336.

Uccellini, L. W., P. J. Kocin, R. A. Petersen, C. H. Wash, and K. F. Brill, 1984: The Presidents' Day cyclone of 18-19 February 1979: Synoptic overview and analysis of the subtropical jet streak influencing the precyclogenetic period. *Mon. Wea. Rev.,* **112,** 31–55.

——, D. Keyser, K. F. Brill, and D. H. Wash, 1985: The Presidents' Day cyclone of February 1979: Influence of upstream trough amplification and associated tropopause folding on rapid cyclogenesis. *Mon. Wea. Rev.,* **113,** 962–988.

Vera, C. S., and P. K. Vigliarolo, 2000: A diagnostic study of cold-air outbreaks over South America. *Mon. Wea. Rev.,* **128,** 3–24.

Volkert, H., 1999: Components of the Norwegian cyclone model: Observations and theoretical ideas in Europe prior to 1920. *The Life Cycles of Extratropical Cycles: Bergen Symposium Book,* C. W. Newton and S. Grønås, Eds., Amer. Meteor. Soc., 15–28.

——, L. Weickmann, and A. Tafferner, 1991: The papal front of 3 May 1987: A remarkable example of frontogenesis near the Alps. *Quart. J. Roy. Meteor. Soc.,* **117,** 125–150.

Wakimoto, R. M., 1982: Investigations of thunderstorm gust fronts with the use of radar and rawinsonde data. *Mon. Wea. Rev.,* **110,** 1060–1082.

Weisman, R. A., K. G. McGregor, D. R. Novak, J. L. Selzler, M. L. Spinar, B. C. Thomas, and P. N. Schumacher, 2002: Precipitation regimes during cold-season central U.S. inverted trough cases. Part I: Synoptic climatology and composite study. *Wea. Forecasting,* **17,** 1173–1193.

Whitaker, J. S., L. W. Uccellini, and K. F. Brill, 1988: A model-based diagnostic study of the rapid development phase of the Presidents' Day cyclone. *Mon. Wea. Rev.,* **116,** 2337–2365.

Zhang, F., and S. E. Koch, 2000: Numerical simulations of a gravity wave event over CCOPE. Part II: Waves generated by an orographic density current. *Mon. Wea. Rev.,* **128,** 2777–2796.

Chapter 3

Back to Norway: An Essay

KERRY EMANUEL

Program in Atmospheres, Oceans, and Climate, Massachusetts Institute of Technology, Cambridge, Massachusetts

(Manuscript received 10 May 2005, in final form 3 April 2006)

Emanuel

ABSTRACT

The advent of the polar front theory of cyclones in Norway early in the last century held that the development of fronts and air masses is central to understanding midlatitude weather phenomena. While work on fronts continues to this day, the concept of air masses has been largely forgotten, superseded by the idea of a continuum. The Norwegians placed equal emphasis on the thermodynamics of airmass formation and on the dynamical processes that moved air masses around; today, almost all the emphasis is on dynamics, with little published literature on diabatic processes acting on a large scale. In this essay, the author argues that a lack of understanding of large-scale diabatic processes leads to an incomplete picture of the atmosphere and contributes to systematic errors in medium- and long-range weather forecasts. At the same time, modern concepts centered around potential vorticity conservation and inversion lead one to a redefinition of the term "air mass" that may have some utility in conceptualizing atmospheric physics and in weather forecasting.

1. Introduction

Fronts and air masses are generally regarded as the two key concepts that emerged from the "Norwegian school," founded by Vilhelm Bjerknes and carried on by his son, Jacob, together with Tor Bergeron, Halvor Solberg, Erik Palmén, and others. Fronts were considered to be boundaries between air masses with distinct thermodynamic properties. According to Bergeron (1928),

> An air mass is a vast body of air whose physical properties are more or less uniform in the horizontal, while abrupt changes are found along its boundaries, i.e. the frontal zones.

Corresponding author address: Kerry Emanuel, Rm. 54-1620, MIT, 77 Massachusetts Avenue, Cambridge, MA 02139.
E-mail: emanuel@texmex.mit.edu

The Norwegians envisioned an intimate relationship among fronts, air masses, and cyclones:

> The cyclone consists of two essentially different airmasses, the one of cold and the other of warm origin. They are separated by a fairly distinct boundary surface which runs through the center of the cyclone. This boundary surface is imagined to continue, more or less distinctly, through the greater part of the troposphere (Bjerknes and Solberg 1922).

Note that the frontal boundary separating air masses was thought to run through the whole depth of the troposphere. This idea was later discredited by Sanders (1955), who showed, using surface and upper air data, that fronts are generally very shallow features, rarely detectable more than a kilometer or two above the surface. Still later, Hoskins and Bretherton (1972) demonstrated on theoretical grounds that true fronts, defined as near discontinuities in long-front velocity, can only form at rigid boundaries or at preexisting discontinuities in the distribution of potential vorticity, such as the tropopause. But I shall argue later in this essay that deep fronts may indeed occur along the equatorward boundary of arctic air masses.

The Norwegians placed a great deal of emphasis on airmass formation. They defined essentially four air masses, based on whether they were of continental or maritime origin and whether they were warm or cold. Although they recognized the influence of radiative processes in the atmosphere itself, they clearly regarded the underlying surface as the basic progenitor of air masses, largely defining their properties. Bergeron visualized air masses as forming within semipermanent circulation systems, such as wintertime continental highs and subtropical anticyclones:

> The air that takes part in the circulation around any such system will become subject to the prolonged influences of the underlying surface, with the result that there will be a tendency for distinct properties to be acquired. Although the vertical structure of any air mass may be modified by differential advection and vertical stretching and shrinking, the more direct modifications are brought about by interactions between the atmosphere and the earth's surface [Bergeron (1928) as paraphrased by Petterssen (1956)].

Air masses were characterized not only by their surface properties but, in particular, by their vertical structure as revealed by radiosondes. Arctic air masses were characterized as having very stable temperature profiles in the lowest layers, while, for example, maritime tropical air was revealed by deep layers of moist adiabatic lapse rates. The thermodynamics of airmass formation were studied in detail and were regarded on an equal footing as the dynamical processes that moved air masses around.

Then came the dynamics revolution, fostered by the theory of baroclinic instability developed by Charney (1947) and Eady (1949), and the subsequent development of quasigeostrophic theory by Charney and Phillips (e.g., Charney and Phillips 1953) and others. In the ensuing firestorm of progress in dynamical theory and in numerical weather prediction, thermodynamics took a back seat. In many beginning courses in atmospheric dynamics, thermodynamics are often developed only so far as demonstrating the conservation of potential temperature. Although Eady, in his 1949 paper, emphasized the central role of latent heat release in the dynamics of extratropical cyclones, it took another quarter century for theorists to take much interest in this issue, and today the phrase "diabatic processes" is virtually synonymous with "latent heat release." Although it is recognized that radiative processes must be included in numerical weather prediction models, they are regarded by forecasters as operating on a long time scale or as influencing only the boundary layer on diurnal time scales.

Almost all contemporary discussions of weather prediction and predictability focus on dynamical error growth, with some attention paid to the incorrect representation of convection and almost none to radiative processes. Meanwhile, the term "air mass" has been shelved together with such antiquities as "polar front" and "weather breeder."

The contemporary view of the physics of fronts and cyclones can be traced back to the work of Ernst Kleinschmidt, Eric Eady, Jule Charney, Brian Hoskins, and Francis Bretherton. This view may be broadly summarized as follows:

> The baroclinic dynamics of quasi-balanced systems outside the Tropics (and to some extent within them) may be thought of in terms of the conservation and invertibility of potential vorticity.

> Isentropic gradients of potential vorticity, which serve as the conduits of Rossby waves, are concentrated at the tropopause and, effectively, at the surface where there is a strong temperature gradient.

> The troposphere itself may be thought of as a region of constant potential vorticity, or nearly constant potential vorticity gradient.

> Most of the dynamics of extratropical weather systems may be conceptualized in terms of the propagation and interaction of Rossby waves at the tropopause and surface, and to a lesser extent, in the tropospheric continuum, perhaps modified by latent heat release.

> Fronts are features of the surface and (deformed) tropopause.

> Five–ten-day forecast errors are due to dynamical error growth; sensitivity can be approximately measured by adiabatic error growth.

With the exception of the important distinction between the troposphere and stratosphere, the concept of air masses is entirely missing from this point of view, as is any accounting of radiative effects. The author has taught an advanced graduate course in quasi-balanced dynamics at the Massachusetts Institute of Technology (MIT) for 15 yr and very much subscribes to the view described briefly above. But there are times and places where being conscious of radiative processes seems necessary for understanding the medium-range evolution of the atmosphere, and there may be some utility in reintroducing the concept of air masses, albeit with a contemporary spin.

2. Evolution of arctic air

The formation of cold air, by radiative cooling, is problematic. Most of the cooling is at the surface itself, and as the air adjacent to the surface cools, the air mass becomes increasingly stable and impervious to vertical mixing. Aside from weak radiative cooling in the in-

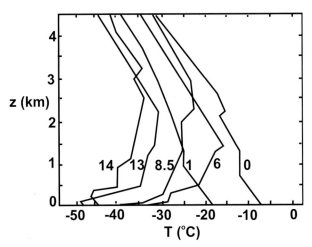

FIG. 1. Successive temperature soundings at Fairbanks, Alaska, in December 1961. Curves labeled with time in days relative to first sounding. From Curry (1983).

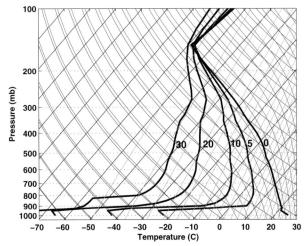

FIG. 2. Evolution of the vertical profile of temperature in a single-column radiative–convective model, beginning with a tropical sounding with 99% of the water vapor removed at each level. Profile labeled in days relative to the initial sounding. Model described in text.

terior, there is no mechanism for propagating the surface cooling upward to affect a deep layer. [The same problem occurs upside down in the ocean, where vertical mixing is the only way to propagate surface warming down into the interior (Sandstrom 1908; Jeffreys 1925).]

In what I view as a landmark paper, Judith Curry drew attention to the peculiar thermodynamics involved in the formation of continental arctic air masses (Curry 1983). She noted that time series of atmospheric soundings, such as those repeated here in Fig. 1, show cooling through the first few kilometers of the atmosphere at rates that exceed those one might expect based on simple radiative models. Figure 1 shows a sequence of soundings from Fairbanks, Alaska, in December. Even 1 km above the surface, the air cools about 30°C over the course of 2 weeks. Of course, some of this might be by advection, but examination of reanalysis data over the Canadian arctic during winter shows that the minimum temperature in an isolated pool of cold air can fall many degrees in a week; this is almost certainly the effect of radiative cooling.

Curry pointed out that the rate of cooling is sensitive to the moisture content of the air, the presence or absence of condensed water (ice crystals, at these temperatures), and the rate of subsidence. Curry began by using a radiative transfer code to calculate the evolution of temperature in a single column, beginning with a dry, nearly moist adiabatic profile with a surface temperature of 0°C. I repeat these calculations here using a different radiative code and starting with a tropical sounding, but removing most of the water vapor so that clouds do not form. [The code uses the radiation scheme of Morcrette (1991), the convection scheme of Emanuel and Živkovic-Rothman (1999), and the fractional cloudiness scheme of Bony and Emanuel (2001).] As Fig. 2 shows, the air cools down very rapidly at the surface, but even at 600 hPa the rate of cooling is large enough to be of concern even for a short-range forecast. If we rerun this

calculation starting from a reasonable moisture profile and allow clouds to develop, the column cools much less rapidly, because of both latent heat release and the insulating effect of clouds. A more interesting calculation is to assume that large-scale subsidence has dried out the column, except near the surface, and allows clouds to develop only in the boundary layer as the air cools. The results of one such calculation are shown in Fig. 3. Clouds form in the boundary layer, which deepens with time owing to the bootstrapping mechanism identified by Curry: Radiative cooling at the cloud top leads to condensation there, which deepens the existing cloud. But as the cloud thickens, this cooling becomes large enough that the boundary layer is destabilized, resulting in convective mixing, which dries the cloud and causes it to break up. [The interested reader should consult Curry (1983), who used a more physically correct representation of ice crystals than the model shown here.]

As Curry recognized, these calculations are not very realistic because they omit the large-scale subsidence that almost certainly accompanies the formation of arctic air. With a single column, it is not possible to directly calculate the subsidence, which requires 2D or 3D dynamics, but one can get a rough feeling of the effects of subsidence by simply specifying a vertical subsidence profile. Here we specify a smooth profile of subsidence, vanishing at the surface and at 100 hPa and reaching a peak value of 0.5 hPa h^{-1} at 750 hPa. To allow condensation to occur in the boundary layer in this calculation, we increase the initial amount of moisture in the lower troposphere. As shown in Fig. 4, the boundary layer grows because of cloud-top cooling, but slowly asymptotes to an equilibrium height of about 750 hPa, at which point this deepening effect is balanced by subsidence. Not only does the cold boundary layer become

a

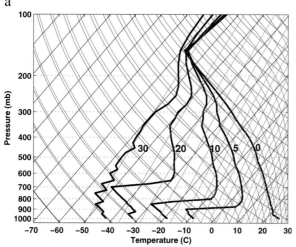

b

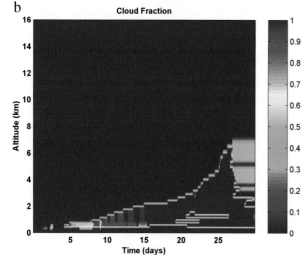

FIG. 3. (a) Evolution of the vertical profile of temperature in a single-column radiative–convective model, beginning with a tropical sounding. In this case, the fraction of water removed increases from 90% at the surface to 99% at 100 hPa. (b) Time–height plot of the fractional cloudiness in the simulation described in (a).

substantially deeper, but the air above the boundary layer cools somewhat faster than it did in the clear case, perhaps because of larger concentrations of water vapor in this simulation. Another interesting feature of this simulation is the development of a new low-level inversion, separating two nearly dry adiabatic layers, at day 30.

While calculations such as Curry's and those presented here cannot be regarded as precise simulations of the formation of arctic air, they do illustrate that the time scale of and depth through which radiative cooling acts are sensitive to the water vapor content of the air mass and the presence of clouds, as well as to the magnitude of the large-scale subsidence. Given that the absolute amounts of water involved are very small, and that some of its sources, such as sublimation of snow, may be difficult to model, it is questionable whether

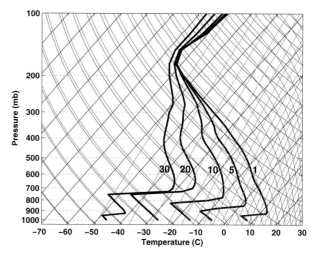

FIG. 4. Evolution of the vertical profile of temperature in a single-column radiative–convective model, beginning with a tropical sounding. In this case, the fraction of water removed increases from 60% at the surface to 99% at 100 hPa. A smooth vertical profile of ω is specified, vanishing at the surface and at 100 hPa and reaching a peak value of 0.5 hPa h^{-1} at 750 hPa.

today's forecast models are capable of accurate simulation of arctic airmass formation on 10-day time scales.

3. A revised airmass classification

The general idea of an air mass is something that once formed, tends to preserve its thermodynamic properties as it is advected around. The Norwegians classified air masses based mostly on the nature of vertical temperature profiles and on moisture content. Here we propose a new classification based on a single quasi-conservative variable, the *saturation potential vorticity* (SPV):

$$\text{SPV} \equiv \rho^{-1}(2\mathbf{\Omega} + \mathbf{\nabla} \times \mathbf{V}) \cdot \mathbf{\nabla} \ln \theta_e^*, \quad (1)$$

where ρ is the air density, $\mathbf{\Omega}$ is the earth's angular velocity vector, $\mathbf{V}$ is the fluid velocity vector, and θ_e^* is the saturation value of the equivalent potential temperature. (I prefer to use the natural log of θ_e^* because it is then proportional to the entropy.) SPV has several very nice characteristics:

It is *always invertible,* provided the flow is balanced, because θ_e^* is a state variable.

It is nearly conserved in very cold air (e.g., arctic air, stratospheric air) because in the cold limit it reduces to the ordinary potential vorticity (PV), as $\theta_e^* \to \theta$ at low temperature.

Neutrality to (slantwise) convection is characterized by SPV = 0, which is equivalent to having moist adiabatic lapse rates along vortex lines (absolute momentum surfaces, in two dimensions). Thus, in much of the tropical and midlatitude free troposphere, where we observe con-

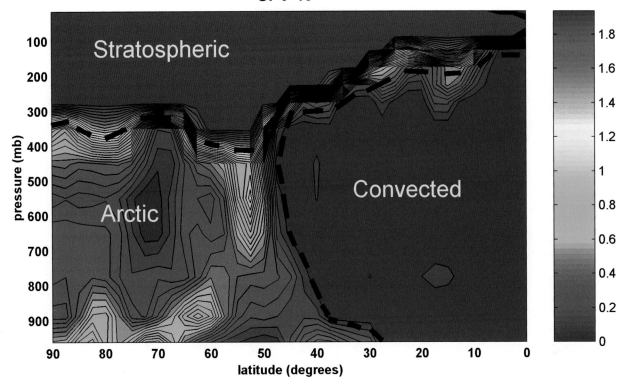

FIG. 5. Cross section of SPV along 90°W at 0000 UTC 8 Mar 2003. Values have been multiplied by 10^4, and all values larger than 2×10^4 have been reset to 2×10^4. The three main air masses are identified.

vective neutrality (Emanuel 1988; Xu and Emanuel 1989), SPV is nearly zero.

Note that although SPV is not materially conserved, it is nearly so in cold air, and in convectively adjusted air it is zero, which is just as good as being conserved.[1]

Based on these characteristics, I define four air masses:

convected: SPV = 0. (Moist adiabatic lapse rates on vortex lines.) Formation time of 1–2 days. Most of the troposphere, most of the time;

stratosphere: High SPV reservoir. Long formation time scales. SPV $\cong$ PV because of low temperatures;

arctic: High SPV because of radiative cooling in the continental interior in winter. Formation time of 4–14 days; and

planetary boundary layer (PBL): Over much of ocean and land during daytime, SPV < 0 because of dry adiabatic (e.g., supermoist adiabatic) lapse rates. Over cold

water and at night over land, SPV may become positive. Formation time of 1–12 h.

An example of a cross section of SPV is shown in Fig. 5; this is taken along 90°W at 0000 UTC 8 March 2003. I have subjectively delineated the boundaries between arctic, stratospheric, and convected air. Note that much of the tropical and midlatitude troposphere has SPV $\simeq$ 0 and that the boundary between the troposphere and stratosphere is usually well delineated, as in the case of PV. The transition zone between convected and arctic air lies between about 30° and 45°N in the section. The actual values of SPV in convected and stratospheric air are well defined by convective and radiative equilibrium, respectively, but as noted is section 2 above, the thermodynamic profiles (and therefore SPV) are highly variable in arctic air because of its sensitive dependence on clouds, water vapor, and subsidence. There is little evidence of PBL air in this section, perhaps because of the low vertical resolution of the reanalysis data used to construct the section.

In summer, arctic air (if it can be said to exist at all) has smaller values of SPV and can only be found at very high latitudes, as illustrated in Fig. 6, which shows a cross section along 100°W at 0000 UTC 7 July 2003. Arctic air can only be found poleward of 70°N, and its SPV is smaller than in the winter section of Fig. 5. The

---

[1] Note that unlike the case of classical PV, all the terms in (1) must be used in calculating SPV; one may not approximate it as the product of the vertical component of vorticity with the vertical gradient of $\ln\theta_e^*$. This is because the latter is typically much less than the vertical gradient of $\ln\theta$.

SPV*10^4

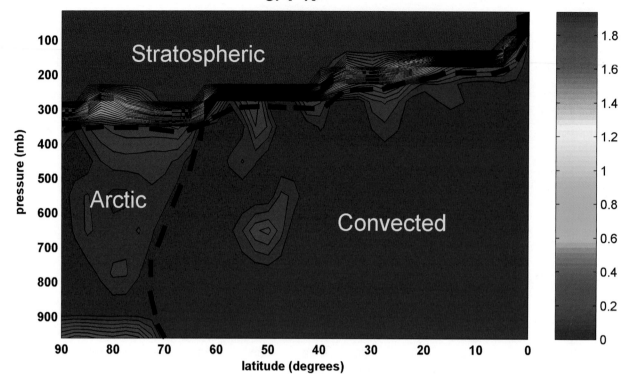

FIG. 6. Cross section of SPV along 100°W at 0000 UTC 7 Jul 2003. Values have been multiplied by 10^4, and all values larger than 2 × 10^4 have been reset to 2 × 10^4.

local patches of relatively high SPV near the tropopause and around 50°N may be real, but they are perhaps artifacts of the model used for the reanalysis.

Note that much of the troposphere often has SPV ≃ 0 even though very little of it is convecting at any one time. Convection is a comparatively fast process and where and when it occurs, it establishes nearly moist adiabatic lapse rates in a matter of hours (perhaps a little longer when the convection is slantwise). But once the convection ceases, it takes radiation and/or subsidence much longer to pull the lapse rates away from moist adiabatic (except in the PBL). A convecting air column over the North Pacific in winter may only take a few days to cross the North American continent, which may not be long enough to change its SPV appreciably.

From the perspective of SPV inversion (remember that SPV can be inverted just like PV, provided a balance approximation is valid), to a first approximation, one only needs to know the distributions of SPV in the artic air and the stratosphere, the distribution of θ_e^* at the top of the boundary layer, and the topology of the boundaries separating the stratospheric, convected, and artic air masses. This suggests a basis for a stripped down, quasi-balanced model integrating only the surface θ_e^*, the tropopause, and the boundary between arctic and convected air. I present one very simple example of this in the next section.

4. The arctic front

The Norwegians talked about a polar front separating polar from tropical air, and they believed that it extended from the surface to the tropopause. Later observational (Reed and Sanders, 1953; Reed 1955; Sanders 1955) and theoretical (Hoskins and Bretherton 1972) work established that true fronts (i.e., near discontinuities in long-front velocity and cross-front temperature gradient) can only form at the surface and at preexisting regions of sharp PV gradients, such as the tropopause. Almost all of the current literature on atmospheric fronts refers to surface and tropopause fronts; the latter, of course, may extend downward even to the lower troposphere. There is a small literature on arctic fronts, and P. Hobbs and coworkers (e.g., Wang et al. 1995) have recognized that these can be quite deep, with regions of strong horizontal temperature contrast extending through much of the troposphere, together with frontogenetical forcing.

The transition between arctic air of high SPV and convected air with SPV = 0 may extend through most or all of the troposphere, as shown in Fig. 5. In principle, deep geostrophic deformation acting on this transition is capable of forming a front extending through the troposphere, much as the Norwegians had imagined. To illustrate this point, we develop a simple, analytic, semigeostrophic model of frontogenesis that closely follows that of Hoskins and Bretherton (1972). The semigeostrophic equations are the geostrophic momentum equations phrased in geostrophic coordinates. For fronts aligned (arbitrarily) along the y axis, and using the Boussinesq, f plane version of the geostrophic momentum equations, the appropriate horizontal, cross-front geostrophic coordinate, X, is defined as

$$X \equiv x + \frac{v'_g}{f}, \tag{2}$$

where x is the physical cross-front coordinate and v'_g is the departure of the meridional component of the geostrophic wind from the background deformation. We assume, as did Hoskins and Bretherton, that on the time scale of front formation, the flow is adiabatic and inviscid, so that both potential temperature and PV are conserved. Conservation of PV can be expressed in geostrophic coordinates as

$$\frac{\partial PV}{\partial \tau} + \frac{dX}{dt}\frac{\partial PV}{\partial X} + w\frac{\partial PV}{\partial Z} = 0, \tag{3}$$

where $\partial/\partial\tau$ is the time derivative holding altitude and X constant, and $\partial/\partial Z$ is the derivative in height holding time and X constant. Given the distribution of PV at any time, the distribution of potential temperature can be found by inverting the PV distribution via

$$\frac{\partial^2\theta}{\partial X^2} + \frac{f^3}{g}\frac{\partial}{\partial Z}\left(\frac{1}{PV}\frac{\partial\theta}{\partial Z}\right) = 0, \tag{4}$$

subject to the time-dependent boundary conditions

$$\frac{\partial\theta}{\partial\tau} + \frac{dX}{dt}\frac{\partial\theta}{\partial X} = 0 \tag{5}$$

on rigid horizontal boundaries (on which w vanishes). The cross-front circulation as represented by a streamfunction ψ may be diagnosed from a Sawyer–Eliassen type equation, which phrased in geostrophic coordinates is

$$\frac{\partial}{\partial X}\left(\frac{PV}{f}\frac{\partial\psi}{\partial X}\right) + \frac{\partial}{\partial Z}\left(\frac{f^2}{g}\frac{\partial\psi}{\partial Z}\right) = Q, \tag{6}$$

where Q is the geostrophic forcing, which is proportional to the product of the geostrophic deformation and the cross-front temperature gradient.

As in Hoskins and Bretherton, we shall consider an idealized, height-independent deformative geostrophic flow given by

$$u_g = -\alpha x \quad \text{and} \quad v_g = \alpha y, \tag{7}$$

where α is the rate of deformation. Owing to symmetry, the deformative part of the geostrophic flow remains constant in time in this problem. Hoskins and Bretherton show that for this flow,

$$\frac{dX}{dt} = -\alpha X. \tag{8}$$

In this application, I start with an initial potential vorticity field that is a function of X alone. Because the geostrophic flow in the x direction does not vary with height, one can see from (3) that PV will never acquire a Z dependence, so that (3) reduces in this case to

$$\frac{\partial PV}{\partial\tau} - \alpha X\frac{\partial PV}{\partial X} = 0, \tag{9}$$

where I have used (8). Equation (9) has a simple analytic solution:

$$PV = F(Xe^{\alpha\tau}), \tag{10}$$

where $F(X)$ is simply the initial PV distribution. As Hoskins and Bretherton point out, time becomes a parameter in this problem. Likewise, the boundary condition for θ [(5)] becomes

$$\frac{\partial\theta}{\partial\tau} - \alpha X\frac{\partial\theta}{\partial X} = 0, \tag{11}$$

which similarly has the solution

$$\theta = G(Xe^{\alpha\tau}), \tag{12}$$

where $G(X)$ is the initial distribution of θ. (A similar condition applies at each boundary.) For simplicity, we take the tropopause here to be a rigid boundary. The geostrophic forcing function Q in (6) becomes, in this case,

$$Q = -2\alpha\frac{\partial\theta}{\partial X}. \tag{13}$$

Thus, given the initial one-dimensional distributions of PV and boundary θ, we immediately have the distributions of these quantities in geostrophic space for all time, from (10) and (12). We can then invert (4) for the interior θ distribution, and (6) and (13) for the streamfunction ψ.

An interesting special case is one in which θ is constant on the upper boundary. We also take the initial lower boundary θ and interior PV both to follow hyperbolic tangent functions, so that they have no gradients as $X \to \pm\infty$. The solution for various quantities at about the time of surface frontal collapse is shown in Fig. 7. Note that the PV itself has formed a front at the surface, and that even though there is no temperature gradient along the tropopause in this case, local extrema of vorticity form there. Both the vorticity and upward vertical motion extend deeper into the troposphere than in the uniform potential vorticity case.

An example of an arctic front as it appears at the 500-hPa level at 1800 UTC 6 January 2004 is shown in Fig. 8. A ribbon of high vorticity extends from the central plains to western Pennsylvania and New York state and through northern New England and the Canadian Maritimes. There was little evidence of a wind shift at the

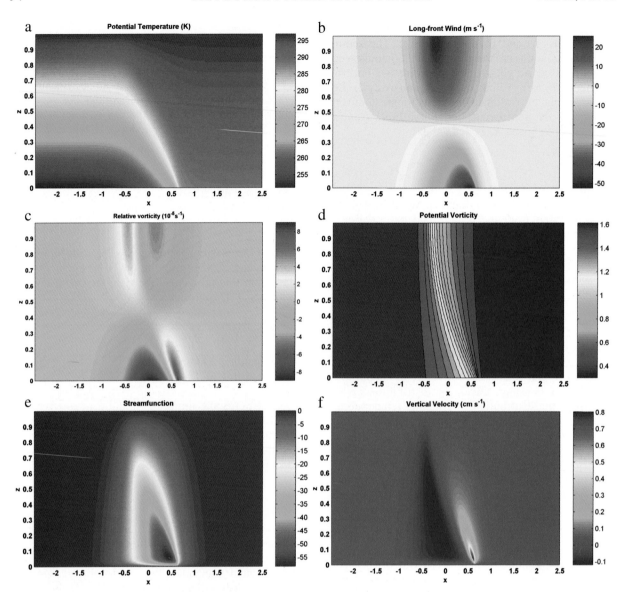

FIG. 7. Solutions of the semigeostrophic model of frontogenesis produced by uniform geostrophic deformation acting on initial hyperbolic tangent profiles of potential vorticity and surface potential temperature. (In this simulation, the potential temperature along the top boundary is constant.) The PV does not vary with altitude in geostrophic coordinates. Solutions are shown at about the time of surface frontal collapse. (a) Potential temperature, (b) long-front wind component (m s^{-1}), (c) vertical component of relative vorticity ($\times 10^{-5}$ s^{-1}), (d) potential vorticity (standard PV units), (e) mass streamfunction, and (f) vertical velocity (cm s^{-1}).

surface, and the only weather possibly associated with this feature were some scattered light snow showers in New England. The wind profiler at Gray, Maine, recorded a wind shift in the altitude range of 500 m–2 km, from west to northwest, sometime between 1100 and 1200 UTC. (The profiler observations did not penetrate above 2 km in the arctic air.)

Clearly, arctic air may develop significant PV gradients at its leading edge, and these may lead to some interesting dynamics not captured by analysis techniques that focus on the surface and the tropopause. Dynamics associated with arctic air are ripe for more comprehensive analysis.

5. Summary

The Norwegian school of meteorology placed roughly equal emphasis on the importance of dynamics and thermodynamics for understanding and forecasting weather. They viewed the development of fronts as being largely a dynamical phenomenon, but they were equally concerned with the thermodynamics of airmass formation. Today, understanding and predicting weather at short to medium range is viewed mostly as a dynamical problem, with some attention paid to surface fluxes and latent heat release, and the idea of airmass formation has gone the way of Edsels and argyle socks. Of course,

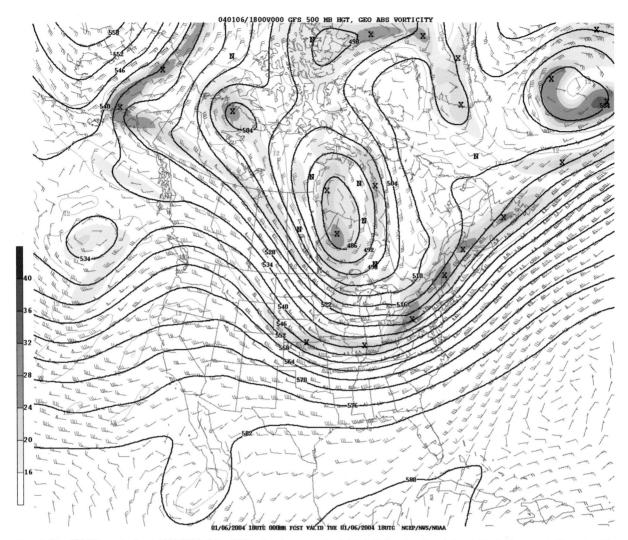

FIG. 8. The 500-hPa analysis at 1800 UTC 6 Jan 2004, showing geopotential height (contours) and the vertical component of geostrophic absolute vorticity (yellow shading).

weather forecast models include processes such as surface fluxes and radiative transfer that the Norwegians considered essential to airmass formation, and much attention is paid to the representation of moist convection, which is critical to the formation of what we here call "convected air." While it is not likely that today's models are making first-order errors in the modification of air masses that are heated from below, which entails the relatively fast and efficient process of convection, little attention is paid to the problem of cooling from below, which is far more problematic. The work of Curry (1983), reviewed here, suggests that radiative cooling over land in winter may sometimes affect deep layers on time scales of 10 days, depending perhaps delicately on such matters as moisture content and the microphysics of ice crystals and their interaction with radiation. I have the impression that medium-range forecasts of what we often refer to as "arctic air outbreaks" are often compromised by incomplete representations of

these physical processes, whereas map discussions almost always focus on the dynamics. The thermodynamics of arctic air is an area ripe for research advances.

The dynamics revolution of the late 1940s did away with the Norwegian concept of air masses and replaced it with the idea of a continuous distribution of properties, modulated by baroclinic wave and frontal processes. But "potential vorticity thinking," advocated, for example, by Hoskins et al. (Hoskins et al. 1985) and widely employed in graduate-level instruction, in effect reintroduced the concept of air masses in a new guise: tropospheric air with low but nearly constant PV and stratospheric air with much larger values of PV. Our current concept of extratropical dynamics holds that what we see on weather maps can be explained by the interaction of two Rossby wave trains: one on the surface temperature gradient and another on the isentropic gradients of PV at the tropopause, where isentropes cross PV contours. (This tropopause zone of intersection is often very narrow, horizontally, giving rise to what might be called "Rossby wave highways.") We supplement this basic view by accounting for PV created by latent heating and for the effects of surface friction.

Soundings in the Tropics, over land in the summer, and over water in the winter often show nearly moist adiabatic structure, though in strongly baroclinic regions one observes moist adiabatic lapse rates on vortex lines (*M* surfaces) rather than in the vertical. Based on this observation, together with the clear utility of PV thinking, I advocate a reclassification of air masses based on the value of a single scalar variable: SPV, defined by (1). This quantity is zero wherever the lapse rate is moist adiabatic on vortex lines (which are nearly vertical in the Tropics), but is large in the stratosphere and in arctic air. It is nearly conserved in cold air where it is nearly equal to conventional PV, and although not conserved in unsaturated warm air, there it tends to be adjusted to near zero by convection. Like PV, it is always invertible, subject to a balance condition. Cross sections, such as those shown in Fig. 5, suggest that the global atmosphere may be approximately described in terms of three or four air masses. "SPV thinking" would proceed along much the same lines as "PV thinking," but replacing surface temperature with surface θ_e, which in convected air is linked to θ_e^* above the boundary layer by the condition of convective neutrality. It would also be concerned with the diabatic formation of arctic air and the dynamics of the SPV transition between arctic and convected air.

Perhaps it is time to bring back the air masses.

Acknowledgments. I am indebted to Dave Schultz for helping me with the analysis of the 8 March 2003 arctic air outbreak in the central United States and to Rob Korty for writing the programs used to construct the SPV cross sections. The manuscript was much improved by the suggestions of several anonymous reviewers. Fred Sanders befriended me when I entered MIT as a freshman in 1973 and remained my friend until his death in 2006. Among the many priceless lessons I have learned from him was to revere our scientific forebears, even while retaining a healthy skepticism of their work.

REFERENCES

Bergeron, T., 1928: Über die dreidimensional verknüpfende Wetteranalysee. *Geofys. Publ.,* **5** (6), 73–78.

Bjerknes, J., and H. Solberg, 1922: Life cycles of cyclones and the polar front theory of atmospheric circulations. *Geofys. Publ.,* **3,** 1–18.

Bony, S., and K. A. Emanuel, 2001: A parameterization of the cloudiness associated with cumulus convection: Evaluation using TOGA COARE data. *J. Atmos. Sci.,* **58,** 3158–3183.

Charney, J. G., 1947: The dynamics of long waves in a westerly baroclinic current. *J. Meteor.,* **4,** 135–163.

——, and N. A. Phillips, 1953: Numerical integration of the quasigeostrophic equations for barotropic and simple baroclinic flows. *J. Atmos. Sci.,* **10,** 71–99.

Curry, J., 1983: On the formation of continental polar air. *J. Atmos. Sci.,* **40,** 2278–2292.

Eady, E. T., 1949: Long waves and cyclone waves. *Tellus,* **1,** 33–52.

Emanuel, K. A., 1988: Observational evidence of slantwise convective adjustment. *Mon. Wea. Rev.,* **116,** 1805–1816.

——, and M. Živkovic-Rothman, 1999: Development and evaluation of a convection scheme for use in climate models. *J. Atmos. Sci.,* **56,** 1766–1782.

Hoskins, B. J., and F. P. Bretherton, 1972: Atmospheric frontogenesis models: Mathematical formulation and solution. *J. Atmos. Sci.,* **29,** 11–37.

——, M. E. McIntyre, and A. W. Robertson, 1985: On the use and significance of isentropic potential vorticity maps. *Quart. J. Roy. Meteor. Soc.,* **111,** 877–946.

Jeffreys, H., 1925: On fluid motions produced by differences of temperature and humidity. *Quart. J. Roy. Meteor. Soc.,* **51,** 347–356.

Morcrette, J.-J., 1991: Radiation and cloud radiative properties in the European Centre for Medium-Range Weather Forecasts forecasting system. *J. Geophys. Res.,* **96,** 9121–9132.

Petterssen, S., 1956: *Motion and Motion Systems.* Vol. I. *Weather Analysis and Forecasting,* 2d ed., McGraw-Hill, 428 pp.

Reed, R. J., 1955: A study of a characteristic type of upper-level frontogenesis. *J. Meteor.,* **12,** 226–237.

——, and F. Sanders, 1953: An investigation of the development of a mid-tropospheric frontal zone and its associated vorticity field. *J. Meteor.,* **10,** 338–349.

Sanders, F., 1955: An investigation of the structure and dynamics of an intense surface frontal zone. *J. Meteor.,* **12,** 542–552.

Sandstrom, J. W., 1908: Dynamische Versuche mit Meerwasser. *Annals in Hydrodynamic Marine Meteorology,* Vol. 36, 6–23.

Wang, P.-Y., J. E. Martin, J. D. Locatelli, and P. V. Hobbs, 1995: Structure and evolution of winter cyclones in the central United States and their effects on the distribution of precipitation. Part II: Arctic fronts. *Mon. Wea. Rev.,* **123,** 1328–1344.

Xu, K.-M., and A. K. A. Emanuel, 1989: Is the tropical atmosphere conditionally unstable? *Mon. Wea. Rev.,* **117,** 1471–1479.

Chapter 4

An Empirical Perspective on Cold Fronts

EDWIN KESSLER

Department of Geography, and School of Meteorology, University of Oklahoma, Norman, Oklahoma

(Manuscript received 28 September 2004, in final form 28 September 2005)

Kessler

ABSTRACT

Oklahoma Mesonetwork data are used to illustrate important atmospheric features that are not well shown by the usual synoptic data. For example, some shifts of wind from south to north that are shown as cold fronts on synoptic charts are not cold fronts by any plausible definition. As previously discussed by Fred Sanders and others, such errors in analysis can be reduced by knowledge of the wide variety of weather phenomena that actually exists, and by more attention to temperatures at the earth's surface as revealed by conventional synoptic data. Mesoscale data for four cases reinforce previous discussions of the ephemeral nature of fronts and deficiencies in the usual analyses of cold fronts. One type of misanalyzed case involves post-cold-frontal boundary layer air that is warmer than the prefrontal air. A second type is usually nocturnal, with a rise of local temperature during disruption of an inversion and a wind shift with later cooling that accompanies advection of a climatological gradient of temperature.

1. Introduction

We want to perceive frontal concepts in a certain way. Frontal concepts are part of the origin of modern meteorological science, and their early promotion and development by J. Bjerknes and others have been recently described and discussed in several fascinating articles in *The Life Cycles of Extratropical Cyclones* (Shapiro and Grønås 1999). Here are gleanings from several of these articles. Vilhelm Bjerknes, the father of Jacob, presented weather forecasting in 1904 as a problem to be addressed through physical laws in an article that calls for no amendment today. In 1919, J. Bjerknes was motivated to develop research toward improved weather forecasting for the fishing and aircraft industries and for other weather-impacted endeavors, and his proposed characteristic structure for extratropical cyclones was based largely on observations along coastal Norway (Bjerknes 1919). The "steering line" and "squall line" in his first model became warm front and cold front in 1922, in an article by Bjerknes and Solberg. This model of the "Norwegian school" became a centerpiece, and in the 1930s it began to take hold in the United States through schools such as the Massachusetts Institute of Technology (MIT), the first to teach meteorology as science. The Weather Bureau began to show fronts in the daily weather report in 1933 and the approach of the Norwegian school was more emphasized after Francis W. Reichelderfer was appointed Chief of Bureau in 1938.

Corresponding author address: Edwin Kessler, 1510 Rosemont Dr., Norman, OK 73072-6337.
E-mail: kess3@swbell.net

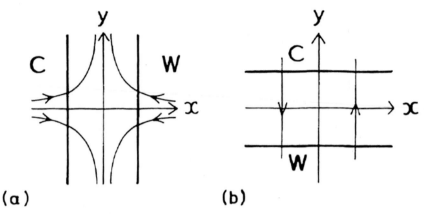

FIG. 1. The (a) confluence and (b) shear mechanism for intensifying an x gradient in temperature. Thick lines show isotherms, and W and C indicate warm and cold air, respectively. Thin lines with arrows denote streamlines. From Hoskins (2003).

Fronts received much attention because they were viewed as originators and/or maintainers of extratropical cyclones, because they are sites of concentrated meteorological activities and because they provided a start toward modeling atmospheric circulations at a time when models were very few. A very early concept visualized development of extratropical cyclones in terms of a kind of wave motion along a front that was circumpolar, but it was soon realized that fronts are highly variable and not always continuous in space and time. With greatly expanded upper-air observations, a consequence of World War II, and a burgeoning industry of civil aviation, there were important new revelations at the same time that the meteorological culture had become somewhat transfixed by elaborations of frontal analysis. While attention to fronts had some qualities of a social or political movement, Reed wrote in 2003 of both frustration and changing attitudes during the 1950s: "Fred [Sanders] and I were kindred spirits who fed each other's discontent with what we regarded as the nearly blind acceptance by many meteorologists of the model of cyclone development advocated by the Norwegian school. It seemed obvious to us—and we were not alone in this view—that fronts often strengthened during cyclogenesis rather than providing sharp preexisting thermal discontinuities on which cyclones formed" (Johnson and Houze 2003, p. 3). The spread of revised thinking is further illustrated by Gedzelman (1985, p. 46) whose review article includes, "Development at the surface is accompanied and even perhaps preceded by development aloft."[1]

_____

[1] The conversations between Reed and Sanders, and Sanders' emphasis on analysis of surface temperature are especially interesting to me in view of a meeting I had when a high school student in the early 1940s with J. P. McAuliffe, then head of the Weather Bureau in Corpus Christi, Texas. Mr. McAuliffe was very well known locally and highly respected. He showed a weather map that he had drawn depicting winds and isotherms, and he commented that fronts were becoming too much emphasized.

2. Some comments on frontal analysis and on frontal development

It is elementary that temperatures at the surface of earth diminish from subtropical latitudes to the poles as a consequence of poles-to-equator variation of received solar radiation. The poleward gradients cause westerly winds that increase with height and these "thermal winds" interact in a very complex way with themselves and with continents and oceans to produce global weather. Earth's atmosphere is very thin in comparison with its horizontal extent, and horizontal temperature contrasts are most influential toward direct solenoidal circulations when the contrast is large over horizontal distances that are not greatly larger than the vertical depth of the atmosphere. Large temperature contrasts over short distances can be important for at least two reasons: First, motion of such a zone, shown as a front, produces a sudden and large local temperature change. Second, variation of air density across the front can stimulate local circulations that can be violent, such as intense thunderstorms, and that can interact significantly with larger scales of motion. So these moving local areas of strong temperature contrast that we call fronts can be important, and it is eminently worthwhile to develop a rational frontal analysis. Contemporary analyses often show more fronts than are justified, and when they are inappropriately shown, they distract unnecessarily from other important features. It is regrettable that contemporary analysis is often so poor that this point needs emphasis and even explanation! (Sanders and Doswell 1995; Sanders 1999).

How is the magnitude of a temperature gradient increased? Three processes come quickly to mind. First, there are effects produced by adjacent sources of heat and cold. For example, the warm Gulf Stream is adjacent to the northern United States and Canada. Second, the north–south climatological temperature gradient in cooler seasons can be concentrated by confluence of the wind whenever the wind gradient has a component par-

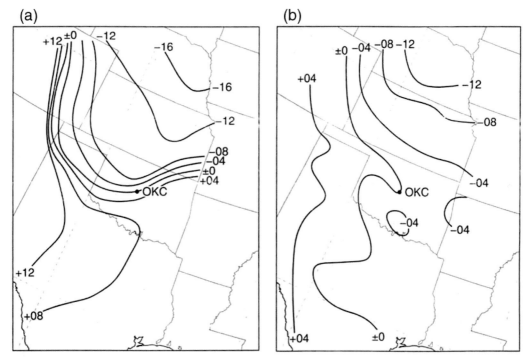

FIG. 2. (a) Mean departure of surface potential temperature from that at Oklahoma City, for four cases of abrupt cooling at the author's nearby farm home, at National Weather Service (NWS) map time nearest the time of the event. (b) Same as in (a), but for four cases of abrupt warming at the farm. From Sanders and Kessler (1999).

allel to the gradient of temperature regardless of the latter's sign. This is illustrated in Fig. 1a, borrowed from Hoskins (in Johnson and Houze 2003). Continued confluence would tend to produce a front oriented north–south. Third, as illustrated in Fig. 1b, the gradient of temperature (or of any other quantity) can be intensified when wind shear bends isopleths together.

There is an important implication for dynamical processes in the difference between Figs. 1a and 1b. In Fig. 1a, the wind is nondivergent, and indeed, the gradient would be intensified whether the wind field is divergent or convergent, so long as the confluent axis (the y axis in the diagram, usually known as the axis of dilatation) is within 45° of the direction of the isotherms. The wind is also nondivergent in Fig. 1b, but it is rotational, and local rotation or vorticity at earth's surface is produced by convergence in the wind field. Thus, the configuration shown in Fig. 1b *follows* either, and probably both, convergence in the surface wind field or an active twisting term in the vorticity equation, possible dynamical results of properties of the three-dimensional flow. Note also in Fig. 1b that a situation of northerly flow is depicted on the left-hand side and southerly flow on the right-hand side. Additional convergence would intensify the temperature gradient and deepen an area of low pressure with the cyclonic circulation such as that commonly seen on surface weather maps. These comments suggest an influence of large-scale circulations on the development of fronts rather than an effect of fronts on large-scale circulations.

We now proceed from this historical sketch of frontal concepts and meager outline of some relevant dynamical and kinematic processes to a few real-world cases that add to examples of problematic analyses presented elsewhere.

3. Four examples

a. Sanders and Kessler (1999): The advection of climatological temperature gradient

Here is a brief review of the paper by Sanders and Kessler (1999, hereafter SK), which examined wind shifts to the north accompanied by abrupt cooling and also examined wind shifts to the north with abrupt warming followed by gradual cooling, all at the present author's farm home in rural central Oklahoma, 46 km south of the weather station at Will Rogers International Airport in Oklahoma City. SK is summarized by Fig. 2. A remarkable dissimilarity existed between the distributions of potential temperature characteristic of the two groups in the illustration, which show averages of the most intense cases among a population twice as large.

Synoptic-scale data with wind shifts accompanied by abrupt cooling and then by continued more gradual cooling showed a strong gradient of potential temperature at the cold front and a smaller northward-directed gradient from about 100 mi north of the front. Wind shifts accompanied by warming followed by gradual cooling

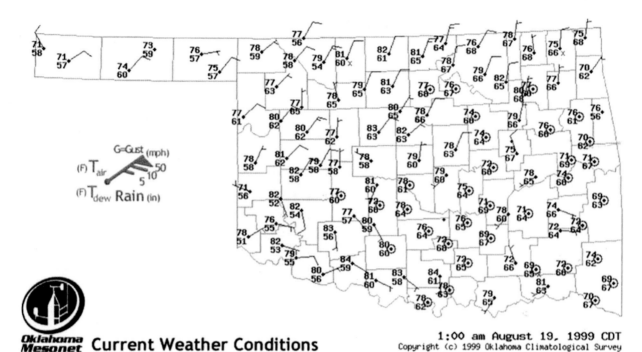

Fig. 3. Data from the Oklahoma Mesonetwork showing the distribution of temperature, dewpoint, and wind at 0100 Central Daylight Savings Time (CDT) 19 Aug 1999. Although official analyses showed a cold front across the state, note that temperatures with northeast winds were several degrees Fahrenheit warmer than prevailed southeast of the "front."

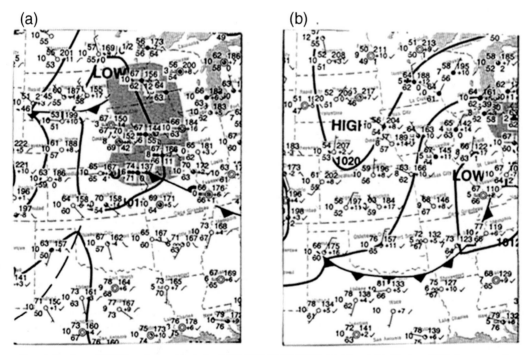

Fig. 4. Portions of NWS surface analysis for 0600 CST (0700 CDT) (a) 18 and (b) 19 Aug 1999, for comparison with Fig. 3. The synoptic-scale data are insufficient to indicate the real nature of the temperature and wind distributions over OK.

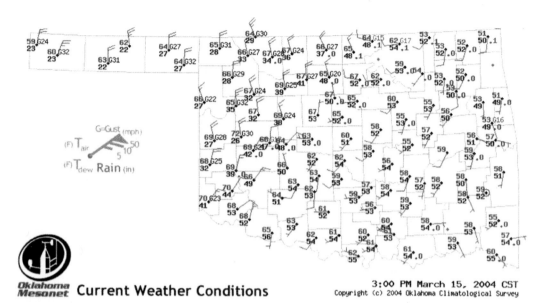

FIG. 5. Data from the Oklahoma Mesonetwork showing the distribution of temperature, dewpoint, and wind at 1500 CST 15 Mar 2004. Although temperatures in flow from the northwest in the northwestern part of Oklahoma were 10°F warmer than in southwesterly flow in the southeastern part of the state, the zone between the differing airflows was shown in NWS analyses as a cold front.

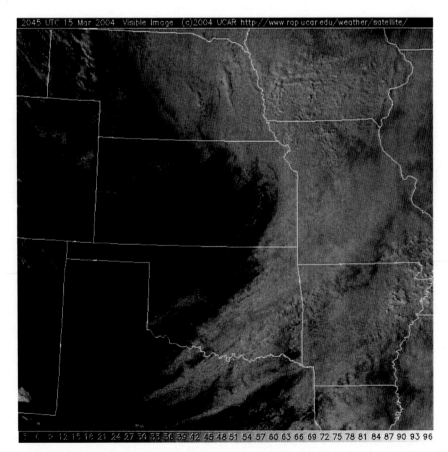

FIG. 6. This photo from a satellite at about the same time as shown in Fig. 5 indicates that the temperature differences in Fig. 5 are attributable, at least in part, to cloud cover in southeastern OK and clear skies in northwestern OK. The same conclusion is indicated by pyrheliometric data of the Oklahoma Mesonetwork.

TABLE 1. Meteorological parameters at the author's OK farm during similar time periods on 18 and 19 Aug 1999. All times are CDT. Observations are for the 5-min period ending at the listed time. Pressure is in inches of mercury, measured at the station. Temperature (Temp) and dewpoint (Dew pt) are in degrees Fahrenheit. Relative humidity (RH) is in percent. Wind direction (Dir) is that from which the wind is blowing (0 = north, 90 = east, . . .). Wind speeds (Spd) are in miles per hour. Rainfall (Rain) is measured in inches, accumulated since 7:00 p.m. Solar radiation (Solar) is in watts per square meter.

| | Year | Month | Date | Hour | Min | Pres | Temp | Dew pt | RH | Wind | | | Rain | Solar |
										Dir	Spd	Max		
WASH	1999	08	18	06	05	28.84	70	57	63	129	1	2	0.00	0
WASH	1999	08	18	06	10	28.84	70	57	64	90	1	2	0.00	0
WASH	1999	08	18	06	15	28.84	70	56	62	101	1	2	0.00	0
WASH	1999	08	18	06	20	28.84	70	56	62	111	2	2	0.00	0
WASH	1999	08	18	06	25	28.84	70	57	63	106	1	2	0.00	0
WASH	1999	08	18	06	30	28.84	69	58	67	104	1	1	0.00	0
WASH	1999	08	18	06	35	28.84	68	59	71	92	1	1	0.00	0
WASH	1999	08	18	06	40	28.84	68	61	79	114	1	2	0.00	0
WASH	1999	08	18	06	45	28.85	67	61	81	95	0	1	0.00	1
WASH	1999	08	18	06	50	28.85	67	59	74	41	2	2	0.00	1
WASH	1999	08	18	06	55	28.85	68	57	67	73	1	2	0.00	3
WASH	1999	08	18	07	00	28.85	69	57	65	98	0	1	0.00	5
WASH	1999	08	18	07	05	28.85	69	57	66	75	0	1	0.00	9
WASH	1999	08	18	07	10	28.85	69	56	62	132	0	1	0.00	19
WASH	1999	08	18	07	15	28.85	70	60	70	149	1	1	0.00	28
WASH	1999	08	18	07	20	28.85	69	61	76	149	1	2	0.00	38
WASH	1999	08	18	07	25	28.85	68	61	78	148	0	1	0.00	51
WASH	1999	08	18	07	30	28.85	67	62	82	148	0	0	0.00	66
WASH	1999	08	18	07	35	28.84	67	62	84	148	0	0	0.00	83
WASH	1999	08	18	07	40	28.84	67	62	83	149	0	0	0.00	99
WASH	1999	08	18	07	45	28.84	68	61	77	102	0	2	0.00	115
WASH	1999	08	18	07	50	28.84	70	62	75	120	1	2	0.00	131
WASH	1999	08	18	07	55	28.84	72	60	68	123	1	3	0.00	146
WASH	1999	08	18	08	00	28.85	73	63	71	79	1	2	0.00	162
WASH	1999	08	18	15	05	28.77	101	59	25	108	2	5	0.00	867
WASH	1999	08	18	15	10	28.77	101	58	24	179	3	7	0.00	863
WASH	1999	08	18	15	15	28.77	102	60	26	274	7	11	0.00	855
WASH	1999	08	18	15	20	28.76	101	60	26	296	5	11	0.00	845
WASH	1999	08	18	15	25	28.76	101	59	26	274	3	5	0.00	829
WASH	1999	08	18	15	30	28.76	101	60	26	254	3	8	0.00	817
WASH	1999	08	18	15	35	28.76	101	59	25	251	7	12	0.00	831
WASH	1999	08	18	15	40	28.76	101	59	25	213	8	13	0.00	838
WASH	1999	08	18	15	45	28.76	101	58	24	208	5	10	0.00	804
WASH	1999	08	18	15	50	28.75	101	58	24	97	3	7	0.00	743
WASH	1999	08	18	15	55	28.75	101	59	25	329	4	6	0.00	698
WASH	1999	08	18	16	00	28.75	101	59	25	228	6	9	0.00	688

had no gradient of potential temperature at the wind shift line, but a gradient to the north that we identify with the climatology of North America. In both situations, the wind continued from the north, and gradual local cooling continued. This is attributable to *advection of a climatological temperature gradient* that is characteristic of North America in winter. Of course, cooling by advection is tempered by solar heating as Arctic air moves southward.

Initial warming is attributable to disruption of a nocturnal inversion in the pre–wind shift air. The wind shift may be, at least in part, a manifestation of a disturbance that propagates into the warm air ahead and is more commonly observed in south Texas.[2] The discussions of propagating wind shifts by Sanders (1999, see his section 5) and by Schultz (2004) may deal with a different phenomenon. There is much more to investigate with detailed observations and perhaps with tracer materials and sensors.

The paper in hand proclaims emphatically that eventual local cooling after a shift of the wind to the north is not, per se, evidence of a front with the wind shift, and a wind shift should not be shown as a cold front unless it is accompanied by at least a first-order discontinuity of temperature with local temperature decline. Regrettably, this has been written and said before without effect, and more than once! Absent such temperature change with the wind shift, it may be shown

---

[2] At Corpus Christi, wind shifts to the north sometimes mark a reversal of flow of tropical air that had previously been emplaced northward in advance of a cold front. In such cases the arrival of polar air may follow the wind shift by a few hours. This bold statement is based in part on personal experience; the phenomenon, including its diurnal properties, should be investigated further.

TABLE 1. (*Continued*)

	Year	Month	Date	Hour	Min	Pres	Temp	Dew pt	RH	Wind Dir	Spd	Max	Rain	Solar
WASH	1999	08	19	02	30	28.75	77	63	63	14	0	1	0.00	0
WASH	1999	08	19	02	35	28.75	77	64	65	21	1	1	0.00	0
WASH	1999	08	19	02	40	28.75	77	64	64	37	1	1	0.00	0
WASH	1999	08	19	02	45	28.75	76	66	70	29	1	1	0.00	0
WASH	1999	08	19	02	50	28.75	74	67	80	5	1	2	0.00	0
WASH	1999	08	19	02	55	28.75	73	67	82	3	1	2	0.00	0
WASH	1999	08	19	03	00	28.75	74	64	72	27	2	3	0.00	0
WASH	1999	08	19	03	05	28.76	75	64	69	27	4	6	0.00	0
WASH	1999	08	19	03	10	28.76	74	65	72	0	11	15	0.00	0
WASH	1999	08	19	03	15	28.77	74	65	73	4	11	14	0.00	0
WASH	1999	08	19	03	20	28.77	75	64	69	357	11	16	0.00	0
WASH	1999	08	19	03	25	28.77	76	64	66	359	11	14	0.00	0
WASH	1999	08	19	03	30	28.77	77	64	64	9	10	14	0.00	0
WASH	1999	08	19	03	35	28.78	77	64	64	7	7	10	0.00	0
WASH	1999	08	19	03	40	28.78	77	64	65	358	6	9	0.00	0
WASH	1999	08	19	03	45	28.78	76	64	65	7	6	10	0.00	0
WASH	1999	08	19	03	50	28.78	76	64	65	6	7	9	0.00	0
WASH	1999	08	19	03	55	28.78	76	64	65	6	6	9	0.00	0
WASH	1999	08	19	04	00	28.78	76	64	65	5	8	11	0.00	0
WASH	1999	08	19	06	50	28.81	75	68	80	24	9	13	0.00	2
WASH	1999	08	19	06	55	28.81	75	68	80	24	9	14	0.00	4
WASH	1999	08	19	07	00	28.81	75	68	80	24	9	13	0.00	7
WASH	1999	08	19	07	05	28.81	75	68	80	23	11	14	0.00	12
WASH	1999	08	19	07	10	28.82	75	68	80	21	11	15	0.00	18
WASH	1999	08	19	07	15	28.82	75	68	80	21	9	14	0.00	26
WASH	1999	08	19	07	20	28.82	75	68	80	21	11	15	0.00	35
WASH	1999	08	19	07	25	28.82	75	68	79	23	11	15	0.00	46
WASH	1999	08	19	07	30	28.83	75	68	79	24	10	14	0.00	60
WASH	1999	08	19	07	35	28.83	75	68	79	22	11	15	0.00	75
WASH	1999	08	19	07	40	28.83	75	68	78	22	12	16	0.00	91
WASH	1999	08	19	07	45	28.83	76	68	78	19	12	17	0.00	106
WASH	1999	08	19	07	50	28.83	76	68	78	21	12	16	0.00	119
WASH	1999	08	19	07	55	28.83	76	68	78	23	11	17	0.00	132
WASH	1999	08	19	08	00	28.84	76	68	77	28	11	17	0.00	146
WASH	1999	08	19	16	00	28.82	94	54	26	38	13	20	0.00	799
WASH	1999	08	19	16	05	28.82	94	54	26	29	14	18	0.00	784
WASH	1999	08	19	16	10	28.81	94	55	27	18	13	19	0.00	775
WASH	1999	08	19	16	15	28.81	94	54	26	35	12	17	0.00	762
WASH	1999	08	19	16	20	28.81	94	53	25	36	12	16	0.00	747
WASH	1999	08	19	16	25	28.81	94	54	26	39	12	18	0.00	676
WASH	1999	08	19	16	30	28.81	94	55	27	32	11	16	0.00	659
WASH	1999	08	19	16	35	28.81	94	55	27	28	11	15	0.00	707
WASH	1999	08	19	16	40	28.81	95	54	26	41	11	16	0.00	692
WASH	1999	08	19	16	45	28.81	95	55	27	39	11	16	0.00	677
WASH	1999	08	19	16	50	28.81	94	55	27	40	12	17	0.00	664
WASH	1999	08	19	16	55	28.81	94	55	27	38	12	19	0.00	648
WASH	1999	08	19	17	00	28.81	94	55	27	37	11	18	0.00	633
WASH	1999	08	19	17	05	28.81	94	55	27	44	11	16	0.00	617
WASH	1999	08	19	17	10	28.81	94	57	29	36	12	18	0.00	600
WASH	1999	08	19	17	15	28.81	94	57	29	39	12	17	0.00	583

as a wind shift, or as suggested in SK, designated as a "TROF", and viewers of the weather map should realize that if the shifted wind persists from the north in the Northern Hemisphere, cooler weather is usually in store for stations to the south.

Another important matter for further study is revealed in SK but not discussed there. Eight cases of wind shifts to the north with marked sudden warming followed by gradual cooling and nine cases of wind shifts to the north with marked sudden cooling are presented in Table 1 in that paper. The times of the warming cases are distributed from 1800 to 0300 CST, and there is just one case during midday hours. The onset times of cooling cases are more concentrated, with seven of the nine occurring between 1100 and 1600 CST. Other experiences of the present author have convinced him that this diurnal cycle is not a statistical aberration. Among questions requiring investigation: Does a larger dataset show that cold frontal passages in Oklahoma occur most often during midday and early afternoons, and, if so, why?

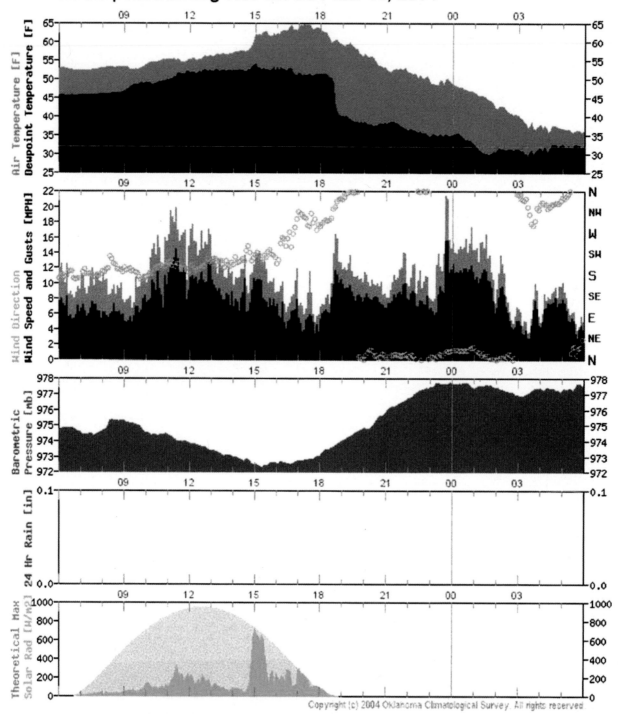

FIG. 7. Detailed evolution of weather parameters in central OK on 15 Mar 2004 is shown in this meteogram for Norman. Notice that the wind was steady from the southwest during the sudden rise of temperature near 1600 CST, but the bottom frame shows that the sun came out during that period. The further rise of temperature until 1830 CST accompanied a wind shift to the northwest. The precipitous decline of dewpoint at 1930 CST accompanied a further shift of wind to the north and a sharp temporary increase of wind speed.

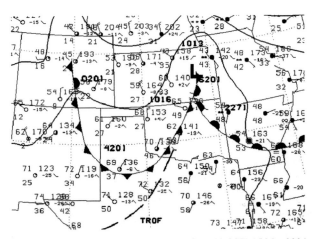

FIG. 8. Sparse synoptic data at 2100 UTC (1500 CST) 15 Mar 2004 neither support the frontal analysis as shown nor provide a clue to the complex events shown in Fig. 7. But the temperatures at three stations ahead of the cold front shown across Oklahoma are cooler than the temperatures at three stations behind.

b. 19 August 1999: A "cold front" with warmer air to the north

A wind shift in central Oklahoma on 19 August 1999 was shown on maps as a cold front although, for several hours, temperatures behind the front were warmer than those that preceded it. Figure 3 shows the distribution of temperature and wind in Oklahoma at 0100 CDT. Note that winds were calm in southeastern Oklahoma and the surface temperatures there were lower than those behind the wind shift line. Temperatures show a weak minimum along the wind shift line in central Oklahoma. With certainty, this cannot be a cold front in the Norwegian sense.

The wind shift occurred in the Oklahoma City area during 0100–0300 CST, and moved southward about 30 miles during that 2-h period. Table 1 shows temperatures and winds at the author's farm, about 24 km south of Norman during selected periods on 18 and 19 August. Notice that temperatures on 18 August were about 5°F cooler during the morning than during the same times on 19 August after the wind shift had passed. During the afternoon of 19 August, temperatures were cooler than on 18 August. As with the cases discussed in section 3a above, the change from warmer to cooler temperatures during the afternoon reflects advection of a *climatological temperature gradient and is not indicative of conditions at the onset of northerly winds.*

The official surface analyses at 0600 CST 18 and 19 August 1999 are shown in Fig. 4. They show that a cold front had passed through all of Oklahoma. But, as discussed above, temperatures on 19 August were about 5°F warmer than at the same time on the previous day.

c. 15 March 2004: Effect of solar heating on frontal analysis

On 15 March 2004, a shift of wind from southwest to northwest was shown on official weather maps as a cold front. The distribution of wind and temperature over Oklahoma at 1500 CST 15 March is shown in Fig. 5. Surface temperatures over northwestern Oklahoma behind the wind shift line were uniformly about 10°F higher than temperatures over southeastern Oklahoma. An explanation for the relative warmth with northwest winds is offered by the satellite-based view in Fig. 6. Cloudy conditions in southeastern Oklahoma reduced solar heating of the subtropical air mass.

Figure 7 is the meteogram for Norman on 15 March. This shows that the local shift in wind direction was gradual and began at about 1500 CST. At 1600 CST, the temperature rose from 57° to 62°F in 10 min when the sun came out and continued to rise slowly to a high of 65°F at 1830 CST while the wind continued to back irregularly to the northwest. Then the wind direction remained nearly invariant until 1930 CST when it resumed backing to the north and the dewpoint fell precipitously—about 12°F in 10 min—and continued thereafter to decline slowly. The fall of dewpoint accompanied a sudden increase of average wind and gusts. Gusts increased from approximately 7 to 15 mph (1 mph = ~0.4470 m s^{-1}) as the dewpoint fell. The importance of data at the mesoscale is shown by the absence of any of these features in the usual synoptic-scale data.

Figure 5 shows a sharp break in dewpoint in a zone-oriented southwest–northeast in northwestern Oklahoma. Even a more thorough study might not tell us clearly about the origins of the various portions of the air mass that overlaid the state, but it is certainly fully clear that the wind shift line was not a cold front although shown as such on weather maps.

What was the origin of the wind shift zone shown in Fig. 5? Perhaps it was a consequence of an irregular distribution of convergence at low altitudes, which at first produced rotation and a low pressure area. Then the convergence may have been concentrated in the western rotational sector and brought to juxtaposition airstreams that had been moving in different directions and continued to do so.

Figure 8 shows a portion of the official analysis for 2100 UTC (1500 CST) 15 March. The data shown are sparse, and we do not know what other data may have been available to the analyst. However, the only feature of the map that supports the cold front as shown is the area of low pressure, and virtually all of the coded indicators of frontal character are incorrect. The best that can be said of the analysis is that the cold front and warm front are drawn in the traditional way. Mesoscale data are teaching us that it would be better to look for clear indications of a cold front before so indicating such a presence, and we should be searching synoptic data for signs of phenomena other than classical fronts. In the present case, such indicators are given by the temperatures shown at three stations behind the marked cold front, all of which are warmer than temperatures at three stations ahead of the front.

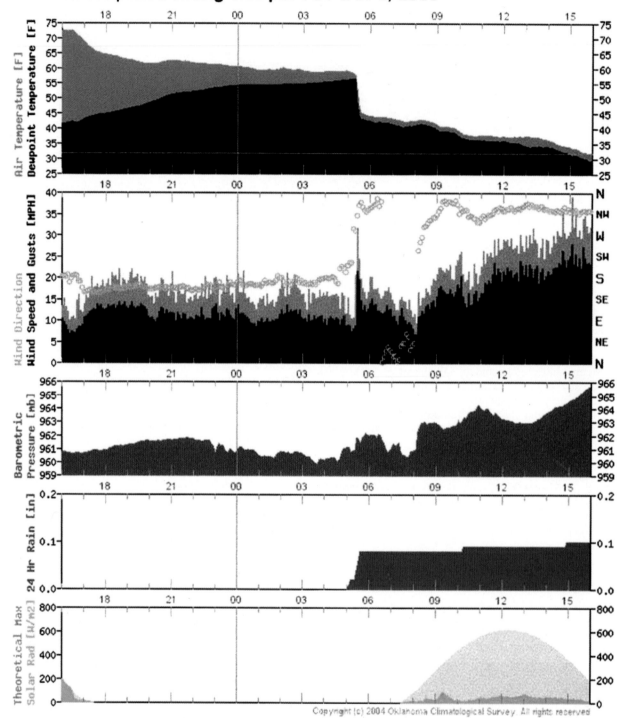

FIG. 9. (a) Meteogram for Norman, OK, on 9 and 10 Dec 2003. The temperature at the cold front is nearly of zero order. Passage of the front was followed by a rain shower with associated wind shifts. (b) Meteogram for Ardmore, OK, 80 miles south of Norman, on 9 and 10 Dec 2003. The cold front, which was very sharp at Norman, is manifested here as a gradual shift of wind direction and a much more protracted temperature decline. The ultimate temperature decline was similar, given an adjustment for latitude.

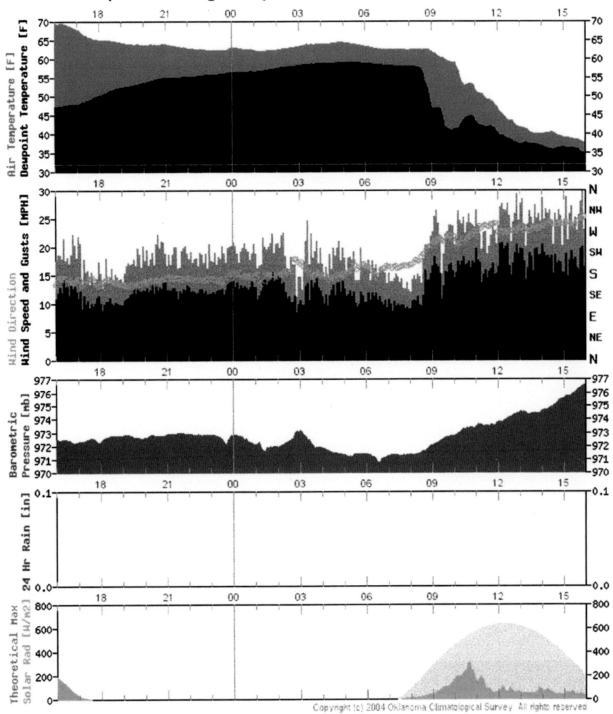

**Ardmore 24-Hour Mesonet Meteogram
for the period ending 4:00 pm CST Dec 9, 2003**

FIG. 9. (*Continued*)

d. 9 December 2003: Rapidly weakening cold front

On 9 December 2003, a real and strong cold front passed through Oklahoma. As shown by the meteograms for Norman and Ardmore, Oklahoma, (Figs. 9a and 9b), its profile was rapidly altered as it moved through the state. Notice that the front appears nearly as a zero-order discontinuity at Norman and barely as a first-order discontinuity at Ardmore, just 80 miles to the south.

Such remarkable changes are not uncommon! Sometimes, for example, a front intensifies and then wanes as it passes through Oklahoma and its speed may be quite variable. Maintenance of a strong front depends on persistence of the strength and location of controlling confluent and/or shear flow in the presence of turbulent diffusion and nonadiabatic processes.

Although a detailed mesoscale analysis has not been attempted in this case, it seems obvious that the analysis of fronts is made especially difficult when mesoscale disturbances are present at or near the front, as shown in this case by wind variations and precipitation near the time of frontal passage. The coincidental presence of such mesoscale disturbances as thunderstorm outflows with associated temperature variations of their own in synoptic data can move the analyst to mark the front in a "wrong" place. As fronts are more intense, diffusion and other nonadiabatic processes are also more important and tend more to offset effects of processes illustrated by Fig. 1. This is probably why the cliff-like structure shown in Fig. 9a rarely exceeds about 20°F in Oklahoma although the change of temperature over a day or two can be very much larger.[3]

4. Concluding remarks

The cases shown here in sections 3a–d are not at all exceptional. Often in the Northern Hemisphere, a cold front is marked on the west side of a cyclonic weather system and a warm front is marked on the east side regardless of whether the temperatures correspond to the frontal associations or not. Such fronts as indicated are usually significant only with respect to the accompanying wind shift and sometimes not significant at all.

There are no well accepted standards for marking fronts on weather maps, and as noted by F. Sanders (2004, personal communication), "What I do see wrong is the practice of identifying ALL wind shifts whether or not accompanied by density contrast, as fronts. This practice deflects attention from real fronts when they occur, from the synoptic circumstances when they occur, and from other interesting and important structures in the surface boundary layer."

To describe all wind shifts to the north as cold fronts is to obscure important atmospheric process of great beauty and complexity and to impede the development of understanding. A front as marked at present on weather maps cannot be relied upon to indicate a definite atmospheric structure, nor can, for example, a denoted cold front even be relied on to indicate falling temperatures at stations passed by the front! And, very regrettably, the large markers of fronts as presented on analyzed weather charts sometimes obscure the very

data that could help the viewer to interpret better the associated weather conditions and processes.

Mesoscale data of the high quality now readily accessible from the Oklahoma Mesonetwork and elsewhere can help greatly to reveal numerous atmospheric phenomena that have been little identified and appreciated in the recent past. Synoptic data at national and international levels are often inadequate guides to analysis, but synoptic-scale analyses can nevertheless be improved through closer attention to details in the synoptic-scale data itself. There must be a flight away from arbitrary labeling of wind shift lines as fronts, and analysts should be on the lookout for varied phenomena. Such an attitude, reinforced in meteorology departments where our future meteorologists are trained, should help greatly to provide more meaningful and accurate weather maps and eventually to promote better weather forecasts.

A tribute to frontal concepts is in order. Fronts were first conceived as cause, they are now seen largely as effect, and they are by no means unimportant. Through their focus, through controversy, through related gatherings of minds, and in company with technical achievements and administrative developments, frontal concepts have been a leading cause of marked improvements to forecasting practice. Let us continue to build from the base provided by our marvelous new technological aids and new means for detailed observations.

Acknowledgments. Thoughtful comments by Charles Doswell, Fred Sanders, David Schultz, very helpful annotations on the original text by an anonymous reviewer, and contributions of illustrations by Derek Arndt and Mark Shafer of the Oklahoma Climatological Survey and Oklahoma Mesonetwork are all gratefully acknowledged.

REFERENCES

Bjerknes, J., 1919: On the structure of moving cyclones. *Mon. Wea. Rev.,* **47,** 95–99.
——, and H. Solberg, 1922: Life cycle of cyclones and the polar front theory of atmospheric circulation. *Geofys. Publ.,* **3,** 3–18.
Gedzelman, S. D., 1985: Atmospheric circulation systems. *Handbook of Applied Meteorology,* D. D. Houghton, Ed., John Wiley and Sons, 3–61.
Hoskins, B., 2003: Back to frontogenesis. *A Half Century of Progress in Meteorology: A Tribute to Richard Reed,* R. H. Johnson and R. A. Houze Jr., Eds., Amer. Meteor. Soc., 49–60.
Johnson, R. H., and R. A. Houze Jr., Eds., 2003: *A Half Century of Progress in Meteorology: A Tribute to Richard Reed.* Amer. Meteor. Soc., 138 pp.
Sanders, F., 1999: A proposed method of surface map analysis. *Mon. Wea. Rev.,* **127,** 945–955.
——, F., and C. Doswell III, 1995: A case for detailed surface analysis. *Bull. Amer. Meteor. Soc.,* **76,** 505–521.
——, F., and E. Kessler, 1999: Frontal analysis in the light of abrupt temperature changes in a shallow valley. *Mon. Wea. Rev.,* **127,** 1125–1133.
Schultz, D. M., 2004: Cold fronts with and without prefrontal wind shifts in the central United States. *Mon. Wea. Rev.,* **132,** 2040–2053.
Shapiro, M., and S. Grønås, Eds., 1999: *The Life Cycles of Extratropical Cyclones.* Amer. Meteor. Soc., 359 pp.

---

[3] Schulz (2004) shows a case with a 30°F cliff, and even larger values occur occasionally.

Chapter 5

Perspectives on Fred Sanders' Research on Cold Fronts

DAVID M. SCHULTZ*

*Cooperative Institute for Mesoscale Meteorological Studies, University of Oklahoma, and
NOAA/National Severe Storms Laboratory, Norman, Oklahoma*

(Manuscript received 24 March 2004, in final form 2 December 2004)

Schultz

Fronts are a real and important feature of our environment, and an effort should be made to better understand them. We hope that this investigation is a contribution in that direction.   —Sanders (1967, p. 4.3)

ABSTRACT

One characteristic of Fred Sanders' research is his ability to take a topic that is believed to be well understood by the research community and show that interesting research problems still exist. Among Sanders' considerable contributions to synoptic meteorology, those concerned with surface cold fronts have been especially influential. After a brief historical review of fronts and frontal analysis, this chapter presents three stages in Sanders' career when he performed research on the structure, dynamics, and analysis of surface cold fronts. First, his 1955 paper, "An investigation of the structure and dynamics of an intense surface frontal zone," was the first study to discuss quantitatively the dynamics of a surface cold front. In the 1960s, Sanders and his students further examined the structure of cold fronts, resulting in the unpublished 1967 report to the National Science Foundation, "Frontal structure and the dynamics of frontogenesis." For a third time in his career, Sanders published several papers (1995–2005) revisiting the structure and dynamics of cold fronts. His 1967 and 1995–2005 work raises the question of the origin and dynamics of the surface pressure trough and/or wind shift that sometimes precedes the temperature gradient (hereafter called a prefrontal trough or prefrontal wind shift, respectively). Sanders showed that the relationship between this prefrontal feature and the temperature gradient is fundamental to the strength of the front. When the wind shift is coincident with the temperature gradient, frontogenesis (strengthening of the front) results; when the wind shift lies ahead of the temperature gradient, frontolysis (weakening of the front) results. A number of proposed mechanisms for the formation of prefrontal troughs and prefrontal wind shifts exist. Consequently, much research remains to be performed on these topics.

---

* Current affiliation: Division of Atmospheric Sciences and Geophysics, Department of Physics, University of Helsinki, and Finnish Meteorological Institute, Helsinki, Finland.

Corresponding author address: Dr. David Schultz, Finnish Meteorological Institute, Erik Palménin Aukio 1, P. O. Box 503, FI-00101 Helsinki, Finland.
E-mail: david.schultz@fmi.fi

1. Introduction

Synoptic meteorology has had a reputation of being less rigorous than other disciplines (e.g., Reed 2003, p. 2), perhaps rightly so in some instances. But Fred Sanders and his colleagues Dick Reed and Chester Newton provided legitimacy to our discipline by merging the application of dynamics and quantitative diagnosis with the study of observed weather systems (e.g., Gyakum et al. 1999). Besides his contributions as a teacher and mentor to many, Fred Sanders made fundamental contributions to synoptic meteorology in the structure, dynamics, and analysis of surface cold fronts. Sanders' research illustrates one of his characteristics that I find most inspiring: the ability to take a weather phenomenon that is considered solved by the research community and show that compelling research problems still exist.

Any study of a cold front, whether it be an idealized simulation or an observational analysis, inevitably will be compared to Sanders' (1955) "An investigation of the structure and dynamics of an intense surface frontal zone." Sanders (1955) was the first, the simplest, and, I would argue, still the best quantitative study of the structure and dynamics of a cold front. Based on Sanders' 1954 Ph.D. thesis, this paper has influenced numerous synoptic and mesoscale meteorologists, theoreticians, and modelers by illustrating the archetype classical cold front. One who was so influenced was Daniel Keyser. Specifically, a goal of Keyser's Ph.D. thesis, eventually published as Keyser and Anthes (1982), was to reproduce the intense, low-level updraft at the leading edge of Sanders' (1955) cold front in a primitive equation model starting with idealized initial conditions (D. Keyser 2003, personal communication).

Sanders' (1955) paper serves as a launching point for further studies of cold fronts, a topic to which Sanders would return at two later times in his career (the 1960s and 1995–2005). In the 1960s, Sanders and his students showed that the pressure trough and wind shift in some fronts lay ahead of the temperature gradient in the warm air, rather than coincident as in zero- and first-order discontinuity models of fronts (e.g., Petterssen 1933; Godson 1951; Saucier 1955, p. 109; Bluestein 1993, 240–248). Such features are hereafter called *prefrontal troughs* or *prefrontal wind shifts*. This body of research by Sanders and his students was consolidated into a report to the National Science Foundation entitled, "Frontal structure and the dynamics of frontogenesis," softbound within a distinctive, dark red cover (Sanders 1967). This report has become a sought-after cult classic among some meteorologists. Although some material eventually appeared in print (Sanders 1983), this report remains largely unpublished.

Later, during 1995–2005, as an outgrowth of his long-standing criticism of the quality of operational frontal analyses and the lack of operational surface isotherm analysis (Sanders and Doswell 1995; Sanders 1999a, 2005; Sanders and Hoffman 2002), Sanders revisited these issues of the nonsimultaneity of the temperature gradient and wind shift by presenting analyses of more nonclassical cold fronts (Sanders 1999b; Sanders and Kessler 1999).

The purpose of this chapter is to collect and review some aspects of cold-frontal structure, evolution, and dynamics in the context of Sanders' work. Other chapters in this volume that discuss other aspects of surface fronts

are Schultz and Roebber (2008) for a model simulation of the Sanders (1955) cold front, Emanuel (2008) for arctic fronts as potential vorticity fronts, Bluestein (2008) for fronts and other surface boundaries over the southern plains, and Bosart (2008) for coastal fronts, cold-air damming, and cool-season fronts adjacent to the eastern slopes of the Rocky Mountains. Sanders' contributions to surface frontal analysis are discussed in more detail by Hoffman (2008) and Kessler (2008).

In section 2 of this chapter, Sanders' career-spanning research on surface cold fronts is put into perspective by briefly reviewing the history of frontal research. In section 3, the structure, kinematics, frontogenesis, and dynamics of a classical cold front are reviewed from the work of Sanders (1955) and others. Section 4 reviews Sanders' (1967) further analyses of cold fronts, whereas section 5 reviews Sanders' 1995–2005 work. Section 6 concludes this chapter with a discussion of the implications for cold fronts from the perspective of two dichotomies: theory versus observations, and research versus operations.

2. Cold fronts: Changing paradigms

Since the first exposition by the Norwegian school of the concept of atmospheric fronts, the attitude of meteorologists towards fronts has gone from great enthusiasm through disappointment to the present air of confusion, consisting of acceptance with little understanding.
—Sanders (1967, p. 1.1)

Discontinuities in surface wind, temperature, and pressure, features we now recognize as fronts, have long been observed. Ficker (1923), Gold (1935), Bergeron (1959), Taljaard et al. (1961), Kutzbach (1979, section 6.7), Davies (1997), Newton and Rodebush Newton (1999), and Volkert (1999) have presented historical reviews of early frontal research. Surprisingly, the basic vertical structure of a cold front was advanced by Loomis as early as 1841 (Fig. 1). Unfortunately, Loomis's (1841) schematic made little impact at the time because the role of baroclinicity in midlatitude cyclones and fronts was not seriously discussed until the late 1800s (Kutzbach 1979, p. 30). Other early frontal studies included Bjerknes (1917) and others by his colleagues, who collectively became known as the Norwegian school (Friedman 1989, 92–94, 122–137, 158–178). Their work culminated in the conceptual model of extratropical cyclone structure and evolution known as the Norwegian cyclone model (Bjerknes 1919; Bjerknes and Solberg 1922; Bergeron 1937; Godske et al. 1957, section 14.3).

By applying the principles of physics and mathematics to the atmosphere, as well as employing detailed observational analysis from a dense network of surface stations over Norway, the Norwegian cyclone model synthesized and built upon earlier work to create a compact modern theory for cyclogenesis and frontogenesis. The Norwegian cyclone model has been quite successful in provid-

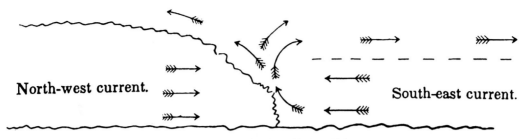

North-west current.   South-east current.

FIG. 1. Cross section through what would later be called a cold front. From Loomis (1841).

ing a common language to facilitate communication among scientists, forecasters, and the public. Even today, the Norwegian cyclone model serves as a first step in describing the structure and evolution of midlatitude cyclones (e.g., Wallace and Hobbs 1977, 126–127; Carlson 1991, chapter 10).

a. Interest wanes in frontal research

> Sometimes I wonder whatever happened to fronts?
> —Sanders (1967, p. 5.1)

Between the two world wars, the Norwegian cyclone model achieved a worldwide following, a period Ball (1960, p. 51) refers to as the "front-happy years." Ball's assertion is supported by Gold's (1935) detailed 51-page article documenting the state of frontal thinking during this period. After World War II, however, the discipline of frontal structure and dynamics waned. As Taljaard et al. (1961, p. 28) stated, "A perusal of the titles of the more than 100 articles in the *Compendium* [*of Meteorology*] would leave the uninitiated reader with the impression that there are no such things as fronts and air masses," despite the *Compendium* being "a survey of the current state of meteorology" (Malone 1951, p. v). Similarly, Sanders' quote above lamented the lack of interest in fronts. He continued, "fronts have passed through a sort of Dark Age of neglect in which only a loyal few worried very much about them" (Sanders 1967, p. 5.1).

Perhaps one impetus for the waning interest in frontal research, as noted by Kirk (1966), Sanders (1967), and Hoskins (1982), was the changing view of fronts as the result of cyclogenesis via the Charney (1947) and Eady (1949) paradigms for baroclinic instability, rather than the seat of the instability as originally envisioned by Bjerknes and Solberg (1922). Reed (2003, p. 3) noted that he and Sanders were "kindred spirits" during their days at MIT in the 1950s because "it seemed obvious to us— and we were not alone in this view—that fronts often strengthened during cyclogenesis rather than providing sharp preexisting thermal discontinuities on which cyclones formed." The driving force behind cyclogenesis was the upper-level short-wave trough, and early observational evidence for the importance of positive vorticity advection aloft in cyclogenesis came from Petterssen (1955) and Petterssen et al. (1955) and was later sup-

ported by Sanders (1987) and Sanders and Auciello (1989) for explosively developing cyclones over the North Atlantic Ocean.

Another possible reason that interest in fronts waned was that, unlike other atmospheric features, a uniform definition of a front could not be agreed upon by meteorologists (e.g., Sanders and Doswell 1995; Sanders 1999a). This point becomes even more apparent when viewed in the context of exercises comparing the range of interpretations of the atmosphere by subjective analyses among different analysts (e.g., Vincent and Borenstein 1980; Uccellini et al. 1992). Such exercises may have frustrated meteorologists by the perceived unscientific nature of frontal research. Or, perhaps, the inactivity in frontal research may be a result of the attitude epitomized by Shapiro et al. (1985, p. 1168): "one can say that surface fronts are presently considered one of the better understood and predictable of mesoscale atmospheric phenomena."

During the initial heyday of mesoscale meteorology in the 1980s, it appeared that a renewed interest in fronts would develop with the rise of mesoscale modeling (Keyser and Uccellini 1987), mesoscale instabilities (e.g., frontal cyclogenesis, conditional symmetric instability), and a cornucopia of field programs (e.g., GALE, FRONTS, ERICA, STORM-FEST, and FASTEX), but this interest was short-lived. For a second time in the history of modern meteorology, interest declined in frontal research. This second decline can be measured by participation at the eleven *Conferences on Mesoscale Processes* from 1983 to 2005, when the number of presentations on fronts declined from a high of 32 (21% of the total number of presentations) in 1990 to 6 (2.5%) in 2005 (Fig. 2).

b. Operational frontal-analysis methods change

> . . . the practice of frontal analysis of surface data spread virtually everywhere outside the tropics, despite disappointment in cyclone behavior which often deviated substantially from the Norwegian rules.
> —Sanders (1983, p. 177)

In the operational forecasting environment, fronts encountered a different fate, but the outcome was the same—eventual disillusionment. The acceptance of the Norwegian cyclone model by the U.S. Weather Bureau

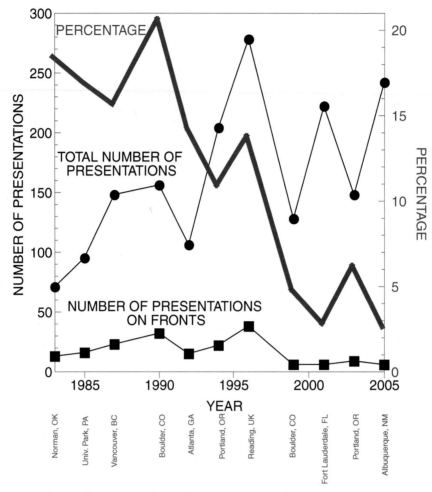

FIG. 2. Presentations (oral and poster) given at the American Meteorological Society *Conferences on Mesoscale Processes* from 1983 to 2005 with the phrase "front/frontogenesis/frontal/baroclinic zone" in the title: number of these "frontal" presentations (solid line with squares), total number of presentations at the conference (solid line with circles), and percentage of these "frontal" presentations out of the total number of presentations (thick red line). Locations of the conferences are listed along the bottom.

(Namias 1983; Newton and Rodebush Newton 1999) led to the Weather Bureau abandoning surface isobar and isotherm analyses in favor of the now-familiar station model and frontal notations on its operational weather maps.

The most striking change in the new map is the substitution of symbols indicating position and movements of air masses for the old familiar concentric ellipses of isobars and isotherms of weather maps in use until now. Isobars are still present, but more widely distributed, so that the map is much less striped-up with these curving lines.
—*Science* (1941)

Thus, the Norwegian frontal analysis supplanted operational isotherm analysis in 1941. Over fifty years later, Fred Sanders became the leading figure in arguing that the inability to trust the Norwegian frontal analyses resulted from omitting the isotherms on analyses that were presumably constructed based on those very same un-

analyzed isotherms! What was gained by this change in 1941 was a more compact description of the present weather using the conceptual model of the Norwegian cyclone. Unfortunately, the Norwegian analysis methods injected greater subjectivity into the analysis of weather maps. In addition, the meteorological community failed to evolve their frontal-analysis techniques given new advances, new meteorological structures and phenomena, and the growing emphasis on mesoscale analysis (e.g., Kocin et al. 1991). [At least two attempted proposals for revised analysis conventions did not become generally accepted (Colby and Seitter 1987; Young and Fritsch 1989).] Some attempts at automating surface isotherm analysis (e.g., Renard and Clarke 1965; Clarke and Renard 1966; Huber-Pock and Kress 1989; and others reviewed in Table 1 in Hewson 1998) may have failed because the horizontal grid spacing of the datasets at the time was too coarse. Later attempts with higher-resolution datasets by Hewson (1998) and McCann and Whistler (2001), however, have been much more successful. Along with the increasing automation in the forecast office, some have reported that map-analysis skills have atrophied or have been lost entirely (e.g., Bosart 1989, 2003; Mass 1991; Sanders and Doswell 1995; McIntyre 1999). In this environment, Sanders and Doswell (1995) made a call for returning to operational isotherm analyses.

The waning scientific interest in frontal research and the reduction in operational isotherm analyses have left the atmospheric science community clinging to an outdated and sometimes incorrect caricature of fronts that evolved from the Norwegian cyclone model, a state not too dissimilar from the period when Sanders (1955) was written (Reed 2003, p. 3). Despite the abundant evidence that fronts are more complicated than those presented by the Norwegian cyclone model, these caricatures of frontal structure and dynamics persist. Many authors have argued that scientific and forecasting progress has been inhibited sometimes by the success of the Norwegian cyclone model (e.g., Sutcliffe 1952; P. Williams 1972; Schwerdtfeger 1981, p. 505; Hoskins 1983, pp. 1 and 14; Mass 1991; Schultz and Trapp 2003, section 7). For example, some surface analysts identify the north–south-oriented boundary equatorward of a surface cyclone as a cold front, even if this feature is only a dryline or a lee trough without a significant temperature gradient (e.g., Hobbs et al. 1990, 1996; Sanders and Doswell 1995; Sanders 1999a). The crusade against these caricatures of fronts and the deterioration of surface analysis techniques has one of its most outspoken proponents in Fred Sanders (Hoffman 2008).

To counter these caricatures, we present the properties of cold fronts as derived from examples that Fred Sanders has published. In the next section, we begin with the archetypal example of a cold front: Sanders' (1955) paper.

3. Sanders (1955): The archetypal cold front

If we are to learn anything about fronts, we must at least be sure our research is done on "real" fronts, and not just regions where someone has drawn a line on a weather map.   —Sanders (1967, p. 1.2)

Keyser (1986) reviewed the characteristics of surface cold fronts from three observational studies: Sanders (1955), Ogura and Portis (1982), and Shapiro (1982). These three fronts represented simple cases from which to build dynamical conceptual models. These three fronts, however, all occurred over the central plains of the United States, all were characterized by weak or nonexistent cyclogenesis (e.g., Fig. 3a), and two of them (Sanders and Shapiro) produced little, if any, precipitation. Thus, generalizing from these three cases requires caution. Patterned after the conclusions in Sanders (1955, p. 552), Keyser (1986, p. 230) identified common structural aspects from the above three studies. Below, we examine

the evidence for these conclusions and, where applicable, extend them to cold fronts in general.

Fronts are strongest at the surface and weaken rapidly with altitude. As shown in Sanders (1955), the horizontal gradient of potential temperature was strongest near the surface and weakened upward (Fig. 3b). Neglecting diabatic effects, Sanders (1955) showed that frontogenesis in this cross section could be expressed as the sum of two terms: confluence and tilting (Fig. 4). He showed that confluence dominated near the earth's surface (Fig. 4a), where the tilting term was small owing to the horizontal gradients of vertical motion being small (Fig. 4b). Farther aloft, tilting was strongly frontogenetical in the warm air directly above the surface front and strongly frontolytical within the frontal zone (Fig. 4b). Above the surface within the frontal zone, frontolysis by tilting dominated the frontogenesis by confluence (Figs. 4b,c), explaining the weakening of the frontal zone away from the ground. Later observational (e.g., Ogura and Portis 1982; Shapiro 1984) and modeling studies (e.g., Hoskins and Bretherton 1972; Keyser et al. 1978; Rutledge 1989; Koch et al. 1995; Thompson and Williams 1997) have confirmed this basic tenet of Sanders (1955), although the release of latent heat in the updraft may offset the frontolysis by tilting (e.g., Rao 1966; Palmén and Newton 1969, p. 261; Bond and Fleagle 1985; Orlanski et al. 1985; Koch et al. 1995; Bryan and Fritsch 2000; Locatelli et al. 2002; Colle 2003). Other studies have shown that the near-vertical isentropes at the leading edge of the front in the boundary layer imply that tilting effects are small (e.g., Pagowski and Taylor 1998; Tory and Reeder 2005; Reeder and Tory 2005; Schultz and Roebber 2008).

A narrow plume of rising warm air exists above the surface frontal position. Sanders (1955) calculated an upward vertical motion exceeding 0.25 m s^{-1} at a height of 1 km above the nose of the front (Fig. 3c). Subsequent direct measurements of updraft plumes of precipitating cold fronts (e.g., Browning and Harrold 1970; Carbone 1982) and other nonprecipitating cold fronts (e.g., Young and Johnson 1984; Shapiro 1984; Shapiro et al. 1985; Bond and Shapiro 1991; Ralph et al. 1999; Neiman et al. 2001) indicate that updrafts of cold fronts can be as strong as several meters per second. Dry, inviscid, idealized, two-dimensional cold fronts (e.g., Hoskins and Bretherton 1972) do not capture such magnitudes or the strong vertical gradients of vertical motion. The addition of Ekman pumping by Blumen (1980) into the analytic solutions of Hoskins and Bretherton (1972) produced greater, but still insufficient, vertical motion and a maximum in the midtroposphere rather than in the lowest few kilometers. The solutions to these two inviscid semi-geostrophic models were calculated only to the time when the surface front collapsed (mathematically, the temperature gradient at the surface becomes infinite). To obtain updraft plumes similar to those observed above, idealized models of fronts can be formulated in one of three ways: (i) numerical solution of a primitive equation model (e.g., Keyser and Anthes 1982, 1986; Tory and Reeder 2005),

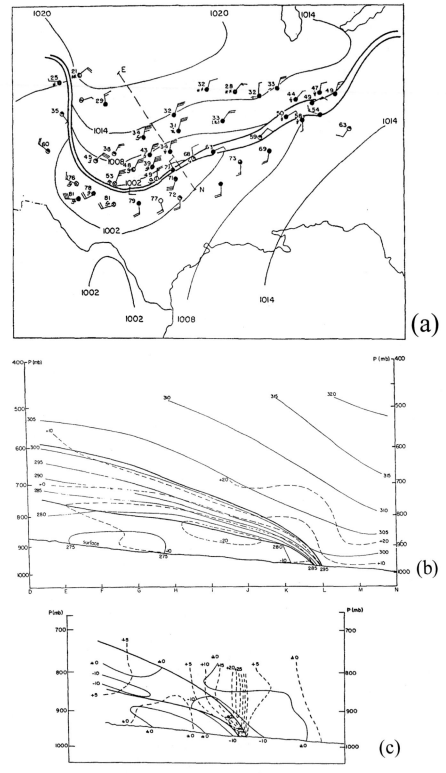

FIG. 3. The 0330 UTC 18 Apr 1953 boundaries of frontal zone (thick solid lines). (a) Surface chart: sea level pressure (thin solid lines every 6 hPa). Plotted reports follow conventional station model. Dashed line E−N indicates position of vertical cross section in (b),(c). (b) Cross section through cold front along E−N in (a): potential temperature (thin solid lines every 5 K) and horizontal wind component normal to cross section (dashed lines every 10 m s^{-1}; positive values represent flow into the plane of the cross section). Distance between adjacent letters on the horizontal axis is 100 km (Sanders 1955, his Figs. 2, 9, 10). (c) Cross section through cold front along E−N in (a): horizontal divergence (light solid lines, 10^{-5} s^{-1}) and vertical velocity (dashed lines every 5 cm s^{-1}).

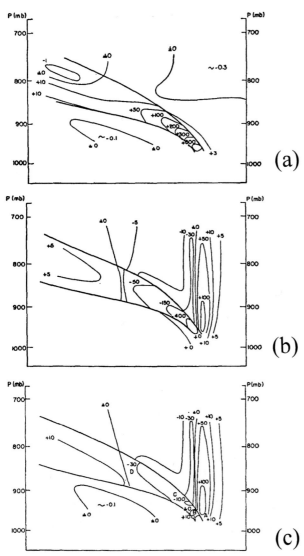

FIG. 4. Cross section through cold front along E–N in Fig. 3a at 0330 UTC 18 Apr 1953: boundaries of frontal zone (thick solid lines). (a) Confluence frontogenesis; (b) tilting frontogenesis; and (c) sum of (a) and (b) [°C (100 km)$^{-1}$ (3 h)$^{-1}$]. From Sanders (1955, his Figs. 11–13).

(ii) numerical solution of a semigeostrophic model including viscosity (e.g., Xu et al. 1998; Xu and Gu 2002), or (iii) analytical solution beyond collapse of an inviscid semigeostrophic model using Lagrangian potential vorticity conservation (e.g., Cullen 1983; Cullen and Purser 1984; Purser and Cullen 1987; Cho and Koshyk 1989; Koshyk and Cho 1992). The third approach produces a strong vertical gradient of vertical motion after the frontal collapse, but the frontal and wind structures appear unrealistic because the effects of boundary layer and surface friction are neglected. These results, as well as results published by others, showed that the strength of the vertical motion plume at the leading edge of modeled cold fronts was very sensitive to the formulation of the boundary layer (e.g., Blumen and Wu 1983; Thompson and

Williams 1997; Chen et al. 1997; Pagowski and Taylor 1998; Chen and Bishop 1999; Tory and Reeder 2005; Reeder and Tory 2005) and the lower boundary conditions (e.g., Xu et al. 1998; Gu and Xu 2000; Xu and Gu 2002).

Keyser and Anthes (1982) and Tory and Reeder (2005) found that the frictional convergence at the surface cold front was a consequence of the depletion of the alongfront momentum by the downward turbulent flux of momentum to the surface. The depletion of momentum caused the winds to become subgeostrophic, deviate toward the pressure trough/front, and generate inflow toward the front, producing the updraft. The strength of the updraft and its acceleration with height above the ground can be explained by the presence of low-level, near-neutral static stability, as noted by Browning (1990) and Bond and Shapiro (1991). The vertical isentropes at the leading edge of the front explain how such strong updrafts can be generated in the presence of otherwise stable prefrontal soundings. These results were supported by observations of cold fronts showing the importance of friction to the strength of the updraft (e.g., Browning and Harrold 1970; Bond and Fleagle 1985; Fleagle and Nuss 1985; Fleagle et al. 1988; Chen and Bishop 1999; Yu and Bond 2002).[1]

The frontal zone, a region of statically stable stratification, tilts rearward over the colder postfrontal air. Sanders (1955) showed that air parcels originating in the prefrontal environment near the earth's surface entered the front, experienced an increasing horizontal temperature gradient, and then were transported aloft in the updraft and rearward into the frontal zone (e.g., trajectory ABCD in Fig. 4c). This rearward tilt of cold-frontal zones (Fig. 3b) is due to the cross-frontal vertical shear of the direct ageostrophic circulation tilting the isentropes rearward with height (R. T. Williams 1972). Specifically, the rear-to-front ageostrophic flow near the surface and the

---

[1] Two caveats to this section require stating. First, the maximum vertical motion observed at the nose of cold fronts and the minimum horizontal scale across the front are sensitive to the resolution of the data. This resolution dependence may explain partially why the vertical motion at the leading edge of the Sanders (1955) front is an order of magnitude smaller than more recent, direct measurements through cold fronts. What controls the minimum scale of fronts remains an unanswered question (e.g., Emanuel 1985a; Boyd 1992). Whereas previous large-scale, idealized model simulations of dry cold fronts did not develop gravity current–like fronts (e.g., Hoskins and Bretherton 1972; Gall et al. 1987; Snyder et al. 1993), simulations of dry cold fronts by Snyder and Keyser (1996) and Chen and Bishop (1999) showed that a gravity current–like leading edge could be produced, given sufficient resolution.

Second, it is important to distinguish between precipitating and nonprecipitating fronts. The addition of moisture to idealized coldfrontal simulations results in narrower ascent plumes with stronger vertical motions (e.g., Sawyer 1956; Hsie et al. 1984; Mak and Bannon 1984; Bannon and Mak 1984; Emanuel 1985b; Thorpe and Emanuel 1985). Also, because of the strong vertical motions that could be produced by the leading edge of the cold outflow from a precipitation system, much stronger vertical motions could be possible on even smaller scales.

front-to-rear ageostrophic flow aloft tilts the frontal zone rearward with height (see also Bluestein 1993, 337–338). In the absence of the ageostrophic circulation, quasigeostrophic frontogenesis would produce unrealistic vertical fronts (Stone 1966; Williams and Plotkin 1968; Williams 1968). In addition, the postfrontal subsidence in the lower to midtroposphere may be responsible for maintaining the static stability of the frontal zone (e.g., Keyser and Anthes 1982, p. 1798; Ogura and Portis 1982, 2781–2782). Longwave radiation from the tops of postfrontal stratocumulus clouds may also enhance the stability of the frontal zone, as reviewed by Keyser (1986, 231–232). Although the overwhelming majority of published cross sections through cold fronts shows rearward-tilting frontal zones, some fronts tilt forward, as discussed by Schultz and Steenburgh (1999), Parker (1999), van Delden (1999), Stoelinga et al. (2002), and Schultz (2005).

Warm air is entrained into the frontal zone near the ground. Because near-surface air parcel trajectories from the warm air were ingested into the frontal zone, Sanders (1955) appears to be the first to note that the front was not a material boundary. Later, others came to the same conclusion (e.g., Ligda and Bigler 1958; German 1959; Brundidge 1965; Blumen 1980; Young and Johnson 1984; Shapiro 1984; Smith and Reeder 1988; Schultz and Mass 1993; Miller et al. 1996; Parker 1999). Despite this evidence, some textbooks still claim that fronts are nearly material surfaces (e.g., Wallace and Hobbs 1977, 116–117).

Determining whether a front is a material surface comprises two issues. First, consider an adiabatic front. In the absence of mixing, Smith and Reeder (1988) noted that some fronts may move along at the advective speed of the cold air and, hence, be considered material surfaces. Yet other fronts, such as those in the presence of alongfront warm advection, may move at a speed faster than the advective speed—in other words, propagation is occurring. Smith and Reeder (1988, p. 1940) said, "In essence, the frontal zone, centered on the position of the maximum surface temperature gradient, advances principally because of the differential alongfront temperature advection in the presence of an alongfront temperature gradient as noted by Gidel (1978)" [see also Sanders (1999a)]. Such fronts cannot be characterized as material surfaces. Thus, whether or not a front is a material surface may depend on other factors such as the alongfront temperature gradient.

Second, because of the no-slip condition at the earth's surface, prefrontal air parcels with zero horizontal velocity are overtaken by a moving front (e.g., Xu and Gu 2002, 104–105). For this reason, fronts cannot be considered material surfaces. It is this near-surface entrainment that leads to the next characteristic of cold fronts.

The postfrontal boundary layer is well mixed or slightly unstable. There are two possible explanations for the well-mixed or slightly unstable postfrontal boundary layer. The first mechanism was proposed by Sanders (1955) and Clarke (1961), who argued that fluxes from the ground in their cases were sufficient to yield this well-mixed postfrontal environment. Subsequent circulations in the planetary boundary layer were then essential for transporting this heat vertically (e.g., Fleagle et al. 1988; Chen et al. 1997). Alternatively, the second mechanism is as follows. The idealized cold front simulations of Keyser and Anthes (1982) and Xu and Gu (2002) showed that cold advection, in conjunction with a no-slip lower boundary condition, results in near-surface warm prefrontal air passing into the frontal zone by entrainment. In the presence of a thermally insulated lower boundary where surface heat fluxes are zero, superadiabatic lapse rates in the postfrontal air result. Upward turbulent heat transport then produces the postfrontal neutral stratification, a mechanism earlier proposed by Brundidge (1965). This second mechanism likely operates in general, whereas the first mechanism becomes nonnegligible under conditions of strong surface heat fluxes. Specifically, mesoscale model simulations of the Sanders (1955) cold front showed that the surface fluxes were not needed to reproduce this well-mixed layer (Schultz and Roebber 2008).

The prefrontal boundary layer is weakly stable. Relative to the postfrontal boundary layer, the prefrontal boundary layer was weakly stable in the three cases examined by Keyser (1986). This statement does not generalize well to other cold fronts, however, as there can be a tremendous range in the static stability of the prefrontal environment. For example, rope clouds associated with cold fronts (e.g., Cochran et al. 1970; Shaughnessy and Wann 1973; Janes et al. 1976; Woods 1983; Seitter and Muench 1985; Shapiro et al. 1985; Bond and Shapiro 1991) are typically associated with prefrontal soundings characterized by surface-based, shallow moist-neutral layers topped by strong inversions and dry air aloft owing to large-scale subsidence (e.g., Shaughnessy and Wann 1973; Woods 1983). Even vertical motions of several meters per second within the shallow moist layer cannot penetrate the inversion, thus limiting the depth of the rope cloud. The appearance of prefrontal boundary layers over the North Pacific Ocean that were nearly moist neutral and saturated (e.g., Bond and Fleagle 1988) may explain the prevalence for rope clouds over the oceans. In contrast, cold fronts moving into deeper surface-based moist layers with substantially weaker capping inversions may lead to deep moist convection (e.g., Koch 1984; Dorian et al. 1988; Bluestein 2008).

Previous research results support the above generalizations of many of the structural aspects of cold fronts. In addition, textbook illustrations of cold fronts show a discontinuity (or near discontinuity) in temperature, a simultaneous wind shift, and coincident pressure trough with a surface cold-frontal passage, a feature predicted from zero- and first-order discontinuity theory (e.g., Petterssen 1933; Godson 1951; Saucier 1955, p. 109; Bluestein 1993, 240–248). [In this regard, Schultz (2004, his Fig. 9) and Schultz and Roebber (2008, their Fig. 2b) showed that the front studied by Sanders (1955) may not

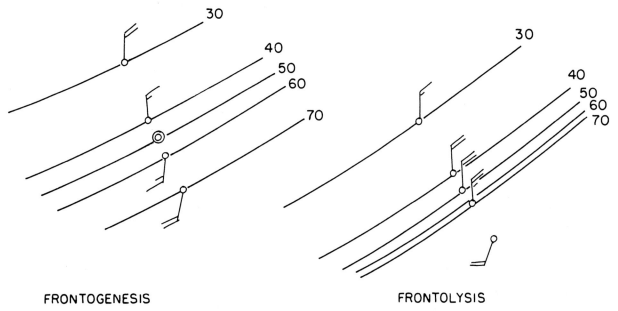

FRONTOGENESIS   FRONTOLYSIS

FIG. 5. Schematic of frontogenetical and frontolytical scenarios. Surface temperatures (solid lines every 10°F) and surface winds (one pennant, full barb, and half barb denote 25, 5, and 2.5 m s^{-1}, respectively). From Sanders (1967, his Fig. 2.9).

be as classical as previously believed. A mesoscale model simulation of the Sanders (1955) cold front presented by Schultz and Roebber (2008) also reveals some potentially interesting aspects of this cold front that deviate from Sanders' (1955) original analysis.] Not all cold fronts, however, may feature the simultaneity of the temperature decrease, wind shift, and pressure trough, as is demonstrated in the next two sections.

4. Sanders (1967): Further studies of cold fronts

Fronts do not just suddenly exist. They form, go through intensifications and weakenings, become diffuse and finally indistinguishable.   —Sanders (1967, p. 4.8)

Sanders resumed his analyses of surface fronts in the 1960s. The National Science Foundation awarded Sanders $117,200 over four years to perform a "description of typical frontal structure in the three-dimensional fields of temperature and wind, diagnosis of the fields of vertical motion and divergence associated with fronts, study of the frontogenetical and frontolytical processes, and study of the effects of friction" (Sanders 1967, p. 1). His 1967 report was principally a collection of ten appendices, comprising excerpts from student theses and papers presented at conferences. The first four appendices analyzed cases of surface cold fronts over Texas and Oklahoma, and these appendices are summarized below; the remaining six appendices dealt with upper-level fronts and other projects unrelated to fronts and are not discussed here.

a. "Detailed analysis of an intense surface cold front" by Jon Plotkin (S.M. thesis, August 1965)

This section presented an analysis of the Texas cold front of 20–21 January 1959 using standard synoptic surface data. One-hour temperature drops associated with this cold front ranged from 18°C (33°F) at Mineral Wells to 3°C (6°F) farther equatorward at Galveston. The changes in the wind consisted of two generally separate features: a change in direction, followed by an increase in speed. The wind shift preceded the temperature drop by as much as an hour at some stations. The temperature drop accompanied, or was close to, the increase in wind speed, with the pressure trough occurring in the warm air.

Confluence was the strongest frontogenetical process acting on the front. The front would inevitably weaken, however, because of the convergence at the wind shift not being coincident with the temperature gradient, along with the frontolytical effect of turbulence. Indeed, divergence quickly followed the frontal passage and the front weakened as it moved equatorward. This conclusion would become a common theme for Sanders: when the temperature gradient and wind shift were coincident, the front would undergo frontogenesis, but when the wind shift preceded the temperature gradient, the front would undergo frontolysis (Fig. 5). This sentiment echoed that of Petterssen (1936, p. 21), who proposed the following rules, "(a) Fronts that move towards a trough increase in intensity. (b) Fronts that leave a trough dissolve." Sanders' conclusions were tentative, however, because of the hourly reporting of the stations and the lack of simultaneous pressure, temperature, and wind

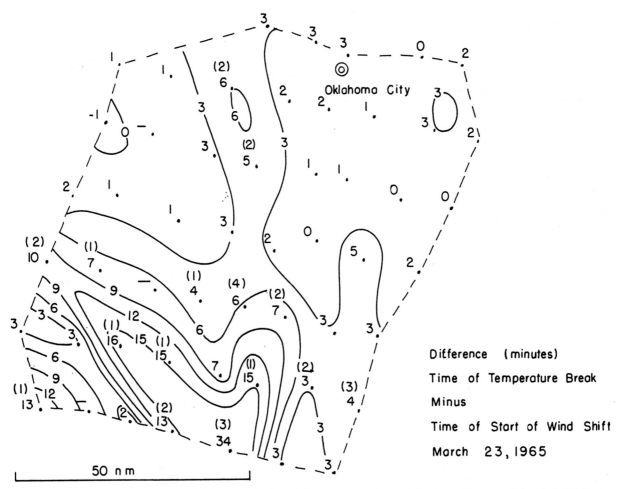

FIG. 6. Difference (solid lines every 3 min) between the time of the temperature drop and the time of the start of the wind shift for the cold front on 23 Mar 1965. Number in parentheses represents difference between the time of the start of the wind shift and the increase in the wind speed. From Sanders (1967, his Fig. 2.13).

data from stations. These limitations were remedied in the next appendix.

b. "Detailed analysis of an intense surface cold front" by Fred Sanders and Jon Plotkin (paper delivered at meeting of the American Meteorological Society, Denver, Colorado, 25 January 1966)

During the 1960s, the National Severe Storms Laboratory beta network (Kessler 1964, 1965) covered south-central Oklahoma with a surface observing station spacing of 16–24 km. This network provided Sanders the opportunity to acquire high temporal and spatial resolution data in the region where strong cold fronts were relatively common. Some of the material from this appendix was later published in Sanders (1983).

On 23 March 1965, a cold front moved through the network. The time between the temperature break and the wind shift ranged from −1 to 34 min (Fig. 6). A composite of the time series at individual stations during the frontal passage showed that the average temperature

drop of 1.4°C (2.5°F) in 1 min occurred around the same time as the wind strengthened (Fig. 7). Within 9 min after the initial temperature drop, the temperature had decreased an average of 7.2°C (13°F). Assuming a two-dimensional and steady-state front, the maximum vorticity and convergence lay ahead of the maximum temperature gradient and frontogenesis by about 1 km (Fig. 8). This appendix abruptly ended, leaving the next appendix to expand on this event in more detail.

c. "Analysis of mesoscale frontogenesis and deformation fields" by R. Throop Bergh (S.M. thesis, May 1967)

Additional analysis of the 23 March 1965 cold front, along with two more fronts (24 March 1964 and 24 April 1965), included horizontal maps of the divergence, vorticity, axes of dilatation, resultant deformation, and frontogenesis at the time the fronts were in the meso-network (e.g., Fig. 9). For these three events, there was no correlation between the strength of the temperature (density) gradient and the frontal speed, as might be

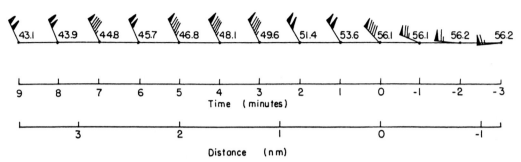

FIG. 7. Average temperature (numbers along top line in °F) and wind (one pennant, full barb, and half barb denote 0.5, 1, and 5 m s^{-1}, respectively) at each station in the network relative to the time of the temperature drop ($t = 0$) for the cold front on 23 Mar 1965. From Sanders (1967, his Fig. 2.14).

expected if the fronts behaved as density currents (e.g., Seitter 1986). Even with the small spacing of the network, most wind shifts occurred less than 5 min before the temperature decrease, although for at least one station this value was as large as 34 min. Despite the huge rates of frontogenesis calculated [0.45°C km^{-1} h^{-1} or 1.5°F (n mi)$^{-1}$ h^{-1}] (e.g., Fig. 9), all three fronts maintained a nearly constant intensity as they moved through the network, suggesting that a balance existed between deformation frontogenesis and turbulent frontolysis.

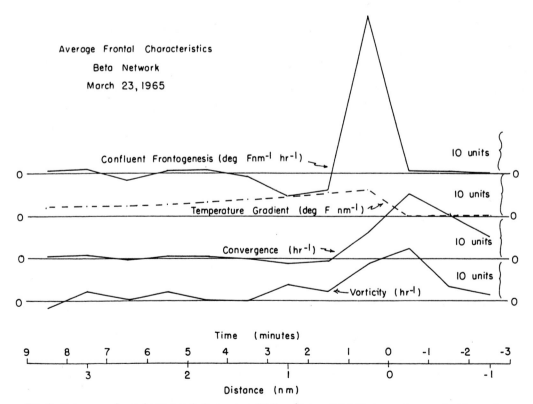

FIG. 8. Average confluent frontogenesis [top row, solid line, °F (n mi)$^{-1}$ h^{-1}], temperature gradient [second row, dashed line, °F (n mi)$^{-1}$], convergence (third row, solid line, h^{-1}), and vorticity (bottom row, solid line, h^{-1}). From Sanders (1967, his Fig. 2.15).

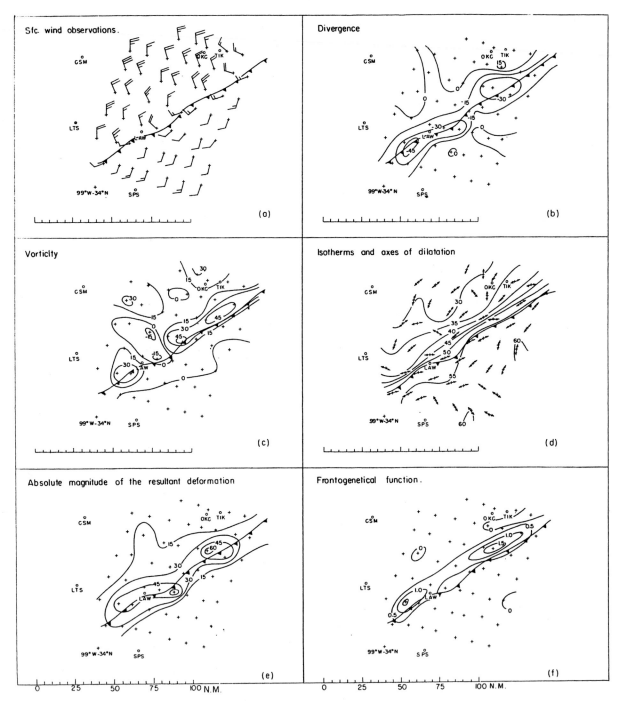

FIG. 9. Cold front on 23 Mar 1965: (a) surface winds (one pennant, full barb, and half barb denote 25, 5, and 2.5 m s^{-1}, respectively); (b) divergence (solid lines every 15 × 10^{-4} s^{-1}); (c) vorticity (solid lines every 15 × 10^{-4} s^{-1}); (d) isotherms (solid lines every 5°F) and direction of axes of dilatation; (e) absolute magnitude of the resultant deformation (solid lines every 15 × 10^{-4} s^{-1}); and (f) confluent frontogenesis [solid lines every 0.5°F (n mi)$^{-1}$ h^{-1}]. From Sanders (1967, his Fig. 3.8).

d. "Mesoscale analysis of complex cold front based on surface and tower data" by William T. Sommers (S.M. thesis, July 1967)

The final analysis of a surface cold front in Sanders (1967) was performed on the front of 8–9 June 1966.

This case differed from the others in that the front appeared to undergo 3 h of frontolysis. No large fronto-lytical deformation was present because of the near co-incidence of the temperature decrease and the wind shift. Thus, diabatic turbulent processes must have been acting to weaken the front. Data from a 444.6-m (1458.5

feet) tower showed that the leading edge of the front, as defined by the wind change, was vertical or forward tilting below 44.5 m (146 feet) AGL, but rearward tilting with height above.

5. Sanders at the turn of the millenium

> Routine analysis does not stop with consideration of the temperature field. —Sanders (1999a, p. 954)

In 1995, Sanders coauthored a critique of current surface analysis techniques with Chuck Doswell (Sanders and Doswell 1995). Returning once again to his roots by trying to raise the quality of operational surface analysis, Sanders was poised to revisit the topic of fronts once again. Three more papers (Sanders 1999a,b, 2005) advanced Sanders' agenda of bringing more science into surface map analysis and are discussed below. A fourth paper (Sanders and Kessler 1999) on interesting frontal passages in rural Oklahoma is discussed in more detail by Kessler (2008). A fifth paper (Sanders and Hoffman 2002) describes a climatology of surface baroclinic zones and is discussed by Hoffman (2008).

Sanders' (1999a) "A proposed method of surface map analysis" picks up where Sanders and Doswell (1995) left off. Sanders (1999a) presented instances where operationally constructed surface maps bore little resemblance to the actual surface frontal positions. To deal with the frequent absence of a relationship between analyzed fronts and surface potential temperature gradients, Sanders (1999a) proposed analysis notation for three features: nonfrontal baroclinic zones, fronts, and baroclinic troughs. These features would be distinguished by the magnitude of the surface potential temperature gradient and the existence of a cyclonic wind shift: fronts would possess both, nonfrontal baroclinic zones would possess only the magnitude of the surface potential temperature gradient, and baroclinic troughs would possess only the cyclonic wind shift.

Because many fronts are associated with a prefrontal wind shift and pressure trough [see Schultz (2005) for a review], Sanders (1999a) proposed a process by which such a prefrontal feature would occur. Quasigeostrophically, in the presence of an alongfront temperature gradient, the speed of movement of the pressure trough would exceed the advective speed of the isotherms in the front by several meters per second; thus, the pressure trough would be propagating relative to the flow. This process, however, has not been rigorously evaluated for observational cases (Schultz 2005).

Sanders' (1999b) "A short-lived cold front in the southwestern United States" analyzed a cold front in the southwestern United States on 26–27 March 1991. [Sanders spotted this case, interestingly, during the Surface Analysis Workshop at the National Meteorological Center (Uccellini et al. 1992).] He found that, during the day, clear skies on the warm side of the front and cloudy skies on the cold side intensified the cross-front

temperature gradient, resulting in an ageostrophic secondary circulation that produced convergence at the front and led to further intensification. Although there was a 6-h period where the temperature drop, pressure trough, and wind shift were coincident, eventually, the pressure trough and wind shift traveled eastward at 17.2 m s^{-1}, which was faster than the 11.8 m s^{-1} advective speed of the surface isotherms. This arrangement resulted in nonsimultaneity of the wind shift and the temperature gradient, leading to mixing within the frontal zone being unopposed by any frontogenetical process and the eventual weakening of the front.

Sanders' (2005) "Real front or baroclinic trough?" examined surface analyses prepared by the Hydrometeorological Prediction Center between 7 February and 29 March 2002. Sanders (2005) found that about 50% of the analyzed fronts were associated with baroclinic zones meeting his criteria of at least 8°C (220 km)$^{-1}$. He used this statistic to argue for a better distinction on operational surface analyses between fronts and baroclinic troughs—the difference between the two being "a substantial temperature change at the time of the cyclonic wind shift" (Sanders 2005, p. 650).

6. Conclusion

> This complexity should not be cause for despair! It is what is there and to deny it cannot benefit forecast accuracy. —Sanders (1999a, p. 947)

Fred Sanders contributed much to the understanding and analysis of surface fronts. With an eye toward reducing the overreliance on the Norwegian cyclone model, Sanders (1955) performed quantitative calculations investigating the dynamics of a surface cold front over the south-central United States. With access to high-resolution data from the National Severe Storms Laboratory beta network, further research by Sanders and his students, culminating in Sanders' (1967) "Frontal structure and the dynamics of frontogenesis," raised the issue of the importance of the relationship between the temperature gradient and the wind field to surface frontogenesis. Sanders (1967) and his classification scheme (Sanders 1999a) provided some basic terminology and the groundwork for my review of prefrontal troughs and wind shifts (Schultz 2005). Although such prefrontal features had been discussed previously in the literature by people other than Sanders, an extensive review of them had not been performed. To my knowledge, Sanders (1999b) was apparently the first to document the regional evolution of a surface cold front and its associated prefrontal features over the southwestern United States, if not over the western United States. Recently, Sanders' outspoken presentations at cyclone workshops (e.g., Gyakum et al. 1999), AMS conferences, and the symposium in his honor have been his attempt to revive these issues of surface analysis and show the complexities of fronts differing from the Nor-

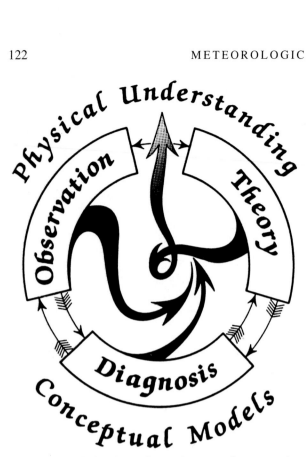

FIG. 10. Physical understanding and conceptual representation through the union of theory, diagnosis, and observation. From Shapiro et al. (1999, his Fig. 1).

wegian cyclone model. Fred Sanders is nothing if not persistent.

Shapiro et al. (1999), building upon earlier ideas by Bergeron (1959), Doswell et al. (1981), and Hoskins (1983), argued that scientific inquiry progresses most effectively when a synergy between theory, observation, and diagnosis occurs to produce physical understanding expressed in the form of conceptual models (Fig. 10). One example of what happens when the elements of this schematic figure work in harmony is illustrated at the beginning of this chapter: Sanders (1955) calculated frontogenesis diagnostics on observations of a cold front, which subsequently inspired the numerical experiments of Keyser and Anthes (1982) to advance the knowledge of the structure and dynamics of the leading edge of cold fronts. A counterexample of what happens when links become severed from each other was presented in section 2. Specifically, the Norwegian cyclone model concept of cold fronts became nearly impervious to modification by new scientific results (theory, observations, and diagnosis) because people failed to appreciate the rich spectrum of cold fronts possible in the real atmosphere and how that spectrum deviated from the reigning Norwegian paradigm.[2] Thus, these new results were not reconciled with the conceptual models

with sufficient veracity to modify the paradigm. Other examples exist where previous researchers have noted the limitations in extending their modeling research because of inadequate verifying observations of cold fronts (e.g., Keyser 1986; Keyser and Pecnick 1987; Blumen 1997; Pagowski and Taylor 1998). Thus, the inattention to fronts and frontal research has stymied more rapid progress because of the lack of links between theory, observations, and diagnosis. This chapter hopes to start a dialog on reconnecting theory, observation, and diagnosis to conceptual models for cold fronts.

Alternative structures and evolutions of cold fronts are often observed by operational forecasters and analysts but have not been placed in a dynamical context. There are opportunities to expand the knowledge reviewed in this chapter into the operational sector. Ultimately, this argument leads to the inevitable conclusion that forecaster training and the manual analysis of the data are important to improved understanding of the atmosphere. Beyond his talents as a scientist, teacher, and mentor, Fred Sanders has also been an outspoken advocate for forecaster education and surface analysis (e.g., Sanders and Doswell 1995; Sanders 1999a). Intuitive forecasters [i.e., forecasters who construct their conceptual understanding on the basis of dynamic visual images, as defined by Pliske et al. (2004)] are good at incorporating a variety of information into the hypothesis formation and hypothesis testing stages of forecasting (e.g., Roebber et al. 2004). Thus, providing improved conceptual models of cold-frontal processes and dynamics leads to improved forecasting skill for intuitive forecasters. Consequently, effective forecaster education, along with an emphasis on weather analysis skills, is required for the best forecasters to excel at their talents (e.g., Doswell et al. 1981; Bosart 2003; Doswell 2004).

> We trust that this paper will not be the last nor most comprehensive to report on these structures. . . .
> —Sanders and Kessler (1999, p. 1132)

Acknowledgments. This chapter benefited tremendously from the encouragement, historical perspective, guidance, and comments provided by Dan Keyser. Discussions with and comments by the following individuals further enriched this manuscript: George Bryan, Phil Cunningham, Charles Doswell, Steve Koch, John Locatelli, Chester Newton, Michael Reeder, Jay Shafer, Mark Stoelinga, Qin Xu, and an anonymous reviewer. Funding was provided by NOAA/OAR/NSSL under NOAA–OU Cooperative Agreement NA17RJ1227. I also thank Howie Bluestein, Lance Bosart, Brad Colman, and Todd Glickman for organizing this symposium. Finally, I owe tremendous gratitude to Fred Sanders for his years of mentorship, encouragement, and collaboration. Through the diversity of his scientific interests, his passion for science, and his personal integ-

---

[2] How such a process works in science in general is discussed in more detail by Kuhn (1970, especially chapters 6–8).

rity, Fred has served as a positive role model throughout my career.

REFERENCES

Ball, F. K., 1960: A theory of fronts in relation to surface stress. *Quart. J. Roy. Meteor. Soc.,* **86,** 51–66.

Bannon, P. R., and M. Mak, 1984: Diabatic quasi-geostrophic surface frontogenesis. *J. Atmos. Sci.,* **41,** 2189–2201.

Bergeron, T., 1937: On the physics of fronts. *Bull. Amer. Meteor. Soc.,* **18,** 265–275.

——, 1959: Methods in scientific weather analysis and forecasting: An outline in the history of ideas and hints at a program. *The Atmosphere and Sea in Motion: Scientific Contributions to the Rossby Memorial Volume,* B. Bolin, Ed., Rockefeller Institute Press, 440–474.

Bjerknes, J., 1917: Über die Fortbewegung der Konvergenz- und Divergenzlinien. *Meteor. Z.,* **34,** 345–349.

——, 1919: On the structure of moving cyclones. *Geophys. Publ., 1* (2), 1–8.

——, and H. Solberg, 1922: Life cycle of cyclones and the polar front theory of atmospheric circulation. *Geophys. Publ., 3* (1), 3–18.

Bluestein, H. B., 1993: *Observations and Theory of Weather Systems.* Vol. 2, *Synoptic–Dynamic Meteorology in Midlatitudes,* Oxford University Press, 594 pp.

——, 2008: Surface boundaries of the southern Plains: Their role in the initiation of convective storms. *Synoptic–Dynamic Meteorology and Weather Analysis and Forecasting: A Tribute to Fred Sanders, Meteor. Monogr.,* No. 55, Amer. Meteor. Soc.

Blumen, W., 1980: A comparison between the Hoskins-Bretherton model of frontogenesis and the analysis of an intense surface frontal zone. *J. Atmos. Sci.,* **37,** 64–77.

——, 1997: A model of inertial oscillations with deformation front-ogenesis. *J. Atmos. Sci.,* **54,** 2681–2692.

——, and R. Wu, 1983: Baroclinic instability and frontogenesis with Ekman boundary layer dynamics incorporating the geostrophic momentum approximation. *J. Atmos. Sci.,* **40,** 2630–2637.

Bond, N. A., and R. G. Fleagle, 1985: Structure of a cold front over the ocean. *Quart. J. Roy. Meteor. Soc.,* **111,** 739–759.

——, and ——, 1988: Prefrontal and postfrontal boundary layer processes over the ocean. *Mon. Wea. Rev.,* **116,** 1257–1273.

——, and M. A. Shapiro, 1991: Research aircraft observations of the mesoscale and microscale structure of a cold front over the eastern Pacific Ocean. *Mon. Wea. Rev.,* **119,** 3080–3094.

Bosart, L. F., 1989: Automation: Has its time really come? *Wea. Forecasting,* **4,** 271.

——, 2003: Whither the weather analysis and forecasting process? *Wea. Forecasting,* **18,** 520–529.

——, 2008: Coastal fronts, cold air damming, and fronts adjacent to higher terrain. *Synoptic–Dynamic Meteorology and Weather Analysis and Forecasting: A Tribute to Fred Sanders, Meteor. Monogr.,* No. 55, Amer. Meteor. Soc.

Boyd, J. P., 1992: The energy spectrum of fronts: Time evolution of shocks in Burgers' equation. *J. Atmos. Sci.,* **49,** 128–139.

Browning, K. A., 1990: Organization of clouds and precipitation in extratropical cyclones. *Extratropical Cyclones, The Erik Palmén Memorial Volume,* C. W. Newton and E. O. Holopainen, Eds., Amer. Meteor. Soc., 129–153.

——, and T. W. Harrold, 1970: Air motion and precipitation growth at a cold front. *Quart. J. Roy. Meteor. Soc.,* **96,** 369–389.

Brundidge, K. C., 1965: The wind and temperature structure of nocturnal cold fronts in the first 1,420 feet. *Mon. Wea. Rev.,* **93,** 587–603.

Bryan, G. H., and J. M. Fritsch, 2000: Diabatically driven discrete propagation of surface fronts: A numerical analysis. *J. Atmos. Sci.,* **57,** 2061–2079.

Carbone, R. E., 1982: A severe frontal rainband. Part I: Stormwide hydrodynamic structure. *J. Atmos. Sci.,* **39,** 258–279.

Carlson, T. N., 1991: *Mid-Latitude Weather Systems.* Harper Collins, 507 pp.

Charney, J. G., 1947: The dynamics of long waves in a baroclinic westerly current. *J. Meteor.,* **4,** 135–162.

Chen, C., and C. Bishop, 1999: Reply. *Mon. Wea. Rev.,* **127,** 258–263.

——, ——, G. S. Lai, and W.-K. Tao, 1997: Numerical simulations of an observed narrow cold-frontal rainband. *Mon. Wea. Rev.,* **125,** 1027–1045.

Cho, H.-R., and J. N. Koshyk, 1989: Dynamics of frontal discontinuities in the semigeostrophic theory. *J. Atmos. Sci.,* **46,** 2166–2177.

Clarke, L. C., and R. J. Renard, 1966: The U.S. Navy numerical frontal analysis scheme: Further development and a limited evaluation. *J. Appl. Meteor.,* **5,** 764–777.

Clarke, R. H., 1961: Mesostructure of dry cold fronts over featureless terrain. *J. Meteor.,* **18,** 715–735.

Cochran, H., N. Thomas, and F. C. Parmenter, 1970: "Rope" cloud. *Mon. Wea. Rev.,* **98,** 612–613.

Colby, F. P., Jr., and K. L. Seitter, 1987: A new analysis technique for fronts. *Extended Abstracts, Third Conf. on Mesoscale Meteorology,* Vancouver, BC, Canada, Amer. Meteor. Soc., 156–157.

Colle, B. A., 2003: Numerical simulations of the extratropical transition of Floyd (1999): Structural evolution and responsible mechanisms for the heavy rainfall over the northeast United States. *Mon. Wea. Rev.,* **131,** 2905–2926.

Cullen, M. J. P., 1983: Solutions to a model of a front forced by deformation. *Quart. J. Roy. Meteor. Soc.,* **109,** 565–573.

——, and R. J. Purser, 1984: An extended Lagrangian theory of semigeostrophic frontogenesis. *J. Atmos. Sci.,* **41,** 1477–1497.

Davies, H. C., 1997: Emergence of the mainstream cyclogenesis theories. *Meteor. Z.,* **6,** 261–274.

Dorian, P. B., S. E. Koch, and W. C. Skillman, 1988: The relationship between satellite-inferred frontogenesis and squall line formation. *Wea. Forecasting,* **3,** 319–342.

Doswell, C. A., III, 2004: Weather forecasting by humans—Heuristics and decision making. *Wea. Forecasting,* **19,** 1115–1126.

——, L. R. Lemon, and R. A. Maddox, 1981: Forecaster training—A review and analysis. *Bull. Amer. Meteor. Soc.,* **62,** 983–988.

Eady, E. T., 1949: Long waves and cyclone waves. *Tellus, 1* (3), 33–52.

Emanuel, K. A., 1985a: What limits front formation? *Nature,* **315,** 99.

——, 1985b: Frontal circulations in the presence of small moist symmetric stability. *J. Atmos. Sci.,* **42,** 1062–1071.

——, 2008: Back to Norway. *Synoptic–Dynamic Meteorology and Weather Analysis and Forecasting: A Tribute to Fred Sanders, Meteor. Monogr.,* No. 55, Amer. Meteor. Soc.

Ficker, H., 1923: Polarfront, Aufbav, Entstehung und Lebensgeschichte der Zyklonen (Polar front, structure, genesis and life cycle of cyclones). *Meteor. Z.,* **40,** 65–79.

Fleagle, R. G., and W. A. Nuss, 1985: The distribution of surface fluxes and boundary layer divergence in midlatitude ocean storms. *J. Atmos. Sci.,* **42,** 784–799.

——, N. A. Bond, and W. A. Nuss, 1988: Atmosphere-ocean interaction in mid-latitude storms. *Meteor. Atmos. Phys.,* **38,** 50–63.

Friedman, R. M., 1989: *Appropriating the Weather: Vilhelm Bjerknes and the Construction of a Modern Meteorology.* Cornell University Press, 251 pp.

Gall, R. L., R. T. Williams, and T. L. Clark, 1987: On the minimum scale of surface fronts. *J. Atmos. Sci.,* **44,** 2562–2574.

German, K. E., 1959: An investigation of the occlusion process near the Earth's surface. M.S. thesis, Dept. of Meteorology, University of Washington, 46 pp.

Gidel, L. T., 1978: Simulation of the differences and Similanties of warm and cold surface frontogenesis. *J. Geophy. Res.,* **83,** 915–928.

Godske, C. L., T. Bergeron, J. Bjerknes, and R. C. Bundgaard, 1957:

Dynamic Meteorology and Weather Forecasting. Amer. Meteor. Soc., 800 pp.

Godson, W. L., 1951: Synoptic properties of frontal surfaces. *Quart. J. Roy. Meteor. Soc.,* **77,** 633–653.

Gold, E., 1935: Fronts and occlusions. *Quart. J. Roy. Meteor. Soc.,* **61,** 107–157.

Gu, W., and Q. Xu, 2000: Baroclinic Eady wave and fronts. Part III: Unbalanced dynamics—Departures from viscous semigeostrophy. *J. Atmos. Sci.,* **57,** 3414–3425.

Gyakum, J. R., L. F. Bosart, and D. M. Schultz, 1999: The Tenth Cyclone Workshop. *Bull. Amer. Meteor. Soc.,* **80,** 285–290.

Hewson, T. D., 1998: Objective fronts. *Meteor. Appl.,* **5,** 37–65.

Hobbs, P. V., J. D. Locatelli, and J. E. Martin, 1990: Cold fronts aloft and the forecasting of precipitation and severe weather east of the Rocky Mountains. *Wea. Forecasting,* **5,** 613–626.

——, ——, and ——, 1996: A new conceptual model for cyclones generated in the lee of the Rocky Mountains. *Bull. Amer. Meteor. Soc.,* **77,** 1169–1178.

Hoffman, E., 2008: Surface potential temperature as an analysis and forecasting tool. *Synoptic–Dynamic Meteorology and Weather Analysis and Forecasting: A Tribute to Fred Sanders, Meteor. Monogr.,* No. 55, Amer. Meteor. Soc.

Hoskins, B. J., 1982: The mathematical theory of frontogenesis. *Annu. Rev. Fluid Mech.,* **14,** 131–151.

——, 1983: Dynamical processes in the atmosphere and the use of models. *Quart. J. Roy. Meteor. Soc.,* **109,** 1–21.

——, and F. P. Bretherton, 1972: Atmospheric frontogenesis models: Mathematical formulation and solution. *J. Atmos. Sci.,* **29,** 11–37.

Hsie, E.-Y., R. A. Anthes, and D. Keyser, 1984: Numerical simulation of frontogenesis in a moist atmosphere. *J. Atmos. Sci.,* **41,** 2581–2594.

Huber-Pock, F., and Ch. Kress, 1989: An operational model of objective frontal analysis based on ECMWF products. *Meteor. Atmos. Phys.,* **40,** 170–180.

Janes, S. A., H. W. Brandli, and J. W. Orndorff, 1976: "The blue line" depicted on satellite imagery. *Mon. Wea. Rev.,* **104,** 1178–1181.

Kessler, E., 1964: Purposes and programs of the National Severe Storms Laboratory, Norman, Oklahoma. National Severe Storms Laboratory Rep. 23, 17 pp. [Available from National Severe Storms Laboratory, 1313 Halley Circle, Norman, OK 73069.]

——, 1965: Purposes and program of the U.S. Weather Bureau National Severe Storms Laboratory, Norman, Oklahoma. *Trans. Amer. Geophys. Union,* **46,** 389–397.

——, 2008: Reflections on meteorology then and now, and with Fred Sanders. *Synoptic–Dynamic Meteorology and Weather Analysis and Forecasting: A Tribute to Fred Sanders, Meteor. Monogr.,* No. 55, Amer. Meteor. Soc.

Keyser, D., 1986: Atmospheric fronts: An observational perspective. *Mesoscale Meteorology and Forecasting,* P. S. Ray, Ed., Amer. Meteor. Soc., 216–258.

——, and R. A. Anthes, 1982: The influence of planetary boundary layer physics on frontal structure in the Hoskins-Bretherton horizontal shear model. *J. Atmos. Sci.,* **39,** 1783–1802.

——, and ——, 1986: Comments on "Frontogenesis in a moist semigeostrophic model." *J. Atmos. Sci.,* **43,** 1051–1054.

——, and M. J. Pecnick, 1987: The effect of along-front temperature variation in a two-dimensional primitive equation model of surface frontogenesis. *J. Atmos. Sci.,* **44,** 577–604.

——, and L. W. Uccellini, 1987: Regional models: Emerging research tools for synoptic meteorologists. *Bull. Amer. Meteor. Soc.,* **68,** 306–320.

——, M. A. Shapiro, and D. J. Perkey, 1978: An examination of frontal structure in a fine-mesh primitive equation model for numerical weather prediction. *Mon. Wea. Rev.,* **106,** 1112–1124.

Kirk, T. H., 1966: Some aspects of the theory of fronts and frontal analysis. *Quart. J. Roy. Meteor. Soc.,* **92,** 374–381.

Koch, S. E., 1984: The role of an apparent mesoscale frontogenetic circulation in squall line initiation. *Mon. Wea. Rev.,* **112,** 2090–2111.

——, J. T. McQueen, and V. M. Karyampudi, 1995: A numerical study of the effects of differential cloud cover on cold frontal structure and dynamics. *J. Atmos. Sci.,* **52,** 937–964.

Kocin, P. J., D. A. Olson, A. C. Wick, and R. D. Harner, 1991: Surface weather analysis at the National Meteorological Center: Current procedures and future plans. *Wea. Forecasting,* **6,** 289–298.

Koshyk, J. N., and H.-R. Cho, 1992: Dynamics of a mature front in a uniform potential vorticity semigeostrophic model. *J. Atmos. Sci.,* **49,** 497–510.

Kuhn, T. S., 1970: *The Structure of Scientific Revolutions.* 2d ed. University of Chicago Press, 210 pp.

Kutzbach, G., 1979: *The Thermal Theory of Cyclones.* Amer. Meteor. Soc., 255 pp.

Ligda, M. G. H., and S. G. Bigler, 1958: Radar echoes from a cloudless cold front. *J. Meteor.,* **15,** 494–501.

Locatelli, J. D., M. T. Stoelinga, and P. V. Hobbs, 2002: Organization and structure of clouds and precipitation on the mid-Atlantic coast of the United States. Part VII: Diagnosis of a nonconvective rainband associated with a cold front aloft. *Mon. Wea. Rev.,* **130,** 278–297.

Loomis, E., 1841: On the storm which was experienced throughout the United States about the 20th of December, 1836. *Trans. Amer. Philos. Soc.,* **7,** 125–163.

Mak, M., and P. R. Bannon, 1984: Frontogenesis in a moist semigeostrophic model. *J. Atmos. Sci.,* **41,** 3485–3500.

Malone, T. F., Ed., 1951: *Compendium of Meteorology.* Amer. Meteor. Soc., 1334 pp.

Mass, C., 1991: Synoptic frontal analysis: Time for a reassessment? *Bull. Amer. Meteor. Soc.,* **72,** 348–363.

McCann, D. W., and J. P. Whistler, 2001: Problems and solutions for drawing fronts objectively. *Meteor. Appl.,* **8,** 195–203.

McIntyre, M. E., 1999: Numerical weather prediction: A vision of the future, updated still further. *The Life Cycles of Extratropical Cyclones,* M. A. Shapiro and S. Grønås, Eds., Amer. Meteor. Soc., 337–355.

Miller, L. J., M. A. LeMone, W. Blumen, R. L. Grossman, N. Gamage, and R. J. Zamora, 1996: The low-level structure and evolution of a dry arctic front over the central United States. Part I: Mesoscale observations. *Mon. Wea. Rev.,* **124,** 1648–1675.

Namias, J., 1983: The history of polar front and air mass concepts in the United States—An eyewitness account. *Bull. Amer. Meteor. Soc.,* **64,** 734–755.

Neiman, P. J., F. M. Ralph, R. L. Weber, T. Uttal, L. B. Nance, and D. H. Levinson, 2001: Observations of nonclassical frontal propagation and frontally forced gravity waves adjacent to steep topography. *Mon. Wea. Rev.,* **129,** 2633–2659.

Newton, C. W., and H. Rodebush Newton, 1999: The Bergen School concepts come to America. *The Life Cycles of Extratropical Cyclones,* M. A. Shapiro and S. Grønås, Eds., Amer. Meteor. Soc., 41–59.

Ogura, Y., and D. Portis, 1982: Structure of the cold front observed in SESAME-AVE III and its comparison with the Hoskins–Bretherton frontogenesis model. *J. Atmos. Sci.,* **39,** 2773–2792.

Orlanski, I., B. Ross, L. Polinsky, and R. Shaginaw, 1985: Advances in the theory of atmospheric fronts. *Adv. Geophys.,* **28B,** 223–252.

Pagowski, M., and P. A. Taylor, 1998: Fronts and the boundary layer—Some numerical studies. *Bound.-Layer Meteor.,* **89,** 469–506.

Palmén, E., and C. W. Newton, 1969: *Atmospheric Circulation Systems.* Academic Press, 603 pp.

Parker, D. J., 1999: Passage of a tracer through frontal zones: A model for the formation of forward-sloping cold fronts. *Quart. J. Roy. Meteor. Soc.,* **125,** 1785–1800.

Petterssen, S., 1933: Kinematical and dynamical properties of the field of pressure, with application to weather forecasting. *Geophys. Publ.,* **10** (2), 1–92.

——, 1936: Contribution to the theory of frontogenesis. *Geophys. Publ.,* **11** (6), 1–27.

——, 1955: A general survey of factors influencing development at sea level. *J. Meteor.,* **12,** 36–42.

——, G. E. Dunn, and L. L. Means, 1955: Report of an experiment in forecasting of cyclone development. *J. Meteor.,* **12,** 58–67.

Pliske, R. M., B. Crandall, and G. Klein, 2004: Competence in weather forecasting. *Psychological Investigations of Competence in Decision Making,* K. Smith et al., Eds., Cambridge University Press, 40–68.

Purser, R. J., and M. J. P. Cullen, 1987: A duality principle in semigeostrophic theory. *J. Atmos. Sci.,* **44,** 3449–3468.

Ralph, F. M., P. J. Neiman, and T. L. Keller, 1999: Deep-tropospheric gravity waves created by leeside cold fronts. *J. Atmos. Sci.,* **56,** 2986–3009.

Rao, G. V., 1966: On the influences of fields of motion, baroclinicity and latent heat source on frontogenesis. *J. Appl. Meteor.,* **5,** 377–387.

Reed, R. J., 2003: A short account of my education, career choice, and research motivation. *A Half Century of Progress in Meteorology: A Tribute to Richard Reed,* Meteor. Monogr., No. 53, Amer. Meteor. Soc., 1–7.

Reeder, M. J., and K. J. Tory, 2005: The effect of the continental boundary layer on the dynamics of fronts in a 2D model of baroclinic instability. II: Surface heating and cooling. *Quart. J. Roy. Meteor. Soc.,* **131,** 2409–2429.

Renard, R. J., and L. C. Clarke, 1965: Experiments in numerical objective frontal analysis. *Mon. Wea. Rev.,* **93,** 547–556.

Roebber, P. J., D. M. Schultz, B. A. Colle, and D. J. Stensrud, 2004: Toward improved prediction: High-resolution and ensemble modeling systems in operations. *Wea. Forecasting,* **19,** 936–949.

Rutledge, S. A., 1989: A severe frontal rainband. Part IV: Precipitation mechanisms, diabatic processes and rainband maintenance. *J. Atmos. Sci.,* **46,** 3570–3594.

Sanders, F., 1955: An investigation of the structure and dynamics of an intense surface frontal zone. *J. Meteor.,* **12,** 542–552.

——, 1967: Frontal structure and the dynamics of frontogenesis. Final Report to the National Science Foundation, Grant GP-1508, 10 pp. with 10 appendixes.

——, 1983: Observations of fronts. *Mesoscale Meteorology—Theories, Observations and Models,* D. K. Lilly and T. Gal-Chen, Eds., Reidel, 175–203.

——, 1987: A study of 500 mb vorticity maxima crossing the east coast of North America and associated surface cyclogenesis. *Wea. Forecasting,* **2,** 70–83.

——, 1999a: A proposed method of surface map analysis. *Mon. Wea. Rev.,* **127,** 945–955.

——, 1999b: A short-lived cold front in the southwestern United States. *Mon. Wea. Rev.,* **127,** 2395–2403.

——, 2005: Real front or baroclinic trough? *Wea. Forecasting,* **20,** 647–651.

——, and E. P. Auciello, 1989: Skill in prediction of explosive cyclogenesis over the western North Atlantic Ocean, 1987/88: A forecast checklist and NMC dynamical models. *Wea. Forecasting,* **4,** 157–172.

——, and C. A. Doswell III, 1995: A case for detailed surface analysis. *Bull. Amer. Meteor. Soc.,* **76,** 505–521.

——, and E. Kessler, 1999: Frontal analysis in the light of abrupt temperature changes in a shallow valley. *Mon. Wea. Rev.,* **127,** 1125–1133.

——, and E. G. Hoffman, 2002: A climatology of surface baroclinic zones. *Wea. Forecasting,* **17,** 774–782.

Saucier, W. J., 1955: *Principles of Meteorological Analysis.* University of Chicago Press, 438 pp.

Sawyer, J. S., 1956: The vertical circulation at meteorological fronts and its relation to frontogenesis. *Proc. Roy. Soc. London,* **A234,** 346–362.

Schultz, D. M., 2004: Cold fronts with and without prefrontal wind shifts in the central United States. *Mon. Wea. Rev.,* **132,** 2040–2053.

——, 2005: A review of cold fronts with prefrontal troughs and wind shifts. *Mon. Wea. Rev.,* **133,** 2449–2472.

——, and C. F. Mass, 1993: The occlusion process in a midlatitude cyclone over land. *Mon. Wea. Rev.,* **121,** 918–940.

——, and W. J. Steenburgh, 1999: The formation of a forward-tilting cold front with multiple cloud bands during Superstorm 1993. *Mon. Wea. Rev.,* **127,** 1108–1124.

——, and R. J. Trapp, 2003: Nonclassical cold-frontal structure caused by dry subcloud air in northern Utah during the Intermountain Precipitation Experiment (IPEX). *Mon. Wea. Rev.,* **131,** 2222–2246.

——, and P. J. Roebber, 2008: The fiftieth anniversary of Sanders (1955): A mesoscale model simulation of the cold front of 17–18 April 1953. *Synoptic–Dynamic Meteorology and Weather Analysis and Forecasting: A Tribute to Fred Sanders,* Meteor. Monogr., No. 55, Amer. Meteor. Soc.

Schwerdtfeger, W., 1981: Comments on Tor Bergeron's contributions to synoptic meteorology. *Pure Appl. Geophys.,* **119,** 501–509.

Science, 1941: A new type of weather map. *Science,* **94** (1 Aug. Suppl.), 10.

Seitter, K. L., 1986: A numerical study of atmospheric density current motion including the effects of condensation. *J. Atmos. Sci.,* **43,** 3068–3076.

——, and H. S. Muench, 1985: Observation of a cold front with rope cloud. *Mon. Wea. Rev.,* **113,** 840–848.

Shapiro, M. A., 1982: Mesoscale weather systems of the central United States. CIRES/NOAA Tech. Rep., University of Colorado, 78 pp. [Available from Cooperative Institute for Research in Environmental Sciences, University of Colorado/NOAA, Boulder, CO 80309.]

——, 1984: Meteorological tower measurements of a surface cold front. *Mon. Wea. Rev.,* **112,** 1634–1639.

——, T. Hampel, D. Rotzoll, and F. Mosher, 1985: The frontal hydraulic head: A micro-α scale (~1 km) triggering mechanism for mesoconvective weather systems. *Mon. Wea. Rev.,* **113,** 1166–1183.

——, and Coauthors, 1999: A planetary-scale to mesoscale perspective of the life cycles of extratropical cyclones: The bridge between theory and observations. *The Life Cycles of Extratropical Cyclones,* M. A. Shapiro and S. Grønås, Eds., Amer. Meteor. Soc., 139–185.

Shaughnessy, J. E., and T. C. Wann, 1973: Frontal rope in the North Pacific. *Mon. Wea. Rev.,* **101,** 774–776.

Smith, R. K., and M. J. Reeder, 1988: On the movement and low-level structure of cold fronts. *Mon. Wea. Rev.,* **116,** 1927–1944.

Snyder, C., and D. Keyser, 1996: The coupling of fronts and the boundary layer. Preprints, *Seventh Conf. on Mesoscale Processes,* Reading, United Kingdom, Amer. Meteor. Soc., 520–522.

——, W. C. Skamarock, and R. Rotunno, 1993: Frontal dynamics near and following frontal collapse. *J. Atmos. Sci.,* **50,** 3194–3211.

Stoelinga, M. T., J. D. Locatelli, and P. V. Hobbs, 2002: Warm occlusions, cold occlusions, and forward-tilting cold fronts. *Bull. Amer. Meteor. Soc.,* **83,** 709–721.

Stone, P. H., 1966: Frontogenesis by horizontal wind deformation fields. *J. Atmos. Sci.,* **23,** 455–465.

Sutcliffe, R. C., 1952: Principles of synoptic weather forecasting. *Quart. J. Roy. Meteor. Soc.,* **78,** 291–320.

Taljaard, J. J., W. Schmitt, and H. van Loon, 1961: Frontal analysis with application to the Southern Hemisphere. *Notos,* **10,** 25–58.

Thompson, W. T., and R. T. Williams, 1997: Numerical simulations of maritime frontogenesis. *J. Atmos. Sci.,* **54,** 314–331.

Thorpe, A. J., and K. A. Emanuel, 1985: Frontogenesis in the presence of small stability to slantwise convection. *J. Atmos. Sci.,* **42,** 1809–1824.

Tory, K. J., and M. J. Reeder, 2005: The effect of the continental boundary layer on the dynamics of fronts in a 2D model of baroclinic instability. I: An insulated lower surface. *Quart. J. Roy. Meteor. Soc.,* **131,** 2389–2408.

Uccellini, L. W., S. F. Corfidi, N. W. Junker, P. J. Kocin, and D. A. Olson, 1992: Report on the surface analysis workshop held at the National Meteorological Center 25–28 March 1991. *Bull. Amer. Meteor. Soc.,* **73,** 459–472.

van Delden, A., 1999: The slope of isentropes constituting a frontal zone. *Tellus,* **51A,** 603–611.

Vincent, D. G., and H. Borenstein, 1980: Experiments concerning variability among subjective analyses. *Mon. Wea. Rev.,* **108,** 1510–1521.

Volkert, H., 1999: Components of the Norwegian cyclone model: Observations and theoretical ideas in Europe prior to 1920. *The Life Cycles of Extratropical Cyclones,* M. A. Shapiro and S. Grønås, Eds., Amer. Meteor. Soc., 15–28.

Wallace, J. M., and P. V. Hobbs, 1977: *Atmospheric Science: An Introductory Survey.* Academic Press, 467 pp.

Williams, P., Jr., 1972: Western Region synoptic analysis—Problems and methods. NOAA/NWS Western Region Tech. Memo NWSTM WR-71, 71 pp. [Available from NOAA/NWS Western Region Headquarters, 125 S. State Street, Rm. 1311, Salt Lake City, UT 84138-1102.]

Williams, R. T., 1968: A note on quasi-geostrophic frontogenesis. *J. Atmos. Sci.,* **25,** 1157–1159.

——, 1972: Quasi-geostrophic versus non-geostrophic frontogenesis. *J. Atmos. Sci.,* **29,** 3–10.

——, and J. Plotkin, 1968: Quasi-geostrophic frontogenesis. *J. Atmos. Sci.,* **25,** 201–206.

Woods, V. S., 1983: Rope cloud over land. *Mon. Wea. Rev.,* **111,** 602–607.

Xu, Q., and W. Gu, 2002: Semigeostrophic frontal boundary layer. *Bound.-Layer Meteor.,* **104,** 99–110.

——, ——, and J. Gao, 1998: Baroclinic Eady wave and fronts. Part I: Viscous semigeostrophy and the impact of boundary condition. *J. Atmos. Sci.,* **55,** 3598–3615.

Young, G. S., and R. H. Johnson, 1984: Meso- and microscale features of a Colorado cold front. *J. Climate Appl. Meteor.,* **23,** 1315–1325.

——, and J. M. Fritsch, 1989: A proposal for general conventions in analyses of mesoscale boundaries. *Bull. Amer. Meteor. Soc.,* **70,** 1412–1421.

Yu, C.-K., and N. A. Bond, 2002: Airborne Doppler observations of a cold front in the vicinity of Vancouver Island. *Mon. Wea. Rev.,* **130,** 2692–2708.

Chapter 6

The Fiftieth Anniversary of Sanders (1955): A Mesoscale Model Simulation of the Cold Front of 17–18 April 1953

DAVID M. SCHULTZ*

Cooperative Institute for Mesoscale Meteorological Studies, University of Oklahoma, and NOAA/National Severe Storms Laboratory, Norman, Oklahoma

PAUL J. ROEBBER

Atmospheric Science Group, Department of Mathematical Sciences, University of Wisconsin—Milwaukee, Milwaukee, Wisconsin

(Manuscript received 14 September 2004, in final form 14 September 2006)

Schultz   **Roebber**

ABSTRACT

Over 50 yr have passed since the publication of Sanders' 1955 study, the first quantitative study of the structure and dynamics of a surface cold front. The purpose of this chapter is to reexamine some of the results of that study in light of modern methods of numerical weather prediction and diagnosis. A simulation with a resolution as high as 6-km horizontal grid spacing was performed with the fifth-generation Pennsylvania State University–National Center for Atmospheric Research (PSU–NCAR) Mesoscale Model (MM5), given initial and lateral boundary conditions from the National Centers for Environmental Precipitation–National Center for Atmospheric Research (NCEP–NCAR) reanalysis project data from 17 to 18 April 1953. The MM5 produced a reasonable simulation of the front, albeit its strength was not as intense and its movement was not as fast as was analyzed by Sanders. The vertical structure of the front differed from that analyzed by Sanders in several significant ways. First, the strongest horizontal temperature gradient associated with the cold front in the simulation occurred above a surface-based inversion, not at the earth's surface. Second, the ascent plume at the leading edge of the front was deeper and more intense than that analyzed by Sanders. The reason was an elevated mixed layer that had moved over the surface cold front in the simulation, allowing a much deeper vertical circulation than was analyzed by Sanders. This structure is similar to that of Australian cold fronts with their deep, well-mixed, prefrontal surface layer. These two differences between the model simulation and the analysis by Sanders may be because upper-air data from Fort Worth, Texas, was unavailable to Sanders. Third, the elevated mixed layer also meant that isentropes along the leading edge of the front extended vertically. Fourth, the field of frontogenesis of the horizontal temperature gradient calculated from the three-dimensional wind

* Current affiliation: Division of Atmospheric Sciences and Geophysics, Department of Physics, University of Helsinki, and Finnish Meteorological Institute, Helsinki, Finland.

Corresponding author address: Dr. David M. Schultz, Finnish Meteorological Institute, Erik Palménin Aukio 1, P. O. Box 503, FI-00101 Helsinki, Finland.
E-mail: david.schultz@fmi.fi

differed in that the magnitude of the maximum of the deformation term was larger than the magnitude of the maximum of the tilting term in the simulation, in contrast to Sanders' analysis and other previously published cases. These two discrepancies may be attributable to the limited horizontal resolution of the data that Sanders used in constructing his cross section. Last, a deficiency of the model simulation was that the postfrontal surface superadiabatic layer in the model did not match the observed well-mixed boundary layer. This result raises the question of the origin of the well-mixed postfrontal boundary layer behind cold fronts. To address this question, an additional model simulation without surface fluxes was performed, producing a well-mixed, not superadiabatic, layer. This result suggests that surface fluxes were not necessary for the development of the well-mixed layer, in agreement with previous research. Analysis of this event also amplifies two research themes that Sanders returned to later in his career. First, a prefrontal wind shift occurred in both the observations and model simulation at stations in western Oklahoma. This prefrontal wind shift was caused by a lee cyclone departing the leeward slopes of the Rockies slightly equatorward of the cold front, rather than along the front as was the case farther eastward. Sanders' later research showed how the occurrence of these prefrontal wind shifts leads to the weakening of fronts. Second, this study shows the advantage of using surface potential temperature, rather than surface temperature, for determining the locations of the surface fronts on sloping terrain.

---

1. Introduction

The December 1955 issue of the *Journal of Meteorology* (now the *Journal of the Atmospheric Sciences*) contained the article "An investigation of the structure and dynamics of an intense surface frontal zone" by Frederick Sanders. This paper, which examined the central U.S. cold front of 17–18 April 1953, was the first, and arguably still the best, quantitative description of the structure and dynamics of an observed surface cold front. Nearly every published case study of a cold front is inevitably compared to this archetypal cold front. Given that it has been over 50 yr since Sanders (1955), we feel that a reexamination of that study is timely, using modern methods of synoptic and mesoscale meteorology and the context of 50 yr of research on fronts. Specifically, in this chapter, we perform a simulation using a mesoscale numerical weather prediction model of the 17–18 April 1953 cold front. The model simulation provides a high-resolution, four-dimensional dataset for diagnosis. As is shown in this chapter, analyzing the model output raises some new and previously underemphasized points about frontal structure and dynamics, in general, and Sanders (1955), in particular. Analysis of this case also foreshadows several other developments in frontal research that would be championed by Sanders later in his career.

The mesoscale model and its configuration are discussed in section 2. Section 3 compares the model output with the observed data analyzed in Sanders (1955) to validate the veracity of the simulation. Section 4 addresses a debate over the causes of well-mixed postfrontal boundary layers. Finally, section 5 concludes this chapter.

2. Model simulation

The fifth-generation Pennsylvania State University–National Center for Atmospheric Research (PSU–NCAR) Mesoscale Model (MM5), version 3, a nonhydrostatic, primitive equation model (Dudhia 1993; Grell et al. 1994), is used to generate a mesoscale model simulation of the 17–18 April 1953 cold front, based on the procedure described by Roebber and Gehring (2000). MM5 was run for 30 h with initial conditions provided by the 1200 UTC 17 April 1953 data from the National Centers for Environmental Precipitation–National Center for Atmospheric Research (NCEP–NCAR) reanalysis project (Kalnay et al. 1996). Lateral boundary conditions during the simulation were generated by linear interpolation of the 6-h NCEP–NCAR reanalysis project data. The model was run in a one-way interactive mode (Zhang et al. 1986) with three domains similar to the outermost three domains in Roebber et al. (2002, their Fig. 2). The outermost domain (D1), with 54-km horizontal grid spacing, was designed to represent synoptic-scale features, with two nested grids to capture the higher-resolution features of the front. Domain 2 (D2) had a horizontal grid spacing of 18 km, and domain 3 (D3) had a horizontal grid spacing of 6 km. The model had 23 vertical σ levels, with $\sigma = (p - p_{\text{top}})/(p_{\text{surf}} - p_{\text{top}})$, where p is pressure, p_{top} is the top of the model (100 hPa), and p_{surf} is the surface pressure. The 23 σ levels had a relative concentration near the earth's surface in order to provide better resolution in the planetary boundary layer. Table 1o contains a list of those 23 σ levels.

An explicit moisture scheme with prognostic equations for cloud water, ice, rainwater, and snow (Reisner et al. 1998) was employed in all domains for grid-resolvable precipitation. The Kain–Fritsch cumulus parameterization scheme (Kain and Fritsch 1993) was used in the two outermost domains (D1 and D2). Radiative processes were handled using a cloud–radiation scheme with a 2-min time step, in which diurnally varying shortwave and longwave radiative fluxes interact with clear air and explicit clouds, whereas the surface fluxes were used in the ground energy-budget calculations (Dudhia 1989). The planetary boundary layer was modeled using the high-resolution Blackadar (1979) scheme (Zhang and Anthes 1982) coupled with a five-layer soil model (Dudhia 1996). The simulation employed four-dimensional data assimilation of the NCEP–NCAR reanalysis project data throughout the simulation on the outermost domain to keep the synoptic-scale error growth small. The assimilation technique (Stauffer and Seaman 1990) adds Newtonian relaxation terms to the prognostic equations for wind, temperature, and water vapor at each model time step.

3. Comparison of model output with the Sanders (1955) analysis

In this section, we compare the results of the model simulation with the analyses by Sanders (1955). One diagnostic for this comparison is the frontogenesis function, F, the Lagrangian rate of change of the magnitude of the horizontal potential temperature gradient because of the three-dimensional wind. This form is the same as that given by Miller [(1948), his Eq. (7)] and the adiabatic form of Bluestein [1986, p. 181, Eq. (9.11); 1993, p. 253, Eq. (2.3.21)], except we use the horizontal, not three-dimensional, potential temperature gradient:

$$F = \frac{d}{dt}|\nabla_H \theta|, \qquad (1)$$

where

$$\frac{d}{dt} = \frac{\partial}{\partial t} + u\frac{\partial}{\partial x} + v\frac{\partial}{\partial y} + w\frac{\partial}{\partial z},$$

$$\nabla_H = \mathbf{i}\frac{\partial}{\partial x} + \mathbf{j}\frac{\partial}{\partial y}.$$

This expression can be expanded to

$$F = -\underbrace{\frac{1}{|\nabla_H \theta|}\left[\frac{\partial\theta}{\partial x}\left(-\frac{\partial u}{\partial x}\frac{\partial\theta}{\partial x} - \frac{\partial v}{\partial x}\frac{\partial\theta}{\partial y}\right) + \frac{\partial\theta}{\partial y}\left(-\frac{\partial u}{\partial y}\frac{\partial\theta}{\partial x} - \frac{\partial v}{\partial y}\frac{\partial\theta}{\partial y}\right)\right]}_{F_{\text{def}}} - \underbrace{\frac{1}{|\nabla_H \theta|}\left[\frac{\partial\theta}{\partial z}\left(-\frac{\partial w}{\partial x}\frac{\partial\theta}{\partial x} - \frac{\partial w}{\partial y}\frac{\partial\theta}{\partial y}\right)\right]}_{F_{\text{tilt}}}. \qquad (2)$$

The form of the frontogenesis function used in Sanders [(1955), his Eq. (1)] was a two-dimensional form of Eq. (2), where all derivatives in the y direction were zero. Term F_{def} represents the horizontal processes intensifying the horizontal potential temperature gradient, also called the deformation term. This term is equivalent to Petterssen (1936) frontogenesis and is the sum of terms 2, 3, 6, and 7 in Bluestein [1986, p. 181, Eq. (9.11); Bluestein 1993, p. 253, Eq. (2.3.21)]. Because Petterssen frontogenesis is related to the vertical circulation associated with a frontal zone (e.g., Keyser et al. 1988), we use F_{def} to locate the active front on a surface map. Term F_{tilt} is called the tilting term and is equivalent to the sum of terms 4 and 8 in Bluestein [1986, p. 181, Eq. (9.11); Bluestein 1993, p. 253, Eq. (2.3.21)]. The tilting term F_{tilt} accounts for the role of the vertical motion w in altering the horizontal potential temperature gradient through tilting of isentropes from horizontal to vertical. Thus, F as given by Eq. (2) can be used to explain how the horizontal gradient of potential temperature is changing instantaneously because of the three-dimensional flow.

a. Surface maps

At 2100 UTC 17 April, the model simulation produced a 993-hPa low pressure center over northeastern New Mexico from which extended a broad baroclinic zone from Oklahoma and Kansas eastward to Indiana and Ohio (Fig. 1a). The model simulation compared well with the sub-996-hPa low center analyzed in the same location at 2130 UTC by Sanders (Fig. 1b), although his analyzed temperature gradient was much sharper, especially near the low center. This is likely explained by the coarse initialization of the surface front 9 h earlier and the weak frontogenetical forcing on the synoptic scale prior to 2100 UTC.

By 0300 UTC 18 April, the 997-hPa modeled low center departed the lee slopes of the Rocky Mountains of northern New Mexico onto the southern plains near the Texas–Oklahoma border (Fig. 1c). Sanders analyzed a sub-1002-hPa low center at 0330 UTC, shifted slightly southward relative to the simulation (Fig. 1d). The front strengthened considerably over Oklahoma in both Sanders' analysis and the model simulation, supported by the increasing Petterssen frontogenesis along the front near the low center (Fig. 1c). The strengthening of the front is not surprising because the deformation associated with the flow field around the low center helps provide forcing for Petterssen frontogenesis. Sanders (1955, his Table 1) showed that the cold front was stronger and narrower closer to the low center [along cross section AH rather than along cross section IP, both cross sections in Sanders (1955)]. Schultz (2004) also argued for the importance of departing lee cyclones toward providing frontogenetical forcing along frontal zones over the southern plains. Despite this frontogenesis, the modeled front remained too weak compared with Sanders' analysis. Specifically, Sanders (1955, his Table 1) determined the front to be 25 km wide with a horizontal potential temperature gradient of 56 K (100 km)$^{-1}$ at 1000 ft (305 m) AGL. By comparison, the model simulation produced a front twice as wide (48 km) with a horizontal potential temperature gradient 4 times weaker [about 16 K (100 km)$^{-1}$] at $\sigma = 0.945$ (about 400 m

TABLE 1. Sigma values for the model simulations.

23-level simulation	36-level simulation
	0.9995
	0.9985
0.998	0.9975
	0.9965
	0.9940
0.991	0.9905
	0.9875
	0.9830
0.973	0.9750
	0.9650
0.945	0.9500
	0.9300
0.910	0.9100
	0.8900
0.870	0.8700
	0.8500
0.825	0.8300
	0.8100
0.775	0.7850
	0.7550
0.725	0.7250
	0.6950
0.675	0.6650
0.625	0.6250
0.575	0.5750
0.525	0.5250
0.475	0.4750
0.425	0.4250
0.375	0.3750
0.325	0.3250
0.275	0.2750
0.225	0.2250
0.175	0.1750
0.125	0.1250
0.075	0.0750
0.025	0.0250

above the surface; Fig. 4a). Even if the simulated frontal zone was the right width, the temperature gradient would still be a factor of 2 too weak. The model resolution or the coarse resolution of the initial fields, for example, might be inadequate to capture the actual intensity.

By 0900 UTC 18 April, the low center moved eastward and continued to weaken into a broad trough elongated along the leading edge of the frontal zone with minimum pressure of 1003 hPa (Fig. 1e). With the passage of the low center to the east, the cold front over Oklahoma and northern Texas moved equatorward more quickly (Fig. 1e). Sanders' analysis, however, had a slightly deeper cyclone (less than 1002 hPa) and the cold front was much farther south, almost to the Texas–Mexico border (Fig. 1f). Although the location of the surface low center was predicted reasonably well throughout the simulation, unfortunately, the cold air was not able to advance equatorward as quickly as occurred in Sanders' analysis, even with applying the four-dimensional data assimilation. This is perhaps not surprising, given the difficulties that operational numerical weather prediction models have experienced in forecasting equatorward-moving cold fronts in the central United States (e.g., Mesinger 1996).

b. Prefrontal wind shift

One interesting aspect of this event is that not all the stations experienced a classic frontal passage with a simultaneous temperature decrease, pressure minimum, and wind shift. Figure 2 presents time series of observations from two nearby stations in Oklahoma. Tinker Air Force Base (TIK) showed a classic frontal signature with simultaneous wind shift, temperature drop, and pressure rise (Fig. 2a). On the other hand, Fort Sill (FSI), about 150 km southwest of TIK, experienced a prefrontal wind shift to northerly 3 h before the frontal passage (Fig. 2b). The reason for this prefrontal wind shift is discussed by Hutchinson and Bluestein (1998), Schultz (2004), and Bosart et al. (2008) where the prefrontal wind shift was attributable to a lee cyclone occurring slightly equatorward of the cold front. The timing and location of the lee cyclogenesis was not precise enough in the model simulation to exactly reproduce the observed time series (not shown). Nevertheless, the model simulation reproduced prefrontal wind shifts at stations in western Oklahoma, although for a much smaller region and less dramatic than were observed (not shown). The sensitivity of the frontal characteristics has particularly severe consequences for using numerical model guidance for operational forecasting of cold-frontal passages, which can be quite sensitive to the timing and location of lee cyclogenesis.

Although Sanders (1955) did not mention the prefrontal wind shift in his study, Sanders (1967, 1983, 1999), Keshishian and Bosart (1987), Hutchinson and Bluestein (1998), Schultz (2004, 2005), and Bosart et al. (2008) later recognized the significance of these types of features. If cyclogenesis occurred along the cold front, the temperature decrease, pressure minimum, and wind shift would occur simultaneously. The front would also be strong because of large Petterssen frontogenesis. After the passage of the cyclone to the east or if the lee cyclogenesis occurred in the warm air, then a prefrontal wind shift, representing the passage of the cyclone, would occur before the temperature drop associated with the front, resulting in frontolysis.

c. Implications for surface analysis: The importance of surface potential temperature

A surface analyst might use the surface temperature field to analyze the cold front across the leading edge of the surface temperature gradient southwest across west Texas and the Texas Panhandle (e.g., Figs. 1a,c,e). In contrast, Sanders' analysis (Figs. 1b,d,f) showed the surface front curling back to the northwest, cutting across the surface isotherms. How is this situation to be interpreted? Given that the elevation across Texas varies from sea level in the east to more than 2000 m in the

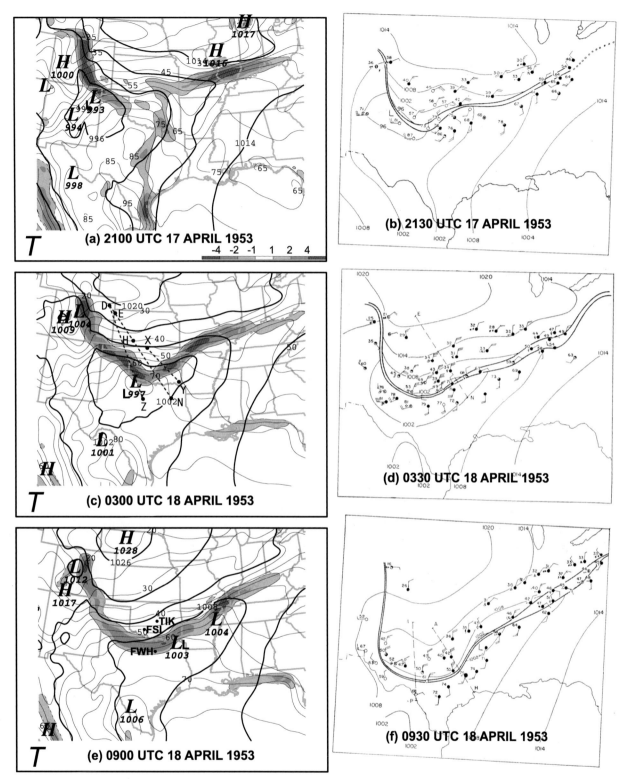

FIG. 1. (left) Surface evolution from a portion of the outermost domain D1 of model simulation: mean sea level pressure (thick solid lines every 6 hPa), temperature at lowest model level [$\sigma = 0.998$, approximately 14 m AGL; thin solid lines every 2.8°C (5°F)], and Petterssen (1936) frontogenesis at the lowest model level ($\sigma = 0.998$) [°C (100 km)$^{-1}$ (3 h)$^{-1}$, shaded according to scale in (a)]. Small sans serif Ls represent locations of the surface low from Sanders' analyses. (a) 2100 UTC 17 Apr, (c) 0300 UTC 18 Apr, and (e) 0900 UTC 18 Apr 1953. The locations of cross sections DZ, EN, and XY are displayed in (c). (right) Surface analyses by Sanders (1955), stretched and rotated to match the size and orientation of those from the model simulation: mean sea level pressure (thin solid lines every 6 hPa), frontal zone (thick solid lines), and station models (standard notation); (d) location of cross section EN. (b) 2130 UTC 17 Apr (his Fig. 1), (d) 0330 UTC 18 Apr (his Fig. 2), and (f) 0930 UTC 18 Apr 1953 (his Fig. 3).

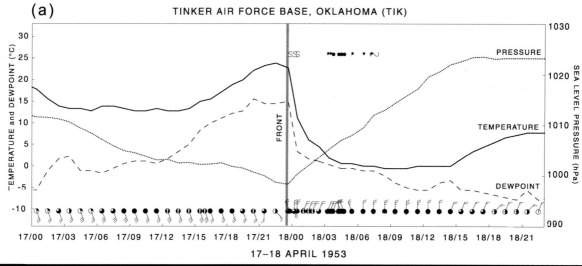

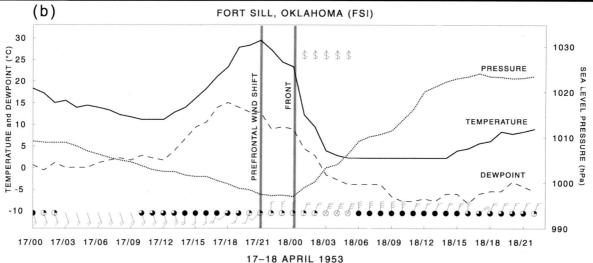

FIG. 2. Time series of observed data during the frontal passages at (a) TIK and (b) FSI. Locations of TIK and FSI are displayed in Fig. 1e. One full barb and half barb on winds denote 5 and 2.5 m s^{-1}, respectively.

west, the change in elevation could possibly be responsible for the temperature decrease. A plot of surface *potential* temperature from the model simulation showed almost no potential temperature gradient along and on the equatorward side of the Petterssen frontogenesis maximum in north-central Texas (e.g., Fig. 3c), suggesting that the decreasing surface temperature heading westward along the front in west Texas (e.g., Fig. 1e) was likely attributable to the increasing elevation to the west across Texas. Furthermore, the surface potential temperature gradient associated with the stable stratification intersecting sloping terrain over central and southern Texas at 0300 and 0900 UTC is apparent by comparing the temperature analyses (Figs. 1c,e) to the potential temperature analyses (Figs. 3b,c). These differences demonstrate the importance of analyzing sur-

face potential temperature rather than surface temperature for situations possessing elevation gradients, a point advocated by Sanders and Doswell (1995), Sanders (1999), and Sanders and Hoffman (2002).

d. Cross section

Figure 4 compares the vertical structure of the simulated front with Sanders' (1955) analysis along his cross section DN (location illustrated in Fig. 1c). Sanders constructed DN because this line segment passed through both the sounding locations of Dodge City, Kansas, and Oklahoma City, Oklahoma. Although there are some basic similarities between the model and analysis, such as the rearward-sloping stable layer repre-

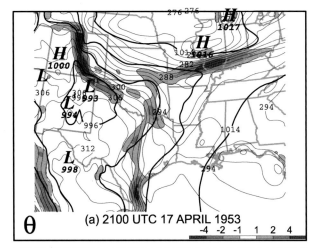

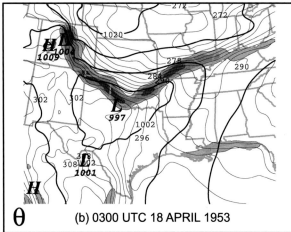

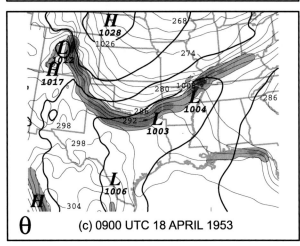

FIG. 3. Surface evolution from a portion of the outermost domain D1 of model simulation: mean sea level pressure (thick solid lines every 6 hPa), potential temperature at lowest model level ($\sigma = 0.998$, approximately 14 m AGL; thin solid lines every 2 K), and Petterssen (1936) frontogenesis at lowest model level ($\sigma = 0.998$) [°C (100 km)$^{-1}$ (3 h)$^{-1}$, shaded according to scale in (a)]: (a) 2100 UTC 17 Apr, (b) 0300 UTC 18 Apr, and (c) 0900 UTC 18 Apr 1953.

senting the frontal zone, many characteristics differ between these two fronts.

First, unlike Sanders' analysis, the strongest horizontal temperature gradient was not found at the earth's surface, but slightly above the surface on the third σ level above the ground ($\sigma = 0.973$). This difference was possibly because of the presence of a prefrontal surface stable layer (Fig. 4a), likely caused by a developing nocturnal inversion. Such a situation resembles those analyzed by Brundidge (1965) and Neiman et al. (1991), and modeled by Reeder and Tory (2005). Specifically, Neiman et al. (1991, their Fig. 8) showed that a cold front moving into a surface-based inversion less than 50 m deep resulted in the strongest horizontal temperature gradient occurring at the top of the inversion rather than at the surface. Although such a surface stable layer was not present in Sanders' analysis, the Fort Worth, Texas, sounding from 0300 UTC 18 April possessed a shallow surface-based inversion that was not present 6 h earlier (Fig. 5). This inversion may be similar to that produced by the model. That Sanders' analysis did not include the Fort Worth sounding (Sanders 1955, his Figs. 5 and 8) suggests that Sanders may have been unaware of the existence of that inversion, a result confirmed by Sanders himself (F. Sanders 2004, personal communication).

A second difference between the model and observations is the lack of clouds in the model. Surface station records indicate broken and overcast skies before and after the frontal passage (e.g., Figs. 1b,d,f and 2) with towering cumulus at the frontal passage. The model does not produce any clouds at the frontal passage and much less than were observed elsewhere (not shown).

A third difference pertains to the depth and strength of the plume of vertical motion at the leading edge of the front. Whereas Sanders' analysis showed a maximum of 0.25–0.30 m s^{-1} between 900 and 800 hPa (Fig. 4c), the simulation showed a much deeper maximum of vertical motion between 900 and 600 hPa, with a maximum exceeding 0.7 m s^{-1} around 700 hPa (Fig. 4a). In turn, this ascent plume led to a cold dome in the midtroposphere. Why was there deeper vertical motion associated with the cold front in the model? The answer is tied to another difference between the model and analysis, discussed in the next few paragraphs.

Fourth, whereas Sanders' analysis showed the isentropes within the frontal zone to be shallow and gently sloping (Fig. 4b), the isentropes at the leading edge of the simulated front were nearly vertical (Fig. 4a). Mesoscale model simulations of other cold fronts have showed similar vertical structures that were not resolved by the corresponding analyses of observations, such as the cold front of 28–29 December 1988 (Koch and Kocin 1991; Chen et al. 1997; Koch 1999).

The steeply sloped isentropes seem to have occurred above the surface position of the front in a layer, between 850 and 500 hPa, of much less stability than analyzed by Sanders (cf. Figs. 4a,b). Observed sound-

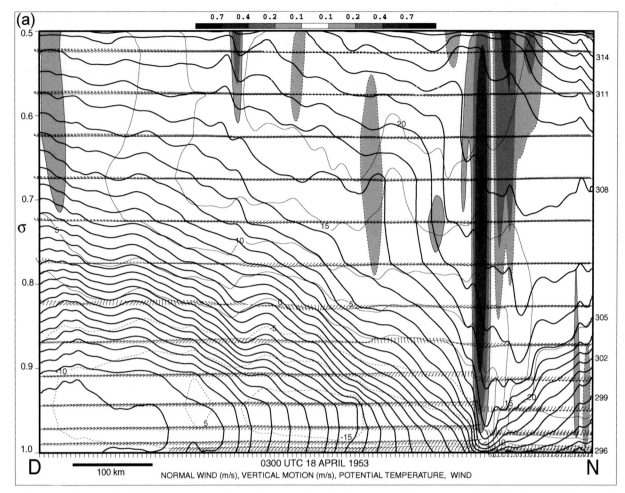

FIG. 4. (a) Cross section DN (location in Fig. 1) from innermost domain D3 of the model simulation at 0300 UTC 18 Apr 1953: potential temperature (black solid lines every 1 K), wind (one pennant, full barb, and half barb denote 25, 5, and 2.5 m s⁻¹, respectively; every 6 km), front-normal wind speed [every 5–20 m s⁻¹, with 10 m s⁻¹ contour intervals thereafter, in gray lines with negative (zero and positive) values contoured in dashed (solid) lines], and vertical motion [m s⁻¹, shaded according to scale; positive (negative) values outlined in solid (dashed) lines]. (b) Same cross section, but analyzed by Sanders (1955, his Fig. 9). Potential temperature (thin solid lines every 5 K), horizontal wind component normal to the cross section (dashed lines every 10 m s⁻¹ with positive values representing flow directed into the plane of the cross section), and boundaries of the frontal zone (thick solid lines). (c) Same cross section, but analyzed by Sanders (1955, his Fig. 10). Horizontal divergence (thin solid lines, labeled in 10⁻⁵ s⁻¹), vertical velocity (dashed lines every 5 cm s⁻¹), and boundaries of frontal zone (thick solid lines).

ings show the existence of this less-stable layer (e.g., Figs. 5 and 6b,c), which can be traced back in time to the higher elevation in the southwestern United States (not shown). Horizontal analyses of stability from the model (as measured by the difference of potential temperature over the 600–800-hPa layer) also suggested that this layer originated over the high terrain farther west during the previous day's diurnal heating (Fig. 6); in other words, this layer was an elevated mixed layer (e.g., Carlson et al. 1983; Benjamin and Carlson 1986; Lanicci and Warner 1991). Thus, these observations support the interpretation of the eastward movement of the elevated mixed layer as a real feature that might have reached near Sanders' cross section DN (Fig. 4b), had the 0300 UTC 18 April Fort Worth sounding been available to him.

The sensitivity of this elevated mixed layer and the location of the cross section selected by Sanders to the resulting structure of the cold front is illustrated with two cross sections on either side of DN. (Locations of cross sections are shown in Fig. 6c.) Whereas DN partly straddles the leading edge of the elevated mixed layer, cross section DZ cuts through the elevated mixed layer and cross section XY lies east of the elevated mixed layer. Cross section DZ shows the steepness of the isentropes at the leading edge of the front (Fig. 7a). Following the strong vertical motion above the leading edge of the front in the lower and midtroposphere, there is a narrow region of strong descent in the mid- and upper troposphere (Fig. 7a). This descent is reminiscent of narrow descent plumes above other cold fronts east of the Rocky Mountains described by Ralph et al. (1999).

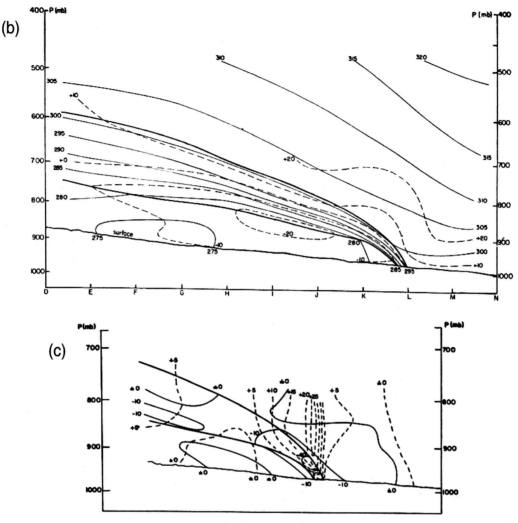

FIG. 4. (Continued)

In contrast, cross section XY, just slightly east of DN, did not have the elevated mixed layer overhead (Fig. 7b). The front possessed a shallow vertical motion plume of 0.15–0.20 m s⁻¹ near 900 hPa, more akin to that analyzed by Sanders (Fig. 7b). This cross section with vertical isentropes at the leading edge of a shallow front is reminiscent of that of Bond and Fleagle (1985), Browning (1990), and Bond and Shapiro (1991), where near-vertical isentropes at the leading edge of a front support relatively strong vertical motions in an environment that is otherwise statically stable. Thus, these three cross sections (DN, DZ, and XY) show the difference that the elevated mixed layer can have on the structure of the cold front and the depth and intensity of the vertical motions.

These aspects of the simulated cold front (steeply sloping isentropes, narrow and strong vertical motion plume, and midtropospheric cold dome) are remarkably similar to those of observations (e.g., Clarke 1961; Garratt 1988; Ryan et al. 1989; May et al. 1990; Hanstrum et al. 1990; Smith et al. 1995; Deslandes et al. 1999) and idealized model simulations (e.g., Reeder 1986; Physick 1988; Reeder et al. 1991; Tory and Reeder 2005; Reeder and Tory 2005) of Australian cold fronts moving into the deep well-mixed surface boundary layer over central Australia. This well-mixed layer, as deep as 3 km, forms because of intense surface heating over the desert. Reeder (1986) found that the presence of the prefrontal well-mixed layer increased the vertical motion at the leading edge of the cold front more than 3 times over that with a more stable prefrontal environment. This result is also consistent with Sawyer–Eliassen semigeostrophic frontal theory (e.g., Eliassen 1990) that predicts stronger and narrower vertical motions in a less stable environment for a given frontogenetic forcing (e.g., Hakim and Keyser 2001). As observational proof of this effect, Keyser and Carlson (1984) computed the secondary circulation at the leading edge of an elevated mixed layer in the presence of a midtropospheric front over the central United States.

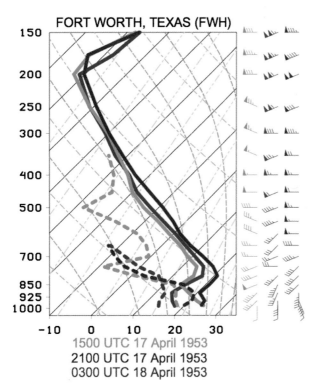

FORT WORTH, TEXAS (FWH)

1500 UTC 17 April 1953
2100 UTC 17 April 1953
0300 UTC 18 April 1953

FIG. 5. Skew T–log$p$ plot of temperature (°C, thick solid lines), dewpoint (°C, thick dashed lines), and wind (one pennant, full barb, and half barb denote 25, 5, and 2.5 m s^{-1}, respectively) profiles at FWH at 1500 UTC 17 Apr (green), 2100 UTC 17 Apr (blue), and 0300 UTC 18 Apr 1953 (pink). Location of FWH is displayed in Fig. 1e.

Finally, another difference between the simulated and analyzed fronts can be seen in the low-level postfrontal cold air. Whereas the postfrontal air in the simulation was slightly superadiabatic near the surface (Figs. 4a and 7), Sanders' analysis showed this layer to be well mixed (Fig. 4b). Observed postfrontal soundings (e.g., Oklahoma City at 0300 UTC 18 April, not shown) all showed near-neutral surface layers, not superadiabatic layers. More on the structure of the postfrontal boundary layer is discussed in section 4.

e. Frontogenesis cross sections

One of the important results from Sanders (1955) was the role of the deformation and tilting terms to frontogenesis. Both the model simulation and Sanders' analysis showed deformation frontogenesis concentrated within the frontal zone with the maximum value near the surface where the horizontal temperature gradient and wind shift were strongest (cf. Figs. 8a,b). The steeply tilting isentropes above the front produced a vertically oriented maximum of deformation frontogenesis that was not present in Sanders' analysis. The model simulation showed relatively weak tilting frontogenesis throughout the cross section with a frontogenesis max-

imum right above the leading edge (Fig. 8c), whereas, in Sanders' analysis, tilting was weakly positive on the warm side of the updraft plume and strongly negative within the frontal zone (Fig. 8d). Consequently, above the leading edge of the front, the deformation term dominated the total frontogenesis in the model simulation (Fig. 8e) much more strongly than in Sanders' analysis (Fig. 8f). In contrast, the tilting term in the model simulation did not play a substantial role toward frontolysis within the frontal zone (Fig. 8e), although it was frontogenetical at the leading edge of the front in Sanders' analysis (Fig. 8f). These results were not unique to this particular cross section from the model simulation; many other cross sections chosen (not shown) showed the relatively weak tilting frontogenesis relative to the deformation frontogenesis. Importantly, these results differ from previously published studies that show the dominance of the tilting term above the surface in cold fronts (e.g., Bond and Fleagle 1985; Koch et al. 1995; Schultz and Steenburgh 1999; Locatelli et al. 2002a; Colle 2003), but agree with results published by others (e.g., Pagowski and Taylor 1998; Tory and Reeder 2005; Reeder and Tory 2005).

The difference lies in the details of the isentropes within the updraft plume. This result is not surprising, because the tilting term is proportional to $\partial\theta/\partial z$, a term that is near zero in the elevated well-mixed layer. Because the isentropes at the leading edge of the front were more vertical in the model simulation (Fig. 8c) and more horizontal in Sanders' analysis (Fig. 4b), the effect of the horizontal gradients of vertical motion on these differing thermal fields resulted in very different fields for tilting frontogenesis. Latent heat release in the updraft may offset the tilting frontogenesis (e.g., Rao 1966; Palmén and Newton 1969, p. 261; Bond and Fleagle 1985; Orlanski et al. 1985; Koch et al. 1995; Bryan and Fritsch 2000; Locatelli et al. 2002a; Colle 2003), but this effect was not occurring in the model simulation because the updraft never became saturated and clouds did not form.

4. Development of the well-mixed postfrontal air

In his review of surface cold fronts, Keyser (1986) noted that the postfrontal boundary layer of cold fronts is characteristically well mixed or slightly unstable. Schultz (2008) discussed two possible explanations for this boundary layer stratification. The first explanation is that fluxes from the ground yielded this well-mixed postfrontal environment (Sanders 1955; Clarke 1961). Subsequent circulations in the planetary boundary layer were then essential for transporting this heat vertically (e.g., Fleagle et al. 1988). The second explanation was first proposed by Brundidge (1965) and later was demonstrated using idealized simulations of cold fronts by Keyser and Anthes (1982) and Xu and Gu (2002, 104–105). These authors showed that cold advection, in conjunction with a no-slip lower boundary condition, resulted in near-surface warm prefrontal air passing into the frontal zone as cold air is transported overtop. In the presence of a thermally insulating lower boundary where surface heat fluxes were zero, superadiabatic lapse rates in the postfrontal air resulted. Upward turbulent heat transport then produced the postfrontal neutral stratification. This entrainment mechanism likely operates in general, whereas the postfrontal flux mechanism becomes important under conditions of strong surface heat fluxes.

In an attempt to determine the relative importance of these effects for the simulation of the Sanders (1955) cold front, a companion simulation was performed where the surface fluxes were set to zero (Fig. 9). Because the fluxes were turned off, the position of the cyclone was slightly different (farther south) than in the control simulation (not shown) and the wind shift was not coincident with the temperature gradient at the leading edge of the front (Fig. 9). Nevertheless, in the no-flux simulation, we focus on the stability of the near-surface postfrontal air where the isentropes were vertical (Fig. 9), implying well-mixed conditions. Because these well-mixed conditions formed in the absence of surface fluxes, the well-mixed postfrontal conditions must have been attributable to entrainment within the frontal zone at the lower boundary, as described by Keyser and Anthes (1982) and Xu and Gu (2002). The inclusion of the surface fluxes in the control simulation resulted in a slight superadiabatic layer (Fig. 4a) where sensible heat flux from the ground to the surface layer of the atmosphere occurred faster than the model's boundary layer parameterization was able to transfer this heat vertically.

One possible reason for the inability of the model to reproduce the observed postfrontal boundary layer is inadequate vertical resolution in the boundary layer. To address the effect that vertical resolution would have on the simulation of the cold front, we performed a new simulation with nonzero surface fluxes and higher vertical resolution (36 levels, compared to the original 23 levels). In the new simulation, 23 of the 36 levels were below $\sigma = 0.7$ (cf. the original simulation where 10 of the 23 levels were below $\sigma = 0.7$). The 36 σ levels for that simulation are listed in Table 1. There was not a significant difference in the postfrontal boundary layer structure, nor the structure of the cold front, in general, between the 23- and 36-level simulations with nonzero surface fluxes. Thus, the number of model vertical levels within the well-mixed layer was sufficient to transport the heat vertically. Consequently, the inability of the model to create the well-mixed layer is likely due to limitations of the Blackadar scheme for this situation.

5. Conclusions

Over fifty years have passed since Sanders (1955) was published. In this chapter, we bring modern methods of numerical weather predication and diagnosis to bear

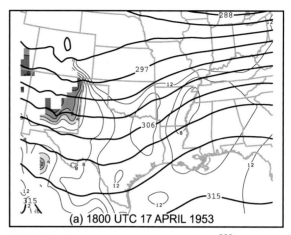

(a) 1800 UTC 17 APRIL 1953

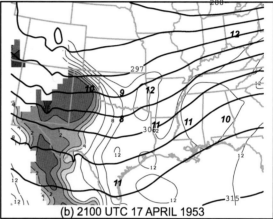

(b) 2100 UTC 17 APRIL 1953

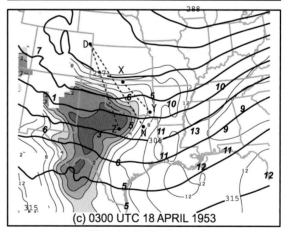

(c) 0300 UTC 18 APRIL 1953

700-hPa HEIGHT (dam)
600–800-hPa POTENTIAL TEMPERATURE DIFFERENCE
(K, shaded < 4 K, contoured < 12 K)

FIG. 6. Midtropospheric static stability from a portion of the outermost domain D1 of model simulation: difference in potential temperature between 600 and 800 hPa (thin solid lines for values 0, 1, 2, 3, 4, 6, 8, 10, and 12 K, shaded for values less than 4 K) and 700-hPa geopotential height (thick solid lines every 3 dam). (b), (c) Boldface italic numbers represent observed values for values less than 14 K. (c) Locations of cross sections DN (Figs. 4 and 9), DZ (Fig. 7a), and XY (Fig. 7b). (a) 1800 UTC 17 Apr, (b) 2100 UTC 17 Apr, and (c) 0300 UTC 18 Apr 1953.

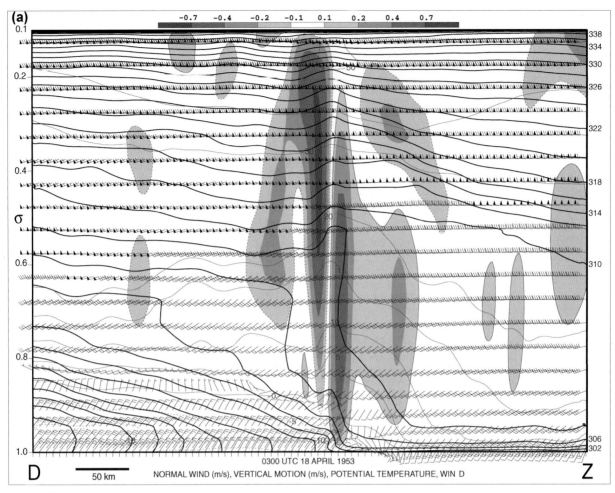

FIG. 7. (a) Same as in Fig. 4a, but for cross section DZ (location in Fig. 6c). (b) Same as in Fig. 4a, but for cross section XY (location in Fig. 6c). Note change in scale of vertical motion.

upon the cold front of 17–18 April 1953. This study is an example of historical weather events that have been simulated using a mesoscale numerical model to raise new issues through a combination of observed data and modeling (e.g., Anthes 1990; Schultz 2001; Locatelli et al. 2002b, 2005). In this way, we use past weather events, together with the present generation of mesoscale numerical models, to look for future research directions based on unresolved issues. To summarize our results, we follow the format of Keyser (1986) and Schultz (2008), who discussed the characteristics of classical cold fronts. These include the following:

Fronts are strongest at the surface and weaken rapidly with altitude. Because the model simulated a prefrontal near-surface stable layer, the strongest part of the front was right above this stable layer, not at the surface. This structure resembled that of other published results (e.g., Brundidge 1965; Neiman et al. 1991). Because some sounding data were not available for Sanders' analysis, it is possible that he might have analyzed this same feature. Nevertheless, above this stable layer, Sanders' conclusion that this cold front weakened above

the surface stable layer was valid in the model simulation.

A narrow plume of rising warm air exists above the surface frontal position. The narrow plume of ascent was reproduced by the model simulation, capturing the magnitude of about 0.25 m s^{-1} analyzed by Sanders (1955) and its height (within the lowest 100 hPa of the surface), *if* the cross section was shifted about 150 km east. Otherwise, an elevated mixed layer would be transported from over the high terrain in the west and into the plane of Sanders' cross section in the model. This very low stability air was associated with ascent 400-hPa deep and a maximum exceeding 0.7 m s^{-1}.

The frontal zone, a region of statically stable stratification, tilts rearward over the colder postfrontal air. Because of the adiabatic cooling in the ascent plume and the elevated mixed layer, the leading edge of the cold advection associated with the cold front was vertical within the lowest kilometer of the earth's surface in the model simulation (not shown). Sanders' (1955) analysis did not show this feature. The limited number of soundings used to construct his vertical cross section

(b)

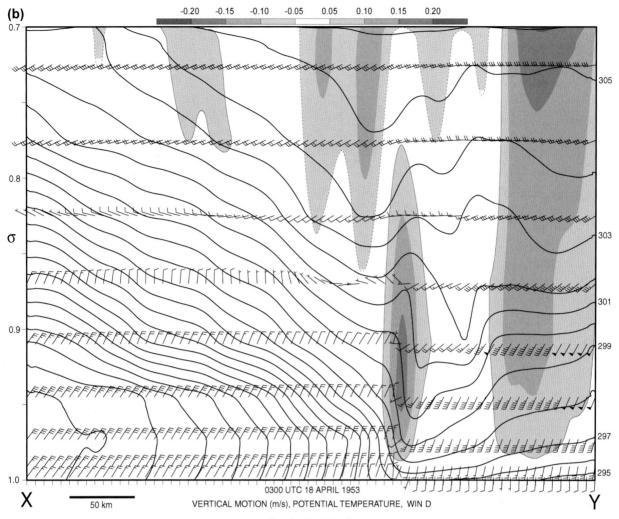

FIG. 7. (*Continued*)

may not have had sufficient horizontal resolution for Sanders to capture this small-scale (~100 km) feature. Alternatively, numerous observations and model simulations (e.g., Australian fronts) show steeply sloping leading edges to cold fronts. Perhaps this feature should be a more generally recognized feature of cold fronts.

Warm air is entrained into the frontal zone near the ground. The postfrontal boundary layer is well mixed or slightly unstable. Two complementary hypotheses exist to explain the stability of the postfrontal boundary layer. One relies on sensible heat flux from the warm ground to the cold postfrontal air. The other relies on the entrainment of near-surface prefrontal warm air into the lower frontal zone. In either case, subsequent vertical mixing redistributes the near-surface energy to produce a well-mixed or slightly unstable layer. One of the advantages of simulating the 17–18 April 1953 cold front is to evaluate and isolate processes like surface fluxes thought to be important through sensitivity experiments. When the surface fluxes were set to zero in an additional simulation, a well-mixed postfrontal

boundary layer still developed, suggesting that the entrainment mechanism (e.g., Brundidge 1965; Keyser and Anthes 1982; Xu and Gu 2002) was acting. To explore this process in more detail, however, would require a large-eddy simulation in a boundary layer model at very high resolution.

The prefrontal boundary layer is weakly stable. At 0300 UTC 18 April [the time of the cross section in Fig. 4 and that in Sanders (1955)], the model simulation showed a near-surface stable layer ahead of the front. However, as discussed by Schultz (2008), a variety of prefrontal boundary layer stabilities are possible.

The research described in this chapter also foreshadows later work by Sanders in two respects. First, the model simulation and a reanalysis of the observed data illustrated the presence of a prefrontal trough attributable to a cyclone departing the lee slopes of the Rockies south of the cold front. Later, Sanders (1967, 1983, 1999) presented case studies of prefrontal troughs associated with cold fronts over the central United States. The second foreshadowing was the advantage of using

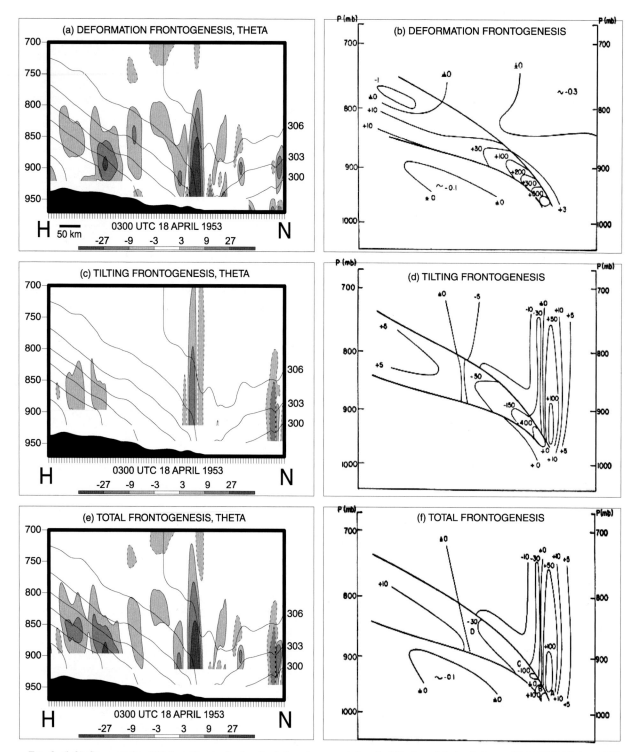

FIG. 8. (left) Cross section HN (see Fig. 1c for location) from innermost domain D3 of model simulation at 0300 UTC 18 Apr 1953: potential temperature (thin solid lines every 3 K): (a) deformation frontogenesis, (c) tilting frontogenesis, and (e) total frontogenesis of horizontal temperature gradient by three-dimensional wind. All frontogenesis fields are in units of °C (100 km)$^{-1}$ (3 h)$^{-1}$, shaded according to scale; positive (negative) values contoured in solid (dashed) lines. (right) Corresponding cross section HN analyzed by Sanders (1955), stretched to match the size and orientation of those from the model simulation: (b) deformation frontogenesis (his Fig. 11), (d) tilting frontogenesis (his Fig. 12), and (f) total frontogenesis of horizontal temperature gradient by three-dimensional wind (his Fig. 13). All frontogenesis fields are in units of °C (100 km)$^{-1}$ (3 h)$^{-1}$.

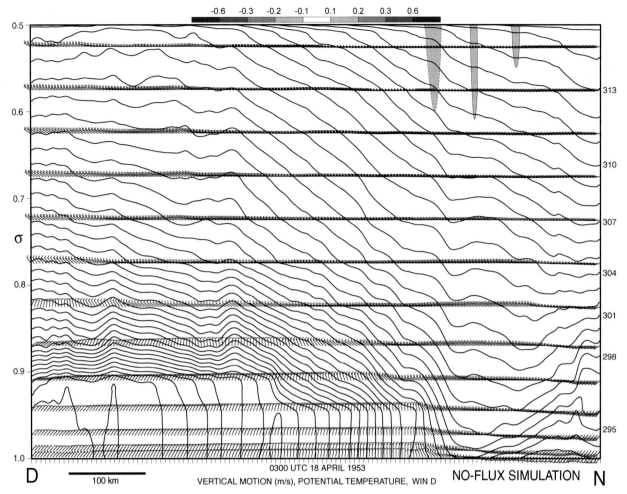

FIG. 9. Same as in Fig. 4a, but for simulation with no surface fluxes and no front-normal wind speed.

surface potential temperature to analyze fronts (e.g., Sanders and Doswell 1995; Sanders 1999; Sanders and Hoffman 2002).

In reexamining the surface and upper-air data that Sanders used to analyze the cold front of 17–18 April 1953, the challenges that he faced in constructing his analyses became very clear to us. Although soundings were available during this period every 6 h, the spatial resolution of the sounding network was no better than it is presently. Thus, the true magnitude and scale of small-scale features like the updraft plume remain unknown, let alone the magnitude and horizontal extent of larger-scale features for which verification of the model simulation was not fully possible (e.g., the elevated mixed layer in Figs. 6b,c). The model simulation examined in this chapter provides some context for the veracity of Sanders' conclusions, simulating the vertical motion at the leading edge of the front within a factor of 2, at worst, of that analyzed by Sanders. Although his conclusions have irrevocably shaped our perceptions of the structure and dynamics of cold fronts, this study presents some different interpretations of Sanders' anal-

ysis. Nevertheless, this study is not meant to criticize what was, at the time, a bold and revealing study. Because this high-resolution, four-dimensional dataset departs significantly in some important respects from the analysis by Sanders, the model simulation lends further insight into the structures and dynamics of cold fronts that remain to be incorporated into our conceptual models of cold fronts.

Acknowledgments. Discussions with and comments by the following individuals greatly enriched this study: Dan Keyser, Michael Reeder, Mark Stoelinga, and Qin Xu. Richard Johnson, Fred Sanders, and an anonymous reviewer provided comments on an earlier version of this chapter. David Ovens helped convert the MM5 output to GEMPAK. The NCDC Customer Service Branch and the National Severe Storms Laboratory supplied the observed data. Funding for Schultz was provided by NOAA/OAR/NSSL under NOAA–OU Cooperative Agreement NA17RJ1227. Funding for Roebber was provided by the National Science Foundation under Grant ATM-0106584.

REFERENCES

Anthes, R. A., 1990: Advances in the understanding and prediction of cyclone development with limited-area fine-mesh models. *Extratropical Cyclones, The Erik Palmén Memorial Volume,* C. W. Newton and E. O. Holopainen, Eds., Amer. Meteor. Soc., 221–253.

Benjamin, S. G., and T. N. Carlson, 1986: Some effects of surface heating and topography on the regional severe storm environment. Part I: Three-dimensional simulations. *Mon. Wea. Rev.,* **114,** 307–329.

Blackadar, A. K., 1979: High resolution models of the planetary boundary layer. *Advances in Environmental Science and Engineering,* J. R. Pfafflin and E. N. Ziegler, Eds., Vol. 1, Gordon and Breach Science Publishers, 50–85.

Bluestein, H. B., 1986: Fronts and jet streaks: A theoretical perspective. *Mesoscale Meteorology and Forecasting,* P. S. Ray, Ed., Amer. Meteor. Soc., 173–215.

——, 1993: *Synoptic-Dynamic Meteorology in Midlatitudes. Vol. II: Observations and Theory of Weather Systems.* Oxford University Press, 594 pp.

Bond, N. A., and R. G. Fleagle, 1985: Structure of a cold front over the ocean. *Quart. J. Roy. Meteor. Soc.,* **111,** 739–759.

——, and M. A. Shapiro, 1991: Research aircraft observations of the mesoscale and microscale structure of a cold front over the eastern Pacific Ocean. *Mon. Wea. Rev.,* **119,** 3080–3094.

Bosart, L. F., A. Wasula, W. Drag, and K. Meier, 2008: Strong surface fronts over sloping terrain and coastal plains. *Synoptic–Dynamic Meteorology and Weather Analysis and Forecasting: A Tribute to Fred Sanders, Meteor. Monogr.,* No. 55, Amer. Meteor. Soc.

Browning, K. A., 1990: Organization of clouds and precipitation in extratropical cyclones. *Extratropical Cyclones, The Erik Palmén Memorial Volume,* C. W. Newton and E. O. Holopainen, Eds., Amer. Meteor. Soc., 129–153.

Brundidge, K. C., 1965: The wind and temperature structure of nocturnal cold fronts in the first 1,420 feet. *Mon. Wea. Rev.,* **93,** 587–603.

Bryan, G. H., and J. M. Fritsch, 2000: Diabatically driven discrete propagation of surface fronts: A numerical analysis. *J. Atmos. Sci.,* **57,** 2061–2079.

Carlson, T. N., S. G. Benjamin, G. S. Forbes, and Y.-F. Li, 1983: Elevated mixed layers in the regional severe storm environment: Conceptual model and case studies. *Mon. Wea. Rev.,* **111,** 1453–1473.

Chen, C., C. H. Bishop, G. S. Lai, and W.-K. Tao, 1997: Numerical simulations of an observed narrow cold-frontal rainband. *Mon. Wea. Rev.,* **125,** 1027–1045.

Clarke, R. H., 1961: Mesostructure of dry cold fronts over featureless terrain. *J. Meteor.,* **18,** 715–735.

Colle, B. A., 2003: Numerical simulations of the extratropical transition of Floyd (1999): Structural evolution and responsible mechanisms for the heavy rainfall over the northeast United States. *Mon. Wea. Rev.,* **131,** 2905–2926.

Deslandes, R., M. J. Reeder, and G. Mills, 1999: Synoptic analyses of a subtropical cold front observed during the 1991 Central Australian Fronts Experiment. *Aust. Meteor. Mag.,* **48,** 87–110.

Dudhia, J., 1989: Numerical study of convection observed during the Winter Monsoon Experiment using a mesoscale two-dimensional model. *J. Atmos. Sci.,* **46,** 3077–3107.

——, 1993: A nonhydrostatic version of the Penn State–NCAR mesoscale model: Validation tests and simulation of an Atlantic cyclone and cold front. *Mon. Wea. Rev.,* **121,** 1493–1513.

——, 1996: A multi-layer soil temperature model for MM5. Preprints, *Sixth Annual PSU/NCAR Mesoscale Model Users' Workshop,* Boulder, CO, National Center for Atmospheric Research, 49–50.

Eliassen, A., 1990: Transverse circulations in frontal zones. *Extratropical Cyclones, The Erik Palmén Memorial Volume,* C. W. Newton and E. O. Holopainen, Eds., Amer. Meteor. Soc., 155–165.

Fleagle, R. G., N. A. Bond, and W. A. Nuss, 1988: Atmosphere-ocean interaction in mid-latitude storms. *Meteor. Atmos. Phys.,* **38,** 50–63.

Garratt, J. R., 1988: Summertime cold fronts in southeast Australia—Behavior and low-level structure of main frontal types. *Mon. Wea. Rev.,* **116,** 636–649.

Grell, G. A., J. Dudhia, and D. R. Stauffer, 1994: A description of the fifth-generation Penn State/NCAR Mesoscale Model (MM5). NCAR Tech. Note NCAR/TN-398+STR, 138 pp. [Available from NCAR, P.O. Box 3000, Boulder, CO 80307-3000.]

Hakim, G. J., and D. Keyser, 2001: Canonical frontal circulation patterns in terms of Green's functions for the Sawyer–Eliassen equation. *Quart. J. Roy. Meteor. Soc.,* **127,** 1795–1814.

Hanstrum, B. N., K. J. Wilson, and S. L. Barrell, 1990: Prefrontal troughs over Southern Australia. Part I: A climatology. *Wea. Forecasting,* **5,** 22–31.

Hutchinson, T. A., and H. B. Bluestein, 1998: Prefrontal wind-shift lines in the plains of the United States. *Mon. Wea. Rev.,* **126,** 141–166.

Kain, J. S., and J. M. Fritsch, 1993: Convective parameterization for mesoscale models: The Kain–Fritsch scheme. *The Representation of Cumulus Convection in Numerical Models, Meteor. Monogr.,* No. 24, Amer. Meteor. Soc., 165–170.

Kalnay, E., and Coauthors, 1996: The NCEP/NCAR 40-Year Reanalysis Project. *Bull. Amer. Meteor. Soc.,* **77,** 437–471.

Keshishian, L. G., and L. F. Bosart, 1987: A case study of extended East Coast frontogenesis. *Mon. Wea. Rev.,* **115,** 100–117.

Keyser, D., 1986: Atmospheric fronts: An observational perspective. *Mesoscale Meteorology and Forecasting,* P. S. Ray, Ed., Amer. Meteor. Soc., 216–258.

——, and R. A. Anthes, 1982: The influence of planetary boundary layer physics on frontal structure in the Hoskins-Bretherton horizontal shear model. *J. Atmos. Sci.,* **39,** 1783–1802.

——, and T. N. Carlson, 1984: Transverse ageostrophic circulations associated with elevated mixed layers. *Mon. Wea. Rev.,* **112,** 2465–2478.

——, M. J. Reeder, and R. J. Reed, 1988: A generalization of Petterssen's frontogenesis function and its relation to the forcing of vertical motion. *Mon. Wea. Rev.,* **116,** 762–780.

Koch, S. E., 1999: Comments on "Numerical simulations of an observed narrow cold-frontal rainband." *Mon. Wea. Rev.,* **127,** 252–257.

——, and P. J. Kocin, 1991: Frontal contraction processes leading to the formation of an intense narrow rainband. *Meteor. Atmos. Phys.,* **46,** 123–154.

——, J. T. McQueen, and V. M. Karyampudi, 1995: A numerical study of the effects of differential cloud cover on cold frontal structure and dynamics. *J. Atmos. Sci.,* **52,** 937–964.

Lanicci, J. M., and T. T. Warner, 1991: A synoptic climatology of the elevated mixed-layer inversion over the southern Great Plains in spring. Part I: Structure, dynamics, and seasonal evolution. *Wea. Forecasting,* **6,** 181–197.

Locatelli, J. D., M. T. Stoelinga, and P. V. Hobbs, 2002a: Organization and structure of clouds and precipitation on the mid-Atlantic coast of the United States. Part VII: Diagnosis of a nonconvective rainband associated with a cold front aloft. *Mon. Wea. Rev.,* **130,** 278–297.

——, ——, and ——, 2002b: A new look at the Super Outbreak of tornadoes on 3–4 April 1974. *Mon. Wea. Rev.,* **130,** 1633–1651.

——, ——, and ——, 2005: Re-examination of the split cold front in the British Isles cyclone of 17 July 1980. *Quart. J. Roy. Meteor. Soc.,* **131,** 3167–3181.

May, P. T., K. J. Wilson, and B. F. Ryan, 1990: VHF radar studies of cold fronts traversing southern Australia. *Beitr. Phys. Atmos.,* **63,** 257–269.

Mesinger, F., 1996: Forecasting cold surges east of the Rocky Mountains. Preprints, *15th Conf. on Weather Analysis and Forecasting,* Norfolk, VA, Amer. Meteor. Soc., 68–69.

Miller, J. E., 1948: On the concept of frontogenesis. *J. Meteor.,* **5,** 169–171.

Neiman, P. J., P. T. May, B. B. Stankov, and M. A. Shapiro, 1991: Radio acoustic sounding system observations of an arctic front. *J. Appl. Meteor.,* **30,** 881–892.

Orlanski, I., B. Ross, L. Polinsky, and R. Shaginaw, 1985: Advances in the theory of atmospheric fronts. *Advances in Geophysics,* Vol. 28B, Academic Press, 223–252.

Pagowski, M., and P. A. Taylor, 1998: Fronts and the boundary layer—Some numerical studies. *Bound.-Layer Meteor.,* **89,** 469–506.

Palmén, E., and C. W. Newton, 1969: *Atmospheric Circulation Systems.* Academic Press, 603 pp.

Petterssen, S., 1936: Contribution to the theory of frontogenesis. *Geophys. Publ.,* **11** (6), 1–27.

Physick, W. L., 1988: Mesoscale modeling of a cold front and its interaction with a diurnally heated land mass. *J. Atmos. Sci.,* **45,** 3169–3187.

Ralph, F. M., P. J. Neiman, and T. L. Keller, 1999: Deep-tropospheric gravity waves created by leeside cold fronts. *J. Atmos. Sci.,* **56,** 2986–3009.

Rao, G. V., 1966: On the influences of fields of motion, baroclinicity, and latent heat source on frontogenesis. *J. Appl. Meteor.,* **5,** 377–387.

Reeder, M. J., 1986: The interaction of a surface cold front with a prefrontal thermodynamically well-mixed boundary layer. *Aust. Meteor. Mag.,* **34,** 137–148.

——, and K. J. Tory, 2005: The effect of the continental boundary layer on the dynamics of fronts in a 2D model of baroclinic instability. II: Surface heating and cooling. *Quart. J. Roy. Meteor. Soc.,* **131,** 2409–2429.

——, D. Keyser, and B. D. Schmidt, 1991: Three-dimensional baroclinic instability and summertime frontogenesis in the Australian region. *Quart. J. Roy. Meteor. Soc.,* **117,** 1–28.

Reisner, J., R. M. Rasmussen, and R. T. Bruintjes, 1998: Explicit forecasting of supercooled liquid water in winter storms using the MM5 mesoscale model. *Quart. J. Roy. Meteor. Soc.,* **124,** 1071–1107.

Roebber, P. J., and M. G. Gehring, 2000: Real-time prediction of the lake breeze on the western shore of Lake Michigan. *Wea. Forecasting,* **15,** 298–312.

——, D. M. Schultz, and R. Romero, 2002: Synoptic regulation of the 3 May 1999 tornado outbreak. *Wea. Forecasting,* **17,** 399–429.

Ryan, B. F., K. J. Wilson, and E. J. Zipser, 1989: Modification of the thermodynamic structure of the lower troposphere by the evaporation of precipitation ahead of a cold front. *Mon. Wea. Rev.,* **117,** 138–153.

Sanders, F., 1955: An investigation of the structure and dynamics of an intense surface frontal zone. *J. Meteor.,* **12,** 542–552.

——, 1967: Frontal structure and the dynamics of frontogenesis. Final Report to the National Science Foundation, Grant GP-1508, 10 pp. + 10 appendixes.

——, 1983: Observations of fronts. *Mesoscale Meteorology—Theories, Observations, and Models,* D. K. Lilly and T. Gal-Chen, Eds., Reidel, 175–203.

——, 1999: A proposed method of surface map analysis. *Mon. Wea. Rev.,* **127,** 945–955.

——, and C. A. Doswell III, 1995: A case for detailed surface analysis. *Bull. Amer. Meteor. Soc.,* **76,** 505–521.

——, and E. G. Hoffman, 2002: A climatology of surface baroclinic zones. *Wea. Forecasting,* **17,** 774–782.

Schultz, D. M., 2001: Reexamining the cold conveyor belt. *Mon. Wea. Rev.,* **129,** 2205–2225.

——, 2004: Cold fronts with and without prefrontal wind shifts in the central United States. *Mon. Wea. Rev.,* **132,** 2040–2053.

——, 2005: A review of cold fronts with prefrontal troughs and wind shifts. *Mon. Wea. Rev.,* **133,** 2449–2472.

——, 2008: Perspectives on Fred Sanders' research on cold fronts. *Synoptic–Dynamic Meteorology and Weather Analysis and Forecasting: A Tribute to Fred Sanders, Meteor. Monogr.,* No. 55, Amer. Meteor. Soc.

——, and W. J. Steenburgh, 1999: The formation of a forward-tilting cold front with multiple cloud bands during Superstorm 1993. *Mon. Wea. Rev.,* **127,** 1108–1124.

Smith, R. K., M. J. Reeder, N. J. Tapper, and D. R. Christie, 1995: Central Australian cold fronts. *Mon. Wea. Rev.,* **123,** 16–38.

Stauffer, D. R., and N. L. Seaman, 1990: Use of four-dimensional data assimilation in a limited-area mesoscale model. Part I: Experiments with synoptic-scale data. *Mon. Wea. Rev.,* **118,** 1250–1277.

Tory, K. J., and M. J. Reeder, 2005: The effect of the continental boundary layer on the dynamics of fronts in a 2D model of baroclinic instability. I: An insulated lower surface. *Quart. J. Roy. Meteor. Soc.,* **131,** 2389–2408.

Xu, Q., and W. Gu, 2002: Semigeostrophic frontal boundary layer. *Bound.-Layer Meteor.,* **104,** 99–110.

Zhang, D., and R. A. Anthes, 1982: A high-resolution model of the planetary boundary layer—Sensitivity tests and comparisons with SESAME-79 data. *J. Appl. Meteor.,* **21,** 1594–1609.

——, H.-R. Chang, N. L. Seaman, T. T. Warner, and J. M. Fritsch, 1986: A two-way interactive nesting procedure with variable terrain resolution. *Mon. Wea. Rev.,* **114,** 1330–1339.

Part II
Analysis and Diagnosis

Chapter 7

Ensemble Synoptic Analysis

GREGORY J. HAKIM AND RYAN D. TORN

University of Washington, Seattle, Washington

(Manuscript received 9 September 2004, in final form 17 May 2005)

Hakim   **Torn**

ABSTRACT

Synoptic and mesoscale meteorology underwent a revolution in the 1940s and 1950s with the widespread deployment of novel weather observations, such as the radiosonde network and the advent of weather radar. These observations provoked a rapid increase in our understanding of the structure and dynamics of the atmosphere by pioneering analysts such as Fred Sanders. The authors argue that we may be approaching an analogous revolution in our ability to study the structure and dynamics of atmospheric phenomena with the advent of probabilistic objective analyses. These probabilistic analyses provide not only best estimates of the state of the atmosphere (e.g., the expected value) and the uncertainty about this state (e.g., the variance), but also the relationships between all locations and all variables at that instant in time. Up until now, these relationships have been determined by sampling in time by, for example, case studies, composites, and time-series analysis. Here the authors propose a new approach, ensemble synoptic analysis, which exploits the information contained in probabilistic samples of analyses at one or more instants in time.

One source of probabilistic analyses is ensemble-based state-estimation methods, such as ensemble-based Kalman filters. Analyses from such a filter may be used to study atmospheric phenomena and the relationships between fields and locations at one or more instants in time. After a brief overview of a research-based ensemble Kalman filter, illustrative examples of ensemble synoptic analysis are given for an extratropical cyclone, including relationships between the cyclone minimum sea level pressure and other synoptic features, statistically determined operators for potential-vorticity inversion, and ensemble-based sensitivity analysis.

1. Introduction

Progress in synoptic and dynamic meteorology is often marked by the advent of new or improved observing systems and novel methods for analyzing and understanding these data. For example, deployment of the

Corresponding author address: Gregory J. Hakim, Department of Atmospheric Sciences, Box 351640, University of Washington, Seattle, WA 98195-1640.
E-mail: hakim@atmos.washington.edu

routine radiosonde network and the emergence of quasigeostrophic (QG) theory provided an opportunity for analysts like Fred Sanders to both better document and understand synoptic-scale weather systems during the latter half of the twentieth century. We propose that emerging techniques in state estimation ("data assimilation") may offer new opportunities to analyze and understand atmospheric phenomena. We group these opportunities under the title "ensemble synoptic analysis" (ESA), which derives from the fact that the methods pertain to ensembles of analyses valid at an instant (or multiple instants) in time. Our goal is to outline the analysis techniques available to ESA and to demonstrate these techniques through illustrative examples.

In deriving basic understanding of atmospheric phenomena, the analysis often revolves around discovering and exploiting relationships between fields and between locations; for example, for extratropical cyclones, geostrophic balance relates wind and pressure, and tropopause disturbances make important contributions to surface development. Up until now, these relationships have been determined by sampling methods involving long periods of time, such as multiple case studies, composites, and time series analysis. A central attribute of ESA that distinguishes it from these other methods is that it uses a probabilistic estimate for the analysis, rather than a single deterministic analysis. Specifically, an ensemble approach is used to generate a sample of analyses valid at an instant (or multiple instants) in time. Although not yet available from operational centers, these probabilistic analyses may reach operational deployment in the future. (A pseudo-operational system has been available online at www.atmos.washington. edu/~enkf/ since December 2004.)

Current three-dimensional variational data assimilation (e.g., 3DVAR) systems rely upon knowledge of established dynamical relationships, such as hydrostatic and geostrophic balance, to specify covariance relationships and to determine the state (analysis). Here we reverse the process so that *data assimilation is used as a tool for revealing dynamical relationships*. The technique used to generate probabilistic analyses is an ensemble Kalman filter (EnKF), which we will describe in more detail in section 2. This technique applies an ensemble of nonlinear forecasts to approximate the extended Kalman filter, which for Gaussian errors and linear operators provides the maximum likelihood state that also has minimum error variance. The EnKF provides the best estimate of the state of the atmosphere (ensemble mean), state error (ensemble variance), and the relationships between all locations and all variables (ensemble covariance).

Although it is tempting to perform ESA on operational ensemble forecasts, the forecast lead times must be chosen such that the memory of the initial perturbations are lost, since the analysis ensembles are currently specified by ad hoc methods that are not designed to sample the probability distribution of the analysis (e.g., total-energy singular vectors and bred-grown modes). As such, the initial ensemble covariance may be inappropriate for use in ESA; therefore, we apply an EnKF to determine ensemble analyses and forecasts.

After a brief overview of the EnKF in section 2, ESA is defined in section 3. Ensemble analyses and select covariance relationships are discussed for an extratropical cyclone in section 4. Section 5 is devoted to using ESA for potential vorticity (PV) inversion, and section 6 is devoted to using ESA for sensitivity analysis. A summary is provided in section 7.

2. Ensemble Kalman filters

Modern state estimation involves the synthesis of observations and a model's estimate of these observations by appropriately weighting these two pieces of information. This weighting depends on the error associated with the observations relative to the model estimate of the observations. Current operational state estimation systems are deterministic, and as such they do not provide the probabilistic, flow-dependent data needed for ESA. Such probabilistic analyses are now being explored by the research community, including those generated by EnKFs (e.g., Evensen 1994; Houtekamer and Mitchell 1998; Hamill and Snyder 2000). A brief overview of state estimation and the EnKF is provided here; the interested reader may find more background information in Daley (1993), Kalnay (2002), and Hamill (2006).

In mathematical terms, the maximum likelihood analysis, assuming Gaussian statistics, is determined by

$$\mathbf{x}^a = \mathbf{x}^b + \mathbf{K}[\mathbf{y}^0 - \mathcal{H}(\mathbf{x}^b)], \qquad (1)$$

where $\mathbf{x}^a$ is the analysis state vector,[1] $\mathbf{x}^b$ is the background state vector, $\mathbf{y}^0$ is the observation vector, and $\mathcal{H}(\mathbf{x}^b)$ is a vector-valued function that returns a column vector of observations, as estimated by the background state. Normally the background is given by a short-term model forecast, and therefore $\mathcal{H}(\mathbf{x}^b)$ is the model's estimate of the observations. This operation can be as simple as linear interpolation from model grid points to observation locations, but it may also be a complicated nonlinear function of the state, such as a radar equation.

The Kalman gain matrix is $\mathbf{K}$, which is given by

$$\mathbf{K} = \mathbf{P}^b\mathbf{H}^\mathrm{T}(\mathbf{H}\mathbf{P}^b\mathbf{H}^\mathrm{T} + \mathbf{R})^{-1}, \qquad (2)$$

where $\mathbf{P}^b$ is the background-error covariance matrix, $\mathbf{R}$ is the observation-error covariance matrix, and superscript T denotes the matrix transpose. The diagonal elements of $\mathbf{P}^b$ and $\mathbf{R}$ contain the error variance in the background and observations, respectively, and the off-diagonal elements indicate the covariance relationships between model state variables and observations, respectively. Matrix $\mathbf{H}$ is the linearization of $\mathcal{H}$ about the background state. The error covariance matrix for the background estimate of the observations is $\mathbf{H}\mathbf{P}^b\mathbf{H}^\mathrm{T}$ (i.e., the same as $\mathbf{R}$ except that it applies to the model's estimate of the observations). Essentially, $\mathbf{K}$ determines the weight given to the new observational information [i.e., the innovation $\mathbf{y}^0 - \mathcal{H}(\mathbf{x}^b)$] relative to the background estimate. Considering just a single observation, so that $\mathbf{R}$ and $\mathbf{H}\mathbf{P}^b\mathbf{H}^\mathrm{T}$ are scalars, observations with large errors relative to the background ($\mathbf{R} \gg \mathbf{H}\mathbf{P}^b\mathbf{H}^\mathrm{T}$) have

---

[1] The vectors $\mathbf{x}^a$ and $\mathbf{x}^b$ are constructed from the multivariate three-dimensional arrays of model gridpoint values by packing all gridpoint values into a one-dimensional array (a column vector) following a chosen procedure. The details of the procedure may be chosen arbitrarily, provided that it is invertible; the three-dimensional arrays must be recoverable from the vector values.

small $\mathbf{K}$; therefore, the analysis is weighted toward the background estimate.

The crucial difference between the EnKF and 3DVAR involves the assumptions for $\mathbf{P}^b$. For 3DVAR, $\mathbf{P}^b$ is typically fixed to assumed relationships (e.g., background errors are approximately time independent and spatially homogeneous and isotropic), whereas for the EnKF, $\mathbf{P}^b$ depends on time and space and is estimated with an ensemble of nonlinear forecasts by

$$\mathbf{P}^b = \frac{1}{M-1}\mathbf{X}^{b\prime}\mathbf{X}^{b\prime\mathrm{T}}. \qquad (3)$$

Here $\mathbf{X}^{b\prime}$ is a matrix containing an ensemble background state estimate in M column vectors, with one ensemble member in each column vector; the superscript prime notation indicates that the ensemble-mean state has been removed from each ensemble member. Because $\mathbf{P}^b$ is flow dependent, the influence of an observation is expected to reduce analysis error more than the 3DVAR flow-independent $\mathbf{P}^b$. The EnKF comes at the added expense of having to integrate the full nonlinear forecast model M times, rather than once, as for 3DVAR. These M forecasts are then all updated with new observations using an ensemble square root filter as described in Whitaker and Hamill (2002). This analysis ensemble is then immediately available for initializing a fresh ensemble forecast and another assimilation step. Note that there is no need to generate synthetic perturbations around a single deterministic analysis as is the current procedure for populating ensembles at operational centers.[2]

Our implementation of an EnKF at the University of Washington utilizes the Weather Research and Forecasting model, version 2.0.2 (WRF; Michalakes et al. 2001), in a perfect model scenario. A "truth" integration is performed first, which is then sampled to generate observations that are assimilated with the EnKF to produce analysis ensembles. Each of the 100 ensemble members utilizes the same model configuration: ~100-km horizontal grid spacing on a 90 × 90 grid, with 28 vertical levels. Model physical parameterizations include warm-rain microphysics, the Medium-Range Forecast (MRF) planetary boundary layer scheme (Hong and Pan 1996), and the convective parameterization scheme of Janjic (1994). Observations consist of 250 randomly spaced surface pressure observations sampled from the truth run, which employs Global Forecast System (GFS) analysis lateral boundary conditions. The uniform observing system used here is, of course, unrealistic but has the advantage of being straightforward to implement and allows for unambiguously defined analysis errors. Surface pressure observations have been shown to effectively constrain tropospheric anal-

---

[2] As a technical matter, an initial ensemble is required to start an EnKF, although the specific choice becomes unimportant after the filter has cycled through several assimilation steps. Here we populate the initial ensemble with forecasts of different lead time verifying at the same time (0000 UTC 24 March 2003).

ysis errors on the synoptic scale (Whitaker et al. 2004) and are useful for quickly generating the ensemble analyses needed for ESA. Although the results depend on this choice of observation network, our main purpose here is to illustrate ESA techniques on a suitable dataset. The filter is initialized at 0000 UTC 24 March 2003, about 4 days prior to the time of interest; the analysis error statistics come into equilibrium with observation errors within about 36 h (i.e., by about 1200 UTC 25 March 2003).

3. Ensemble synoptic analysis

ESA is explored here using *linear* relationships and Gaussian statistics. ESA is not necessarily limited by these assumptions, but it seems logical to explore these prior to more complicated nonlinear relationships and non-Gaussian statistics. Given two multivariate ensemble samples of data that have had the ensemble mean removed (indicated by superscripted primes), the individual ensemble members are stored in column vectors of the matrices $\mathbf{X}'$ and $\mathbf{Y}'$, which are size $n \times M$ and $N \times M$, respectively. Each of the M columns represents an ensemble member, which has $n - 1$ and $N - 1$ degrees of freedom for $\mathbf{X}'$ and $\mathbf{Y}'$, respectively. A specific example to be considered more thoroughly in section 5 applies to potential vorticity inversion, where $\mathbf{X}'$ represents the potential vorticity at n grid points, and $\mathbf{Y}'$ represents the state variables that are recovered from the PV inversion (e.g., u, v, and T at all the grids points gives $N = 3n$).

A linear relationship between $\mathbf{X}'$ and $\mathbf{Y}'$ may be expressed as

$$\mathbf{Y}' = \mathbf{L}\mathbf{X}', \qquad (4)$$

where $\mathbf{L}$ is an $N \times n$ linear operator that maps $\mathbf{X}'$ into $\mathbf{Y}'$; that is, the relationship between $\mathbf{X}'$ and $\mathbf{Y}'$. The goal here is to recover $\mathbf{L}$ given the samples for $\mathbf{X}'$ and $\mathbf{Y}'$, which is a standard problem in statistics. The solution

$$\mathbf{L} = \mathbf{Y}'\mathbf{X}'^{-1} \qquad (5)$$

assumes that $\mathbf{X}'$ is invertible. When $\mathbf{X}'$ is singular or not square, a pseudoinverse may still exist (e.g., Golub and Van Loan 1996, p. 257), which will be further explained below; hereafter, references to inverse matrices are understood to include the pseudoinverse. Prior to exploring the properties of this solution, we consider an alternative expression for (5). Right multiplying (4) by $\mathbf{X}^{\mathrm{T}}$ gives

$$\mathbf{Y}'\mathbf{X}'^{\mathrm{T}} = \mathbf{L}\mathbf{X}'\mathbf{X}'^{\mathrm{T}} \qquad (6)$$

or, equivalently,

$$\mathrm{cov}(\mathbf{Y}', \mathbf{X}') = \mathbf{L}\,\mathrm{cov}(\mathbf{X}', \mathbf{X}'). \qquad (7)$$

This reveals that

$$\mathbf{L} = \mathrm{cov}(\mathbf{Y}', \mathbf{X}')\mathrm{cov}(\mathbf{X}', \mathbf{X}')^{-1}, \qquad (8)$$

where cov is shorthand notation for a covariance matrix

of the indicated arguments. This expression shows that the operator **L** may be understood as a linear regression of **Y′** on **X′**. Although (8) provides a useful interpretation for **L**, it is not computationally efficient.

Returning to (5), a formidable calculation might be anticipated given the size of the state vector for the problems we consider ($N \sim 10^6$). A major simplification to this potentially large calculation is available because of the small size of the ensembles that we consider ($M \sim 100$). Using the singular value decomposition (SVD) for **X′** (e.g., Golub and Van Loan 1996),

$$\mathbf{X'} = \mathbf{USV}^\mathsf{T}, \tag{9}$$

which applies to any matrix; the inverse of **X′** is given by

$$\mathbf{X'}^{-1} = \mathbf{VS}^{-1}\mathbf{U}^\mathsf{T}. \tag{10}$$

Here we have used the fact that **U** and **V** are orthogonal matrices (transpose is the inverse), and **S** is diagonal. An even better expression for the inverse derives from using (9) to eliminate **U** in favor of **X′**,

$$\mathbf{X'}^{-1} = \mathbf{VS}^{-2}\mathbf{V}^\mathsf{T}\mathbf{X'}^\mathsf{T}, \tag{11}$$

so that from (5) we find

$$\mathbf{L} = \mathbf{Y'VS}^{-2}\mathbf{V}^\mathsf{T}\mathbf{X'}^\mathsf{T}. \tag{12}$$

Thus in order to determine **L** from the known data **X′** and **Y′**, only matrices **V** and **S** need to be calculated. From (9) we find that

$$\mathbf{X'}^\mathsf{T}\mathbf{X'} = \mathbf{VS}^2\mathbf{V}^\mathsf{T}, \tag{13}$$

so that **V** and **S** are the eigenvector and eigenvalue matrices, respectively, of $\mathbf{X'}^\mathsf{T}\mathbf{X'}$. Of greatest importance is that **V**, **S**, and $\mathbf{X'}^\mathsf{T}\mathbf{X'}$ are all $M \times M$, or in the present application, 100×100, so they are trivial to manipulate numerically. However, this simplification also reflects the fact that there are only 100 independent degrees of freedom available for the analysis, and this is one factor that will impose a limit on the insight that may be gained from ESA.

These ideas are illustrated for an extratropical cyclone using two examples: potential vorticity inversion and sensitivity analysis. Before discussing the results, it will prove helpful to first consider a brief overview of the case and of the ensemble-determined covariance relationships.

4. ESA applied to an extratropical cyclone

An extratropical cyclone is chosen for analysis because of its familiarity and the fact that Fred Sanders made important contributions to our understanding of these features (e.g., Sanders and Gyakum 1980; Sanders 1986, 1987). The (randomly) chosen case occurred on 28–29 March 2003. During this time, a cyclone moved from the central United States to Michigan as it deepened. An overview of the vertical structure of the system on 0600 UTC 29 March is provided in Fig. 1. The developing low is marked by well-defined warm and cold fronts that are characterized by abrupt horizontal potential temperature gradients and troughs in the surface pressure field (Fig. 1a). A broad and meridionally extensive 500-hPa trough is located to the west of the surface low, as may be expected for a developing surface cyclone (Fig. 1b). What is less clear is the role played by smaller-scale shortwave disturbances, which are more obvious in the 500-hPa PV distribution (Fig. 1b, red lines). As will be shown below, the southern disturbance is apparently important to the surface development of the low over Michigan despite its distant location. On the dynamic tropopause,[3] an elongated region of potentially cold air is located in the region of the 500-hPa trough, as well as a stronger gradient in potential temperature near Texas (Fig. 1c).

Figure 1 also introduces the first example regarding the importance of considering analyses probabilistically by showing all ensemble members (gray lines) in addition to the ensemble mean (solid black lines). All ensemble members must be regarded as equally likely realizations of the state, so that considering just one is potentially misleading (e.g., Leith 1974).

Having considered the ensemble distribution and mean value for selected parameters, we now explore aspects of the ensemble analysis covariance matrix, which is at the heart of ESA for Gaussian statistics. To simplify the presentation and to remain focused on the surface cyclone, covariance relationships are considered for a single grid point: the point of lowest surface pressure, which will be referred to as the cyclone central pressure. The sample of cyclone central pressure values is normalized by the ensemble standard deviation (2.7 hPa), so that the covariances carry the value of the covarying field. Defining this ensemble sample of M normalized surface cyclone central pressure values by the *row vector* **y′** and the full $N \times M$ ensemble state matrix by **X′**, the expression we evaluate is

$$\mathbf{c} = -\mathrm{cov}(\mathbf{X'}, \mathbf{y'}) = -\mathbf{X'y'}^\mathsf{T}. \tag{14}$$

The prime notation indicates that the ensemble mean has been removed, and the negative sign in (14) is applied to the covariances so that the fields reflect a deeper cyclone. Note that **c** is an $N \times 1$ column vector, with each value reflecting the linear relationship between the metric of interest (cyclone central pressure) and all other state variables. Figures 2–4 represent selected parts of vector **c**.

Figure 2 shows the covariance of the cyclone central pressure with the surface pressure and wind fields. When the surface cyclone central pressure is lower than the ensemble mean, as it is for some ensemble members, the pressure at nearby points is also lower; that is, these points covary strongly, as may be expected. The pressure covariance field also shows interesting patterns

---

[3] Defined here by the 1.5×10^{-6} m² K kg⁻¹ s⁻¹ (PVU) surface.

Surface Pressure & Temperature

500 hPa Height and Ertel PV

Tropopause θ & Wind

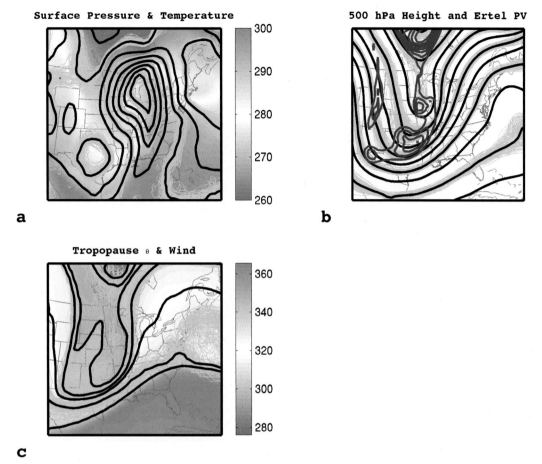

FIG. 1. Ensemble-mean fields at 0600 UTC 29 Mar 2003 of (a) surface pressure (black lines every 4 hPa) and potential temperature (colors); (b) 500-hPa geopotential height (black lines, every 60 m) and Ertel potential vorticity (red lines every 0.25 PVU, starting with 1 PVU); and (c) dynamic tropopause potential temperature (black lines every 10 K and colors). Gray lines show all 100 ensemble members at twice the contour interval of the ensemble-mean field.

away from the low center. Specifically, near the cold front there is a dipole with positive values west of the cold front and negative values to the east. This pattern suggests a phase shift in the location of the pressure trough associated with the cold front such that the surface cold front is displaced eastward when the surface low is deeper; this is intuitively appealing because a deeper low should be associated with stronger cyclonic circulation. The wind field covariance reflects this inference, with a local maximum in wind speed along the front. Note also that the wind vectors have a component directed toward lower pressure, qualitatively consistent with the effect of surface friction on geostrophic flow (e.g., Holton 2004).

At 500 hPa, the geopotential height field covariance with the cyclone central pressure shows that when the surface low is deeper by 2.7 hPa, there is a dipole response at 500 hPa, with lower heights upstream (-10 m) and higher heights downstream (10 m; Fig. 3). This pattern is appealing based on quasigeostrophic reasoning in that the upper-level wave is amplified above the surface low and the dominant wavelength shortened.

Interestingly, the 500-hPa trough located over Texas also covaries strongly with the surface low, suggesting that it also contributes to deepening the cyclone despite being located much farther away than the trough over Illinois. The wind field covariance shows clear qualitative evidence of balance, with the vectors following the height field covariance and the magnitude generally proportional to the geopotential height covariance gradient. Balance issues will be examined more closely in the next section on PV inversion.

Moving up to the tropopause (Fig. 4), the potential temperature field shows that where the 500-hPa heights are lower (higher) the tropopause is colder (warmer). A particularly interesting result is the band of warmer air along the subtropical jet stream that extends from Mexico to North Carolina. It is unclear why this region is dynamically related to the surface low over Michigan, although one possibility is that, on the large scale, the surface low is near the left exit region of a planetary-scale jet. The strong horizontal potential temperature gradients both south and north of the cyclone are en-

Covariance of Surface Low with Surface Pressure & Wind

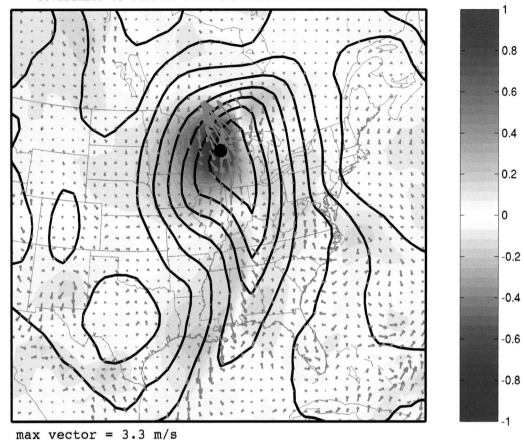

max vector = 3.3 m/s

FIG. 2. Ensemble-based covariance fields between the normalized cyclone central pressure and the surface pressure (colors, hPa) and wind (vectors) at 0600 UTC 29 Mar 2003. The ensemble-mean surface pressure is given in solid lines every 4 hPa, and the sample of cyclone central pressure values are normalized by the ensemble standard deviation. The location of lowest pressure in the surface cyclone position is denoted by the black dot.

hanced for a deeper cyclone, which suggests a greater tendency to form upper-level fronts.

To summarize this section, the covariance relationships are generally qualitatively consistent with expectations for an extratropical cyclone, although they also suggest some interesting relationships that may not be obvious using other analysis techniques, for example, the relative importance of the two upper-level disturbances to the surface low and the relationship between the surface low and the subtropical jet. We proceed to use these statistical relationships to perform a more complicated diagnostic: piecewise PV inversion.

5. Statistical potential vorticity inversion

Typically, PV inversion involves the specification of balance constraints between wind, mass, and temperature; boundary conditions; and numerical approximations (e.g., Hoskins et al. 1985; Davis and Emanuel 1991). Here we determine the inversion operator *statistically* using ESA methods outlined in section 3. In the earlier notation the linear operator $\mathbf{L}$ is estimated by

$$\mathbf{L} = \mathbf{X'P'}^{-1}, \qquad (15)$$

where $\mathbf{X'}$ is an ensemble state matrix, $\mathbf{P'}$ is an ensemble PV matrix, and the matrix inverse is calculated as described in section 3. Note also that the symbol $\mathbf{X'}$ represented independent variables in section 3, but here the state vector is dependent on the potential vorticity. A complete specification of $\mathbf{X'}$ and $\mathbf{P'}$ will be given below when individual experiments are discussed. Piecewise PV inversion may then proceed given the operator $\mathbf{L}$ and some subsample of the PV,

$$\mathbf{x'} = \mathbf{Lp'}, \qquad (16)$$

where $\mathbf{x'}$ is a column vector for the state that is attributable to the column vector of PV, $\mathbf{p'}$. For example, $\mathbf{p'}$ may consist of one column of $\mathbf{P'}$, with all entries but one set to zero, in which case $\mathbf{x'}$ is a (discrete) Green's function. Repeating this example for all rows of $\mathbf{P'}$ but one set to zero yields an ensemble sample for the Green's function, which may be used to estimate a mean solution and variance (i.e., error).

Because the Ertel PV has nonlinear terms, (15) may

Covariance of Surface Low with 500 hPa Height & Wind

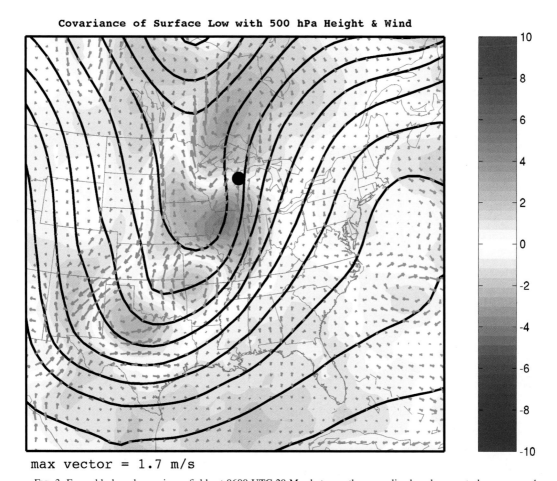

max vector = 1.7 m/s

FIG. 3. Ensemble-based covariance fields at 0600 UTC 29 Mar between the normalized cyclone central pressure and the 500-hPa geopotential height (colors, m) and wind (vectors). The ensemble-mean 500-hPa geopotential height field is given by solid lines every 60 m. The sample of cyclone central pressure values are normalized by the ensemble standard deviation, and the surface cyclone position is denoted by the black dot.

seem like an inappropriate estimator for this field. Recall that we have removed the ensemble mean from matrices **X′** and **P′**, so that **L** represents a linearization about the ensemble mean. As with all piecewise Ertel PV inversions, some linearization is necessary, and the ensemble linearization is particularly useful because it requires no space or time averaging.

Recall that *M,* the number of ensemble members (here 100), places a restriction on exactly how much information can be extracted by ESA, and this plays an important role here. One way to think about the problem is that there are only 100 degrees of freedom in the ensemble, so we cannot expect to constrain all $O(10^6)$ degrees of freedom in the numerical model given a small ensemble. Consider a simple example in three-dimensional Cartesian coordinates (x, y, z) for a two-member ensemble with vectors lying in the (x, y) plane. Because there is no information normal to the plane, the z direction is "invisible" to the ensemble statistics; in mathematical terms, it does not lie in the span of the ensemble—no linear combination of ensemble vectors in the (x, y) plane project off the plane. For PV inversion, our

results are limited to the space spanned by the 100 singular vectors of the potential vorticity field (**V** in the notation of section 3), which means that there will be many directions that are invisible. A second way to think about this problem has to do with noise in the covariance calculation at distances far from a particular point. Close to this particular point, the covariance relationships should be better than those far from this point where, by random chance, a small ensemble may suggest a strong relationship. These spurious relationships far from the point introduce noise into the calculation that may adversely affect the inversion. Mathematically this means that covariances between fields physically separated by large distances are potentially spurious; one possible solution often used in the EnKF literature is to simply zero out these entries (e.g., Houtekamer and Mitchell 1998; Hamill et al. 2001). This procedure increases the rank of the covariance matrix (effectively, the size of the ensemble), but tests of this idea are beyond the scope of the present analysis. Our goal here is simply to illustrate proof of concept; a more thorough analysis will be published elsewhere.

Covariance of Surface Low with Tropopause θ & Wind

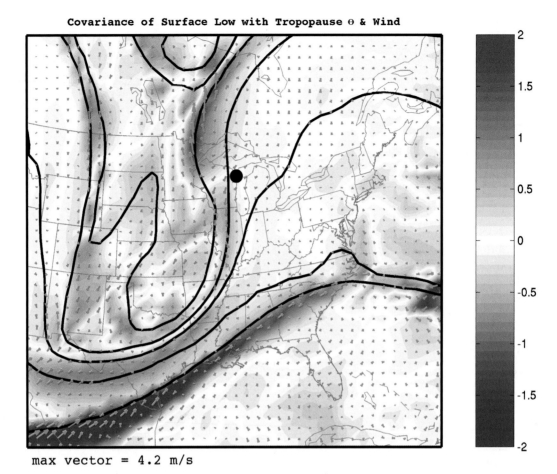

max vector = 4.2 m/s

FIG. 4. Ensemble-based covariance fields at 0600 UTC 29 Mar between the normalized cyclone central pressure and the dynamical tropopause potential temperature (colors, °C) and wind (vectors). The ensemble-mean tropopause potential temperature is given by solid lines every 10°C. The sample of cyclone central pressure values are normalized by the ensemble standard deviation, and the surface cyclone position is denoted by the black dot.

Mindful of the problems described above, piecewise PV inversion may be applied to large-scale features over the entire grid or to smaller features in a window around a local region. We choose the latter application by limiting the data to lie close to a point at 500 hPa for two examples: a potential vorticity blob at 500 hPa and an east–west cross section through the same location.

The first example is illustrated in Figs. 5 and 6 for piecewise PV inversion of a blob of PV located mainly over Oklahoma. Inversion (15) is calculated over 21 × 21 grid of points independently for the geopotential height, u, and v fields; that is, three operators are calculated. A piecewise inversion is then accomplished by applying these operators on a subset of the 21 × 21 PV field by choosing a vector that is zero everywhere but for the grid points indicated by the red lines in Fig. 5. The results show that the height field reaches a local minimum near the center of the PV anomaly, and the wind field has cyclonic circulation around the PV blob, with maximum speeds near the edges of the blob, and decays toward zero at larger distances. These patterns are qualitatively in accord with QG theory, which has

essentially a Laplacian relationship between the geopotential height field and the potential vorticity field, so that the geopotential height minimum is near the PV maximum. Because no assumptions are made regarding balance, ageostrophic wind vectors may be computed by the difference between the geostrophic wind determined from the inverted height field and the inverted full wind field. The results show anticyclonic circulation about the low geopotential height region; as expected (e.g., Holton 2004, p. 67), the flow is subgeostrophic around the low (Fig. 6).

The present example is more exploratory for the technique rather than definitive of its quantitative properties; nevertheless, a brief analysis of the sensitivity of the results is provided. Increasing the grid size to 31 × 31 points yields a geopotential height minimum of −74 m as compared with −75 m for the 21 × 21 grid and a maximum wind speed of 17 m s^{-1} as compared with 21 m s^{-1} for the 21 × 21 grid. Further increasing the grid to 41 × 41 gives a height minimum of −75 m and a maximum wind speed of 16 m s^{-1}; moreover the plotted fields are very similar to those shown in Fig. 5. The

Statistical Piecewise PV Inversion

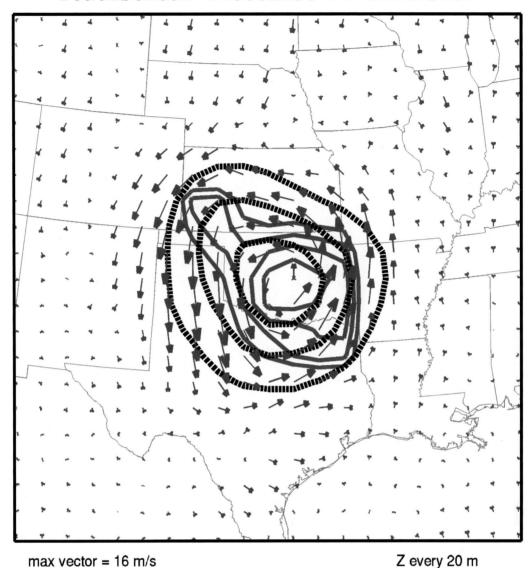

max vector = 16 m/s   Z every 20 m

FIG. 5. Piecewise Ertel PV inversion for a region of 500-hPa PV given by the red contours (every 0.5 PVU, starting with 1 PVU) at 0600 UTC 29 Mar. Dashed black lines give the inverted 500-hPa geopotential height field every 20 m, and the vectors give the inverted wind field.

results start to degrade for grid sizes larger than 61 × 61, at which point spurious noise in the covariances probably starts to pollute the local signal in the ensemble. A test of the results with respect to the number of ensemble members indicates that 75 members give results very similar to Fig. 5, and the results do not start to degrade considerably until fewer than 50 members are used (not shown). A convergence test in the space of PV singular vectors shows that 25 singular vectors achieve about the same results as approximately 75 ensemble members, which reflects the optimal dimensional reduction of SVD (not shown).

The above inversion procedure is repeated for the second example, which applies to a zonal cross section through the same PV blob as in Fig. 5. Piecewise PV inversion on PV in the 450–550-hPa layer of the cross section yields a familiar pattern (e.g., Hoskins et al. 1985, their Fig. 15). A local minimum in geopotential height is found near the PV maximum, with warm air above and cold air below the anomaly (Fig. 7). This distribution is qualitatively as expected from hydrostatic balance, although hydrostatic balance is not assumed in the calculation. The wind field normal to the cross section shows the strongest winds at the edges of the PV anomaly, which decrease with increasing distance. Note that because boundary conditions are not specified, there

Ageostrophic Velocity

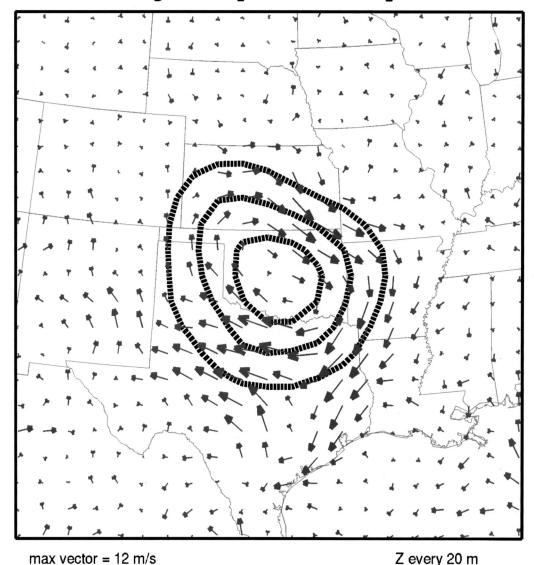

max vector = 12 m/s   Z every 20 m

FIG. 6. Ageostrophic velocity field (vectors) for the piecewise PV inversion shown in Fig. 5. Geopotential height is given by dashed lines every 20 m. Note the area of low geopotential is associated with anticyclonic ageostrophic circulation; the total wind around the low is subgeostrophic.

is no ambiguity in the solution near the surface. This is an attractive property given the strong boundary condition influence on piecewise PV inversion (Hakim et al. 1996).

These two examples suggest that ESA PV inversion is a potentially useful tool. Advantages of this technique include inverting the unapproximated Ertel PV and freedom from balance assumptions, boundary conditions, map factors, etc. As such, irregular grids and boundaries are no more difficult than regular domains. Furthermore, the inversion is linear and thus superposition is rigorous and straightforward, and the implementation is very easy. The foremost disadvantage of the technique is the

nonuniqueness of the inversion operator due to ensemble size, which may need to be large when considering piecewise PV inversions for anomalies extending over a large number of grid points. This problem may be overcome by larger ensembles or, more realistically, by techniques designed to increase the effective ensemble size (e.g., covariance localization) (Houtekamer and Mitchell 1998; Hamill et al. 2001).

We note that although the operator **L** has a null space because of ensemble subsampling there is a more fundamental space that may prove useful for studying, or even defining, balance dynamics and balance models. This space, the orthogonal complement of **L**, consists

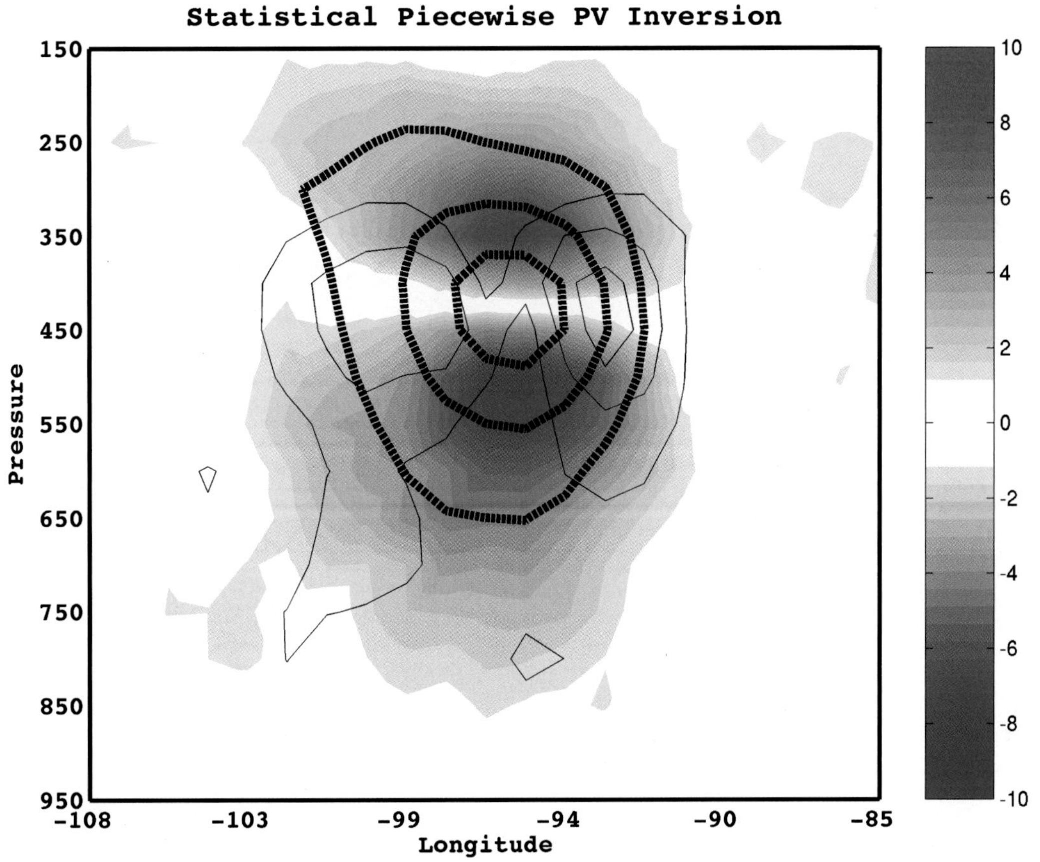

Statistical Piecewise PV Inversion

FIG. 7. East–west cross section of piecewise Ertel PV inversion for a blob of PV bounded in the vertical by the 450–550-hPa layer and in the horizontal by values greater than about 1 PVU. Thick solid black lines give the geopotential every 20 m, and thin solid black lines give the wind speed normal to the cross section every 4 m s^{-1}. Colors give the potential temperature field (°C).

of the subspace of the model state **X**, which cannot be recovered by PV inversion, even for full-rank ensembles. Consider a hydrostatic incompressible model having n grid points for model velocities u and v and thermodynamic variable T for a total of $N = 3n$ degrees of freedom. A full-rank PV matrix has at most n degrees of freedom, leaving $2n$ degrees of freedom orthogonal to the space spanned by the PV inversion operator. What is the dynamical basis for this large model subspace? In the case of linearized dynamics around a state of rest, these directions correspond to inertia–gravity waves and the n modes resolved by the PV correspond to Rossby waves. What is interesting is that the ensemble approach permits a natural extension of this separation to finite amplitude states, with the orthogonal complement corresponding to unbalanced modes linearized about the ensemble-mean state. This interpretation suggests a statistical definition for balance: Balance dynamics are defined by the state subspace that covaries with potential vorticity.

We close this section by noting that there is nothing in principle that limits ESA PV inversion to the wind and mass field as is typically the case; one could also determine, for example, the precipitation field "attributable" to a particular PV anomaly.

6. Statistical dynamical sensitivity analysis

The final ESA example concerns sensitivity analysis, which addresses problems such as determining how changes to an initial condition affect a subsequent forecast. This analysis is equivalent to determining the factors most important to the dynamics of a weather system. For the cyclone examined here, an application of these ideas concerns the factors that control the cyclone central pressure at 0600 UTC 29 March, which is the metric chosen for the analysis. A 24-h time interval is selected for examination, and sensitivity is defined by the covariance between the chosen metric and the 0600 UTC 28 March analysis. We shall refer to the ensemble solution as "the control." Specifically, the control ensemble-mean analysis at 0600 UTC 28 March is perturbed by

$$\mathbf{x}_0' = \alpha \, \text{cov}(\mathbf{X}_0', y') = \alpha \mathbf{X}_0' y'^{\text{T}}, \qquad (17)$$

where $\mathbf{x}_0'$ is the perturbation state vector at 0600 UTC

Initial P′ (lines) & Θ′ (color)

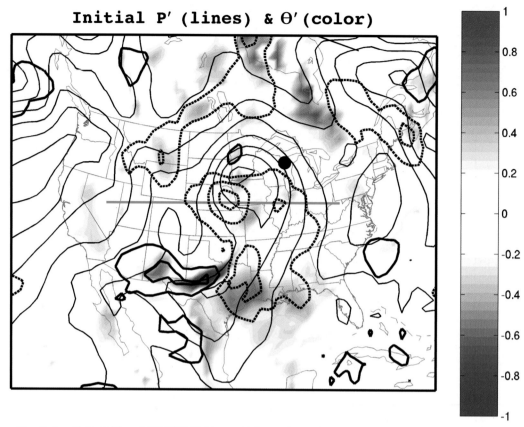

FIG. 8. Sensitivity fields at 0600 UTC 28 Mar pertaining to the cyclone central pressure at 0600 UTC 29 Mar as determined by ESA for $\alpha = -2.5$; the predicted change in the cyclone central pressure at 0600 UTC 29 Mar is -6 hPa. Thick black lines show the perturbation surface pressure field every 0.25 hPa, and the colors show the surface potential temperature perturbations. Thin solid lines show the control ensemble-mean surface pressure at 0600 UTC 28 Mar every 4 hPa.

28 March ($N \times 1$), $\mathbf{X}_0'$ is the control ensemble state matrix at 0600 UTC 28 March ($N \times M$; ensemble mean removed), and $\mathbf{y}'$ is the normalized control ensemble sample of cyclone central pressure at 0600 UTC 29 March ($1 \times M$; mean removed). As in prior calculations [see (14)], normalization of the central pressure is by the ensemble standard deviation. Parameter α controls the amplitude of the initial perturbations.

One possible factor that may be expected to affect the 0600 UTC 29 March cyclone central pressure is the cyclone central pressure at 0600 UTC 28 March; that is, a deeper initial cyclone may be expected to produce a deeper ending cyclone. In fact, the results show that the positioning of the cyclone relative to other flow features is more important than the initial amplitude (Fig. 8), with the pressure covariance field indicating a displacement of the low center farther south. An east–west cross section through this feature exhibits a nearly barotropic structure, with essentially no vertical tilt and largest amplitude at the surface and the tropopause (Fig. 9). This structure stands in contrast to adjoint sensitivity analyses, which often give structures that are highly tilted in the vertical (e.g., Langland et al. 1995). A sec-

ond disturbance is found farther to the south, near a weak frontal wave in Texas. The surface covariances indicate that a displacement of this feature northeastward along the front is an important factor for the intensity of the parent cyclone. Moreover, the surface temperature covariance field indicates that a stronger cross-front temperature gradient near the frontal wave is also linked to the intensity of the main surface low 24 h later. Overall, this particular case leaves the impression that changes in the phase relationships of existing features are the most important control on the central pressure of the surface cyclone.

With $\alpha = -2.2$, a perturbed ensemble-mean initial condition defined by (17) is evolved in the full WRF model. The WRF solution for this initial condition gives a cyclone central pressure that is 4.6 hPa lower than in the control case at 0600 UTC 29 March, as compared with 6 hPa lower as predicted by the ensemble covariance (Fig. 10). A more stringent test of the method derives from the following covariance considerations at 0600 UTC 29 March. The linearized ensemble dynamics obey

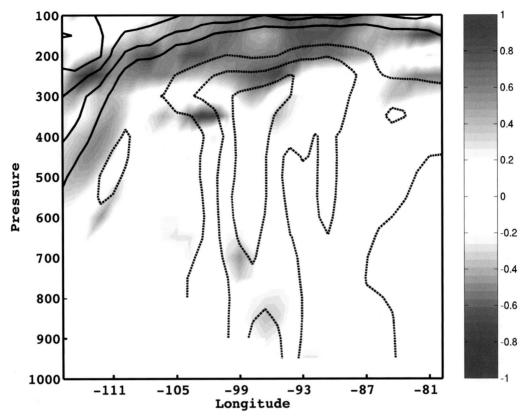

FIG. 9. Cross section near 40°N through the sensitivity geopotential height field (black lines every 2 m) and potential temperature field (colors).

$$X'_t = WX'_0,\qquad(18)$$

where X'_t is the ensemble control solution at 0600 UTC 29 March, X'_0 is the ensemble control initial condition, and W represents the tangent-linear WRF model.[4] Right multiplying (18) by y^T gives

$$X'_t y'^T = WX'_0 y'^T.\qquad(19)$$

Using (17), we find that

$$x'_t = Wx'_0 = \alpha\ \mathrm{cov}(X'_t,\ y').\qquad(20)$$

Equation (20) says that the perturbed solution at 0600 UTC 29 March given perturbed initial condition (17) is equal to the control ensemble covariance, scaled by α. Thus, unlike adjoint sensitivity analysis, ESA sensitivity analysis predicts the changes in both the chosen metric and in the full solution fields. Comparing Figs. 10 and 2 we find very similar fields, as predicted by (20).

Finally, the above perturbation procedure is repeated for other values of α to determine the range of linear dynamics for this event in the chosen metric. There is

a good linear relationship between predictions based on the covariance calculation and WRF solutions over the range -8 to 8 hPa in perturbation cyclone central pressure (Fig. 11, light blue dashed line). Note, however, that the linear relationship falls below the main diagonal, which indicates that the WRF response in the chosen metric is systematically smaller than the covariance prediction. One possible explanation for this result is that the ensemble has too much variance in the estimate of the forecast cyclone central pressure. Renormalizing the ensemble sample of cyclone central pressure values [i.e., y' in (17)] with this correction yields a very good linear response (Fig. 11, red dashed line). This plot also clearly indicates that linear perturbation dynamics are violated when the initial perturbations are scaled to produce deviations larger than about 8 hPa in cyclone central pressure at 0600 UTC 29 March. Note also the asymmetry between positive and negative perturbations, which suggests that the cyclone can be deepened by at most ~ 10 hPa, whereas it may be weakened by nearly 15 hPa.

7. Summary

ESA is proposed as a new tool for investigating atmospheric phenomena that is based upon probabilistic analyses, which are estimated from an ensemble. Unlike deterministic analyses, the probabilistic samples dis-

---

[4] Note that W is not known in matrix form but rather only as a computer program. Nevertheless, the action of the model may be represented symbolically as in (18). Furthermore, we do not have the tangent linear model for WRF; however, for small-amplitude perturbations, the difference between two nonlinear runs provides a good approximation to the true tangent linear model.

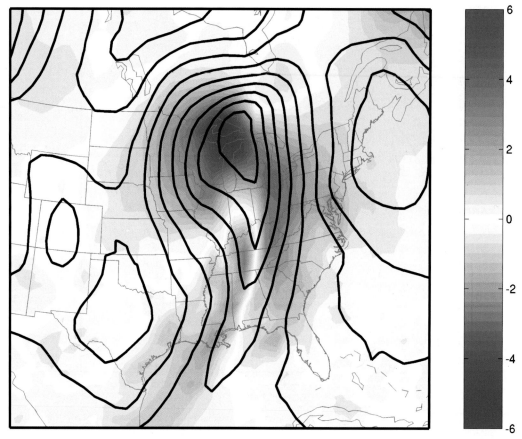

Fig. 10. Difference surface pressure field at 0600 UTC 29 Mar relative to the control for the perturbed initial condition for $\alpha = -2.5$. Pressure perturbations (colors, hPa) are defined as the difference between the perturbed solution at 0600 UTC 29 Mar, as determined by integration of the perturbed initial condition in the WRF model, and the control solution at 0600 UTC 29 Mar. The surface pressure field for the perturbed solution is given by thin solid lines every 4 hPa.

cussed here admit, and quantify, the inherent errors in atmospheric analyses. These ensemble analyses also seamlessly integrate with ensemble prediction, since it is not necessary to create analysis perturbations. Our goal here has been to propose analysis methods that utilize the wealth of information in these probabilistic analyses and, in doing so, to illustrate that state estimation may be used as a tool for atmospheric dynamics. The analysis ensemble is determined by a research-based ensemble Kalman filter consisting of 100 members and is applied to a case of cyclogenesis.

The ensemble covariance relationships qualitatively confirm known linkages between surface cyclones and upper-level disturbances. The results also suggest perhaps less obvious relationships between the cyclone and the surface cold front, a second upper-level shortwave, and the subtropical jet stream. These novel insights underscore the potential for this technique to provide useful new perspectives into phenomena that are less well understood than extratropical cyclones.

An application of ESA to potential vorticity inversion suggests the viability of this approach, which is ap-

pealing given the short algorithm, and freedom from traditional assumptions, such as balance relationships. Furthermore, the analysis motivates a new definition for balance dynamics in terms of the state subspace that covaries with potential vorticity.

A second ESA example addressed dynamical sensitivity analysis and factors that control the intensity of the surface cyclone. Unlike adjoint sensitivity analysis, this technique does not require an adjoint model and is very easy to perform numerically; both this calculation and the PV inversion calculation are easily performed with short software programs. Moreover, the method implicitly accounts for errors in the initial and end states, and for a given initial perturbation, it predicts the solution response in both the chosen metric and the full solution field. Continuing this comparison, the ESA results for this case show that the initial condition sensitivity is highly organized around existing flow features, and the perturbation fields exhibit little vertical tilt. Adjoint-based sensitivity analyses often indicate widespread regions of sensitivity and complicated vertical and horizontal structures (e.g., Langland et al.

WRF Calculated Surface Pressure Difference

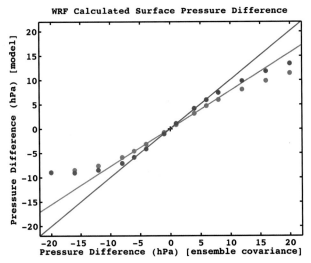

FIG. 11. Cyclone central pressure differences as determined by perturbed integrations of the WRF model (ordinate) against those predicted by the control ensemble covariance statistics (abscissa). The dashed blue line applies to the ensemble covariance perturbations and the dashed red line to the renormalized ensemble covariance perturbations (see text for details). Values along the main diagonal indicate agreement between the covariance predictions and solutions from WRF integrations.

1995; Zou et al. 1998). We caution that this comparison is based on one case, and further research will be needed to compare these two techniques.

Results for ESA sensitivity analysis show that the primary factors affecting the strength of the surface cyclone involve a phase shift of existing flow features, including a frontal wave well removed from the primary cyclone. These features were determined by the covariance between the chosen metric (surface cyclone central pressure) and the full model state 24 h earlier. To check these covariance predictions, the full WRF model was used to integrate covariance-perturbed initial conditions. The solutions show a remarkably good correspondence with the covariance predictions of the cyclone central pressure.

Although the example illustrations given here all apply to a case of cyclogenesis, that choice was mainly based on familiarity and recognition of Fred Sanders' contributions to our understand of this phenomenon. In fact, ESA may be applied to any problem where ensemble samples of data are available.

Acknowledgments. We thank Tom Hamill and an anonymous reviewer for their helpful comments on an earlier version of the manuscript. This research was supported by National Science Foundation Grants ITR-0205648, ATM-0228804, and CMG-0327658 awarded to the University of Washington.

REFERENCES

Daley, R., 1993: *Atmospheric Data Analysis*. Cambridge University Press, 471 pp.

Davis, C. A., and K. A. Emanuel, 1991: Potential vorticity diagnosis of cyclogenesis. *Mon. Wea. Rev.*, **119**, 1929–1953.

Evensen, G., 1994: Sequential data assimilation with a nonlinear quasigeostrophic model using Monte Carlo methods to forecast error statistics. *J. Geophys. Res.*, **99** (C5), 10 143–10 162.

Golub, G. H., and C. F. Van Loan, 1996: *Matrix Computations*. Johns Hopkins University Press, 698 pp.

Hakim, G. J., D. Keyser, and L. F. Bosart, 1996: The Ohio Valley wave-merger cyclogenesis event of 25–26 January 1978. Part II: Diagnosis using quasigeostrophic potential vorticity inversion. *Mon. Wea. Rev.*, **124**, 2176–2205.

Hamill, T. M., 2006: Ensemble-based atmospheric data assimilation: A tutorial. *Predictability of Weather and Climate*, T. Palmer and R. Hagedorn, Eds., Cambridge University Press, 702 pp.

——, and C. Snyder, 2000: A hybrid ensemble Kalman filter–3D variational analysis scheme. *Mon. Wea. Rev.*, **128**, 2905–2919.

——, J. S. Whitaker, and C. Snyder, 2001: Distance-dependent filtering of background error covariance estimates in an ensemble Kalman filter. *Mon. Wea. Rev.*, **129**, 2776–2790.

Holton, J. R., 2004: *An Introduction to Dynamic Meteorology*. Elsevier Science and Technology Books, 560 pp.

Hong, S.-Y., and H.-L. Pan, 1996: Nonlocal boundary layer vertical diffusion in a medium-range forecast model. *Mon. Wea. Rev.*, **124**, 2322–2339.

Hoskins, B. J., M. E. McIntyre, and A. W. Robertson, 1985: On the use and significance of isentropic potential vorticity maps. *Quart. J. Roy. Meteor. Soc.*, **111**, 877–946.

Houtekamer, P. L., and H. L. Mitchell, 1998: Data assimilation using an ensemble Kalman filter technique. *Mon. Wea. Rev.*, **126**, 796–811.

Janjic, Z. I., 1994: The step-mountain eta coordinate model: Further developments of the convection, viscous sublayer and turbulence closure schemes. *Mon. Wea. Rev.*, **122**, 927–945.

Kalnay, E., 2002: *Atmospheric Modeling, Data Assimilation and Predictability*. Cambridge University Press, 364 pp.

Langland, R. H., R. L. Elsberry, and R. M. Errico, 1995: Evaluation of physical processes in an idealized extratropical cyclone using adjoint sensitivity. *Quart. J. Roy. Meteor. Soc.*, **121**, 1349–1386.

Leith, C. E., 1974: Theoretical skill of Monte Carlo forecasts. *Mon. Wea. Rev.*, **102**, 409–418.

Michalakes, J., S. Chen, J. Dudhia, L. Hart, J. Klemp, J. Middlecoff, and W. Skamarock, 2001: Development of a next generation regional weather research and forecast model. *Developments in Teracomputing: Proceedings of the Ninth ECMWF Workshop on the Use of High Performance Computing in Meteorology*, W. Zwieflhofer and N. Kreitz, Eds., World Scientific, 269–276.

Sanders, F., 1986: Explosive cyclogenesis in the west-central North Atlantic Ocean, 1981–84. Part I: Composite structure and mean behavior. *Mon. Wea. Rev.*, **114**, 1781–1794.

——, 1987: Study of 500-mb vorticity maxima crossing the east coast of North America and associated surface cyclogenesis. *Wea. Forecasting*, **2**, 70–83.

——, and J. R. Gyakum, 1980: Synoptic-dynamic climatology of the "bomb." *Mon. Wea. Rev.*, **108**, 1589–1606.

Whitaker, J. S., and T. M. Hamill, 2002: Ensemble data assimilation without perturbed observations. *Mon. Wea. Rev.*, **130**, 1913–1924.

——, G. P. Compo, X. Wei, and T. M. Hamill, 2004: Reanalysis without radiosondes using ensemble data assimilation. *Mon. Wea. Rev.*, **132**, 1190–1200.

Zou, X., Y.-H. Kuo, and S. Low-Nam, 1998: Medium-range prediction of an extratropical oceanic cyclone: Impact of initial state. *Mon. Wea. Rev.*, **126**, 2737–2763.

Chapter 8

Surface Potential Temperature as an Analysis and Forecasting Tool

ERIC G. HOFFMAN

Department of Chemical, Earth, Atmospheric, and Physical Sciences, Plymouth State University, Plymouth, New Hampshire

(Manuscript received 30 August 2004, in final form 7 December 2005)

Hoffman

ABSTRACT

In the last decade, Fred Sanders was often critical of current surface analysis techniques. This led to his promoting the use of surface potential temperatures to distinguish between fronts, baroclinic troughs, and nonfrontal baroclinic zones, and to the development of a climatology of surface baroclinic zones. In this paper, criticisms of current surface analysis techniques and the usefulness of surface potential temperature analyses are discussed. Case examples are used to compare potential temperature analyses and current National Centers for Environmental Prediction analyses.

The 1-yr climatology of Sanders and Hoffman is reconstructed using a composite technique. Annual and seasonal mean potential temperature analyses over the continental United States, southern Canada, northern Mexico, and adjacent coastal waters are presented. In addition, gridpoint frequencies of moderate and strong potential temperature gradients are calculated. The results of the mean potential temperature analyses show that moderate and strong surface baroclinic zones are favored along the coastlines and the slopes of the North American cordillera. Additional subsynoptic details, not found in Sanders and Hoffman, are identified. The availability of the composite results allows for the calculation of potential temperature gradient anomalies. It is shown that these anomalies can be used to identify significant frontal baroclinic zones that are associated with weak potential temperature gradients. Together the results and reviews in this paper show that surface potential temperature analyses are a valuable forecasting and analysis tool allowing analysts to distinguish and identify fronts, baroclinic troughs, and nonfrontal baroclinic zones.

1. Introduction

Surface analysis has long been a key component of weather analysis and forecasting. Before 1941, operational surface analyses in the United States consisted of isobar and isotherm plots (Namias 1983). In 1941, the U.S. Weather Bureau changed its surface analysis practices and adopted the now ubiquitous and familiar analyses with station models, isobars, and fronts (Namias

1983). Current surface analysis techniques continue to this day to be based upon the Norwegian conceptual model (NCM) for cyclone development first espoused in a series of papers published in the early 1920s (e.g., Bjerknes 1919; Bjerknes and Solberg 1922). According to the NCM and nearly every current introductory meteorology textbook (e.g., Ahrens 2005), fronts are defined to be boundaries between air masses of differing densities and follow a conceptual evolution from frontal wave through occlusion when associated with a midlatitude cyclone. It is also stated in these texts that these boundaries are not discontinuities but rather transition zones that are represented by a zone of enhanced tem-

Corresponding author address: Dr. Eric G. Hoffman, Dept. of Chemical, Earth, Atmospheric, and Physical Sciences, Plymouth State University, 313 Boyd Hall, Plymouth, NH 03264.
E-mail: ehoffman@plymouth.edu

perature gradients. In the theoretical/dynamical framework (e.g., Hoskins and Bretherton 1972) fronts are defined to be associated not only with large horizontal thermal gradients but also coincident with large values of cyclonic vorticity and static stability.

On observational surface charts, the frontal line is drawn to mark the warm edge of the frontal temperature gradient. However, it must have been very soon after this analysis was adopted that it became obvious that surface thermal (i.e., density) boundaries do not always conform to the NCM and "textbook" ideas about fronts and air masses. As Sanders (1999) has pointed out, Petterssen (1940) considered temperature near the surface of the earth to be "unrepresentative [of the air mass]." In their text, Wallace and Hobbs (1977, section 3.2.2) begin the discussion of fronts with a temperature and baroclinic zone definition, but then move on to suggest that

> there are large regions of the globe in which it is extremely difficult to locate fronts on the basis of gradients in surface temperature:
> - over the oceans where the surface temperature never departs by more than a few degrees from the temperature of the underlying water,
> - in mountainous terrain where large differences in station elevation introduce spurious temperature gradients.

These two textbooks seem to suggest that temperature is not the defining characteristic of fronts. However, Wallace and Hobbs (1977) state in the very next sentence "it is essential to discriminate between the temperature gradients associated with fronts and those due to other influences." Analysts and forecasters have been in a quandary ever since. Are surface fronts defined by thermal gradients or are surface temperatures "unrepresentative" and "spurious"?

Beginning in the early 1990s, Fred Sanders and others became critical of the surface analyses and surface analysis techniques that have arisen from the above quandary. These criticisms focused on the discrepancies between NCM fronts and the observed features on surface analyses. Mass (1991) pointed out several deficiencies in the application of NCM ideas. Specifically, Mass (1991) was concerned that NCM frontal symbols are used to depict several different types of surface boundaries including cyclone-related frontal zones, shallow topographically induced boundaries (e.g., cold-air damming, coastal fronts, and lee troughs), and shallow diabatically produced boundaries (e.g., the differential temperatures across a surface snow-cover boundary). In addition to these discrepancies and in response to the above quandary, fronts are being identified through analysis of surface features that are sometimes, but not always, collocated with surface temperature boundaries (i.e., density contrasts). Sanders and Doswell (1995) list these features as pressure troughs, pressure tendencies, wind shifts, dewpoint differences, clouds, and precipitation. Sanders and Doswell (1995) suggest that these are secondary and perhaps sometimes tertiary indications of the location of a surface thermal boundary (i.e., a front). Uccellini et al. (1992) showed quite convincingly that even a group of noted research and operational meteorologists use different weightings of these secondary features to produce large differences in the location of surface frontal features (see their Fig. 2). Sanders and Doswell (1995), therefore, suggest that surface analysis should focus on the thermal boundaries. In practice, however, the location of these secondary features, especially wind shifts, clouds, and precipitation, may be important to forecasters.

To rectify the confusion in surface analysis related to the use of the NCM and the secondary frontal characteristics, Sanders (1999) proposed an alternate method of surface analysis. This method centers on the routine analysis of surface potential temperatures. The term "front" is reserved by Sanders for those surface features that have either a moderate or strong surface potential temperature gradient [defined by Sanders to be 8 K (220 km)$^{-1}$ and 8 K (110 km)$^{-1}$, respectively] and an associated pressure trough and wind shift. Surface features that have significant wind shifts or pressure troughs without a moderate or strong potential temperature gradient are labeled baroclinic troughs (BTs). All areas of strong potential temperature gradients not associated with significant pressure troughs are termed nonfrontal baroclinic zones (NBZs). Inspired by these thoughts at their initial exposition in an invited talk by Sanders at the *10th Cyclone Workshop,* at Val Morin, Quebec, Canada (November 1997), routine analyses of surface potential temperature and their gradient have been produced at the University at Albany since November 1998. (These analyses are currently available online at http://www.atmos.albany.edu/index.php?d=wx_surfupprdata, with links to the current surface potential temperature analysis, recent analyses, and an animation.) An example of this plot is shown in Fig. 1. Station models of observed weather features are plotted with the potential temperature (°C) replacing temperature, along with contours of potential temperature (every 4°C) and shading where the gradient exceeds 3.5°C (100 km)$^{-1}$ [3.5 × 10^{-5}°C m^{-1}, ~8 K (220 km)$^{-1}$] and 7°C (100 km)$^{-1}$ [7 × 10^{-5}°C m^{-1}, ~8 K (110 km)$^{-1}$] as proposed in Sanders (1999). In the following section, advantages and disadvantages of using analyses of this type will be discussed with examples provided from actual cases.

As with any meteorological field, understanding and wise use of these surface potential temperature maps depends upon knowledge of the climatology of features on these maps. Sanders and Hoffman (2002, hereafter SH02) constructed a climatology of surface baroclinic zones using these daily potential temperature analyses. In SH02, the frequencies of strong and moderate potential temperature gradients are represented in counts

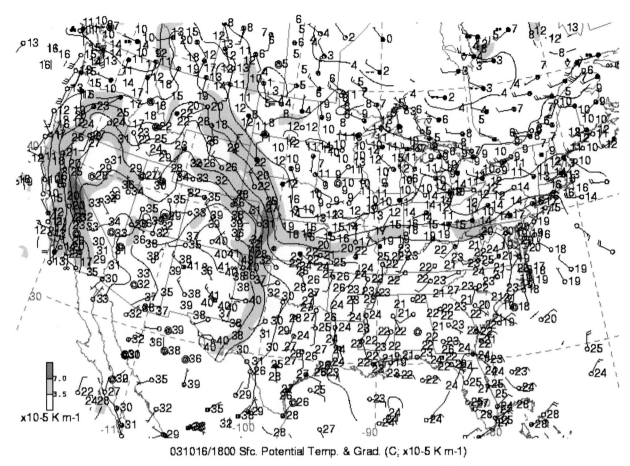

x10-5 K m-1

031016/1800 Sfc. Potential Temp. & Grad. (C; x10-5 K m-1)

FIG. 1. Example surface potential temperature map: potential temperature (solid, every 4°C) and potential temperature gradient ($\times 10^{-5}$°C m^{-1}, light shading > 3.5, dark shading > 7). Conventional station models except for potential temperature (°C) in place of temperature.

per state. SH02 showed that 1) moderate and strong potential temperature gradients occur mainly along the Atlantic and Pacific coasts of North America and along the eastern slopes of the North American cordillera; 2) the National Centers for Environmental Prediction (NCEP) fronts are often not associated with moderate or strong potential temperature gradients; and 3) a great many baroclinic zones are not analyzed in current surface analyses.

SH02's use of frequency of baroclinic zones per state is a simple method that gave a general idea of the large-scale distribution of surface potential temperature. However, it is desirable to examine actual mean surface potential temperature maps and produce contours of gridpoint frequencies for strong and moderate potential temperatures. Calculations of this sort will allow quantitative assessment of mean surface potential temperatures and will show additional geographic and subsynoptic details in the mean and gradient frequencies. Therefore, in section 3, the SH02 climatology is revisited and contour mean surface potential temperature maps are presented along with contour maps of moderate and strong potential temperature gradient fre-

quencies. Last, the paper will conclude with some summary remarks about the relationship between fronts and potential temperature gradients as well as some thoughts about the application of the surface potential temperature technique.

2. Advantages and disadvantages of surface potential temperature analyses

Sanders (1999) suggested that the primary reason for recommending use of potential temperature to analyze the thermal field is to alleviate some of the "spurious thermal gradients" (Wallace and Hobbs 1977) owing to variations in station elevation. A schematic cross section of potential temperatures is given in Fig. 2 (see also Sanders 1999, his Fig. 1). Figure 2 demonstrates that when the boundary layer is relatively well mixed, the potential temperature gradient along sloping terrain is nearly identical to the horizontal gradient just above the terrain (Fig. 2a). However, when the atmosphere becomes stably stratified, as is often the case during nocturnal inversions and with cold air masses, the potential temperature gradient along the sloping terrain may ac-

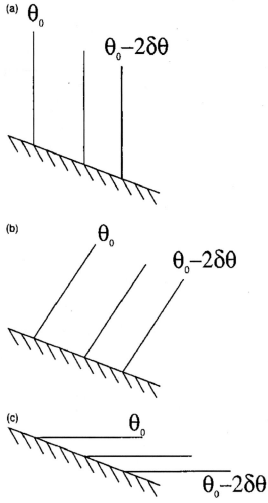

FIG. 2. Schematic vertical cross section of potential temperature over sloping terrain for (a) a well-mixed boundary layer, (b) moderate stratification, and (c) strong stratification. From Sanders (1999, his Fig. 1.).

the horizontal kinematic wind field near the earth's surface would result in nearly adiabatic temperature changes. Therefore, in places where dramatic potential temperature changes occur in the absence of strong advection and frontogenesis/frontolysis, diabatic processes can be inferred. Another parallel between potential temperature and PV may also be useful for diagnosing and understanding the full three-dimensional flow. Low-level potential temperature anomalies are associated with low-level PV anomalies as demonstrated by Bretherton (1966). Hoskins et al. (1985) revealed that cyclogenesis can be understood as the interaction between upper-level and low-level PV anomalies. Therefore, the potential temperature maps can be used to understand and diagnose cyclogenesis from a "PV-thinking" perspective.

While there are many reasons to use surface potential temperature, there are some disadvantages in choosing this analysis method. The primary disadvantage of surface potential temperature analyses is that thermal boundaries are not the only important surface boundary that forecasters and analysts have to consider. As was pointed out in Sanders and Doswell (1995), forecasting atmospheric convection can depend critically on surface moisture boundaries (e.g., the dryline). Such moisture boundaries will not be represented on an analysis of this sort. Clearly this disadvantage can easily be overcome by using an additional analysis of a moisture variable such as mixing ratio (see the University at Albany Web site http:www.atmos.albany.edu/deas/data.html for an example) or by analyzing equivalent potential temperature. From a theoretical perspective other important surface boundaries would be areas of enhanced cyclonic vorticity and stability. Sanders (1999) and the analyses presented here (see Fig. 1) qualitatively account for the cyclonic vorticity by plotting the observed winds. The spatial and temporal observations of lower-tropospheric stability from soundings preclude including stability in this type of analysis. Automated frontal analyses produced by the Met Office take into account these quantities in a more rigorous fashion (e.g., Hewson 1997, 1998).

The automated method for producing these surface potential temperature analyses by SH02 utilizes a GEMPAK Barnes objective analysis technique (Koch et al. 1983). There are some disadvantages to using this technique to produce the analyses, but these do not represent a disadvantage of using surface potential temperature analyses in general. While giving robust results, SH02 noted that the GEMPAK Barnes automated analysis will often underestimate the value of gradients especially in coastal regions where there are larger numbers of onshore observations weighting the immediate offshore grid points toward overland values. The Barnes analysis scheme can also underestimate the value of potential temperature gradients in regions where the density of observing stations is sparse (Steenburgh and Blazek 2001). Internal GEMPAK routines estimate the optimal grid resolution based on the average station density for

tually represent the vertical stratification rather than the actual horizontal gradient (Figs. 2b,c). Despite these limitations, using surface potential temperature gradients reduces the occurrence of spurious gradients. In practice, when stable cold air masses are adjacent to sloping terrain (as in Fig. 2c) the location of the gradient along the slope can reveal the depth of the low-level cold air and may represent the horizontal penetration of the cold air up the sloping surface.

Another advantage of using potential temperature is its conservation for frictionless, unsaturated, adiabatic motions. While surface flows are perhaps least expected to conform to those conditions, it is precisely this nonconservation that can be viewed as advantageous in much the same way that nonconservation of quasigeostrophic or Ertel's potential vorticity (PV) can be used to infer the influence of diabatic heating on upper-level PV anomalies. Presumably, temperature advection and frontogenetical/frontolytical processes diagnosed from the analysis area (continental U.S. and adjacent coastal waters) and the analyses produced are consistent with this grid resolution. However, it should be noted that in some areas this resolution is not dense enough (e.g., the northeastern United States) and in others it is too dense (e.g., the Intermountain West).

If a feature in the potential temperature analysis moved into a region where the gradient is underestimated, the analysis of the gradient would become temporally discontinuous. This could be partially solved by using a first-guess field derived from either a model forecast or from automated time–space conversion techniques that could provide a more spatially and temporally consistent analysis. The GEMPAK Barnes analysis employed here is not tied to a prediction system or a first-guess field of any kind. The Barnes analysis was primarily chosen for its simplicity and availability in the GEMPAK software. However, automated analyses that employ first-guess fields use mathematical techniques that constrain the analysis toward the guess field and not the observations. In the case of modeling this is an important consideration. In this case, the primary intention of these analyses is to represent the *observed* potential temperature features and therefore, it is not constrained in any way (other than by the Barnes Gaussian weighting function) to a model forecast, which in its approximate representation of the atmosphere has its own set of limitations. The analysis is free to represent everything that appears in the observed data (including on occasion a bad data point or two) and analysts and forecasters that use these analyses must understand the limitations associated with grid resolution and data density noted above. Other choices of automated analysis, such as the Local Analysis and Prediction System (Albers et al. 1996), for objectively analyzing surface data might be equally judicious and could produce equally good results.

In summary, the main advantages of using a surface potential temperature analysis are 1) a reduction in the number of temperature gradients observed because of differences in elevation, and 2) the ability to infer diabatic processes from the nonconservation of surface potential temperature. Disadvantages include the lack of consideration of a moisture variable or other weather phenomena and interpretation of potential temperature gradients in strongly stable environments along sloping terrain. The limitations associated with a choice of an automated Barnes analysis scheme not tied to a first-guess field for these particular sets of surface potential analyses need to be understood by analysts and forecasters. In the following section, examples of surface potential temperature analyses are presented to demonstrate the usefulness of these analyses.

Surface potential temperature analysis examples

As with any application, the best way to evaluate a tool or technique is to use it. It has been the experience

in the University at Albany and Plymouth State University map rooms that use of the surface potential temperature analyses often reduces confusion that can be generated by the NCEP frontal analyses. Examples of NCEP and corresponding surface potential temperature analyses are given below to further illustrate the advantages and disadvantages in using potential temperature analyses. It is not the intention of the author to disparage the analysts at NCEP. It seems appropriate to compare any new analysis method to the current method and perhaps some day the analysts at NCEP will be able to incorporate these ideas to make identification of surface frontal features easier.

1) 28–29 JANUARY 2002

During this 2-day period weak cyclonic disturbances traveled northeast along a quasi-stationary surface boundary extending northeast to southwest from the Great Lakes to the Texas panhandle. This boundary separated a slow moving continental polar (cP) air mass over the northern plains from the maritime tropical (mT) air mass over the southeastern United States. The NCEP surface analysis from 1200 UTC 28 January 2002 and the corresponding surface potential temperature analysis are shown in Fig. 3. Comparison of the two figures shows that the stationary/cold front on the NCEP analysis extending southwest from the low center near the Ontario–Quebec border to Oklahoma and then bending back along the eastern slopes of the Rockies has a moderate-to-strong potential temperature gradient only on its southern and western ends. Through much of the Midwest the front is not associated with even a moderate potential temperature gradient and Sanders (1999) would consider this feature to be a baroclinic trough. The cold front extending from the low pressure center located in Wyoming across Utah and Nevada is associated with a moderate gradient as is most of the undulating stationary front over New England associated with cold-air damming (Bell and Bosart 1988). The frontal system depicted along the Carolina coasts and across northern Florida is also associated with a moderate potential temperature gradient. All of these would be considered fronts using the Sanders methodology. However, a large area of moderate-to-strong potential temperature gradient lying across south-central Canada is not represented on the NCEP analyses. This area would be considered a NBZ using the Sanders methodology and users of the NCEP analyses would be unaware that a significant decrease in temperature exists as one moves poleward in south-central Canada. A quick inspection of the analyses reveals that is probably due to differential cooling between the clear skies over central Canada and cloudy skies over southern Canada and the northern United States.

Twelve hours later at 0000 UTC 29 January 2002 there are some noticeable changes in the two analyses (potential temperatures in Fig. 4a and NCEP in Fig. 4b).

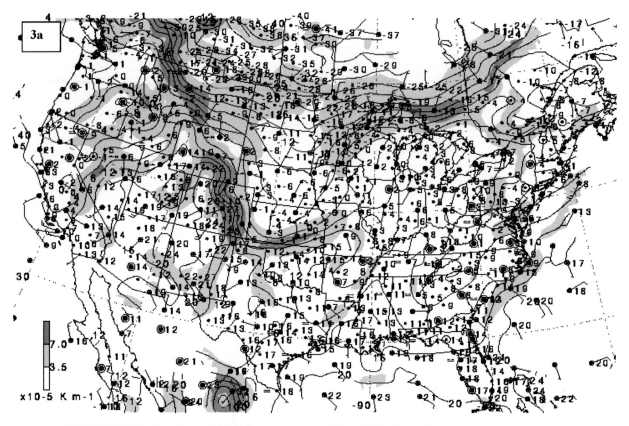

020128/1200 Sfc. Potential Temp. & Grad. (C; x10-5 K m-1)

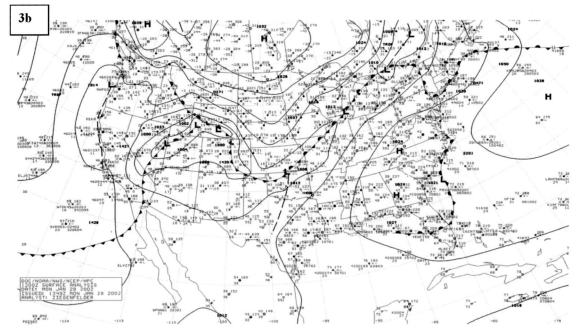

FIG. 3. Surface potential temperature and NCEP analyses for 1200 UTC 28 Jan 2002: (a) same as in Fig. 1 and (b) isobars (solid, every 4 mb), fronts, and station models.

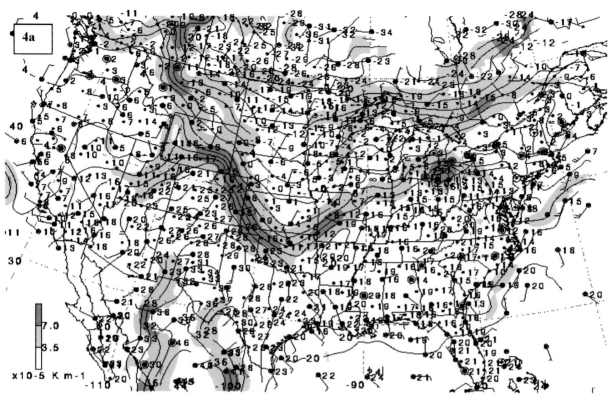

020129/0000 Sfc. Potential Temp. & Grad. (C; x10-5 K m-1)

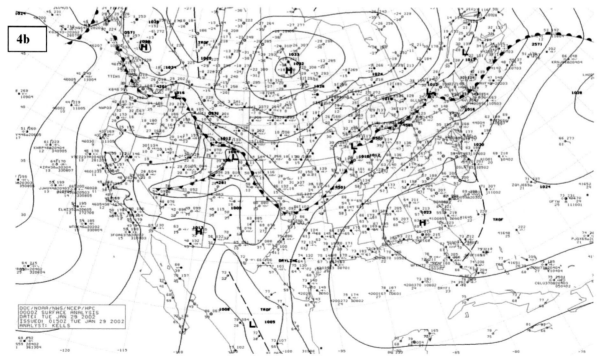

Fig. 4. Same as in Fig. 3, but for 0000 UTC 29 Jan 2002.

The primary feature continues to be the frontal boundary extending across the midwestern United States. We can see in Fig. 4a that the entire length of the front is now associated with a moderate-to-strong potential temperature gradient as the air on the warm side of the front has warmed considerably during the day (in fact the transformation had occurred by 1800 UTC, not shown). However, this same heating has eroded the moderate potential temperature gradient in the southeastern United States along the Atlantic coast at 1200 UTC. NCEP seems to be indicating this transformation with a trough line. The NBZ over southern Canada is still evident especially through Ontario and is still unanalyzed by NCEP. Therefore this example illustrates that significant NBZs can be identified using the potential temperature analyses. In addition, it illustrates that the cross-frontal potential temperature gradient in the Midwest evolves on time scales shorter than 12 h whereas on the NCEP analysis it is still marked with the same symbol.

2) 19–20 FEBRUARY 2002

In this case a moderately strong low pressure system developed over Kansas and Oklahoma and moved northeastward toward the Great Lakes. The NCEP and potential temperature analyses from 1200 UTC 19 February 2002 are shown in Fig. 5. The NCEP analysis has a very complicated system of stationary, cold, warm, and occluded fronts over the central United States (Fig. 5b). However, the accompanying surface potential analysis shows no such structure (Fig. 5a). The NCEP cold front extending southward from the low along the Nebraska–Iowa border corresponds to an area of moderate potential temperature gradient as does some portion of the stationary arctic boundary farther north across the Dakotas. These would be considered fronts using the Sanders method. The other boundaries analyzed by NCEP, including a fully occluded low, would be designated as baroclinic troughs. Meanwhile the most significant feature on the potential temperature map (Fig. 5a) is the moderate-to-strong gradient in the southeastern United States along the Atlantic coast that turns and becomes aligned along the western slopes of the Appalachians indicating vigorous a cold-air damming episode along the East Coast. This would be characterized by Sanders as an NBZ and is completely missing from the NCEP analysis. Another NBZ can be identified over the northeastern United States. The cold potential temperatures over northern New York, Vermont, and southern Canada are caused by overnight radiative cooling associated with clear skies, while the surrounding areas that were cloudy remained warmer. By 0000 UTC 20 February 2002 the NCEP analysis (Fig. 6b) continues to show a complicated frontal system over the lower Mississippi River valley without appreciable potential temperature gradients (Fig. 6a). The NBZs in the southeastern United States along the Atlantic coast and in the northeastern United States have dissipated as daytime

heating has erased the temperature gradients caused by differential radiative cooling.

These two examples demonstrate the most compelling reason for the use of potential temperature analyses. Comparison of the potential temperature analyses and the NCEP analyses allows an analyst to adhere to Wallace and Hobbs's admonition and easily differentiate between thermal gradients associated with fronts (i.e., Sanders' fronts) and those that are not (i.e., NBZs). The potential temperature analyses also show quite nicely that many fronts as analyzed by NCEP are not associated with surface thermal gradients (i.e., BTs). These two examples also show that fronts, baroclinic troughs, and nonfrontal baroclinic zones are strongly influenced by the diurnal heating cycle. Therefore, as might be expected, these features evolve on mesoscale time scales, and do not maintain their characteristics for days at a time as might be inferred from a classic interpretation of the NCM and synoptic scaling.

3. An update to the 1-yr climatology of surface baroclinic zones

To use surface potential temperature analyses, it is useful for an analyst or forecaster to have an understanding of the climatology of the features that can be found on the charts. Therefore, SH02 constructed a climatology of surface baroclinic zones from 1 yr (August 1999–July 2000) of surface potential temperature maps that were produced at the University at Albany. The maps are constructed using a GEMPAK Barnes analysis (Koch et al. 1983) on $0.5° \times 0.5°$ latitude–longitude grid extending from 20° to 55°N and from 135° to 65°W. SH02 subjectively examined the number frequency of strong or moderate potential temperature gradients per state. The results showed that strong potential temperature gradients are found mostly along the U.S. coastlines and along the eastern slopes of the North American cordillera. The seasonal and diurnal variability of these results was also demonstrated. While the number frequency per state approach was more than sufficient to demonstrate the large-scale horizontal structure and frequencies of strong potential temperature gradients, it would be beneficial for forecasters and analysts to visualize the composite structure of the surface potential temperature field and to generate number frequencies for the analyzed grid points themselves. Such a composite would allow users to gain a more quantitative appreciation for the average values of the surface potential temperatures and frequencies of strong and moderate potential temperature gradients. In addition, such a composite would demonstrate any subsynoptic detail that was not available in the original analysis. Therefore, composite surface potential temperature and potential temperature gradients grids were constructed and the results are shown here.

A 1-yr composite analysis of surface potential temperatures was undertaken using the same time period (1

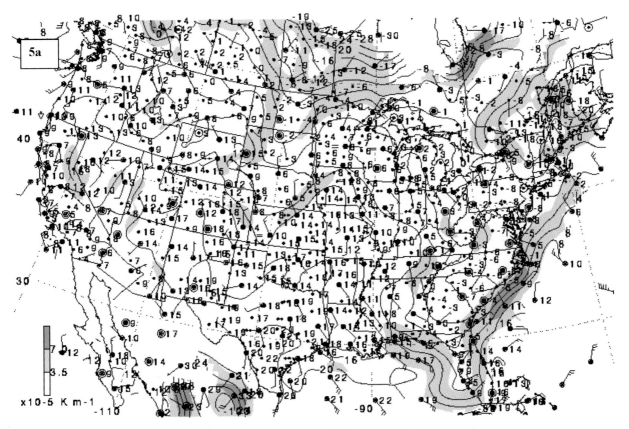

020219/1200 Sfc. Potential Temp. & Grad. (C; x10-5 K m-1)

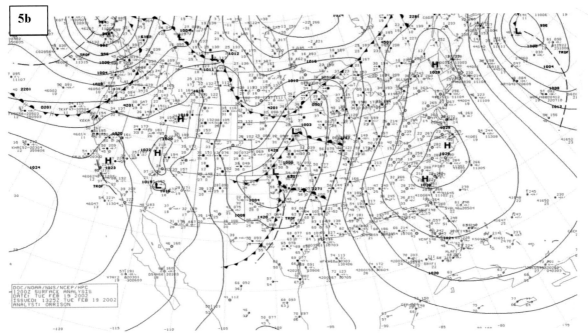

FIG. 5. Same as in Fig. 3, but for 1200 UTC 19 Feb 2002.

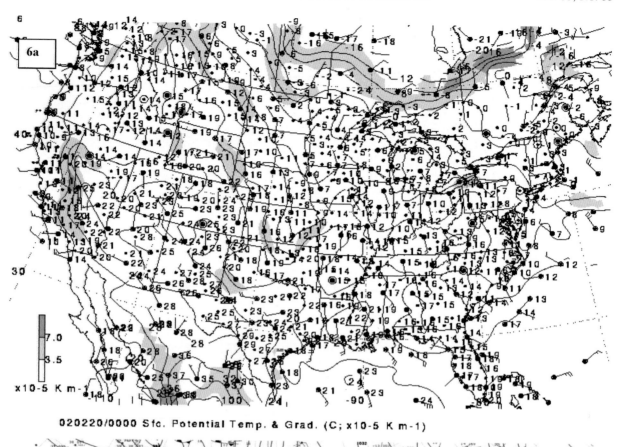

020220/0000 Sfc. Potential Temp. & Grad. (C; x10-5 K m-1)

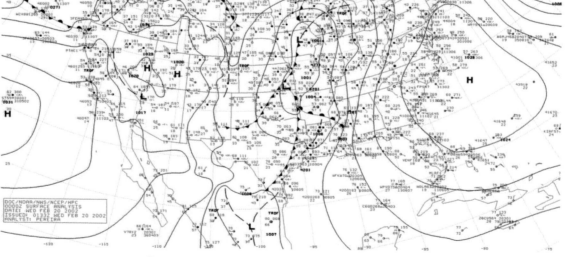

FIG. 6. Same as in Fig. 3, but for 0000 UTC 20 Feb 2002.

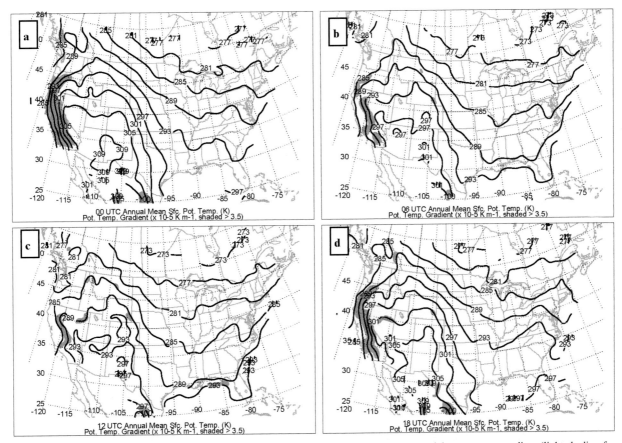

FIG. 7. Annual mean surface potential temperature (K, solid, contour interval is 4 K) and potential temperature gradient (light shading for values > 3.5 × 10⁻⁵ K m⁻¹, dark shading for values >7 × 10⁻⁵ K m⁻¹) for (a) 0000, (b) 0600, (c) 1200, and (d) 1800 UTC.

August 1999–31 July 2000) and method as in SH02. However, because the original gridded analyses were not saved, they had to be reconstructed from the original surface data files available from the University at Albany tape archive. All available surface data [aviation routine weather report (METAR), land-synoptic, ship-synoptic, and Coastal-Marine Automated Network (C-MAN) reports] are used to construct these analyses. Grids of potential temperature were calculated for four time periods per day (0000, 0600, 1200, and 1800 UTC). In the automated GEMPAK analysis routine, if there are less than three observations within the Barnes search radius for a grid point, the grid point is assigned a missing value. Because of the variability in data availability from ship observations over the oceans and inconsistency in data reporting over Mexico, many grid points in these areas frequently get assigned a missing value. Therefore, these missing values must be accounted for in the compositing routine.

The grids for the entire period (1 August 1999–31 July 2000) were used to construct the composite and frequency maps. The composites were constructed by calculating the average potential temperature and potential temperature gradient at each individual grid point. Missing data were not included. However because

of the variability mentioned above, some grid points had considerably fewer observations contributing to the average. Therefore, only those grid points with greater than ⅔ of the observations for the time period were included in the composite (>244 for the annual composite and >60 for the seasonal composites). The other composite grid points were assigned missing values. The same constraints were used for the number frequencies of moderate and strong potential temperature gradients.

a. Mean surface potential temperatures

The 1-yr composite surface potential temperature analysis for 0000, 0600, 1200, and 1800 UTC over the United States, southern Canada, and adjacent nearshore waters is shown in Fig. 7. For all times, the highest potential temperature values occur over the southwestern United States and northern Mexico. Enhanced potential temperature gradients exist along the Pacific coast and the Sierra Nevada Mountains in California as well as along the eastern slopes of the Rockies primarily during the daylight hours at 1800 and 0000 UTC (see Figs. 7a,d). In the central and eastern parts of the United States, the basic north–south temperature gradient is evident with the exception of the small thermal ridge in

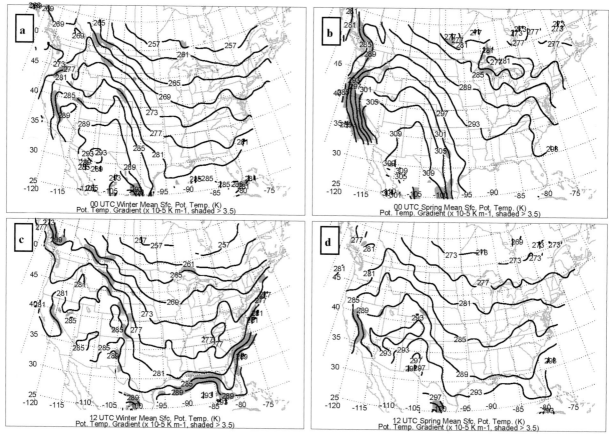

FIG. 8. Seasonal mean surface potential temperature, with same contours, contour interval, and shading as in Fig. 7 but for 0000 UTC in (a) winter and (b) spring and 1200 UTC in (c) winter and (d) spring.

the vicinity of the Appalachians. As was pointed out in SH02 the enhanced potential temperature gradients along the sloping terrain can be mostly attributed to the average static stability of the atmosphere and the influence of elevation on the potential temperature gradient (refer to Fig. 2b). In the southeastern United States along the Atlantic and Gulf coasts of the United States a third area of enhanced potential temperature gradient is shown primarily during the nighttime and early morning periods (0600 and 1200 UTC, Figs. 7b,c). These coastal gradients are mostly due to the land–sea thermal contrasts associated with the warm ocean waters of the Gulf of Mexico and the Gulf Stream.

Seasonal and diurnal variability in the mean potential temperatures are shown in Figs. 8 and 9. At 0000 UTC in both winter and spring (Figs. 8a,b) moderate-to-strong mean potential temperature gradients are evident along the Sierra Nevada Mountains and the eastern slopes of the Rockies. These gradients are due to daytime heating in reduced air density at higher elevations. In California, the mean potential temperature gradient is enhanced further by the proximity of the Sierra Nevada and coastal mountains to the cool California Current. In the spring at 0000 UTC (Fig. 8b), further enhancement of potential temperature gradients between

warm land and cool waters can be observed in the vicinity of the Great Lakes with the mean potential temperature gradient attaining moderate values north of Lake Superior. At 1200 UTC (Figs. 8c,d), overnight cooling in the mountains reduces the gradients found along the mountains slopes. The opposite diurnal land–sea thermal contrast becomes evident at 1200 UTC as the potential temperature gradient becomes enhanced in the southeastern United States along the Atlantic and Gulf coasts between the radiatively cooled land and the warm ocean waters (Figs. 8c,d). Naturally, this enhancement is most evident in the winter composite map (Fig. 8c).

As might be expected, the summer and autumn mean potential temperature maps (Figs. 9a–d) continue to show the same features as were seen in the winter and spring. Notably, the highest mean potential temperature gradients can be found along the northern California and southern Oregon coasts at 0000 UTC in the summer where the mean gradient is actually categorized as strong (Fig. 9a). In the summer, off the coast of the northeastern United States, the mean potential temperature gradient is enhanced, especially along the Maine coast (Fig. 9a). As might be expected the enhanced mean potential temperature gradients in the southeastern Unit-

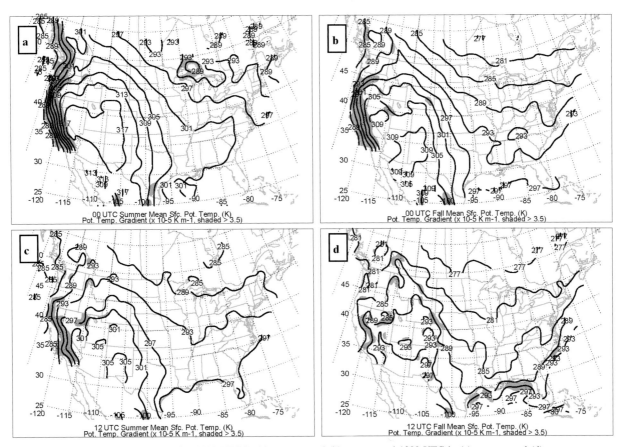

FIG. 9. Same as in Fig. 7, but for 0000 UTC in (a) summer and (b) autumn and 1200 UTC in (c) summer and (d) autumn.

ed States along the Atlantic and Gulf coasts at 1200 UTC are weakest in summer and strongest in winter (cf. Figs. 8c and 9c). All of these results compare directly to results from SH02. However, SH02 surmised from the number frequencies by state that the numbers of strong potential temperature gradients lying along the eastern slopes of the North American cordillera were greatest at 1200 UTC because of increased stability overnight. Yet, Figs. 7, 8, and 9 seem to suggest that in the mean the area of enhanced potential temperature gradient is largest at 0000 UTC. Perhaps this can be reconciled by comparing the gridpoint frequencies of moderate-to-strong gradients to the per state frequencies in SH02.

b. Frequency of moderate and strong potential temperature gradients

Figure 10 shows the number of moderate potential temperature gradients for the winter and spring seasons at 0000 and 1200 UTC. As might be expected, high frequencies of moderate gradients occur in the same locations where the mean potential temperature gradient is enhanced. Moderate potential temperature gradients are found nearly every day at 0000 UTC along the northern California coasts and the western slopes of the Sierra

Nevada Mountains (>75 per 91 days). Along the eastern slopes of the Rockies, it can be seen that the frequency of moderate gradients is higher at 1200 UTC (Figs. 10c,d) than at 0000 UTC (Figs. 10a,b). This corresponds directly with the result in SH02 despite the overall area of the mean potential temperature gradient being larger at 0000 UTC (see above). Interestingly, both the winter and spring frequency maps indicate that at 0000 UTC (Figs. 10a,b) moderate gradients become more frequent away from the slopes of the southern Rockies over eastern Colorado, western Kansas, and in the Oklahoma and Texas panhandles. Another feature not readily apparent in the mean potential temperature maps is the relative higher frequency of moderate potential temperature gradients at 1200 UTC in the winter (Fig. 10c) along the western slopes of the Appalachians. It could be surmised that this is due to the blocking effect of these mountains on cold fronts approaching from the west (Schumacher et al. 1996; O'Handley and Bosart 1996).

In the summer and autumn (Figs. 11a–d), the persistence of moderate gradients along the West Coast is most evident. Spring (Fig. 10b) and summer (Fig. 11a) also show high frequencies of moderate potential temperature gradients around the northern Great Lakes. The relatively high number of moderate potential temperature gradients in summer over Maine mentioned in SH02

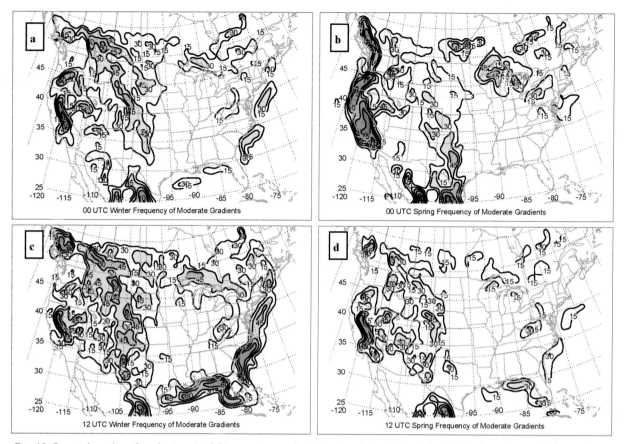

FIG. 10. Seasonal number of moderate potential temperature gradients (contour interval is 15, shaded > 30) for 0000 UTC in (a) winter and (b) spring and 1200 UTC in (c) winter and (d) spring.

is evident (Fig. 11a) as well as a region of higher frequency lying on the north side of the climatological location of the Gulf Stream in the oceans south of New England (Fig. 11a). At 1200 UTC during the autumn season the frequency map has a chaotic look over the Rockies (Fig. 11d). This seems to indicate that radiative cooling takes place in favored locations during the autumn creating a large number of small-scale enhanced potential temperature gradient features in persistent locations.

The occurrence of strong potential temperature gradients (Fig. 12) in this analysis is much lower than the occurrence of moderate potential temperature gradients throughout all the seasons. In the summer at 0000 UTC (Fig. 12b), the only location where strong gradients occur with regularity is along the length of the Pacific coast. While at 1200 UTC (Fig. 12d), strong gradients can only be found with regularity along the west slopes of the Sierra Nevada Mountains and the Coastal Range of northern and central California. In the winter, strong gradients at 0000 UTC (Fig. 12a) occur with regularity (more than 15 times in ~90 days) along the eastern slopes of the Canadian Rockies and in the sloping terrain surrounding the elevated great basin of Nevada and southwestern Oregon (Fig. 12a). At 1200 UTC strong

gradients become more frequent in several locations: 1) on the eastern slopes of the Rockies extending farther southward than in winter, 2) in the Fraser River Valley in British Columbia, 3) on the northern side of Lake Superior, and 4) in the southeastern United States along the Atlantic and Gulf coasts (Fig. 12b). At 0000 UTC in the summer (Fig. 12b), strong potential temperature gradients are as frequent as moderate potential temperature gradients along the west coast of the United States.

It was noted in SH02 that the automated Barnes analysis typically underrepresents the magnitude of the potential temperature gradient along coastlines because of the typical higher density of observations over land. SH02 manually corrected the gradients along the coast thereby increasing the frequency of strong potential temperature gradients. No such correction was performed in this analysis, which may be leading to the lower number of strong potential temperature gradients in this study.

c. Subsynoptic features

Several interesting subsynoptic details can be gleaned from the mean potential temperatures and the frequency of moderate and strong potential temperature gradients

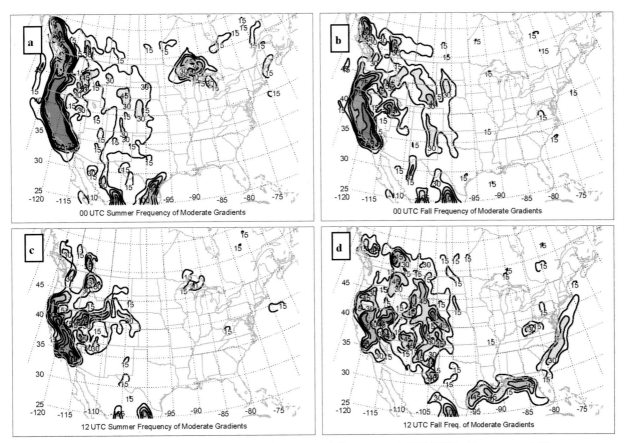

FIG. 11. Same as in Fig. 10, but for 0000 UTC in (a) summer and (b) autumn and 1200 UTC in (c) summer and (d) autumn.

that were not necessarily apparent in SH02. For example, it can be seen that the high frequency of strong gradients observed by SH02 in Michigan is mostly due to land–water contrasts between Lake Superior and the upper peninsula but not as significantly between Lakes Michigan, Huron, and the lower peninsula. Another observation is that the frequency of strong and moderate potential gradients extends meridionally along the entire west coast of the United States at 0000 UTC in all seasons, but at 1200 UTC there is a distinct minimum in northern Oregon and Washington. The relatively high number of strong and moderate potential temperature gradients in Texas observed in SH02 can be seen to be associated with both the southwestern edge of cold air masses at 0000 UTC in the spring in the Texas Panhandle (Figs. 10a,c) and the land–sea contrast along the Gulf coast at 1200 UTC in the autumn and winter (Figs. 10b and 11b). Looking at the results in the New England region in SH02, one might wonder about the minimum number of per state moderate and strong gradients for Connecticut and Rhode Island. It is clear from the analysis above (Fig. 10c) that, at least on the scale of these analyses, strong and moderate potential temperature gradients do not occur often in southern New England but can be found along the Gulf of Maine and farther south on the Atlantic coastline. These are just some of the

additional subsynoptic features of the mean surface potential temperatures and frequency of moderate and strong gradients. Meteorologists using surface potential temperature analysis are encouraged to determine the significant relationships for their particular forecast areas.

d. The relationship of fronts to baroclinic zones

One important result from SH02 was that the frequency of fronts on current NCEP analyses and those analyzed by Morgan et al. (1975) from the daily weather map series, does not appear to generally coincide with the frequency of strong potential temperature gradients. The results here suggest that this statement might even include moderate potential temperature gradients. Perhaps the most striking feature on Figs. 10 and 11 is the relative infrequency of occurrence of even moderate potential temperature gradients in the central United States away from the slopes of the Rockies and the immediate coastlines. Synoptic time scaling suggests that in a ~90-day season one might expect an extratropical cyclone and its associated fronts to cross the central United States every 3–5 days. If fronts associated with these cyclones regularly possessed even moderate potential temperature gradients, one would expect the minimum

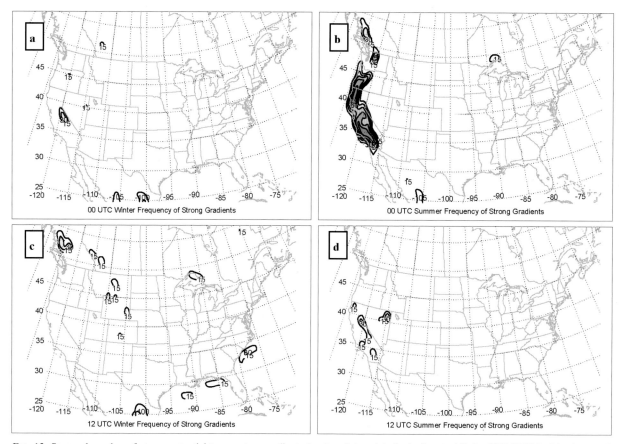

FIG. 12. Seasonal number of strong potential temperature gradients (contour interval is 5, shading > 15) for 0000 UTC in (a) winter and (b) summer and 1200 UTC in (c) winter and (d) summer.

value on Figs. 10 and 11 to be between 20 and 30. Instead, in the central United States, moderate potential temperature gradients occur less than 15 times in ~90 days or about ½ of what might be expected. This seems to match well with SH02 result that less than half of all cold fronts analyzed by NCEP are associated with moderate potential temperature gradients. As shown in SH02, the frequency of frontal occurrence identified by Morgan et al. (1975) is much higher than the observed frequency of moderate and strong potential temperature gradients.

This result is most intriguing. It led Dr. Fred Sanders to consider the relationship of fronts, frontogenesis, and surface baroclinic zones in his presentation at the January 2004 symposium honoring his career at the American Meteorological Society's annual meeting. One could argue that because moderate and strong baroclinic zones are infrequent over the interior central United States, Sanders' (1999) arbitrary choice of 4 and 8 K $(220 \text{ km})^{-1}$ ($\sim$3.5 and 7×10^{-5} K m^{-1}) for values of moderate and strong gradients, respectively, would not accurately represent an anomalous temperature gradient in this region. Perhaps a baroclinic trough associated with a weak temperature gradient (i.e., only 2×10^{-5} K m^{-1}) that is oriented north–south, or about 90° from

the background gradient (cf. Fig. 7), could actually be considered a front. Just such a front occurred in the vicinity of the Mississippi Valley during 1 March 2000 (Fig. 13b). The potential temperature map (Fig. 13a) shows a weak potential temperature gradient oriented north–south associated with the frontal feature on the NCEP analysis (Fig. 13b). In addition, moderate and strong gradients can be seen along the western slopes of the Appalachians and in the southeastern United States along the Atlantic and northeastern Gulf coasts in response to significant cold-air damming. Because the potential temperature gradient associated with the surface trough and wind shift is anomalous for this region, it may be useful in the future, to also plot potential temperature gradient anomalies to identify those areas where the gradients, though weak, may be significant. Figure 14 shows just such a plot where the potential temperature gradient anomaly is calculated as a difference between the observed values and the mean spring mean potential temperature gradient. Figure 14 clearly shows that the NCEP analyzed front in the central United States is associated with a decent potential temperature gradient anomaly. It also shows that the potential temperature gradient along the Appalachians is also anomalously strong. It seems, therefore, that anomalous

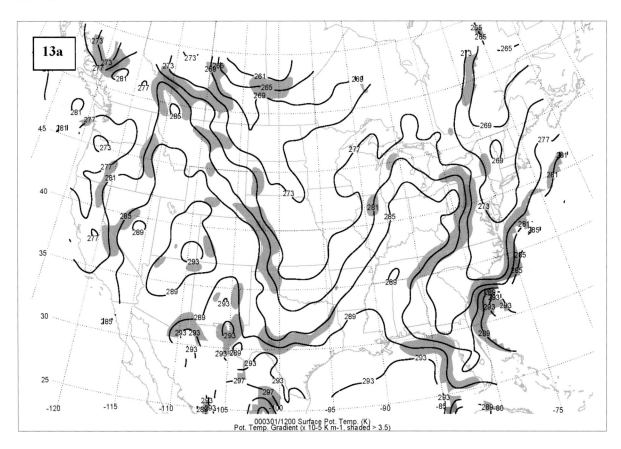

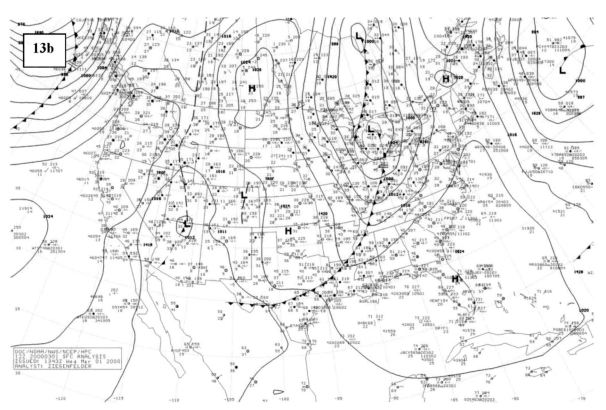

Fig. 13. Same as in Fig. 3, but with station models omitted in (a) and for 1200 UTC 1 Mar 2000.

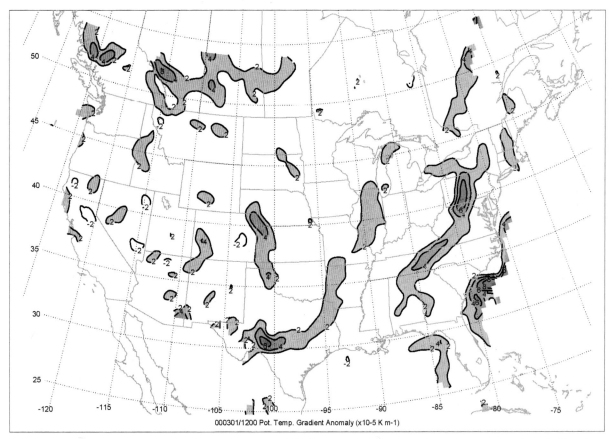

Fig. 14. Surface potential temperature gradient anomaly at 1200 UTC 1 Mar 2000 (solid lines, contour interval is 2×10^{-5} K m^{-1}, shading for values > 2).

gradients could be used to identify what could be termed a front even though it would not be classified as one according to Sanders' (1999) criteria.

4. Concluding discussion

Fred Sanders spent a lifetime dedicated to studying various aspects of the atmosphere and in the last 10 yr focused a good deal of attention on, and was critical of, current surface analysis methods. This led Sanders (1999) to propose a new method for surface analysis that makes use of surface potential temperatures and their gradients to distinguish between fronts, baroclinic troughs, and nonfrontal baroclinic zones. This method was put into practice at University at Albany in the late 1990s and shown to have value for understanding the complexity and richness of surface observations and atmospheric structure. The distinct advantage of using surface potential temperatures is to alleviate spurious thermal gradients associated with varying elevations and it has been shown that these maps allow analysts to quickly distinguish between surface troughs with and without significant thermal gradients as well as identify other nonfrontal baroclinic zones.

A climatology of surface baroclinic zones using these

surface potential temperature analyses has been constructed by SH02 and revisited here. This climatology shows that surface baroclinic zones over the United States are most frequently found along the coasts and slopes of the North American cordillera. A quantitative estimate of the annual and seasonal mean surface potential temperatures was presented and the results compare favorably to SH02. Mean potential temperature maps and gridpoint frequencies of moderate and strong potential temperature gradients gave additional subsynoptic detail not found in SH02. SH02 demonstrated that most fronts analyzed using current practices do not correspond to regions of moderate surface baroclinic zones. However, this paper has shown that some fronts that would not qualify as fronts using the Sanders' (1999) criteria may still have an anomalously strong gradient.

In the opinion of this author, the atmosphere is capable of producing a spectrum of surface boundaries. These boundaries can possess many different characteristics (e.g., thermal, moisture, pressure, wind shift, etc.). Attaching the name "front," "baroclinic trough," "airstream boundary," or "dryline" to a boundary does not make it any more or less important to the forecast. However, it is important to understand and identify the characteristics that distinguish these boundaries. Therefore, while surface potential temperature analysis is valuable to distinguish those boundaries associated with (and without) thermal gradients, it cannot be used in a vacuum. Surface potential temperature analysis should be considered as a part of the entire forecast process that should include analysis of the all the important elements and available data.

Acknowledgments. Fred Sanders' many contributions have left an indelible mark on the field of meteorological science. I am grateful that he took an interest in my studies when I was in graduate school. Our many interesting discussions on the topic of frontal analysis have been a wonderful inspiration. In addition, I thank all of the faculty, staff, and graduate students at the University at Albany for all their help in acquiring data and allowing me to continue to use their computing resources even though I am now located at Plymouth State University. In particular, the work presented here would not have come to fruition without the encouragement of my colleagues and mentors, Dr. Lance Bosart, and Dr. Daniel Keyser. Last I also thank the two anonymous reviewers for helping me improve this manuscript.

REFERENCES

Ahrens, C. D., 2005: *Essentials of Meteorology: An Invitation to the Atmosphere.* 4th ed. Brooks/Cole, 473 pp.

Albers, S. C., J. A. McGinley, D. L. Birkenheuer, and J. R. Smart, 1996: The Local Analysis and Prediction System (LAPS): Analyses of clouds, precipitation, and temperature. *Wea. Forecasting,* **11,** 273–287.

Bell, G. D., and L. F. Bosart, 1988: Appalachian cold-air damming. *Mon. Wea. Rev.,* **116,** 137–161.

Bjerknes, J., 1919: On the structure of moving cyclones. *Geof. Publ.,* **1,** 1–9.

——, and H. Solberg, 1922: Life cycle of cyclones and the polar front theory of atmospheric circulation. *Geofys. Publ.,* **3,** 1–18.

Bretherton, F. P., 1966: Critical layer instability in baroclinic flows. *Quart. J. Roy. Meteor. Soc.,* **92,** 325–334.

Hewson, T. D., 1997: Objective identification of frontal wave cyclones. *Meteor. Appl.,* **4,** 311–315.

——, 1998: Objective fronts. *Meteor. Appl.,* **5,** 37–65.

Hoskins, B. J., and F. P. Bretherton, 1972: Atmospheric frontogenesis models: Mathematical formulation and solution. *J. Atmos. Sci.,* **29,** 11–37.

——, M. E. McIntyre, and A. Robertson, 1985: On the use and significance of isentropic potential vorticity maps. *Quart. J. Roy. Meteor. Soc.,* **111,** 877–946.

Koch, S. E., M. DesJardins, and P. S. Kocin, 1983: An iterative Barnes objective map analysis scheme for use with satellite and conventional data. *J. Climate Appl. Meteor.,* **22,** 1487–1503.

Mass, C. F., 1991: Synoptic frontal analysis: Time for a reassessment? *Bull. Amer. Meteor. Soc.,* **72,** 348–363.

Morgan, G. M., Jr., D. G. Brunkow, and R. C. Beebe, 1975: Climatology of surface fronts. Illinois State Water Survey Circular 122, Dept. of Registration and Education, Urbana, IL, 46 pp.

Namias, J., 1983: The history of polar front and air mass concepts in the United States—An eyewitness account. *Bull. Amer. Meteor. Soc.,* **64,** 734–755.

O'Handley, C., and L. F. Bosart, 1996: The impact of the Appalachian Mountains on cyclonic weather systems. Part I: A climatology. *Mon. Wea. Rev.,* **124,** 1353–1373.

Petterssen, S., 1940: *Weather Analysis and Forecasting.* 1st ed. McGraw-Hill, 503 pp.

Sanders, F., 1999: A proposed method of surface map analysis. *Mon. Wea. Rev.,* **127,** 945–955.

——, and C. A. Doswell, 1995: A case for detailed surface analysis. *Bull. Amer. Meteor. Soc.,* **76,** 505–521.

——, and E. G. Hoffman, 2002: A climatology of surface baroclinic zones. *Wea. Forecasting,* **17,** 774–782.

Schumacher, P. N., D. J. Knight, and L. F. Bosart, 1996: Frontal interaction with the Appalachian Mountains. Part I: A climatology. *Mon. Wea. Rev.,* **124,** 2453–2468.

Steenburgh, W. J., and T. R. Blazek, 2001: Topographic distortion of a cold front over the Snake River plain and central Idaho mountains. *Wea. Forecasting,* **16,** 301–314.

Uccellini, L. W., S. F. Corfidi, N. W. Junker, P. J. Kocin, and D. A. Olson, 1992: Report on the Surface Analysis Workshop held at the National Meteorological Center, 25–28 March 1991. *Bull. Amer. Meteor. Soc.,* **73,** 459–473.

Wallace, J. M., and P. V. Hobbs, 1977: *Atmospheric Science: An Introductory Survey.* Academic Press, 467 pp.

Chapter 9

Dynamical Diagnosis: A Comparison of Quasigeostrophy and Ertel Potential Vorticity

JOHN W. NIELSEN-GAMMON

Department of Atmospheric Sciences, Texas A&M University, College Station, Texas

DAVID A. GOLD*

Department of Atmospheric Sciences, Texas A&M University, College Station, Texas

(Manuscript received 5 December 2005, in final form 28 July 2006)

Nielsen-Gammon   **Gold**

ABSTRACT

Advances in computer power, new forecasting challenges, and new diagnostic techniques have brought about changes in the way atmospheric development and vertical motion are diagnosed in an operational setting. Many of these changes, such as improved model skill, model resolution, and ensemble forecasting, have arguably been detrimental to the ability of forecasters to understand and respond to the evolving atmosphere. The use of nondivergent wind in place of geostrophic wind would be a step in the right direction, but the advantages of potential vorticity suggest that its widespread adoption as a diagnostic tool on the west side of the Atlantic is overdue. Ertel potential vorticity (PV), when scaled to be compatible with pseudopotential vorticity, is generally similar to pseudopotential vorticity, so forecasters accustomed to quasigeostrophic reasoning through the height tendency equation can transfer some of their intuition into the Ertel-PV framework. Indeed, many of the differences between pseudopotential vorticity and Ertel potential vorticity are consequences of the choice of definition of quasigeostrophic PV and are not fundamental to the quasigeostrophic system. Thus, at its core, PV thinking is consistent with commonly used quasigeostrophic diagnostic techniques.

1. Introduction

Much of Fred Sanders' scientific legacy involves the diagnosis of midlatitude weather systems through the use of the simplest possible techniques. Among these

---

* Current affiliation: Silver Lining Tours, Houston, Texas.

Corresponding author address: John W. Nielsen-Gammon, Dept. of Atmospheric Sciences, Texas A&M University, 3150 TAMUS, College Station, TX 77843-3150.
E-mail: n-g@tamu.edu

techniques is the use of an analytical quasigeostrophic model as a comprehensive, self-consistent, but directly understandable representation of midlatitude disturbances (Sanders 1971). More recently, the formal superiority of **Q** vectors as a diagnostic tool for vertical motion compared with vorticity and temperature advection, led to the elucidation of a simple technique for estimating **Q** vectors by eye from weather maps (Sanders and Hoskins 1990).

The purpose of this article is to consider the present state of diagnosis, quasigeostrophic and otherwise, for weather forecasting and dynamical understanding, and in particular to investigate the validity (for optimists) and sources of error (for pessimists) of quasigeostrophic (QG) potential vorticity (PV) diagnosis compared with diagnosis using Ertel PV. Section 2 discusses the current practice of diagnosis in the context of present-day meteorology. In view of the expanding role of PV-based diagnosis of weather systems, section 3 reviews the mathematical distinctions between the pseudopotential vorticity (pseudo PV) and Ertel PV and examines those differences in the context of a rapidly deepening midlatitude cyclone. Concluding remarks are provided in section 4.

In this article, we aspire to be as wise and provocative as Fred Sanders. We suspect, however, that the former is much more elusive than the latter.

2. Diagnosis: Keeping up with the changing meteorological landscape

In recent decades, major changes have taken place in the diagnosis of midlatitude weather systems. Four agents of these changes are the following: 1) the widespread availability of desktop computing power, 2) the improved spatial resolution of numerical weather prediction (NWP) models, 3) the improved accuracy of NWP models, and 4) the development of qualitative and quantitative diagnostic techniques associated with potential vorticity.

With computing power comes the ability to compute forcing for vertical motion. We argue that this computation in some ways makes the inference of vertical motion more difficult, and in particular inhibits diagnosis of the dynamical causes and sensitivities of the vertical motion field.

With improved spatial resolution comes the representation of finer and finer scales of motion in model output. Display and diagnostic approaches based on computing geostrophic winds from heights or pressures are seriously degraded and require massive amounts of smoothing. We argue that using the nondivergent wind in place of the geostrophic wind is fully consistent with quasigeostrophic theory (in fact, is theoretically preferred) and allows more aspects of the balanced dynamics to be retained and diagnosed.

With improved model accuracy comes a change in the nature of the value that a forecaster can add to a numerical forecast. We argue that there are substantial opportunities to add value to ensemble forecasts, but that those opportunities are lost because of a lack of commonly accepted tools (beyond ensemble spread) to diagnose forecast uncertainty easily.

With PV diagnostics, we argue, comes a tool whose proper application may successfully address most of these issues. Here we contrast traditional and PV-based approaches to diagnosis, and in later sections we explore the compatibility between QG-based PV and Ertel PV.

a. Computer power and the purpose of quasigeostrophic diagnosis

The computing power now available on individual workstations is vastly superior to what was available, for example, in 1990, and in turn is undoubtedly vastly inferior to what will be available 16 years hence. One consequence of this fingertip computational resource is the ability to easily compute diagnostic quantities that in the past would have been prohibitively time consuming in an operational environment. For example, following the notation of Holton (1992, chapter 6), the quasigeostrophic omega equation written in traditional form is

$$\left(\sigma\nabla^2 + f_0^2\frac{\partial^2}{\partial p^2}\right)\omega = f_0\frac{\partial}{\partial p}\left[\mathbf{V}_g \cdot \boldsymbol{\nabla}\left(\frac{1}{f_0}\nabla^2\Phi + f\right)\right]$$
$$+ \nabla^2\left[\mathbf{V}_g \cdot \boldsymbol{\nabla}\left(-\frac{\partial\Phi}{\partial p}\right)\right] \quad (1)$$

and in $\mathbf{Q}$-vector form on an f plane is

$$\left(\sigma\nabla^2 + f_0^2\frac{\partial^2}{\partial p^2}\right)\omega = -2\boldsymbol{\nabla}\cdot\mathbf{Q}, \quad (2a)$$

where

$$\mathbf{Q} = \left(-\frac{R}{p}\frac{\partial\mathbf{V}_g}{\partial x}\cdot\boldsymbol{\nabla}T\right)\mathbf{i} + \left(-\frac{R}{p}\frac{\partial\mathbf{V}_g}{\partial y}\cdot\boldsymbol{\nabla}T\right)\mathbf{j}. \quad (2b)$$

A list of symbols is provided in Table 1. As is well known but still sometimes misunderstood, the right-hand sides of (1) and (2a) are mathematically identical at the level of quasigeostrophic theory. This can be shown using the equations for geostrophic and hydrostatic balance:

$$\mathbf{V}_g = \frac{1}{f_0}\mathbf{k} \times \boldsymbol{\nabla}\Phi \quad \text{and} \quad (3)$$

$$-\frac{\partial\Phi}{\partial p} = \frac{RT}{p}. \quad (4)$$

By making certain simplifying assumptions, or by using other simplified forms of the equation as in Sutcliffe (1947) and Trenberth (1978), it is possible to "solve" the omega equation by inspection of weather maps. The most common simplification is the assumption that the left-hand side of the equation is roughly proportional to $-\omega$ (see Durran and Snellman 1987 for an evaluation of this assumption). Other simplifications specific to the traditional form are that the vertical derivative of the vorticity advection is proportional to the vorticity advection itself in the midtroposphere and that the negative Laplacian of temperature advection is proportional to the temperature advection itself. A final simplification is that the forcing important to deep vertical motion can be adequately diagnosed using one or two standard levels or a layer. With assumptions such

TABLE 1. List of symbols.

Symbol	Meaning	Value/definition
Q	Ertel potential vorticity	Eq. (7)
$\check{Q}$	Scaled Ertel potential vorticity	Eq. (20)
$\mathbf{Q}$	$\mathbf{Q}$ vector	Eq. (2b)
R	Gas constant for dry air	$287 \text{ J kg}^{-1} \text{ K}^{-1}$
T	Temperature	
$\mathbf{V}$	Horizontal wind	
$\mathbf{V}_g$	Geostrophic wind	
f	Coriolis parameter	
f_0	Reference Coriolis parameter	$1.0 \times 10^{-4} \text{ s}^{-1}$
g	Gravitational acceleration	9.8 m s^{-2}
$\mathbf{i}$	Unit vector in the x direction	
$\mathbf{j}$	Unit vector in the y direction	
$\mathbf{k}$	Unit vector in the $-p$ (upward) direction	
p	Pressure	
p_0	Reference pressure	1000 hPa
q	Pseudopotential vorticity	Eq. (12c)
q_{com}	Pseudopotential vorticity as commonly computed	Eq. (22)
Φ	Geopotential height	
ζ	Vertical component of relative vorticity in isobaric coordinates	
ζ_{ag}	Vertical component of ageostrophic relative vorticity	
ζ_g	Geostrophic relative vorticity	
ζ_θ	Vertical component of relative vorticity in isentropic coordinates	
κ	R/c_p	2/7
θ	Potential temperature	$T\left(\dfrac{p_0}{p}\right)^\kappa$
$-\hat{\theta}_p$	Basic-state stratification	$-\dfrac{\partial \hat{\theta}}{\partial p}$
σ	QG stability	$-\dfrac{R\hat{T}}{p\hat{\theta}}\hat{\theta}_p$
ψ	Streamfunction	$u = -\dfrac{\partial \psi}{\partial y}, \quad v = \dfrac{\partial \psi}{\partial x}$
ω	Omega	$\dfrac{dp}{dt}$
∇^2	(Horizontal) Laplacian	$\dfrac{\partial^2}{\partial x^2} + \dfrac{\partial^2}{\partial y^2}$
$J(a, b)$	Jacobian operator	$\dfrac{\partial a}{\partial x}\dfrac{\partial b}{\partial y} - \dfrac{\partial a}{\partial y}\dfrac{\partial b}{\partial x}$
$\dfrac{d}{dt}$	Lagrangian derivative	$\dfrac{\partial}{\partial t} + u\dfrac{\partial}{\partial x} + v\dfrac{\partial}{\partial y} + \omega\dfrac{\partial}{\partial p}$
$\dfrac{d_h}{dt}$	Horizontal Lagrangian derivative	$\dfrac{\partial}{\partial t} + u\dfrac{\partial}{\partial x} + v\dfrac{\partial}{\partial y}$
$\dfrac{d^{(0)}}{dt}$	Lagrangian derivative to leading order	$\dfrac{\partial}{\partial t} + u^{(0)}\dfrac{\partial}{\partial x} + v^{(0)}\dfrac{\partial}{\partial y}$
a	An arbitrary variable	
$a^{(0)}$	Approximate value of a to lowest order	
$\hat{a}$	Basic-state value of a, constant in x, y, and t	
a'	Departure of a from basic-state value	

as these, one or two conventional weather maps suffice for the adept forecaster to subjectively infer the vertical motion pattern associated with synoptic-scale motions in the atmosphere.

Computer power has made many of these simplifications fade from practice. Computers can compute, essentially instantaneously, the entire right-hand side of the omega equation, and perhaps soon it will become operationally routine to solve the omega equation directly for the vertical motion. Presently, a standard diagnostic quantity displayed on National Weather Service workstations is the 700–400-hPa mean $\mathbf{Q}$-vector divergence, multiplied by -2.

Certainly computing the right-hand side in this way has its advantages. Meteorologists need not estimate the sum of two quantities that partially cancel (as in the traditional omega equation) or neglect some nonnegligible forcing terms (as in certain other forms of the omega equation). Nevertheless, this technique creates some problems of its own.

One problem is the introduction of a scale inconsistency in the solution of the omega equation. In all but the $\mathbf{Q}$-vector form, both the forcing and the response involve a horizontal Laplacian. The Laplacian operator is a scale-selective operator, and the Laplacian of temperature advection is much "noisier" than the temperature advection itself. Other forcing terms (including the $\mathbf{Q}$-vector forcing term) are of similarly high order and thus are similarly noisy. However, the noisy fields are equated in the omega equation to the (three dimensional) Laplacian of the vertical motion field, and while the Laplacian of vertical motion will be just as noisy as the forcing, the vertical motion itself will be smooth. To a large extent, the neglect of the Laplacian on one side of the equation was balanced by neglecting the Laplacian on the other. With no such compensation now taking place, the meteorologist must somehow infer a smooth omega field from a noisy forcing field.

The smoothing problem will go away if computers routinely solve the omega equation for vertical motion, but a separate problem, already serious, will become even worse. This problem is the loss of the ability to predict the impact of errors in analyses and forecasts. When subjective evaluation of the omega equation was routine, meteorologists' skill in diagnosing vertical motion was matched by their skill in diagnosing errors in vertical motion. A meteorologist could easily and directly infer how a change in the intensity or location of a temperature gradient, for example, would affect the distribution of vertical motion, because he or she was using aspects of one to infer the other. Computer display of a computed quantity such as $\mathbf{Q}$-vector divergence obscures the direct relationship between the forcing and the weather patterns, such as the deformation of a temperature gradient, that give rise to it. This problem can be eased somewhat by adoption of the $\mathbf{Q}$-vector partitioning method of Keyser et al. (1992), which allows $\mathbf{Q}$-vector convergence to be treated as having a system-scale component and a frontal-scale component. Other partitionings are also possible (e.g., Jusem and Atlas 1998). Otherwise, the meteorologist is left with the vague hope that changes in the intensity or location of, say, a trough upstream of a region of upward forcing will lead directly to proportional changes in the intensity or location of the vertical motion.

Weather systems are sensitive to details in the temperature and wind patterns. A slight change in the orientation of the wind relative to a temperature gradient can change the magnitude of vertical motion substantially and may even change its sign. Awareness of the subtleties of the weather patterns and their situation-specific consequences to the forecast can be lost when these connections are not made evident to the meteorologist through the process of diagnosis.

b. Improved spatial resolution and quasigeostrophic diagnosis

1) CONTAMINATION OF THE GEOSTROPHIC WIND

The resolution of NWP models regularly improves. Along with this improvement has come a slower but similarly important increase in the ability to deliver high-resolution model output to the forecaster. Both of these factors expose the forecaster to much more detail in NWP output than was available in 1990. No longer are the models constrained by resolution to simulate weather phenomena that comply with the basic assumptions of quasigeostrophic theory. Both gravity waves and mesoscale convective systems now routinely appear in operational NWP output, and routine operational simulations of individual thunderstorms at very short range may not be far away.

Apart from the challenge of making sense of so much more information, there is the added difficulty of extracting useful geostrophic wind information for quasigestrophic diagnosis. Difficulty arises because the pressure gradient is instrumental as a driver of atmospheric motion at all scales, so every form of motion except inertial will have a fundamental signal in the geopotential height field.

To see how the multiple signals contaminate the calculation of geostrophic wind, consider the horizontal inviscid equation of motion:

$$\frac{d\mathbf{V}}{dt} + f\mathbf{k} \times \mathbf{V} + \boldsymbol{\nabla}\Phi = 0. \qquad (5)$$

The slowly evolving flow will follow an approximate geostrophic balance between the second and third terms. Embedded within this flow will be rapidly varying motions (in a Lagrangian sense) for which the second term is smaller than the other two terms. Using (3), the latter statement is equivalent to

$$|f\mathbf{k} \times \mathbf{V}| \ll |f\mathbf{k} \times \mathbf{V}_g|, \quad \text{or} \qquad (6a)$$

$$|\mathbf{V}| \ll |\mathbf{V}_g|, \qquad (6b)$$

where the geostrophic wind is defined using the local value of f. Even if the rapidly varying portion of the wind is relatively weak, the computed geostrophic wind associated with these motions will be large and can easily overwhelm the larger-scale geostrophic wind.

An example of this phenomenon is shown in Figs. 1a,b. The images are from a 12-h forecast made by the fifth-generation Pennsylvania State University–National Center for Atmospheric Research (PSU–NCAR) Mesoscale Model (MM5). The model was triply nested, with only the inner 4-km domain shown here. No cumulus parameterization was used on the inner domain. The forecast featured a broad line of explicit convective activity extending from the northern to the southern extremities of Texas (see updrafts in Fig. 1b). The height contours (Fig. 1a) are broadly oriented west-southwest–east-northeast, but the individual contours are very jagged, particularly in the vicinity of the convection and near gravity waves over New Mexico and northeast Texas. The geostrophic wind, here sampled every 100 km, is so affected by the small-scale height variability that no coherent flow pattern is apparent over the Gulf or over south-central Texas. In contrast, the full wind (Fig. 1b) is smooth and coherent over the entire domain.

With high-resolution model output, additional steps must be taken to remove or filter the small, rapid motions prior to utilizing many diagnostic tools. Coarse sampling of the high-resolution grid can lead to severe aliasing problems. A better approach, and one that is presently in common use, is to apply a smoothing function to the model output prior to or during the diagnostic calculations.

One way smoothing is done is by vertically averaging the forcing. As noted earlier, the **Q**-vector convergence is often averaged over the 400–700-hPa layer. This is justified on theoretical grounds (the forcing over a deep layer will contribute most strongly to the vertical motion response), but also has the salutary effect of filtering any phenomena possessing small vertical scales; such phenomena often possess small horizontal scales as well. But in practice, even this filtering is often not enough, and a second, explicit horizontal smoothing is applied.

One benefit of such filtering is to reduce the forcing noise discussed in section 2a. But this benefit comes at a cost: the loss of some smaller-scale features that would otherwise be amenable to quasigeostrophic or semigeostrophic diagnosis. For example, **Q** vectors can be used to diagnose frontogenesis and vertical circulations (Hoskins and Pedder 1980; Keyser et al. 1988), and semigeostrophic **Q** vectors are applicable even when the cross-front length scale is small, but smoothed **Q** vectors will yield a blurry view of the frontogenetical forcing. Another potentially important feature whose diagnosis is inhibited by smoothing is the PV streamer (e.g., Massacand et al. 2001). In principle, high-resolution NWP models deliver balanced dynamical forecasts of improved fidelity, but because the high resolution is in-

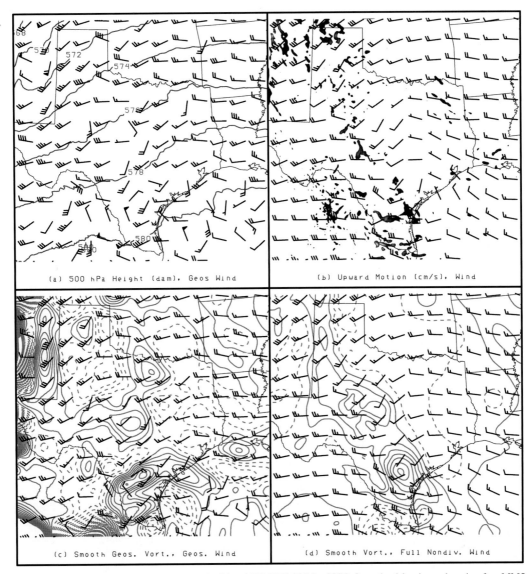

FIG. 1. The 12-h 500-hPa forecast output (valid at 1200 UTC 29 Mar 2006) from the 4-km inner domain of an MM5 forecast model run. Winds sampled every 100 km. (a) Geopotential height (dam) and geostrophic winds (conventional; 1 long barb = 10 kt). (b) Horizontal winds and vertical motion (values greater than 0.1 m s^{-1} shaded; maximum 2 m s^{-1}). (c) Smoothed geostrophic vorticity (<320-km wavelengths suppressed, contour interval 0.2×10^{-4} s^{-1}, negative contours dashed) and smoothed geostrophic wind (<80-km wavelengths suppressed). (d) Smoothed vorticity [as in (c)] and nondivergent wind. Smoothing is by a Gaussian smoother that reduces the amplitude of the specified wavelength to $1/e$.

compatible with conventional geostrophy-based diagnosis, the Bosartian baby is being thrown out with the bathwater.

2) NONDIVERGENT WIND AND QUASIGEOSTROPHIC DIAGNOSIS

Other approaches to quasigeostrophic diagnosis are more readily amenable to high-resolution analyses or model output. The statement that, to lowest order in QG theory, the Coriolis force balances the pressure gradient force in (3) is generally taken to be a constraint on the

horizontal wind under QG dynamics. We will call this option 1. However, (3) can instead be interpreted as a constraint on the geopotential height field (option 2). In option 1, the wind is approximated so that the Coriolis force balances the pressure gradient force, but it would be equally valid to approximate the pressure gradient force to achieve this balance. Or, as a third option, both the pressure and wind are approximated so that their approximate form satisfies QG dynamics. Each of these three options arises naturally from different approaches to deriving the QG set of equations.

The conventional formal approach to deriving the QG

equations proceeds by expanding all dependent variables in series in Rossby number and retaining the leading-order terms. This approach is consistent with the third option, in which no QG variables exactly match their real-world counterparts. Warn et al. (1995) show that this approach leads to a formal inconsistency as one attempts to carry the expansion beyond leading order to obtain what should be a more accurate equation set: the higher-order terms, nominally small, grow exponentially in time under realistically complex flow conditions. This problem is circumvented by choosing a variable to be taken as given rather than subject to formal expansion. For example, one could choose the pressure distribution as given, as in option 1, or the nondivergent wind as given, as in option 2. This single unexpanded variable is called a "slaving" variable, and the other variables are said to be "slaved" to it.

Warn et al. (1995) describe the mathematical procedure and discuss the appropriate characteristics of a QG slaving variable. This variable must be a "slow" variable, in the sense that it projects to leading order onto the Rossby wave modes rather than primarily onto the gravity wave modes. A balance relation (be it geostrophic or some higher-order approximation) serves to slave the other variables to the evolution of the slaving variable. As part of the expansion, the diagnostic (balance) relationship between the slaving variable and the other variables is expanded asymptotically.

By using the observed geopotential height field to determine the geostrophic winds in QG diagnosis, meteorologists are choosing height as the slaving variable. Leading-order (geostrophic) wind, next-order (ageostrophic) wind, and QG vertical motion are diagnosed from the full-height field. Warn et al. (1995) argue on theoretical grounds that height is not a good choice for a slaving variable because it is not slow: gravity waves produce a strong signal in the height field, complicating the equations at higher order. (We have shown an example of this gravity wave contamination in Fig. 1.) As an alternative, Warn et al. (1995) propose potential vorticity as the slaving variable. PV does not project at all onto linear gravity waves, giving it theoretical and practical advantages as a slaving variable.

Vallis (1996) presents asymptotic expansions of the stratified, rotating Boussinesq equations using potential vorticity as a slaving variable. [Muraki et al. (1999) carry out a different quasigeostrophic expansion that also uses PV as a slaving variable.] In this context, the hydrostatic Ertel PV, defined as

$$Q = -g(\zeta_\theta + f)\frac{\partial \theta}{\partial p}, \qquad (7)$$

can be written to leading order as

$$Q = [f + \nabla^2 \psi^{(0)}](-g\hat{\theta}_p) + f_0\left[-g\frac{\partial \theta'^{(0)}}{\partial p}\right], \qquad (8)$$

where $-g\hat{\theta}_p$ is the horizontally and temporally invariant

vertical derivative of basic-state potential temperature, a prime indicates a departure from this basic state, and the superscript (0) indicates a leading-order dependent variable. Because the leading-order variables are in geostrophic and hydrostatic balance, (8) can be rewritten in terms of geopotential height as

$$\frac{Q}{-g\hat{\theta}_p} - f = \frac{1}{f_0}\nabla^2\Phi'^{(0)} - \frac{f_0 p_0^\kappa}{R}\frac{\frac{\partial}{\partial p}\left[p^{1-\kappa}\frac{\partial\Phi'^{(0)}}{\partial p}\right]}{\hat{\theta}_p}. \qquad (9)$$

The crucial concept here is that Q, as the slaving variable, is still the Ertel PV. Equation (9) does not define an approximate form of Q. Instead, (9) implicitly defines the leading-order quasigeostrophic geopotential height field. To obtain it from observed data, one would find the full Ertel PV using (7), compute the left-hand side of (9), and invert (9) using suitable boundary conditions. One could then use (1) or (2) to diagnose vertical motion.

Mohebalhojeh (2002) investigated various possible quasigeostrophic expansions and truncations and found that the slaving of Ertel PV introduces errors in PV inversion. Smaller errors were found with the use of the linearized PV or vorticity itself as a slaving variable.

The choice of PV or linearized PV requires a global inversion, which guarantees full three-dimensional consistency in variables such as temperature. This choice is perhaps best when "PV thinking" (Hoskins et al. 1985) is to be employed. Issues regarding PV thinking and QG diagnosis are explored in more detail in section 3.

The choice of vorticity as a slaving variable has particular advantages for diagnosis of vertical motion. A simple two-dimensional Laplacian inversion suffices to obtain the streamfunction (and horizontal nondivergent velocities) from the vorticity:

$$\psi = \nabla^{-2}\zeta. \qquad (10)$$

The streamfunction, multiplied by f_0, is then substituted for the geopotential height in (1). For (2), the leading-order perturbation temperature may be obtained from the vertical derivative of the streamfunction. Hydrostatic consistency may be enforced through boundary conditions or integral constraints, but it is not necessary to do so since only horizontal temperature variations appear in the omega equation [(1)]. The resulting diagnosis will only be weakly contaminated by inertia–gravity waves, and unlike geopotential height the contamination asymptotes to zero as the wavelength becomes shorter.

In practice, the inversion is simplest if carried out over a global domain or with boundary conditions obtained from a global model. Otherwise, boundary conditions give rise to the harmonic wind, which has zero vorticity and zero divergence. For diagnostic purposes, the harmonic wind should be treated as part of the nondivergent wind. Boundary conditions on a limited domain may be obtained from a line integral of normal

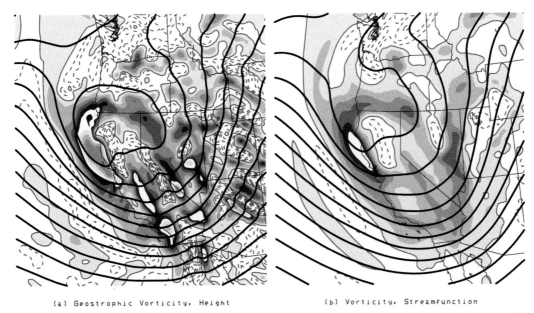

(a) Geostrophic Vorticity, Height   (b) Vorticity, Streamfunction

FIG. 2. The 12-h 500-hPa forecast output (valid at 1200 UTC 29 Mar 2006) of an upper-level trough from the 36-km domain of an MM5 forecast model run. (a) Geopotential height (contour interval 3 dam) and geostrophic vorticity (negative values dashed, positive values shaded, interval 0.5×10^{-5} s^{-1}). (b) Streamfunction (contour interval 3×10^6 m^2 s^{-1}) and vorticity [as in (a)].

velocity around the perimeter. Alternatively, if nondivergent wind rather than streamfunction is desired, the divergence may be inverted with homogeneous boundary conditions to obtain the velocity potential, at which point the irrotational wind may be computed and subtracted from the total wind field to obtain the nondivergent wind.

An example of the use of nondivergent wind for diagnosis is shown in Figs. 1c,d. Figure 1c is the traditional geostrophic vorticity (smoothed to suppress wavelengths less than 320 km) and geostrophic wind (smoothed to suppress wavelengths less than 80 km). Figure 1d is the full vorticity (smoothed as in Fig. 1c) and nondivergent wind (unsmoothed). The vorticity in Fig. 1c is dominated by small-scale features that are nonetheless large enough to survive the smoothing. The smoothed geostrophic winds are still rather noisy. Little useful diagnostic information can be gained from this figure. In contrast, Fig. 1d shows a coherently structured trough with well-defined areas of positive and negative vorticity advection. The unsmoothed nondivergent wind is much smoother than the smoothed geostrophic wind. Further smoothing of Fig. 1c, to more strongly suppress the gravity wave and convective signals, would also have yielded a more coherent (but broader) vorticity maximum in south-central Texas, but as Fig. 1d demonstrates, such violent smoothing is unnecessary.

As a simple plotted field, the streamfunction is inherently more useful than the geopotential height. Unlike height, a given streamfunction gradient corresponds to the same wind speed at all latitudes. Furthermore, streamfunction is of value in the Tropics, where height

becomes almost irrelevant. These advantages, when combined with the superiority of streamfunction over geopotential height for quasigeostrophic diagnosis, suggest that perhaps a greater use of streamfunction is warranted and that such grids should be more regularly produced, either centrally or on local workstations.

Figure 2 is an example of another trough at the same time as Fig. 1, but simulated on the outer 36-km domain of the MM5. Unlike Figs. 1c,d, no fields have been smoothed. Figure 2a features a dynamically significant vorticity maximum off the northern California coast, but over land it is difficult to isolate vorticity features not caused by topography. Topographic gravity waves are particularly prominent in California, southern Nevada, Utah, and northern Arizona. The height contours also display the characteristic wiggles of gravity waves. Figure 2b shows the full vorticity and streamfunction. Little topographic contamination is evident, and the true structure of the upper-level trough and areas of significant vorticity advection are easily discerned.

Excluding gravity waves and other fast motions does not fully eliminate finescale structures associated with high resolution. The ubiquitous enstrophy cascade causes potential vorticity variations, and thus vorticity and stratification variations, to develop finer and finer scales, ultimately producing filaments (vorticity) and fronts (stratification) (Pedlosky 1987, 164–177). Transport, even if by the nondivergent wind, of such finescale and often dynamically inert features will inevitably produce advection fields that are dominated by high-amplitude, small-scale dipoles. Horizontal smoothing can minimize such dipoles, but the smoothing should be designed to eliminate the dynamically inert scales and retain the dynamically significant ones, for reasons discussed in section 2c.

c. Improved NWP accuracy and the purpose of quasigeostrophic diagnosis

Sections 2a and 2b presented arguments that meteorologists have suffered a decline in their ability to diagnose NWP models in an operational setting, despite advances in diagnostic techniques and model accuracy. The essence of this decline is in the ability to relate aspects of the diagnosis to details of particular atmospheric features.

This decline coincides with a shift in the value that can be added to the numerical forecast by human forecasters. Decades ago, forecasters routinely made local forecasts based on their own forecasts of larger-scale weather conditions, applying QG theory, forecasting rules, and their own experience. As NWP models gradually gained accuracy, bias correction (correcting NWP models' known errors), initial error correction (detecting errors in the analysis and determining the forecast errors that will result), and downscaling (converting a large-scale forecast to likely local weather conditions) became increasingly important.

Further gains in accuracy and rapid evolution of NWP models have combined to make bias correction increasingly rare as a forecasting technique. As initial analyses improve through appropriate use of satellite and other information, the opportunities for initial error correction decrease, both because of a reduction in the size of the possible benefit and an inability of the forecaster to examine all the data that goes into the analysis. Downscaling has also begun a gradual decline in importance as models begin to produce explicit forecasts at local scales. Automated techniques for bias correction and downscaling make the forecaster even more superfluous.

Meanwhile, as deterministic weather forecasts are becoming increasingly automated, a different type of weather forecast, the probabilistic weather forecast, is becoming increasingly recognized for its value. Precipitation probabilities have long been a standard forecasting product, but all forecasted parameters have situation-specific uncertainties associated with them, and quantifying those uncertainties (and expressing them in a useful way) has value for many if not most forecast customers.

With ensemble forecasts and other computer-based probabilistic forecast techniques still in relative infancy, there is still a large amount of value that human forecasters might add to a probabilistic forecast. For example, Hurricane Wilma (2005) drifted slowly in the Caribbean before being picked up by midlatitude westerlies associated with a developing trough and carried rapidly across Florida and into the Atlantic. The onset of its rapid northeastward motion was directly tied to the relative positions of the hurricane and trough. The uncertainty in the long-range forecast of Wilma's position was obvious when forecast models spanned the scenarios, but on occasion all models selected a single scenario. Forecasters understanding the dynamical source of the uncertainty can improve upon probabilistic forecasts from a limited ensemble set.

Most current operational ensembles are based either on model perturbations or on initial condition perturbations designed to produce rapid ensemble expansion over a particular interval. Until such ensembles can incorporate the possibilities of rapid uncertainty growth at all time windows within the forecast and can span the range of plausible models, there will be room for improvement in the prediction of forecast uncertainty.

For meteorologists to fill that gap, their diagnostic techniques must be useful for estimating forecast uncertainty. It is difficult to imagine a diagnostic technique less well suited for estimating forecast uncertainty than the practice of plotting deep-layer-smoothed vertical motion forcing. The obfuscation of the relationship between the forcing and the underlying weather patterns, discussed above, makes it difficult to understand the nature of uncertainties in the forecasted or inferred vertical motion field. While error growth usually begins at smaller scales and gradually expands to larger ones (e.g., Tan et al. 2004), the smoothing of the forcing field makes those smaller scales invisible and their potential importance unknown.

Adjoint models (Errico 1997) are extremely useful for determining forecast sensitivity to initial conditions. However, they must be run separately for each particular forecast aspect of interest. They are well suited to individual forecasts of critical importance but are not easily applied to general situations with many forecast variables.

Thus, there exists an unmet need for diagnostic tools that allow meteorologists to infer the uncertainty associated with particular deterministic forecasts. Such tools would be applied to individual ensemble members to understand the dynamical causes of ensemble divergence and the possibility of other sources of forecast divergence unsampled by the ensemble.

d. Potential vorticity diagnostic techniques

The seminal article by Hoskins et al. (1985) introduced a new approach for understanding and diagnosing atmospheric motion using potential vorticity. As with QG theory, the approach can be applied qualitatively (PV thinking) or quantitatively (PV inversion). The use of PV in an operational environment has become increasingly common in Europe, but has yet to gain a significant foothold in the United States. Time will tell whether the situation is analogous to the relatively late adoption of the Norwegian cyclone model forecasting techniques in the United States compared with Europe.

Both QG theory and PV diagnosis can be used to study all aspects of synoptic-scale development. Despite this, the two approaches seem more complementary than alternative. For example, diagnosis of the total synoptic-scale vertical motion seems in practice more suited to traditional QG theory and diagnosis of vertical motion associated with individual features seems more suited to PV thinking, even though total QG vertical motion forcing can be written in terms of PV advection (Hoskins et al. 1985) and diagnosed in a piecewise fashion (e.g., Dixon et al. 2003).

Fundamentally, the distinction between present applications of QG theory and PV diagnosis seems to involve differences between an Eulerian and a Lagrangian world view. Nowhere is this more apparent than in the equation known to QG theory as the height tendency equation and to PV diagnosis as the equation for conservation of pseudo-PV.

In conventional QG form, height tendencies are "forced" by horizontal vorticity advection and the vertical derivative of temperature advection:

$$\frac{\partial \psi^{(0)}}{\partial t} \propto \frac{\partial}{\partial t}\left\{\nabla^2 \psi^{(0)} + f_0 \frac{\partial}{\partial p}\left[\frac{\theta'^{(0)}}{\hat{\theta}_p}\right]\right\}$$

$$= -\mathbf{V}^{(0)} \cdot \nabla[f + \nabla^2 \psi^{(0)}]$$

$$- f_0 \frac{\partial}{\partial p}\left\{\mathbf{V}^{(0)} \cdot \nabla\left[\frac{\theta'^{(0)}}{\hat{\theta}_p}\right]\right\}. \qquad (11)$$

To apply this equation, one examines the three-dimensional spatial pattern of advection fields. The local advection patterns encapsulate the dynamics by governing the local height tendencies. This approach is fundamentally Eulerian: the evolution of individual air parcels or assemblages of air parcels is irrelevant, and no conservation principles are explicitly used.

The PV version of this equation is

$$\frac{d^{(0)}q}{dt} = 0 \quad \text{or} \qquad (12a)$$

$$\frac{\partial q}{\partial t} = -\mathbf{V}^{(0)} \cdot \nabla q, \quad \text{with} \qquad (12b)$$

$$q - f = \nabla^2 \psi^{(0)} + f_0 \frac{\partial}{\partial p}\left[\frac{\theta'^{(0)}}{\hat{\theta}_p}\right], \qquad (12c)$$

where q is the pseudo-PV (Charney and Stern 1962). Note that (11) and (12b) are formally equivalent. Qualitative application uses (12c) to relate height perturbations to the strength of PV anomalies and (12a) to determine how these anomalies will be transported, reconfigured, and amplified by the winds. The conservation of PV and the evolution of individual air parcels are fundamental to the diagnosis.

In doing a PV diagnosis, one could compute the pseudo-PV advection and use (12b) as the primary diagnostic tool, but such an approach would rightly be seen as obscuring the fundamental processes, even at a qualitative level. PV thinking is simplest when the atmosphere is idealized as containing an assemblage of point vortices and PV discontinuities such as at the tropopause, because the motion of a limited number of air parcels determines the evolution of the entire troposphere. But vortices and discontinuities have infinite gradients and therefore infinite advection, so (12b) is ill posed in precisely that situation that is simplest to diagnose. Furthermore, dynamically inert PV filaments are associated with strong PV advection dipoles, but a qualitative diagnosis using (12a) and (12c) recognizes them as inert and easily ignores them. If one wanted a quantitative diagnosis of height tendencies using PV, one would not stop with the computation of advection on the right-hand side of (12b) but instead would proceed with inversion, solving quantitatively for the height tendencies (e.g., Nielsen-Gammon and Lefevre 1996).

For these reasons, PV advection is simply not a useful approach for PV diagnosis, when compared with direct examination of PV and wind on the one hand, and PV advection inversion on the other hand. Both direct examination and inversion avoid many of the problems associated with increased model resolution, and piecewise PV inversion (along with piecewise PV advection inversion) take advantage of modern computing power while retaining the direct conceptual relationship between the dynamical feature, expressed as a PV anomaly, and its impact on its surroundings.

For someone accustomed to conventional QG diagnosis, eyeballing PV amplification and PV superposition is an alien way of doing things. I have seen some meteorologists, when presented with arguments in favor of the addition of PV thinking to their diagnostic arsenal, begin by computing Ertel-PV advection. Such an approach imbues PV diagnosis with many of the same shortcomings of QG diagnosis, but with no compensating benefits. The proper application of PV thinking involves inferring the transport and amplification of areas of PV rather than the advection of PV. Again, it is a Lagrangian approach rather than an Eulerian one.

In addition to retaining a connection with the underlying dynamics (section 2a) and dealing more easily with small-scale structure and gravity waves (section 2b), PV diagnosis also holds promise for the issue raised in section 2c, namely diagnosis of forecast uncertainty. Interactions among three or more significant PV anomalies is likely to be fundamentally less predictable than interactions between two such anomalies (e.g., Fehlmann and Davies 1997). In cases of split flow, the trajectory of a PV anomaly approaching the split determines the future path of the entire associated weather system. In situations of downstream development and Rossby wave propagation, the PV gradients govern the path of wave energy and the ultimate sensitivity of a particular location to weather events happening upstream (e.g., Nielsen-Gammon 2001).

The next section explores the mathematical relationship between Ertel PV and the QG framework.

3. Pseudo-PV and Ertel PV

The preceding section has argued that the changing landscape of model output and forecast accuracy has led to changes in the practice of operational diagnosis of weather systems, with some of those changes being detrimental to the forecasting process.

Alternative approaches, involving the use of streamfunction in place of geopotential height and the expanded application of PV diagnostics, were proposed. The remainder of this article quantifies the differences between the alternative approaches and the QG approach in the context of a rapid cyclogenesis event over the North Pacific Ocean. This section examines the mathematical differences between Ertel PV and pseudo-PV and compares them in the upper and lower troposphere during the cyclogenesis event.

a. Pseudo-PV

As noted earlier, QG diagnosis of height tendencies involves the pseudo-PV, which may be computed from geopotential heights or winds. Conventional PV diagnosis is occasionally performed using the pseudo-PV, taking advantage of the fact that the linearity of the inversion operator makes attribution of particular flow features to particular PV anomalies straightforward. Most qualitative applications of PV diagnosis make use of the Ertel PV, presented either in isobaric or isentropic coordinates. Isentropic display reduces PV evolution to two-dimensional advection, making conceptual understanding of that evolution particularly straightforward. Dynamic tropopause maps (Morgan and Nielsen-Gammon 1998) further reduce PV evolution to quasi-2D advection on a single surface. Given the superior conservation properties of Ertel PV, it is desirable to know to what extent Ertel PV and pseudo-PV are interchangeable for diagnostic purposes.

The Ertel PV is defined in (7). We are interested here in its approximate form in the QG system. As discussed in section 2, there is no single correct approximate form, as choices can be made regarding which variable is the slaving variable and which variables are expanded in the series in Rossby number (Warn et al. 1995). One choice is to retain the full PV, and then to leading order (8) and (9) define the relationship between PV and other variables. Alternatively, one can define a linearized PV as an approximation to the Ertel PV at leading order (Mohebalhojeh 2002):

$$\frac{Q^{(0)}}{-g\hat{\theta}_p} - f = \frac{1}{f}\nabla^2\Phi' - \frac{f_0 p_0^\kappa}{R}\frac{\frac{\partial}{\partial p}\left(p^{1-\kappa}\frac{\partial\Phi'}{\partial p}\right)}{\hat{\theta}_p}. \quad (13)$$

The only difference between (9) and (13) is that one defines an approximation to the geopotential height in terms of Ertel PV, and the other defines an approxi-

mation to the Ertel PV in terms of the geopotential height. However, neither (9) nor (13) are the pseudo-PV.

The PVs in (9) and (13) are both approximately conserved following the three-dimensional flow. The pseudo-PV (Charney and Stern 1962) arises from a desire for a quantity that is approximately conserved following the horizontal motion alone, since the wind itself is purely horizontal to leading order in QG. The QG evolution equation for (8), to leading order and ignoring diabatic and frictional processes, is

$$\frac{\partial Q}{\partial t} + J[\psi^{(0)}, Q] + f_0\omega\frac{\partial(-g\hat{\theta}_p)}{\partial p} = 0. \quad (14)$$

The last term represents advection of basic-state stratification by the vertical motion, and is not negligible. Vertical motion is eliminated by use of the QG thermodynamic equation,

$$\frac{\partial\theta'^{(0)}}{\partial t} + J[\psi^{(0)}, \theta'^{(0)}] + \omega\hat{\theta}_p = 0, \quad (15)$$

turned on its head:

$$\omega = -\frac{1}{\hat{\theta}_p}\left\{\frac{\partial\theta'^{(0)}}{\partial t} + J[\psi^{(0)}, \theta'^{(0)}]\right\}$$
$$= \frac{\partial}{\partial t}\left[\frac{\theta'^{(0)}}{(-\hat{\theta}_p)}\right] + J\left\{\psi^{(0)}, \left[\frac{\theta'^{(0)}}{(-\hat{\theta}_p)}\right]\right\}. \quad (16)$$

Substitution into (14) yields

$$\frac{\partial}{\partial t}\left[Q - f_0 g\frac{\theta'^{(0)}}{(-\hat{\theta}_p)}\frac{\partial\hat{\theta}_p}{\partial p}\right]$$
$$+ J\left\{\psi^{(0)}, \left[Q - f_0 g\frac{\theta'^{(0)}}{(-\hat{\theta}_p)}\frac{\partial\hat{\theta}_p}{\partial p}\right]\right\} = 0. \quad (17)$$

The addition of another term involving temperature has created a quantity conserved following the 2D flow:

$$Q - f_0 g\frac{\theta'^{(0)}}{(-\hat{\theta}_p)}\frac{\partial\hat{\theta}_p}{\partial p} = [f + \nabla^2\psi^{(0)}](-g\hat{\theta}_p)$$
$$+ f_0\left[-g\frac{\partial\theta'^{(0)}}{\partial p}\right] - f_0 g\frac{\theta'^{(0)}}{(-\hat{\theta}_p)}\frac{\partial\hat{\theta}_p}{\partial p}, \quad (18)$$

which, after combining the two terms on the right-hand side and dividing through by $-g\hat{\theta}_p$ becomes the familiar pseudo-PV q:

$$\frac{Q}{-g\hat{\theta}_p} - f_0\frac{\theta'^{(0)}}{\hat{\theta}_p^2}\frac{\partial\hat{\theta}_p}{\partial p} \equiv q = (f + \nabla^2\psi^{(0)}) + f_0\frac{\partial}{\partial p}\left[\frac{\theta'^{(0)}}{\hat{\theta}_p}\right]. \quad (19)$$

b. Differences between pseudopotential vorticity and Ertel potential vorticity

Charney and Stern (1962) find that variations of pseudo-PV on horizontal surfaces are approximately proportional to variations of Ertel PV on isentropic surfaces (see also Hoskins et al. 1985). Yet (19) includes a term (the second term on the left-hand side) involving vertical stratification that destroys this proportionality. The difference arises because Charney and Stern (1962) assume that vertical variations of basic-state stratification are small, whereas in the real atmosphere, in the vicinity of the tropopause, such variations are not small enough to ignore. Thus, the pseudo-PV differs by both additive and multiplicative factors from Ertel PV, even at leading order.

To facilitate direct comparison comparison between Ertel PV and pseudo-PV, we define a scaled Ertel PV:

$$\check{Q} \equiv \frac{Q}{-g\hat{\theta}_p} = \frac{(f + \zeta_\theta)}{\hat{\theta}_p}\frac{\partial\theta}{\partial p}. \tag{20}$$

This scaling differs from that of Thorpe and Bishop (1995) through the lack of a factor f_0 in the denominator and from that of Juckes (1999) because it involves the local basic-state pressure-coordinate stratification rather than the stratification at an air parcel's original level. A consequence of the latter is that this scaled Ertel PV is not conserved.

To make clear the term-by-term relationship between the scaled Ertel PV and the pseudo-PV, (20) is expanded algebraically:

$$\check{Q} = (f + \zeta_g) + \zeta_{ag} + \frac{f_0}{\hat{\theta}_p}\frac{\partial\theta'}{\partial p} + \frac{f - f_0 + \zeta}{\hat{\theta}_p}\frac{\partial\theta'}{\partial p}$$

$$+ \frac{(\zeta_\theta - \zeta)}{\hat{\theta}_p}\frac{\partial\theta}{\partial p}. \tag{21}$$

The most common form of pseudo-PV uses height as the slaving variable:

$$q_{com} = (f + \zeta_g) + f_0\frac{\partial}{\partial p}\left(\frac{\theta'}{\hat{\theta}_p}\right). \tag{22}$$

Equation (22) differs from (19) in that the full heights and temperatures, rather than a balanced approximation thereto, are used to compute q. The relationship between Ertel PV and the common pseudo-PV can then be written as

$$\check{Q} - q_{com} = \underbrace{\{\zeta_{ag}\}}_{AV} + \underbrace{\frac{(f - f_0 + \zeta)}{\hat{\theta}_p}\frac{\partial\theta'}{\partial p}}_{+NL}$$

$$+ \underbrace{\frac{(\zeta_\theta - \zeta)}{\hat{\theta}_p}\frac{\partial\theta}{\partial p}}_{+BC} - \underbrace{\left\{-f_0\theta'\frac{\partial}{\partial p}\left(\frac{1}{-\hat{\theta}_p}\right)\right\}}_{-TP}. \tag{23}$$

The four terms on the right-hand side are, respectively, the ageostrophic vorticity (AV) term, the nonlinear (NL) term, the baroclinic (BC) term, and the scaled temperature perturbation (TP) term. This grouping of terms is not unique but is useful for present purposes. The meaning of the curly brackets is described below.

The AV term is the departure of vertical vorticity from its geostrophic value. The particular form of this term depends upon the choice of a slaving variable (see section 2c). The AV term appears as it does in (23) solely because (22) defined pseudo-PV in terms of geostrophic vorticity, using height as the slaving variable. If the more general definition of pseudo-PV in (19) were used, the AV term would be replaced by the departure of vertical vorticity from its zero-order balanced value, with several options available for defining "balance." Because geostrophy is an especially inaccurate balance approximation, the AV term is largest when geostrophy is used. At the other extreme, if vertical vorticity or the streamfunction is chosen as a slaving variable as recommended in section 2c, the AV term vanishes completely. We write the AV term in curly brackets because its existence depends on the choice of a specific QG system.

The NL term is the product of departures from basic-state vertical vorticity (i.e., from the Coriolis parameter) and departures from basic-state stratification. It will be small when the Rossby number is small and the PV anomalies are weak. Conversely, the NL term will tend to be large within and near strong, localized PV anomalies.

The BC term is a consequence of the difference between isobaric vorticity and isentropic vorticity. Its size will depend upon the slope of isentropic surfaces and the magnitude of vertical shear. It is largest where there are strong horizontal temperature gradients and weak stratification.

The TP term is the equivalent of the second term on the left-hand side of (19). As discussed earlier, it arises as a replacement for the vertical advection of PV. It is not necessarily small, even under QG scaling, and will tend to be largest near the tropopause where rapid vertical variations in basic-state stratification are found.

The complete derivation in (15)–(19) obscures the physical interpretation of the TP term. For clarity, we now provide a simplified derivation. Consider an approximate equation for conservation of PV:

$$\frac{d_h Q}{dt} + \omega\frac{\partial\hat{Q}}{\partial p} \approx 0, \tag{24}$$

where the substantial derivative follows the horizontal motion only and the vertical advection is of the basic-state PV only. The basic-state PV is a product of basic-state vorticity and basic-state stratification:

$$\hat{Q} = -gf_0\hat{\theta}_p. \tag{25}$$

A similar equation governs θ to leading order:

$$\frac{d_h\theta'^{(0)}}{dt} + \omega\hat{\theta}_p = 0. \tag{26}$$

Rearranging (26) to stand for the proposition that vertical motion is proportional to the change in potential temperature at a given level, scaled by the stratification:

$$\omega = \frac{d_h}{dt}\left[\frac{\theta'^{(0)}}{-\hat{\theta}_p}\right], \qquad (27)$$

one can equate the vertical advection of PV to

$$\omega\frac{\partial\hat{Q}}{\partial p} = -\omega g f_0 \frac{\partial\hat{\theta}_p}{\partial p} = -g f_0 \frac{\partial\hat{\theta}_p}{\partial p}\frac{d_h}{dt}\left[\frac{\theta'^{(0)}}{-\hat{\theta}_p}\right]. \qquad (28)$$

Since each of the factors on the right-hand side is constant on a horizontal (pressure) surface, they can be brought within the time derivative. Combining (28) and (24),

$$\frac{d_h Q}{dt} \approx \frac{d_h}{dt}\left[\frac{g f_0}{-\hat{\theta}_p}\frac{\partial\hat{\theta}_p}{\partial p}\theta'^{(0)}\right], \qquad (29)$$

which says that the change in PV due to vertical advection is equal to the leading-order change in the appropriately scaled perturbation θ due to vertical motion. Since

$$\frac{d_h}{dt}\left[Q - \frac{g f_0}{-\hat{\theta}_p}\frac{\partial\hat{\theta}_p}{\partial p}\theta'^{(0)}\right] \approx 0, \qquad (30)$$

we can scale the term in brackets:

$$\frac{d_h}{dt}\left[\breve{Q} - \frac{f_0}{-\hat{\theta}_p^2}\frac{\partial\hat{\theta}_p}{\partial p}\theta'^{(0)}\right] \approx 0 \qquad (31)$$

and apply the chain rule:

$$\frac{d_h}{dt}\left[\breve{Q} - f_0\theta'^{(0)}\frac{\partial}{\partial p}\left(\frac{1}{\hat{\theta}_p}\right)\right] \approx 0 \qquad (32)$$

to obtain a form similar to the potential temperature term in (19) and (23).

We now provide a physical interpretation of the TP term. Near the tropopause, vertical motions can dramatically alter the PV at a given level, even within quasigeostrophy [(14)]. Downward motion, for example, would lower the tropopause and cause the PV locally to increase. At the same time, downward motion would cause θ to increase at that level as well [(15)]. If the change in θ is scaled by the stratification [(16) and (27)] and by the vertical gradient of basic-state PV [(17) and (28)], the increase in scaled θ due to downward motion will be exactly equal to the increase in PV due to downward motion. Thus, the difference between PV and scaled θ will remain constant. Since both PV and θ are otherwise conserved following the horizontal motion, absent diabatic, and frictional processes, the difference between PV and scaled θ is conserved following the horizontal motion even in the presence of vertical motion. The pseudo-PV is, to lowest order, the difference between PV and scaled θ.

What the TP term does, therefore, is subtract from PV a quantity that to leading order is proportional to the change in PV caused by vertical displacements of air parcels in (28). While the PV at a given level can change substantially through vertical motion near the tropopause, the pseudo-PV, to leading order, cannot. As noted by Hoskins et al. (1985) and others, pseudo-PV is not an approximation to PV but rather a different quantity altogether.

The signs in (23) are meant to suggest that the AV, NL, and BC terms are quantities that must be added to pseudo-PV to recover Ertel PV, while the TP term has been added to Ertel PV to obtain pseudo-PV and therefore must be subtracted back out from pseudo-PV to recover Ertel PV.

The TP term has been placed in curly brackets in (23) because, like the AV term, the existence of the TP term depends upon one's choice of QG equation set. It may be satisfactory to work with the linearized PV in (13) within the QG framework, from which the height field (and all other variables) can be recovered through the familiar inversion of a Laplacian. Indeed, the Laplacian in (13) is simpler than that for pseudo-PV: it takes the same basic form as the Laplacian in the QG omega equation with variable vertical stratification. This linearized PV is conserved according to QG dynamics in the same way that θ is conserved: in addition to diabatic (and frictional) processes, it changes through vertical motion acting on the vertical gradient of the basic state. Finally, the linearized quasigeostrophic PV is easier to relate to Ertel PV, because the TP difference term becomes zero.

c. Sample comparison at 300 hPa

We now examine the magnitudes of the terms in (23) in a practical setting: a rapidly deepening extratropical cyclone over the North Pacific. The data source is the National Centers for Environmental Prediction (NCEP)–NCAR reanalysis (Kalnay et al. 1996), reduced to a 2.5° latitude–longitude grid at 50-hPa vertical resolution.

This event was chosen to present a stringent but not severe test for QG dynamics. The event is synoptic in spatial scale and there are no topographic gravity waves, but the rapid intensification of the storm may stress the validity of the QG system. The quarter wavelength of the developing system is 1000 km and the characteristic wind speed in the upper troposphere is 50 m s^{-1}, giving a Rossby number of 0.5.

At upper levels (Fig. 3a), a broad upper-level low is centered over the Sea of Okhotsk. Well to its south, a strong jet with wind speeds approaching 90 m s^{-1} exits Japan, forming a trough near 155°E and a broad ridge near the date line. In the lower troposphere (Fig. 3b), an already strong frontal wave cyclone at (49°N, 168°E) is in the process of rapidly intensifying (not shown) within the upper-level trough–ridge couplet. Warm and cold fronts are evident in the temperature and height fields, with the warm front extending eastward from the low and the intense cold front curving southwestward.

Figure 4a shows the Ertel PV [(7)] of the trough–ridge system at the 325-K isentropic level. This surface

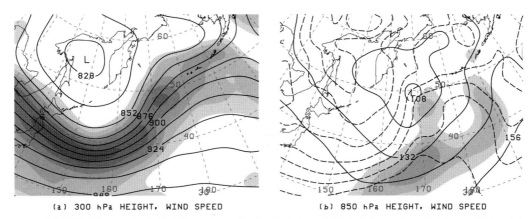

FIG. 3. (a) 300-hPa geopotential height (contours, dam) and wind speed (shaded above 30 m s⁻¹ at 10 m s⁻¹ intervals). (b) 850-hPa geopotential height (solid contours, dam), wind speed (shaded above 20 m s⁻¹ at 5 m s⁻¹ intervals), and temperature (dashed contours, interval 4°C). The time of this and all subsequent maps is 0000 UTC 1 Apr 1987.

intersects the dynamic tropopause, which has been variously defined as 1.5–3 PVU (see Morgan and Nielsen-Gammon 1998). The strong gradient of PV along the tropopause is especially prominent in the trough, from Manchuria to 170°E. As shown by the thick contours, the 325-K surface slopes fairly steeply in the tropo-sphere just south of the tropopause, but north of the tropopause the 325-K surface lies in the stratosphere and is relatively flat.

Figure 4b shows the scaled Ertel PV [(20)] at the 300-hPa level. On a constant pressure surface, the scaling of PV simply involves multiplying it by a constant.

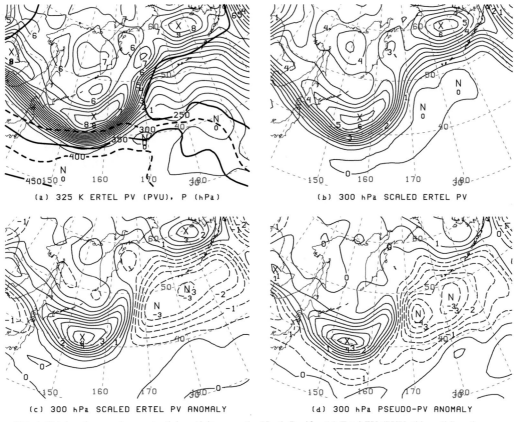

FIG. 4. Total and anomalous potential vorticity over the North Pacific. (a) Ertel PV (PVU, thin solid) and pressure (hPa, thick alternating solid/dashed) on the 325-K isentropic surface. (b) Ertel PV on the 300-hPa isobaric surface, scaled according to (20) (×10⁻⁴ s⁻¹). (c) Same as in (b), but for anomalous scaled Ertel PV (time mean PV removed). (d) Anomalous pseudo-PV (×10⁻⁴ s⁻¹) on the 300-hPa isobaric surface, computed from the analyzed geopotential heights.

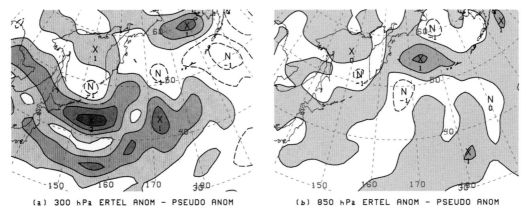

(a) 300 hPa ERTEL ANOM - PSEUDO ANOM   (b) 850 hPa ERTEL ANOM - PSEUDO ANOM

FIG. 5. Difference between anomalous scaled Ertel PV and anomalous pseudo-PV (contour interval 0.5×10^{-4} s^{-1}, negative values dashed, positive values shaded): (a) 300 and (b) 850 hPa.

Making use of a constant pressure surface that intersects the tropopause alters the PV pattern in the trough very little but makes the ridge clearer in this case (Fig. 4b), because 300 hPa is far below the tropopause in the ridge. As is common, the upper-tropospheric jet stream (Fig. 3a) is collocated with the strong horizontal PV gradient at the tropopause.

By subtracting the 7-day-average fields about this date, we obtain the scaled PV anomalies (Fig. 4c). Anomalies are commonly used in PV diagnostics because they isolate the synoptic-scale aspects of the flow from the planetary-scale tropospheric jets and waves. Without experience, it can be difficult to distinguish the dynamically significant raw PV variations from the insignificant ones; computing anomalies isolates the significant variations. For example, the most prominent PV anomalies are seen to be the trough centered at 155°E and the ridge centered at 177°W. While the trough is associated with a prominent maximum in total PV, the ridge is associated with a relatively minor minimum. By contrast, the maxima and minima in total PV near the Kamchatka Peninsula are but minor blips in the PV anomaly field. The ridge and trough anomalies owe most of their strength to horizontal displacements of the strong tropopause PV gradient seen in Figs. 4a,b. The comparative magnitudes of the anomalies serve to illustrate that the lateral displacements of the tropopause PV gradient itself are far more important than most of the isolated maxima and minima north or south of the tropopause PV gradient.

To compute pseudo-PV [(22)], it is necessary to specify a basic-state stratification $g\hat{\theta}_p$. The choice of stratification is somewhat arbitrary, but should generally be related to the specific synoptic situation. Here, $g\hat{\theta}_p$ is determined as a smooth curve fit to the horizontally averaged stratification within the broad area of interest. The resulting pseudo-PV anomalies, computed using the analyzed height and potential temperature field over the same 7-day window, are shown in Fig. 4d.

The pseudo-PV anomalies are strikingly similar to the scaled Ertel-PV anomalies (Fig. 4c) in both pattern and magnitude, but some specific differences are apparent. The structure of the ridge is more complex, the PV maximum in the Bering Sea is weaker, and there is a band of anticyclonic pseudo-PV wrapping around the base of the trough. For this system on this scale, the differences between pseudo-PV and Ertel PV are primarily matters of degree.

The pattern of differences, computed by subtracting the pseudo-PV anomalies in Fig. 4d from the scaled Ertel-PV anomalies in Fig. 4c, is shown in Fig. 5a. In general, there is a tendency for the scaled Ertel-PV anomalies to be of slightly higher amplitude than the pseudo-PV anomalies, since positive differences are prevalent in the trough and negative differences are prevalent in the ridge. However, the differences have a complex spatial structure. For example, the scaled Ertel-PV and pseudo-PV anomalies have similar values within the core of the jet stream (shown in Fig. 3a), but the scaled Ertel-PV anomalies are more positive than the pseudo-PV anomalies on either side of the jet.

The four terms contributing to the difference between the scaled Ertel-PV anomalies and the pseudo-PV anomalies are shown in Fig. 6. In Fig. 6, the zero contours have been suppressed and the contour interval is smaller than in Fig. 5 to more clearly show the differences. As in Figs. 4c,d, anomalies from a 7-day time mean are shown.

The AV term (Fig. 6a) is largest within the trough, where it forms a positive–negative dipole aligned with the tropopause. The jet stream, which follows the tropopause PV gradient, curves cyclonically around the base of the trough, so the winds are subgeostrophic. Consequently, on both sides of the jet, the actual shear vorticity is smaller in magnitude than the geostrophic shear vorticity. Thus, the ageostrophic vorticity is negative on the cyclonic side and positive on the anticyclonic side. Within the trough, because of subgeostrophy, the actual curvature vorticity is also weaker than its geostrophic counterpart, further reinforcing the negative values of the AV term.

The NL term (Fig. 6b) is the other large contributor

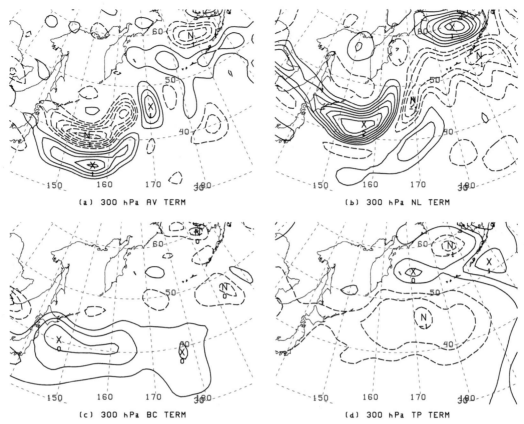

FIG. 6. Difference terms [from (23)] between anomalous scaled Ertel PV and anomalous pseudo-PV. The meanings of individual terms are described in the text following (23). Contour interval is (a), (b), (d) 0.2×10^{-4} s^{-1} and (c) 0.05×10^{-4} s^{-1}. Negative contours are dashed and the zero contour is suppressed.

to the difference between scaled Ertel-PV anomalies and pseudo-PV anomalies. Within the core of any PV anomaly, the NL term should be positive, because both the anomalous vorticity and anomalous stratification have the same sign (Thorpe 1986). Beyond the lateral edges of PV anomalies, the anomalous stratification retains its sign and weakens while the anomalous vorticity changes sign and weakens, so the NL term should be zero or negative. Similarly, below or above strong PV anomalies, the anomalous vorticity retains its sign and weakens while the anomalous stratification changes sign and weakens, so the NL term should be zero or negative there, too. Thus, even in the simple case of a vertically isolated vortex such as that depicted by Thorpe (1986), where the anomalous Ertel PV is zero below and above the core of the vortex, the anomalous pseudo-PV should be positive (Davis 1992).

To facilitate understanding the NL anomaly pattern in Fig. 6b, the full NL term (with time mean included) is plotted in Fig. 7a. Along the core of the jet stream, where the PV gradient is strong, both the vorticity and perturbation stratification are nearly zero, so the full NL term is close to zero there. The southward displacement of the jet stream brings large positive values of the NL term southward (Fig. 7a) where the time-mean NL term

is near zero (not shown), producing a large positive NL anomaly in the base of the trough (Fig. 6b). Meanwhile, in the downstream ridge, the northward displacement of the jet moves near-zero full NL values into an area along 50°N (Fig. 7a) where the time-mean NL term is positive (not shown), resulting in a negative NL anomaly (Fig. 6b).

The midtropospheric full NL pattern is shown in Fig. 7b. Since this level is well beneath the strong tropopause PV anomalies, the idealized situation discussed two paragraphs ago would predict that negative NL values will predominate. Indeed, the full NL term is strongly negative beneath the isolated PV anomaly to the northeast and near (48°N, 166°E) where the center of the tilting extratropical cyclone is located. Beneath the trough and ridge, the NL term is actually near zero, because the positive and negative PV anomalies aloft are underlain by negative and positive, respectively, surface θ anomalies (cf. Fig. 3b) so anomalous stratification (and, to some extent, anomalous vorticity) is weak. The band of high NL values immediately south of the upper-level trough is associated with a tropopause fold, where both vorticity and stratification are high.

The BC term (Fig. 6c) is the weakest of the four difference terms and can be safely ignored at this level.

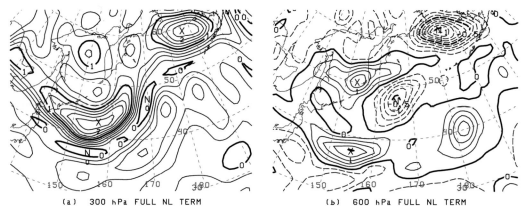

(a) 300 hPa FULL NL TERM   (b) 600 hPa FULL NL TERM

FIG. 7. The full NL term, time mean included, at (a) 300 hPa (contour interval 0.2×10^{-4} s^{-1}) and (b) 600 hPa (contour interval 0.1×10^{-4} s^{-1}).

The TP term (Fig. 6d) is not as small as BC but is much weaker than the other two terms. The term would be expected to be largest near the climatological tropopause where the vertical derivative of basic-state PV is large, because the term exists to cancel the vertical advection of basic-state PV. At the 300-hPa level, the factor multiplying perturbation potential temperature in the TP term in (23) is negative, meaning this term will be negative where the temperature is anomalously warm, in this case within and upstream of the ridge.

A physical interpretation of this pattern is difficult because potential temperature and Ertel PV are affected by both horizontal and vertical advection. But to the extent that anomalously warm temperatures have been produced by subsidence near the tropopause, subsidence would also have increased the Ertel PV there. To obtain a quantity (pseudo PV) that is not affected by subsidence, the Ertel PV must be reduced where subsidence, diagnosed by anomalously high potential temperature, is found. The magnitude of this reduction is given by the negative values of the TP term. Conversely, the TP term is positive where temperatures are cold, implying ascent.

The difference field in Fig. 5a is recovered as AV + NL + BC − TP.

d. Sample comparison at 850 hPa

The 300-K surface (Fig. 8a) slopes from the 850-hPa level to the stratosphere above 350 hPa. In addition to the tropopause PV gradient, an area of Ertel PV greater than 1 PVU is present beneath the upper-level ridge. The steep slope of the 300-K surface reflects the baroclinic zones associated with the developing cyclone. At the location of the 850-hPa low pressure center (Fig. 3b), the 300-K surface is in the upper troposphere.

A constant pressure map, in this case at 850 hPa, isolates the lower-tropospheric scaled Ertel PV (Fig. 8b). In addition to the high PV beneath the ridge, there is a PV maximum near the cyclone center (Fig. 8b). Because there is no strong background PV gradient at this level,

the scaled Ertel-PV anomaly field (Fig. 8c) looks very much like the total scaled PV (Fig. 8b). As at 300 hPa, the anomalous pseudo-PV is very similar in overall structure to the anomalous scaled Ertel PV.

The anomaly difference field (Fig. 5b) shows that the differences between anomalous pseudo-PV and anomalous-scaled Ertel PV are smaller than the PV fields themselves. Even more than at 300 hPa, the difference field tends to be opposite in sign to the anomalous pseudo-PV, indicating that the scaled Ertel PV tends to be of smaller amplitude than the pseudo-PV.

The terms contributing to the anomaly difference are again primarily the AV and NL terms. The AV term (Fig. 9a) is strongly negative near the 850-hPa low center, where the winds are strongly cyclonic and subgeostrophic. The NL term is strongly positive there (Fig. 9b), indicating that the high PV at the low center includes anomalously high stratification as well as anomalously high vorticity. Of the two terms, the AV term has a larger magnitude.

The BC term is shown in Fig. 9c with the same contour interval as the rest of Fig. 9. Recall that the BC term represents the difference between isentropic and isobaric vertical vorticity, and will be large where the isentropic slopes and vertical wind shear are large, such as within strong baroclinic zones. Because fronts are much more common at 850 than 300 hPa, the BC term is much larger in Fig. 9c than in Fig. 6c. Here the term is negative along the cold front, and weakly negative along the warm front as well (see Fig. 3b for frontal locations).

The sign of the term is consistent with thermal wind balance: shear along a sloping isentropic surface should be more anticyclonic than shear along a flat surface within a baroclinic zone. This fact is illustrated by means of a cross section through the cold front and upper-level jet (Fig. 10), which depicts the full (not merely anomalous) BC term. Proceeding right to left along a sloping isentropic surface below the jet, one invariably encounters stronger winds (and therefore ex-

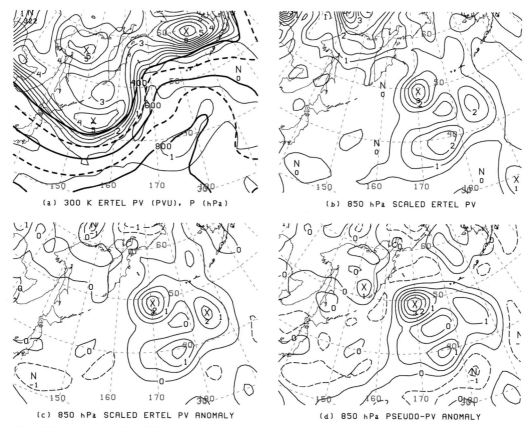

(a) 300 K ERTEL PV (PVU), P (hPa)

(b) 850 hPa SCALED ERTEL PV

(c) 850 hPa SCALED ERTEL PV ANOMALY

(d) 850 hPa PSEUDO-PV ANOMALY

FIG. 8. Total and anomalous PV, as in Fig. 4, but for (a) the 300-K isentropic surface (pressure contour interval is 100 hPa) and (b), (c), (d) the 850-hPa isobaric surface.

periences more anticyclonic shear vorticity) than if one moved horizontally along an isobaric surface. Above the jet the isentropic surfaces slope downward from right to left, but one still encounters stronger anticyclonic wind shear than one would along a horizontal surface. While the BC term attains a smaller magnitude than the AV and NL terms at 850 hPa, it is large where the perturbation PV is weak and therefore causes a change in sign between the scaled Ertel-PV anomalies and the pseudo-PV anomalies.

The TP term (Fig. 9d) is insignificant at low levels, where the vertical gradient of background PV is weak.

e. Summary

The snapshots of PV in the upper and lower troposphere illustrate some common aspects of the pseudo-PV compared with scaled Ertel PV. First, a major contributor to the difference between Ertel and pseudo-PV is the AV term. This term makes the conventional pseudo-PV "too large" within cyclonic circulations, effectively tending to make the amplitude of positive pseudo-PV anomalies larger than the amplitude of their scaled Ertel-PV counterparts. The AV term also causes pseudo-PV to overestimate the shear vorticity along the margins of cyclonically curved jets, and it would also cause

pseudo-PV to underestimate the shear vorticity in anticyclonically curved situations. However, as noted in sections 2b and 3b, the AV term is large only because we have chosen to use height rather than streamfunction in computing pseudo-PV. There is no theoretical reason for preferring the pseudo-PV computed from the height field to the pseudo-PV computed from the actual vorticity.

Partially cancelling the effect of the AV term in some places is the NL term. This term will be positive and particularly strong within strong PV anomalies and zero or negative in areas adjacent horizontally and vertically to strong PV anomalies. In the examples shown above, the largest NL contributions were in the cores of the upper-level trough and lower-level cyclone.

The remaining two terms, the BC term and the TP term, were in general small. One exception is strong frontal zones, where the BC term leads to substantially different scaled Ertel-PV and pseudo-PV values, with pseudo-PV more positive than scaled PV. The smallness of the TP term, which serves the role of neutralizing the effect of vertical advection of basic-state PV, is good news because it means that pseudo-PV is not much different from a linearized QG approximation to the Ertel PV. One could argue that there is little need to work with the "pseudo" version of PV created by eliminating

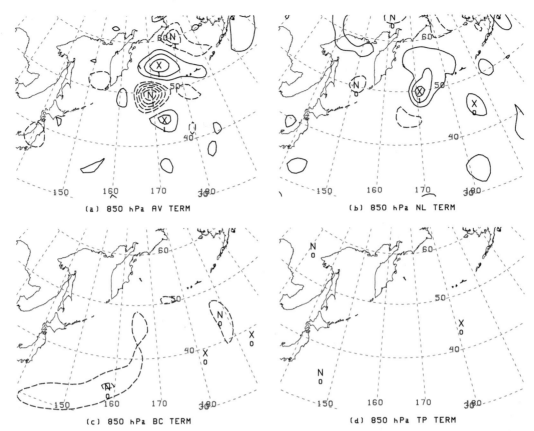

Fig. 9. Difference terms between anomalous scaled Ertel PV and anomalous pseudo-PV, as in Fig. 6, but for 850 hPa. Contour interval in (a)–(d) is 0.2×10^{-4} s^{-1}.

the effect of vertical motion, because the smallness of the TP term demonstrates that neglect of the vertical advection of PV is not as significant as neglect of the nonlinearity and ageostrophic effects.

Two of the four difference terms, the AV and TP terms, are in a sense optional terms. The former's existence depends on the method by which vorticity is computed, and the latter's existence depends on whether the meteorologist prefers to work with a QG PV that can be advected in three dimensions or a pseudo-PV that can only be advected horizontally.

Most importantly for our purposes, the differences between pseudo-PV and scaled Ertel PV are either obvious, small, or both. An intuitive understanding of QG diagnostics in terms of the height tendency equation and pseudo-PV can be translated directly into the Ertel-PV framework. Relatively high and low values of Ertel PV imply relatively high and low values of pseudo-PV, and differences are mostly smaller than the quantities themselves. Furthermore, since it is perfectly legitimate to treat the Ertel PV as a slaving variable and construct QG diagnosis around Ertel PV, as discussed in section 2, there is no theoretical or practical bar to replacing pseudo-PV anomalies with Ertel-PV anomalies in the context of the height-tendency equation. Instead, there are quite definite advantages, the primary one being that

Ertel PV is conserved in an absolute sense rather than an approximate one. One can "think" in QG and "see" Ertel PV, and in a sense have the best of both worlds. The extent to which this qualitative equivalence applies to quantitative piecewise diagnosis of heights and height tendencies is explored in Neilsen-Gammon and Gold (2008).

4. Concluding remarks

Several problems with the traditional diagnosis of vertical motion and development have been brought on by increases in computer power and model resolution. Possible solutions to these problems involve choosing a diagnostic approach that is less sensitive to small-scale noise but that retains a direct connection between conceptual understanding and the real-world situation. One step in this direction, entirely justified by the quasigeostrophic formalism, involves the use of nondivergent wind rather than height as an input parameter for diagnostic calculations such as **Q** vectors. A more comprehensive approach would involve a shift to the use of PV transport as the fundamental qualitative driver of vertical motion and development.

The switch from quasigeostrophic diagnosis to PV thinking, as applied to development, involves transi-

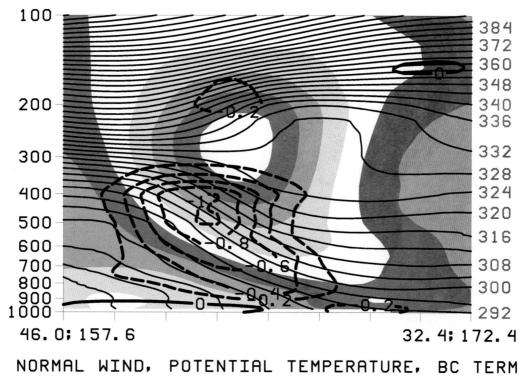

FIG. 10. Cross section through cold front and jet stream: Potential temperature (K; thin contours), component of wind into the section (shading interval 10 m s⁻¹), full (time-mean plus anomaly) BC (thick contours, contour interval 0.1×10^{-4} s⁻¹, negative contours dashed). See text for details.

tioning from pseudo-PV (the fundamental quantity of the QG height-tendency equation) to Ertel PV. The two quantities are closely related, although pseudo-PV is not really a potential vorticity. Apart from a scaling factor that depends on the background stratification, the difference between the two quantities can be written as four terms. Two of those terms are not fundamental to the QG system and can be made to vanish with a different definition of quasigeostrophic PV. The remaining terms involve nonlinearity (the product of perturbation vorticity and stratification) and the difference between vorticity on flat surfaces and sloping isentropic surfaces. As applied to a case of rapid and intense cyclogenesis over the North Pacific, all of the difference terms are small compared with the PV itself, so intuition regarding pseudo-PV and QG height tendencies can generally be applied equally well to Ertel PV.

If one wishes to track the evolution of the system dynamics through PV, a dynamical framework using Ertel PV has the advantage that Ertel PV will be exactly conserved in dry inviscid dynamics. Nonconservation of Ertel PV may be directly attributed to diabatic and frictional processes, whereas the nonconservation of pseudo-PV, linearized PV, etc., is ambiguous.

The use of PV as a diagnostic tool in an operational setting would not generally involve quantitative diagnostic inversions such as are performed in a research setting. Qualitative dynamical reasoning, as in Hoskins

et al. (1985), Nielsen-Gammon (1995), and Morgan and Nielsen-Gammon (1998) should suffice. For example, propagation of Rossby wave energy and subsequent downstream development (e.g., Orlanski and Sheldon 1993) requires a PV gradient along which such a wave can propagate. One might be able to rule out rapid cyclogenesis if there is no clear PV waveguide to channel energy from an upstream development. In another example, the formation of an upper-level cutoff low from the polar jet might only have a local effect on the weather, but if it wanders far enough south to induce winds along the subtropical jet, a downstream wave train might result. Potenial vorticity helps the forecaster see what the atmosphere is capable of doing, thereby helping not just the forecast itself but also the forecast of what can go wrong (i.e., the forecast uncertainty).

Acknowledgments. Interactions with expert diagnosticians such as Fred Sanders, Randall Dole, Kerry Emanuel, Lance Bosart, Daniel Keyser, Michael Morgan, Brian Hoskins, and Alan Thorpe directly fostered the development of the ideas presented here. Chris Davis and an anonymous reviewer greatly helped to improve the clarity of the manuscript. This work was partially supported by the National Science Foundation through Grants ATM-9521383 and ATM-0089906.

REFERENCES

Charney, J. G., and M. E. Stern, 1962: On the stability of internal baroclinic jets in a rotating atmosphere. *J. Atmos. Sci.,* **19,** 159–172.

Davis, C. A., 1992: Piecewise potential vorticity inversion. *J. Atmos. Sci.,* **49,** 1397–1411.

Dixon, M. A. G., A. J. Thorpe, and K. A. Browning, 2003: Layerwise attribution of vertical motion and the influence of potential-vorticity anomalies on synoptic development. *Quart. J. Roy. Meteor. Soc.,* **129,** 1761–1778.

Durran, D. R., and L. W. Snellman, 1987: The diagnosis of synoptic-scale vertical motion in an operational environment. *Wea. Forecasting,* **2,** 17–31.

Errico, R. M., 1997: What is an adjoint model? *Bull. Amer. Meteor. Soc.,* **78,** 2577–2591.

Fehlmann, R., and H. C. Davies, 1997: Misforecasts of synoptic systems: Diagnosis by PV retrodiction. *Mon. Wea. Rev.,* **125,** 2247–2264.

Holton, J. R., 1992: *An Introduction to Dynamic Meteorology.* 3d ed. Academic Press, 511 pp.

Hoskins, B. J., and M. A. Pedder, 1980: The diagnosis of middle latitude synoptic development. *Quart. J. Roy. Meteor. Soc.,* **106,** 707–719.

——, M. E. McIntyre, and A. Robertson, 1985: On the use and significance of isentropic potential vorticity maps. *Quart. J. Roy. Meteor. Soc.,* **111,** 887–946.

Juckes, M., 1999: The structure of idealized upper-tropospheric shear lines. *J. Atmos. Sci.,* **56,** 2830–2845.

Jusem, J. C., and R. Atlas, 1998: Diagnostic evaluation of vertical motion forcing mechanisms by using Q-vector partitioning. *Mon. Wea. Rev.,* **126,** 2166–2184.

Kalnay, E., and Coauthors, 1996: The NCEP/NCAR 40-Year Reanalysis Project. *Bull. Amer. Meteor. Soc.,* **77,** 437–471.

Keyser, D., M. J. Reeder, and R. J. Reed, 1988: A generalization of Petterssen's frontogenesis function and its relation to the forcing of vertical motion. *Mon. Wea. Rev.,* **116,** 762–780.

——, B. D. Schmidt, and D. G. Duffy, 1992: Quasigeostrophic vertical motions diagnosed from along- and cross-isentrope components of the Q vector. *Mon. Wea. Rev.,* **120,** 731–741.

Massacand, A. C., H. Wernli, and H. C. Davies, 2001: Influence of upstream diabatic heating upon an Alpine event of heavy precipitation. *Mon. Wea. Rev.,* **129,** 2822–2828.

Mohebalhojeh, A. R., 2002: On shallow-water potential-vorticity inversion by Rossby-number expansions. *Quart. J. Roy. Meteor. Soc.,* **128,** 679–694.

Morgan, M. C., and J. W. Nielsen-Gammon, 1998: Using tropopause maps to diagnose midlatitude weather systems. *Mon. Wea. Rev.,* **126,** 2555–2579.

Muraki, D. J., C. Snyder, and R. Rotunno, 1999: The next-order corrections to quasigeostrophic theory. *J. Atmos. Sci.,* **56,** 1547–1560.

Nielsen-Gammon, J. W., 1995: Dynamical conceptual models of upper-level mobile trough formation: Comparison and application. *Tellus,* **47A,** 705–721.

——, 2001: A visualization of the global dynamic tropopause. *Bull. Amer. Meteor. Soc.,* **82,** 1151–1167.

——, and R. J. Lefevre, 1996: Piecewise tendency diagnosis of dynamical processes governing the development of an upper-tropospheric mobile trough. *J. Atmos. Sci.,* **53,** 3120–3142.

——, and D. A. Gold, 2008: Potential vorticity diagnosis in the quasigeostrophic and nonlinear balance systems. *J. Atmos. Sci.,* **65,** 172–188.

Orlanski, I., and J. Sheldon, 1993: A case of downstream baroclinic development over western North America. *Mon. Wea. Rev.,* **121,** 2929–2950.

Pedlosky, J., 1987: *Geophysical Fluid Dynamics.* 2d ed. Springer-Verlag, 710 pp.

Sanders, F., 1971: Analytic solutions of the nonlinear omega and vorticity equations for a structurally simple model of disturbances in the baroclinic westerlies. *Mon. Wea. Rev.,* **99,** 393–407.

——, and B. J. Hoskins, 1990: An easy method for estimation of Q-vectors from weather maps. *Wea. Forecasting,* **5,** 346–353.

Sutcliffe, R. C., 1947: A contribution to the problem of development. *Quart. J. Roy. Meteor. Soc.,* **73,** 370–383.

Tan, Z.-M., F. Zhang, R. Rotunno, and C. Snyder, 2004: Mesoscale predictability of moist baroclinic waves: Experiments with parameterized convection. *J. Atmos. Sci.,* **61,** 1794–1804.

Thorpe, A. J., 1986: Synoptic-scale disturbances with circular symmetry. *Mon. Wea. Rev.,* **114,** 1384–1389.

——, and C. H. Bishop, 1995: Potential vorticity and the electrostatics analogy: Ertel–Rossby formulation. *Quart. J. Roy. Meteor. Soc.,* **121,** 1477–1495.

Trenberth, K. E., 1978: On the interpretation of the diagnostic quasigeostrophic omega equation. *Mon. Wea. Rev.,* **106,** 131–137.

Vallis, G. K., 1996: Potential vorticity inversion and balanced equations of motion for rotating and stratified flows. *Quart. J. Roy. Meteor. Soc.,* **122,** 291–322.

Warn, T., O. Bokhove, T. G. Shepherd, and G. K. Vallis, 1995: Rossby number expansions, slaving principles, and balanced dynamics. *Quart. J. Roy. Meteor. Soc.,* **121,** 723–739.

Chapter 10

Finescale Radar Observations of a Dryline during the International H₂O Project (IHOP_2002)

CHRISTOPHER C. WEISS
Department of Geosciences, Texas Tech University, Lubbock, Texas

HOWARD B. BLUESTEIN
School of Meteorology, University of Oklahoma, Norman, Oklahoma

ANDREW L. PAZMANY
ProSensing, Inc., Amherst, Massachusetts

BART GEERTS
College of Engineering, University of Wyoming, Laramie, Wyoming

(Manuscript received 25 February 2004, in final form 9 March 2006)

Weiss   Bluestein   Geerts

ABSTRACT

A case study of a double dryline on 22 May 2002 is presented. Mobile, 3-mm-wavelength Doppler radars from the University of Massachusetts and the University of Wyoming (Wyoming cloud radar) were used to collect very fine resolution vertical-velocity data in the vicinity of each of the moisture gradients associated with the drylines. Very narrow (50–100 m wide) channels of strong upward vertical velocity (up to 8 m s⁻¹) were measured in the convergence zone of the easternmost dryline, larger in magnitude than reported with previous drylines. Distinct areas of descending motion were evident to the east and west of both drylines. Radar data are interpreted in the context of other observational platforms available during the International H₂O Project (IHOP_2002). A variational ground-based mobile radar data processing technique was developed and applied to pseudo-dual-Doppler data collected during a rolling range-height indicator deployment. It was found that there was a secondary (vertical) circulation normal to the easternmost moisture gradient; the circulation comprised an easterly component near-surface flow to the east, a strong upward vertical component in the convergence zone, a westerly return flow above the convective boundary layer, and numerous regions of descending motion, the most prominent approximately 3–5 km to the east of the surface convergence zone.

Corresponding author address: Christopher C. Weiss, Dept. of Geoscience, Texas Tech University, Box 42101, Lubbock, TX 79409.
E-mail: chris.weiss@ttu.edu

1. Introduction

Fred Sanders was interested in the physical processes responsible for the formation of deep convection (Sanders and Blanchard 1993) and the forecasting of deep convection (Sanders and Garrett 1975; Sanders 1986). Independent of his interest in convection, he pursued his interest in the analysis, nature, and behavior of surface boundaries, along which ascending motion on the mesoscale sometimes initiates convective clouds (e.g., Sanders and Doswell 1995; Bluestein 2008). In addition, the second author remembers well his interest in synthesizing all available sources of data, including radar data, in studying the behavior of convective systems (e.g., Miller and Sanders 1980). The second author recalls finding Fred Sanders pouring over sequences of radar images that covered much of the 16th floor corridor of the Green Building at the Massachusetts Institute of Technology while he studied the case just referenced. It is in the spirit of Fred's enthusiasm for convection and surface boundaries that we present this recent study of the behavior of a dryline and its relation to convection initiation (CI), which is based on a new, mobile observing system and a new analysis technique.

The processes responsible for the initiation of deep moist convection have intrigued researchers and forecasters alike for decades. Even in cases in which the large-scale environment is conducive to the development of severe thunderstorms (i.e., when there is high potential instability and strong vertical wind shear over a broad area), CI occurs only locally. CI tends to occur in regions where there is mesoscale convergence of winds in the boundary layer, and therefore forced ascent. Surface boundaries are examples in which there can be mesoscale ascent (Bluestein 2008); therefore, much of the ambiguity surrounding the convection initiation problem can be diminished through a better understanding of the finescale motions near these boundaries.

The dryline, in the southern plains of the United States, is an example of a boundary that frequently initiates intense thunderstorms, some of which are responsible for the production of large hail, damaging winds, and tornadoes during the spring months. The dryline can be thought of as the intersection between the top of a surface-based layer of virtually cool, moist air originating over the Gulf of Mexico and the sloping terrain east of the Rocky Mountains. The prediction of CI along this boundary is particularly complex owing to the large fluctuations in the position of, movement of, and convergence along, the boundary.

During the spring months of 2002, the International H$_2$O Project (IHOP) was conducted over the central and southern plains. The mission of IHOP was the improved characterization of the four-dimensional distribution of water vapor and its application to improving the understanding and prediction of convection (Weckwerth et al. 2004). A multitude of surface and airborne observing platforms were employed through the course of the experiment.

In conjunction with these platforms, W-band (3-mm wavelength; 95 GHz) radars from the University of Massachusetts (UMass) (Bluestein and Pazmany 2000) and the University of Wyoming [Wyoming cloud radar

(WCR)] collected data in a number of drylines during IHOP (Weiss et al. 2006). The narrow half-power beamwidth of the radars (0.18° and 0.7° for UMass and WCR, respectively) allowed for very fine spatial resolution, and thereby the potential existed to resolve scales of motion near the dryline that had been previously unresolved. A better knowledge of these finescale motions will ultimately lead to a more accurate understanding of the success or failure of CI in the dryline convergence zone (DCZ).

Another goal of this project was to develop a retrieval technique to synthesize, from time series of range-height indicator (RHI; i.e., vertical cross sections) data, the two-dimensional wind components in the plane normal to the dryline. Radar velocity observations are inherently one-dimensional, measuring the component of motion toward/away from the radar line of sight. However, using certain assumptions and scanning strategies, the two-dimensional wind field can be recovered.

The purpose of this paper is to describe the analysis technique applied to, and the analysis of, Doppler wind data in drylines on 22 May 2002 during IHOP. Section 2 provides an overview of the development and motion of the drylines. A summary of scanning strategies used by the W-band radar system is presented in section 3. Section 4 details the analysis of data collected with a vertically pointed antenna. Section 5 describes the development and testing of a variational radar synthesis technique, which is applied to data from the drylines, in section 6. Finally, a summary is given in section 7.

2. Overview of dryline motion and development

The early evolution of the 22 May 2002 dryline was typical of that seen over the southern plains. The dryline was located[1] over the western Oklahoma and Texas Panhandles at 1500 UTC (Fig. 1a). Because of the increase in sensible heating during the late morning and early afternoon the boundary layer deepened significantly (not shown), thereby permitting the distinctive eastward advancement of the dryline by 1800 UTC (Fig. 1b). Sharp decreases in dewpoint were noted with dryline passage over Kansas [e.g., Elkhart, Kansas (EHA), and Liberal, Kansas (LBL)]. Farther to the south, surface dewpoints fell in a more gradual manner [e.g., Amarillo, Texas (AMA)], which was less indicative of a distinct dryline passage. This region will be shown to be an intermediate zone between two sharp moisture gradients/drylines later in the day.

By late afternoon the dryline ceased its eastward advancement and stalled over the eastern Oklahoma and

---

[1] The dryline was analyzed at the leading edge of the sharpest surface dewpoint gradient, when reporting surface stations were available. Data from other sources [e.g., University of Wyoming King Air in situ specific humidity measurements; Weather Surveillance Radar-1988 Doppler (WSR-88D) finelines] were used to provide the best estimate of dryline position in areas void of adequate surface measurements.

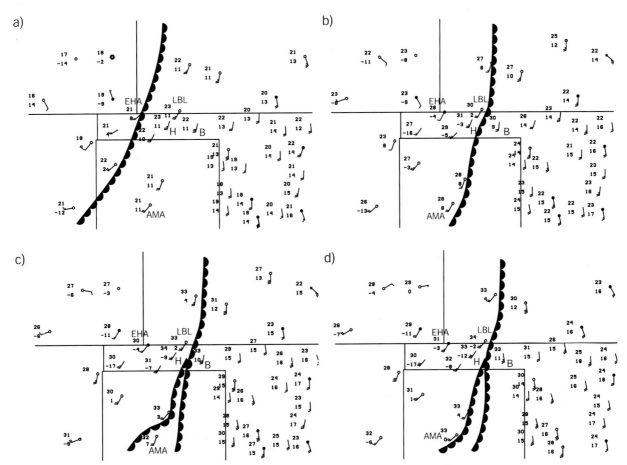

FIG. 1. High plains surface map valid at (a) 1500, (b) 1800, (c) 2100, and (d) 2300 UTC 22 May 2002. Temperature (°C) and dewpoint (°C) for each station are shown. Full (half) wind barbs denote 5 (2.5) m s⁻¹. The scalloped lines indicate the position of (known) drylines. The labels EHA, LBL, AMA, H, and B denote the locations of Elkhart, KS, Liberal, KS, Amarillo, TX, Hooker, OK, and Beaver, OK, respectively.

Texas Panhandles. The dewpoint gradient sharpened considerably during this time, and by 2100 UTC (Fig. 1c) a dewpoint difference of 18°C was evident over the 60 km between Hooker, Oklahoma, and Beaver, Oklahoma. Convergence was also evident on the boundary at this time. The magnitude of this convergence varied considerably as winds were gusty on both sides of the dryline. As seen earlier, the dewpoint at AMA continued to decrease in a steady manner until 2200 UTC, when there was a sharp decrease from 7° (2200 UTC) to 0°C (2300 UTC; Fig. 1d).

It is seen in WSR-88D radar reflectivity data from Amarillo (KAMA) at about 2300 and 0000 UTC (Figs. 2a,b) that there was a double-fineline (i.e., thin maximum in reflectivity) structure over the northern Texas Panhandle.[2] The passage of the westernmost fineline

coincided with the sharp dewpoint decrease observed at AMA at 2300 UTC (Figs. 1c,d). Earlier in the afternoon, AMA was situated between the finelines and experienced a gradual decrease in dewpoint. This behavior was in accord with that observed in other studies (e.g., Hane et al. 1997; Crawford and Bluestein 1997). The two finelines were oriented such that they merged just to the north of the 10-cm wavelength, dual-polarization National Center for Atmospheric Research (NCAR) S-band dual polarization (SPOL) radar (http://www.atd.ucar.edu/rsf/spol/spol.html) at Homestead, Oklahoma (Fig. 3a). It is apparent in visible satellite imagery (Fig. 4) that there was a wedge-shaped area of cumulus cloud cover in the region between the radar finelines over the Oklahoma and northern Texas Panhandles (Figs. 2a,b and 4). A time series of in situ dewpoint measurements taken at 800 m AGL aboard the University of Wyoming King Air (UWKA) (Geerts and Miao 2005) supported the coexistence of separate moisture gradients with each fineline (Figs. 3a,b). It is noted that the domain of UMass operations (white box in Fig. 3a) on this day was to the south of the intersection point.

---

[2] Finelines are often used as evidence of confluence in the dryline convergence zone, as the insect concentration, and therefore radar reflectivity, locally increases in these regions. Because the region of confluence is where the strongest frontogenesis is also occurring, the sharp specific humidity gradient at the dryline is often correlated with a thin reflectivity fineline.

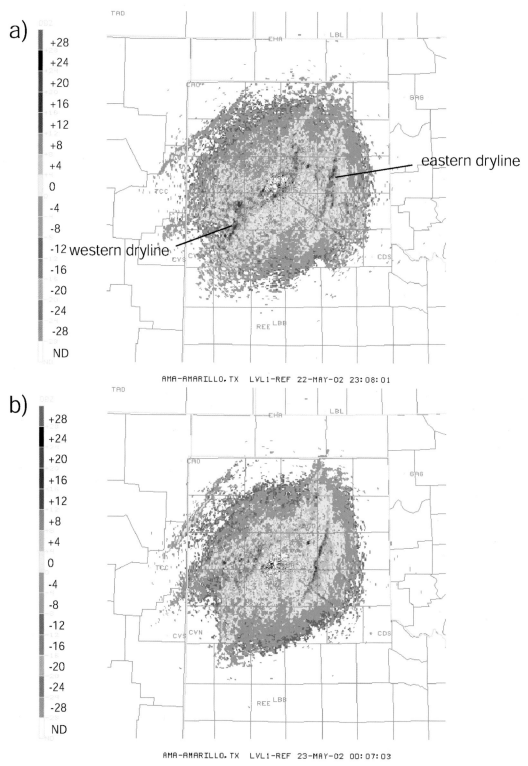

FIG. 2. WSR-88D 0.5° reflectivity (dBZ) at Amarillo valid at (a) 2308 UTC 22 May 2002 and (b) 0007 UTC 23 May 2002. Reflectivity scale provided to the left. The locations of the eastern and western dryline are shown in (a). (c) 0.5° reflectivity (dBZ) from the SMART-R at 2254 UTC 22 May 2002. The black arrow indicates the path of the UMass vertical antenna deployment. Note that these data were collected approximately 20 min after the termination of the UMass vertical antenna deployment. The locations of the eastern and western dryline are shown. Range markers are indicated in black.

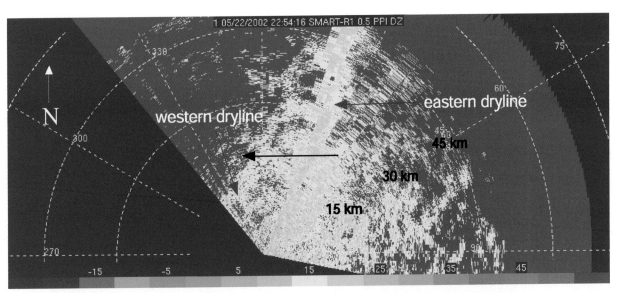

FIG. 2. (*Continued*)

Therefore, the data collection encompassed both dry-lines. Both dryline boundaries demonstrated retrogression by 0000 UTC (Figs. 2a,b).

Deep convection had initiated along the dryline over northwestern Kansas and eastern Nebraska by late in the afternoon. This region was more directly influenced by a longwave trough over the western United States (not shown), which likely produced vertical motion and midlevel cooling that aided in the development of the convection. Farther to the south, no sustained deep convection was initiated along the dryline.

3. W-band radar characteristics

The primary datasets used for this study were collected with the UMass W-band radar. As mentioned above, the very narrow beamwidth of 0.18° permitted radial velocity measurements with high spatial resolution. Consequently, at a range of 1 km from the radar, the effective footprint was approximately 3 m wide. Because the area scanned was void of precipitation, the principal scattering source for the radiation returned to the W-band radar was most likely insects (Wilson and Schreiber 1986; Russell and Wilson 1997; Geerts and Miao 2005). Because the wavelength of the radiation transmitted by the radar was comparable to the size of the targets, Mie scattering was the dominant source of returned power. The minimum detectable signal for the W-band radar was -35 dBZ$_e$ at a range of 1 km from the radar. The average reflectivity in convergence zones at this range (shown later) was about -20 dBZ$_e$, representing a returned power over 30 times the minimum detectable signal.

The following three scanning strategies were utilized in the 22 May case study:

1) Vertical antenna—antenna pointed at 86° (maximum elevation allowed by the positioner) and driven across the boundary. The resulting time series of vertical velocity data were corrected for vehicle motion and converted to a spatial profile using recorded GPS data.

2) Stationary RHI (SRHI)—stationary data collection in which the antenna was rotated from approximately 0° to 86° in elevation. Multiple vertical sectors of radial velocity data were obtained in this manner. Although useful for tracking reflectivity and diagnosing radial velocity, the u and w components could not be retrieved independently with such a collection strategy.

3) Rolling RHI (RRHI)—0°–86° RHIs collected with the radar platform in motion. The radial velocity was adjusted for platform motion. The principles of pseudo-dual-Doppler analysis (e.g., Hildebrand et al. 1996) could be applied to data taken in such a manner to retrieve the individual u and w wind components (described in section 5).

4. Analysis of vertical antenna scans across the dryline

From 2221 to 2235 UTC, the UMass W-band radar executed a westward-moving vertical antenna deployment across the double dryline along U.S. Highway 270 near Elmwood, Oklahoma, in the Oklahoma Panhandle (Fig. 5). The objective of this deployment was to obtain a time series of vertical velocity in the near-dryline environment. The vehicle maintained a nearly constant speed of 27 m s^{-1} during the traverse.

The Shared Mobile Atmospheric Research and Training Radar (SMART-R; Fig. 2c; Biggerstaff et al. 2005) and SPOL (Fig. 3a) both detected a fineline associated with the eastern DCZ. This boundary was oriented in a north-northeast to south-southwest direction. Therefore, the traverses were not precisely normal, but rather

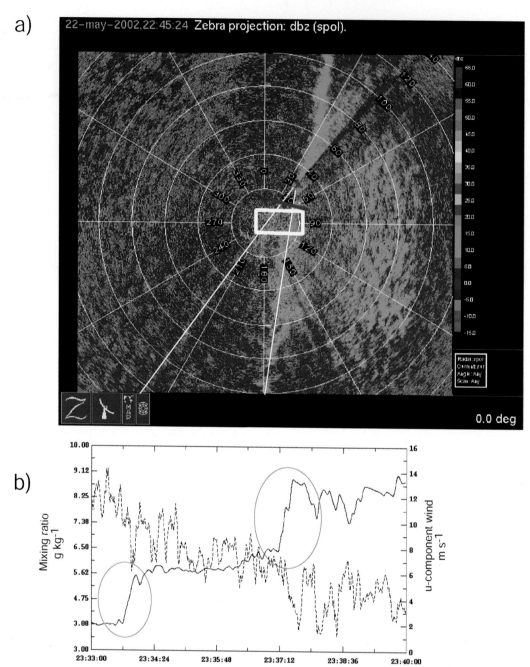

FIG. 3. (a) Reflectivity from SPOL (location indicated in Fig. 4). Reflectivity scale (dBZ) indicated to the right. The straight white lines indicate the axes of two separate drylines. The white box represents the domain of operations for the UMass W-band radar. The red line denotes the flight track of UWKA during 2333–2340 UTC. (b) Traces of in situ specific humidity (g kg^{-1}, solid trace) and u-component wind (m s^{-1}; dashed trace) taken aboard the UWKA for the flight leg indicated in (a). Time (UTC) is indicated along the bottom axis; scales for specific humidity and u-component wind are indicated on the left and right axes, respectively. The two regions of sharp moisture gradient are circled in green.

formed a small angle from normal to the boundary. A secondary, less distinct fineline was evident to the west of the eastern (i.e., targeted) dryline. Though the western dryline was not recognized at the time of data collection, the UMass W-band radar transected this secondary feature just before the termination of the vertical-antenna data collection leg (Fig. 2c). Both finelines were collocated with a specific humidity change of 2–2.5 g kg^{-1} over a distance of approximately 1 km (Fig. 3b) at the UWKA flight level (~800 m AGL), thereby confirming that the identified reflectivity finelines were indeed drylines.

FIG. 4. *Geostationary Operational Environmental Satellite* (*GOES-8*) visible satellite image at 2233 UTC 22 May 2002. Black boundaries denote the state borders. The black dot indicates the location of SPOL.

Both drylines backscattered a detectable signal toward the UMass radar. The time section of reflectivity for this data leg (Fig. 6) shows both drylines clearly as a maximum in reflectivity. The eastern dryline returned a signal in excess of -15 dBZ$_e$, while the western dryline returned less power (a maximum of approximately -20 dBZ$_e$). The reflectivity maxima were associated with a local concentration of boundary layer scatterers that are thought to be composed primarily of insects (Wilson and Schreiber 1986). These insects are assumed to be passive, and therefore their motion is identical to that of their advecting velocity. Convergent regions like those shown in Fig. 6 accumulate these insects and therefore are more reflective (assuming the insect size distribution is the same everywhere). The relatively lower reflectivity of the western dryline in UMass and SMART-R data was presumably caused by a decreased insect concentration, either due to weaker convergence and/or a more limited source of insects.

There was a nearly vertical slope to both drylines in the lowest 1–1.5 km AGL (Fig. 6), consistent with previous observations of the forward-propagating dryline (e.g., Hane et al. 1997). Above the depth of the convective boundary layer (CBL) to the east of each dryline ($\sim$1.5 km AGL for the eastern dryline; $\sim$2 km AGL for the western dryline), the drylines tilted significantly

with height to the east (Fig. 6). Minima in reflectivity were observed in the eastern dryline interface at approximately 1.5 km AGL (Fig. 6). One of these areas (D$_1$ in Fig. 6) was immediately to the east of the surface position of the eastern DCZ. The other position (D$_2$ in Fig. 6) was about 4 km to the east of the DCZ. Even though the UMass velocity data (Fig. 7) were limited in areas D$_1$ and D$_2$, the fringe regions about the reflectivity-void regions indicated weak descent. Similar decreases in maximum echo altitude west of the eastern dryline (e.g., west of B and C in Figs. 6 and 7) were clearly correlated with the bodies of downdrafts. Because the source for the scatterers (i.e., insects) is the surface, the scatterer concentration is nearly zero at higher altitudes (e.g., above the boundary layer). Therefore, downward motion across the dryline interface represents transport from a region where there is a dearth of insects and is therefore associated with a lack of radar reflectivity (Geerts and Miao 2005).

Observations from the WCR were considered to confirm the regions of suspected descent, and for general intercomparison with the UMass W-band radar. The WCR is an airborne W-band radar with smaller antennas, each therefore with a larger beamwidth than that of the UMass W-band radar. The flight leg (Fig. 5) crossed directly over the path of the UMass W-band

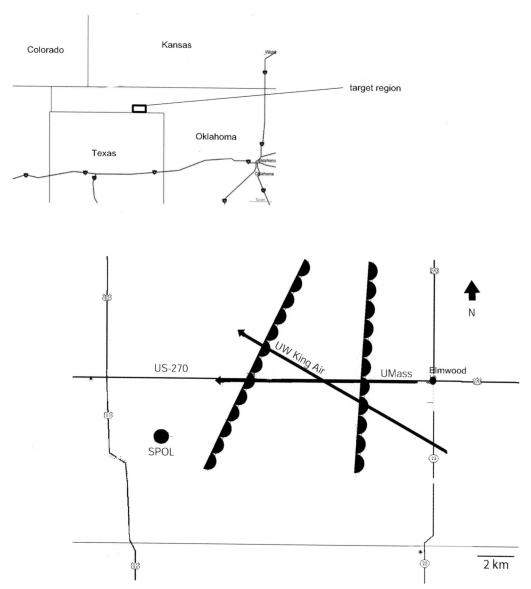

FIG. 5. Maps depicting the vertical antenna UMass W-band deployment. Distance scale located in the lower right-hand corner of the zoomed-in map. The thick lines denote the path of the UMass vertical antenna deployment from right to left (east to west) and the path of the UWKA from right to left (east-southeast to west-northwest). The scalloped lines indicate the position of known drylines. The dot labeled SPOL denotes the position of the SPOL radar near Homestead.

radar and ended approximately 10 min before the completion of the UMass data collection. The flight leg was flown normal to the eastern dryline and therefore formed an angle to the UMass ground leg. The UWKA confirmed downward motion in the suspected subsiding regions D_1 and D_2 (Fig. 8a) marked by reflectivity voids to the east of the eastern DCZ in Fig. 6. A sharp decrease in maximum echo altitude of nearly 1 km is noted by both the UMass and WCR platforms at position D_2 (consider data from 2.0 km AGL at position D_1 and from 1.5 km AGL at position D_2 in Figs. 7 and 8a).

The region of subsiding air approximately 4 km to the east of the dryline (D_2 in Figs. 8a and 8b) was ~3 km wide. Local vertical velocity maxima in excess of

-4 m s^{-1} were observed in this corridor. Though impossible to reconstruct trajectories from these snapshots, UMass and WCR velocity data both confirm that the subsidence extended downward from the CBL top to at least 1 km AGL. The position of this descending air was consistent with that found in the airborne radar study of Weiss and Bluestein (2002). The lowered maximum altitude of returned power in this region (~1 km AGL) suggested that the source region for this downward-moving air was at least in part from above the insect-rich CBL. One can therefore infer that this air had somewhat lower specific humidity. In situ measurements taken aboard the UWKA at 700 m AGL (Fig. 8c) confirm a small local decrease in dewpoint below

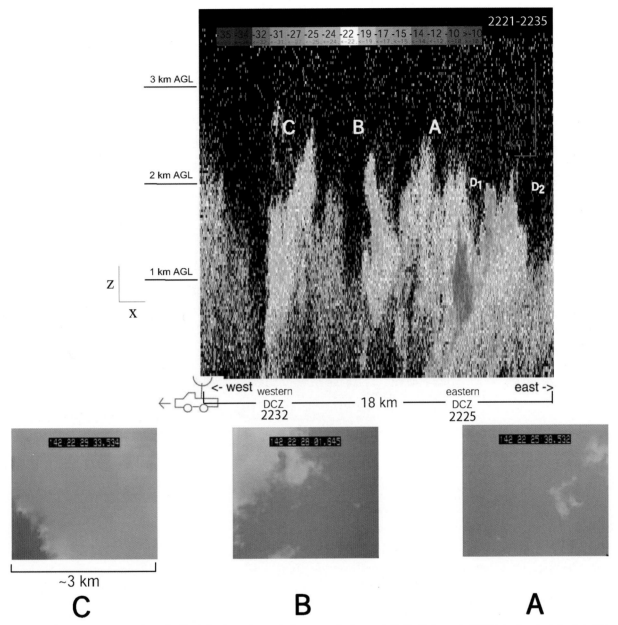

FIG. 6. East–west cross section of reflectivity from the vertical antenna deployment. Reflectivity scale (dBZ$_e$) is shown at the top. The 1-km scales for the horizontal and vertical direction are shown in the upper right-hand corner. Domain size is approximately 18 km wide (east–west) × 3.4 km high. Letters A, B, and C are the locations of cloud cover discussed in the text. The UMass vehicle was in motion toward the west (left). Labels D$_1$ and D$_2$ are referred to in the text. Images of video from the W-band boresighted video camera are shown. The time of dryline passage is indicated below where the eastern and western DCZ were crossed.

the evidence of strongest subsidence. Flight-level equivalent potential temperature drops of 2 K and gust-probe-measured downward vertical velocity of 1–2 m s^{-1} supported the assertion that subsidence was present. We speculate that these elevated regions of descent to the east of this stationary dryline may be microcosms of a larger CBL-scale process for the forward-propagating dryline, where the downward transport of low specific humidity air from above the CBL to the surface may be one mechanism to propagate the dryline eastward

(e.g., through the late morning and early afternoon hours).

UMass W-band measurements of the eastern DCZ indicated a maximum upward vertical velocity of approximately 8 m s^{-1} (Fig. 7). However, the most intense upward motion was evident only over a very narrow region approximately 50–100 m wide. The WCR also sensed strong upward motion, with a maximum w of 6.2 m s^{-1} (Fig. 8a). Although there were small differences in the exact location where the eastern DCZ was

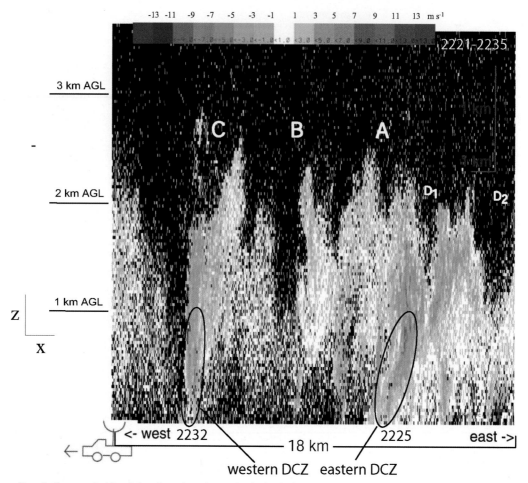

FIG. 7. Same as in Fig. 6, but the colors denote vertical velocity (m s⁻¹). Orange colors indicate upward motion; green colors indicate downward motion. Velocity scale is indicated at the top. Labels D_1 and D_2 are referred to in the text. The lower portions of the eastern and western DCZ are denoted by ovals.

crossed between the ground and flight tracks (Fig. 5), the difference in maximum w was at least partly attributable to the larger beamwidth (UMass: 0.18°; WCR: 0.7°) and faster cross-track platform motion (UMass: 26.8 m s⁻¹; UWKA/WCR: approximately 80–85 m s⁻¹) (LeMone et al. 2007). Both of these factors ultimately increased the size of the resolution volume for the WCR compared with UMass, though the faster WCR platform velocity allowed for timelier dryline sampling.

Both the UMass and WCR detected upward vertical velocity associated with the convergence zone of the western dryline. The WCR indicated a maximum ascent of approximately 3–4 m s⁻¹ (Fig. 8a), while UMass measured a maximum w of ~5 m s⁻¹ in narrow regions (Fig. 7). Again, the effective beamwidth may have contributed to this difference. Also, the difference in slope and decreased distance between the eastern and western drylines[3] (compared with observations from the UMass

---

[3] Recall that the UWKA intercepted the western dryline north of the UMass intercept (Fig. 5).

radar) suggest that the WCR measurements may have been made on a distinctly different portion of the western dryline. Regardless, it is seen in data from both platforms that there was a wide region of descent approximately 3–4 km wide centered about 4–5 km east of the western dryline. A decrease in dewpoint of 1°C was seen here as well as in UWKA in situ data (Fig. 8c).

Reflectivity data from SPOL (not shown) indicated that the eastern dryline had begun to retrograde at approximately 2230 UTC. WCR data from two passes centered around 2230 UTC (nearly identical in position; separated by ~25 min) reveal a widening of the distance between the two drylines, suggesting a different retrogression speed and/or onset for each dryline. Furthermore, the easternmost dryline had established a much more pronounced eastward tilt with height in the second WCR pass. The inclined slope of the interface appears qualitatively similar to that observed in propagating density currents.

A boresighted video camera was mounted on the

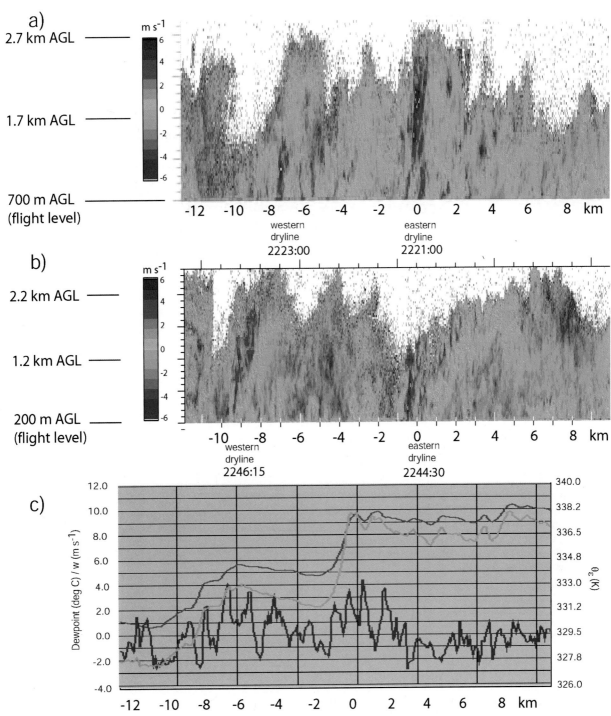

FIG. 8. WCR vertical velocity (m s⁻¹, scale to left) during (a) 2218–2224 and (b) 2241–2248 UTC. (c) Trace of in situ dewpoint (red trace; °C; scale included to left), θ_e (green trace; K; scale included to right), and vertical velocity (blue trace; m s⁻¹; scale included to left) measurements at flight level (700 m AGL) aboard the UWKA during 2218–2224 UTC. The horizontal distance scale in all three images represents the distance (km) east (positive) or west (negative) of the eastern dryline (positioned at 0 km). The vertical distance scale is in km AGL [note the different flight altitudes in (a) and (b)]. Labels D_1 and D_2 are referred to in the text.

W-band antenna to assist the radar operator (the first author) in the proper placement of the narrow beam during field operations. At times when the antenna was pointed vertically, the video served to identify regions of cloud cover directly above the instrument. Three such regions were found during the deployment. The first area (labeled A in Figs. 6 and 7), immediately to the west of the eastern DCZ, was populated by very shallow

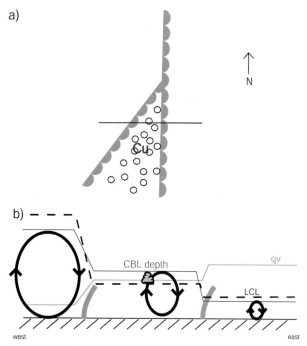

FIG. 9. A (a) plan view and (b) east–west cross section schematic of "wedge sector" cumulus. Circles in (a) denote boundary layer convection. In (b), the "qv" and "CBL depth" traces indicate the surface specific humidity and convective boundary layer depth, respectively. The dashed line denotes the location of the LCL.

cumulus clouds. The second area, likely associated with the ascending branch of a horizontal convective roll (HCR; LeMone 1973), was found halfway between the eastern and western DCZ (labeled B in Figs. 6 and 7), and contained more vigorous convection (Fig. 6). The third area of cloud cover, by far the widest (~2 km), was associated with the DCZ of the western dryline (labeled "C" in Figs. 6 and 7). No cumulus cloud field was seen outside the intermediate region between the drylines.[4] This observation was consistent with the "wedge" shape of cumulus convection seen on satellite images (Fig. 4).

It is hypothesized here that the wedge zone represented an optimal combination (for boundary layer cumulus development) of the highest specific humidity to the east of the eastern dryline and the deepest CBL to the west of the western dryline (Fig. 9). Signal-to-noise ratio (SNR) data (Fig. 10a) from the NCAR Integrated Sounding System (ISS)/Multiple Antenna Profiler Radar (MAPR; Cohn et al. 2001) indicated a rapid increase in the boundary layer depth to 3.5–4.0 km AGL (considering the altitude of the 0-dB SNR surface) by 2200 UTC as the Homestead profiling site sampled the wedge region (Demoz et al. 2006). Upward vertical velocity was evident through this boundary layer depth (Fig. 10b), supporting the notion that surface-based air parcels were freely ascending to this level. Remotely-

sensed soundings from the Atmospheric Emitted Radiance Interferometer (AERI; Feltz et al. 2003) in the wedge region at 2158 UTC confirm a level of free convection (LFC) height of approximately 2.7 km (4.0 km) (Fig. 11a). Considering the depth of the convective ascent noted by the MAPR, it is clear that lifted condensation level (LCL) was reached easily in the wedge region. Another AERI sounding taken at 0013 UTC (Fig. 11b), after the eastern dryline retreated westward through the profiling site, reveals the expected lowering of the LCL (to approximately 2.1 km AGL). MAPR data near the same time indicate a boundary layer depth (again, using the 0-dB SNR surface) of 1.8–2.4 km. With relatively weak upward velocity detected (compared to the wedge region) in the boundary layer, these findings largely confirm the visual observation that the LCL was not reached to the east of the eastern dryline. As the western dryline did not propagate through the profiling site, no MAPR information is available to confirm the depth of the boundary layer in the dry air. However, UWKA in situ data (Fig. 3b) and surface observations (Fig. 1) do identify another sharp decrease in moisture in association with the western dryline, likely limiting cumulus development.

5. The variational analysis of rolling RHI data

a. Motivation and technique development

One goal of this study was to synthesize, with radial velocity from a single radar, the secondary circulation normal to the plane of the dryline. In general, a time series of stationary RHIs are often used to capture the vertical evolution of atmospheric boundaries. With this technique, the individual components of motion in the plane of the RHI cannot be retrieved without making assumptions about one of the components (e.g., w is constant over the domain).

If the platform is in motion during the collection of the RHIs (the RRHI technique), then the two components of motion can be retrieved using pseudo-dual-Doppler principles (Hildebrand et al. 1996). An arbitrary point in space will be seen a number of times from varying "look" angles using such a strategy (Fig. 12). An assumption inherent to this technique is that of *stationarity*—that nothing varies at the point over the time elapsed between the "looks." One can expect this assumption to be more valid for larger scales of motion with limited translation.

Because a substantial number of overlapping radial velocity measurements were made using the RRHI scanning strategy, an analysis technique had to be developed to accommodate the overdetermined system. For this study, a weak-constraint variational (Sasaki 1970) wind synthesis technique was developed for rolling RHI data taken in an east–west plane. A cost function is defined that is a function of the dependent analysis variables. The cost function (J) penalizes the analysis for depar-

---

[4] A narrow band of cirrus clouds, however, was seen.

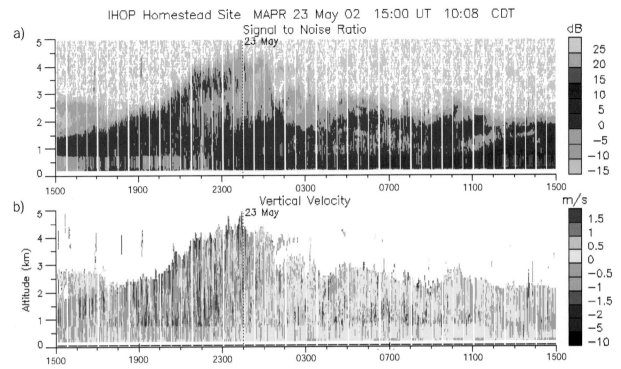

FIG. 10. (a) SNR (dB) and (b) vertical velocity (m s^{-1}) data from the MAPR, positioned at Homestead. The data in the display are valid from 1500 UTC 22 May 2002 to 1500 UTC 23 May 2002.

tures from radial-velocity observations and kinematic or dynamic constraints. The cost function chosen to be minimized for this case was

$$J = \sum_{\text{domain}} (J_{\text{obs}} + \beta J_{\text{continuity}}) \tag{1a}$$

$$J_{\text{obs}} = \sum_{n=1}^{m(x,z)} (c_1^n u + c_2^n w - \mathbf{V}_r^n)^2, \tag{1b}$$

$$J_{\text{continuity}} = \left(\frac{\partial u}{\partial x} + \frac{\partial w}{\partial z} + \kappa w\right)^2, \tag{1c}$$

$$c_1^n = \cos(\alpha^n), \quad \text{and} \tag{1d}$$

$$c_2^n = \sin(\alpha^n), \tag{1e}$$

where Eq. (1b) represents the contribution to the cost function from observational discrepancy and Eq. (1c) denotes the contribution to the cost function from anelastic mass continuity violation. In Eqs. (1a)–(1e) u and w are the analysis values, $\mathbf{V}_r$ is the observed radial velocity (corrected for platform motion), c_1 and c_2 represent geometric coefficients mapping velocities from Cartesian space to that of the radial velocity vectors, κ is the correction to mass continuity for vertical density stratification (assumed constant here), m is the total number of observations per grid point of which n is a specific observation, and α is the elevation angle for each observation. The parameter β represents the relative impact of departures from radial velocity observations and violation of mass continuity in the calculation of J. For the analyses presented below, β was

constant and set proportional to the square of the grid spacing [$(\Delta x)^2$]. The formulation was similar to that developed by Gao et al. (1999) (less a background and smoothness constraint) and Dowell and Bluestein (2002) (neglecting variations in the y direction).

The variations of J with respect to u and w were set equal to zero, yielding two coupled Euler–Lagrange equations (not shown). The two equations were repeatedly solved in turn until the solutions for u and w over the entire domain converged.

b. Testing

A series of observational system simulation experiments (OSSEs) was created to examine the behavior of the variational synthesis technique presented. Specifically, it was desired to elucidate the relation between analysis error and scanning strategy. To this end, a prescribed constant flow field of $u = 10$ m s^{-1} was sampled with a UMass "pseudoradar."[5] The platform velocity was due westward, and the scanning plane was oriented east–west. The platform velocity and vertical antenna rotation rate (hereafter, scan rate) were allowed to vary. Data were "stored" at a frequency of 10 Hz, matching the specifications of the signal processor in the UMass

---

[5] The synthesis procedure also tested successfully with OSSEs derived from large-eddy simulation (LES) output, where large gradients in the horizontal and vertical wind components were present (not shown).

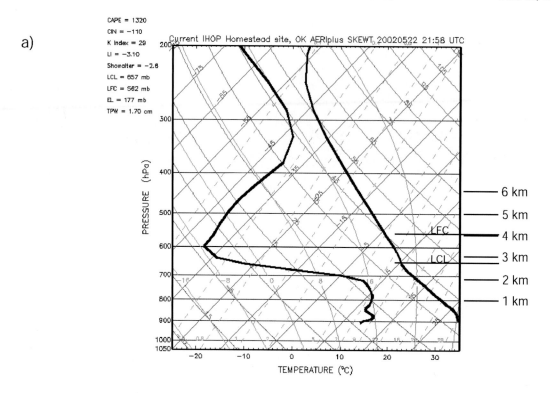

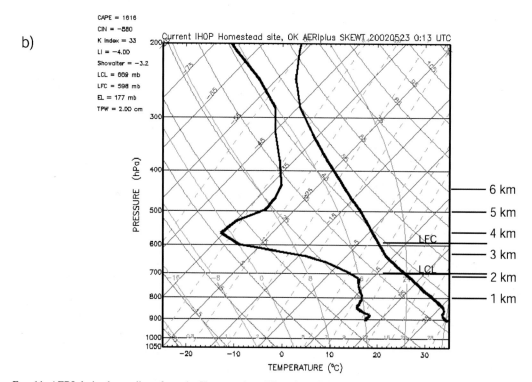

FIG. 11. AERI-derived soundings from the Homestead profiling site valid at (a) 2158 and (b) 0013 UTC. Height (km AGL), LCL, and LFC are identified to the right of each panel.

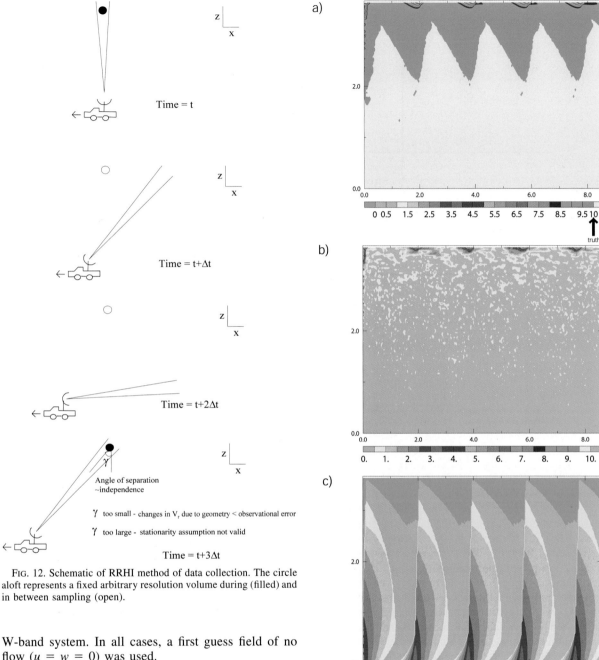

FIG. 12. Schematic of RRHI method of data collection. The circle aloft represents a fixed arbitrary resolution volume during (filled) and in between sampling (open).

W-band system. In all cases, a first guess field of no flow ($u = w = 0$) was used.

Initial simulations were performed with a platform velocity of 15 m s^{-1} and a scan rate of 1.5° s^{-1}—scanning parameters identical to that of the IHOP data collection on 22 May 2002. The u-component analysis winds verified very well against the truth field (Fig. 13a), with a domain total RMS error of 0.10 m s^{-1}. It is noted that almost all of the error accrued in periodic minima in u along the upper portion of the domain. It is in this region that the difference in viewing angle amongst the looks (hereafter, the look angle difference) was small.

Observational error is a natural consequence of any radar platform. These errors can vary in origin (e.g.,

FIG. 13. (a) The u-component wind velocity (shaded; m s^{-1}) from a homogeneous-flow OSSE with a platform velocity of 15 m s^{-1}, a scan rate of 1.5° s^{-1}, and no observational error. (b) The u-component absolute wind velocity error (shaded; m s^{-1}) from the same OSSE as in (a), but with an imposed observation error standard deviation of 1.0 m s^{-1}. (c) Maximum look angle difference (shaded; degrees) for the analyses in (a) and (b). Distance labels (km) are provided on each axis.

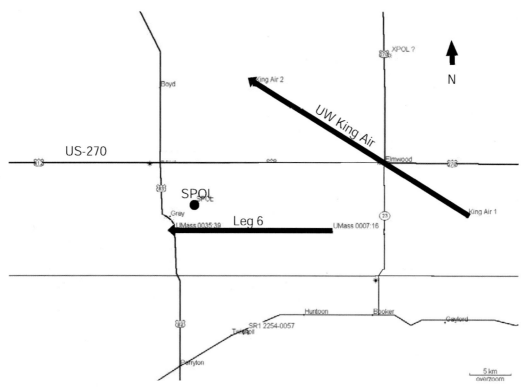

FIG. 14. A map of the rolling RHI UMass W-band deployment (0007–0036 UTC). Distance scale located in the lower-right-hand corner. The thick lines denote the path of the UMass rolling RHI deployment from right to left (east to west), and the path of the UWKA (2345–2351 UTC) from right to left (east-southeast to west-northwest). The dot labeled SPOL denotes the position of the SPOL radar at Homestead.

system noise, sidelobe contamination, anomalous propagation, and beam spreading). To simulate this effect, Gaussian (i.e., random and normal) errors were added to the time series of radial velocity data obtained with the pseudoradar. To mimic the error characteristics of the UMass radar, the standard deviation of the applied error was 1.0 m s^{-1}. These imperfect data were then processed in the same manner as above. The synthesis technique continued to perform very well (Fig. 13b), with an analysis RMS error of 0.379 m s^{-1}. Much like the observation-error-free case, the largest errors were contained in the upper portion of the domain, where maximum look angle differences were subcritical. Since the analysis was biased toward the first-guess values in these areas, an evident underapplication of the observation increments occurred in these locations. Consistent with this conclusion, an analysis error equal and the opposite in magnitude was found if the first guess was changed to $u = 20$ m s^{-1}.

The OSSEs confirm that, given a time series of rolling RHI data, the variational synthesis technique provides an accurate depiction of the true two-dimensional flow, if a sufficient amount of independence (i.e., large look angle difference) exists amongst the observations. To ensure that this condition is met, it is desirable to obtain data with as fast a scan rate as possible, taking into consideration the beamwidth and sample averaging for

the upper limit. A slower platform motion allows for more looks at any arbitrary point, which, considering the random nature of the observational error, permits a better determination of the true radial velocity. However, concerns about the stationarity assumption must be weighed against this benefit. Naturally, one cannot fully resolve temporal scales of motion faster than the time between the first and last looks. For the later looks at each grid point, in particular, the trade-off between stationarity and look angle separation becomes significant.

6. Analysis of rolling RHI scans across the dryline

During 0007–0036 UTC, the UMass W-band radar executed a westward-moving RRHI deployment across the eastern dryline (Fig. 14). The geometry of the RRHI scanning strategy permitted the overlap of rays (Fig. 15) and therefore was compatible with the variational processing technique described above. Assuming stationarity of the wind field for the time period between the looks (discussed below), we used the technique to synthesize the u and w wind components in the plane approximately normal to the dryline.

The dryline was beginning its nocturnal retrogression as the RRHI data were being collected (Fig. 16). The retrogression was not uniform along the dryline as evidenced by the change in the radar fineline location,

a)

1.0 km AGL

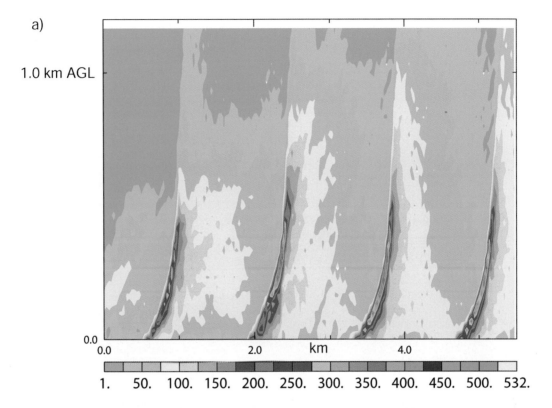

b)

1.0 km AGL

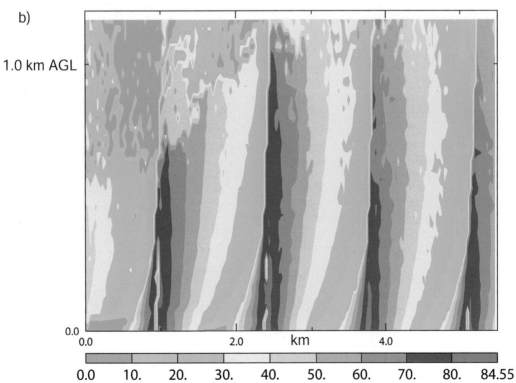

FIG. 15. (a) Look count and (b) maximum look angle difference (degrees) for the 22 May 2002 dryline analyses with no data window cutoff imposed (discussed in text).

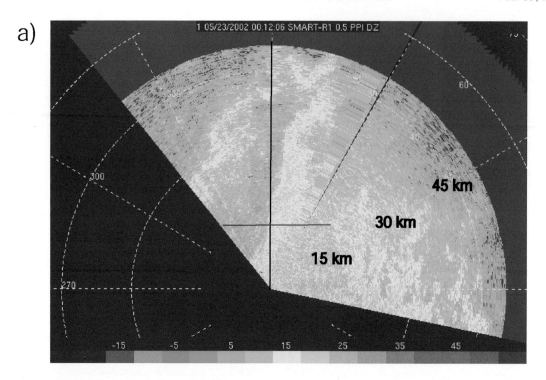

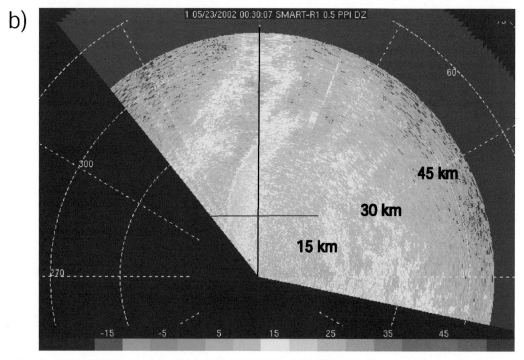

FIG. 16. SMART-R 0.5° reflectivity (dBZ) valid at (a) 0012 and (b) 0030 UTC. The north radial is highlighted in black to show retrogression more clearly. The approximate path of the UMass rolling RHI (right to left) is shown in red. Range markers are indicated in black.

since there was evidence of wave activity along the dryline interface. Data from various radar platforms permitted an estimated retrogression speed between 2 and 5 m s^{-1} during the period of the traverse. The motion

of the dryline and evident heterogeneities in dryline position upwind of the analysis plane made the stationarity assumption more restrictive; therefore, for each point in the analysis domain, a data cutoff window of

60 s (from first observation) was introduced. This window size appeared to balance best the concerns of stationarity (for a window too large) and observations that were too dependent (for a window too small). Reduced total look counts and maximum look angle differences were a natural consequence of the imposed data window (Fig. 17). The UMass radar platform traveled at a nearly constant velocity of 13 m s^{-1} toward the west as RHI sweeps were taken (with a scan rate of 1.5 degrees per second) from the rear horizon up through ~86° above the rear horizon. The raw time series of data were postprocessed to account for truck velocity and pitch (using digital elevation models) before the data were analyzed.

In a composite reflectivity image for the traverse (Fig. 18), the pronounced eastward tilt of the dryline interface with height during retrogression can be seen. As with the vertical antenna deployment, the DCZ appeared as a maximum in reflectivity, presumably due to the local increase in insect concentration in this region. The domain chosen for analysis was the lowest 1 km AGL, where there were no data voids.

It is seen from the analyses using the variational technique that the upper and lower branches of the dryline secondary circulation were resolved quite clearly (Fig. 19a). The near-surface inflow to the DCZ from the east approached $u = -6$ m s^{-1} in some areas of the CBL. Near the top of the CBL, strong westerly component winds (i.e., the return flow) were evident, a combination of air parcels from the moist CBL that had ascended in the DCZ (Hane et al. 1997) and parcels from the dry side that had advected up and over the moist CBL. Westerly winds upward of $u = 15$ m s^{-1} were identified in this region.

The DCZ showed up clearly in the w-component field (Fig. 19b) as a maximum w of 8–10 m s^{-1}. As with the stationary dryline earlier (section 4), the channel of maximum vertical velocity was only ~100 m wide. The eastward tilt of the DCZ with height was again present, much more so than with the stationary dryline. A small area of descent was evident at ~500 m AGL approximately 3 km to the east of the surface position of the DCZ (labeled "D" in Fig. 19b). The position of this descending motion was similar to that shown for the vertical antenna deployment earlier and the airborne Doppler case study of Weiss and Bluestein (2002), both stationary dryline cases.

To examine the effect of the β parameter [Eq. (1a)] on the analysis, the value of β at each point was multiplied by the number of looks at that point. The influence of the mass continuity constraint in the analysis was consequently increased, producing the smoother field depicted in Fig. 20. One can still see very clearly the discontinuity in the u-component (Fig. 20a) and w-component fields at the dryline interface (Fig. 20b) and the rotor circulation on the head of the dryline secondary circulation (DSC). From Figs. 20a and 20b it is clear that the easterly component winds at the surface

extended to the west of the area of maximum upward motion.

7. Summary and discussion

This study was driven by the desire to resolve the finescale structure of a near-dryline environment. This aim was accomplished utilizing instruments capable of high-resolution observations in clear air—the U-Mass W-band and UWKA/WCR—and by developing a pseudo-multiple-Doppler radar processing technique to decompose radial velocity vectors into the individual components of motion.

On the afternoon of 22 May 2002 during IHOP, the UMass W-band was deployed on a double dryline event in the Oklahoma Panhandle. With the antenna pointed vertically, the radar was driven westward across both dryline boundaries. The DCZ was well resolved; a maximum upward vertical velocity $w \sim 8$ m s^{-1} was measured in a narrow channel of the eastern DCZ, approximately 50–100 m wide. In the past, details such as these were hard to discern. For example, Atkins et al. (1998) reported upward motion on the order of 1–2 m s^{-1} using NCAR Electra Doppler Radar (ELDORA) data with a relatively coarse horizontal (along-track) resolution of 600 m. Parsons et al. (1991) used lidar technology (with horizontal resolution of 200 m) to measure a maximum positive vertical velocity of $w \sim 5$ m s^{-1} on a retrograding dryline in west Texas. The larger magnitudes of vertical velocity seen in the current study are likely in part due to the narrow beamwidth of the antenna, which reduced the cross-dryline width over which velocity discontinuities were measured in the DCZ, and allowed for measurements very near the surface without contamination from the ground. Both of these effects allowed for a more accurate determination of the magnitude and depth of convergence in the DCZ.

Areas of subsidence were noted away from the DCZ of both drylines. One such area was found in both UMass W-band and WCR data approximately 4–5 km east of the eastern DCZ. This position was consistent with a similar finding by Hane et al. (1993), in which the area of descent coincided with a moisture gradient at the surface. In this case, it does not appear that the subsidence was sufficient to reach the surface. Small decreases in dewpoint and equivalent potential temperature were identified by UWKA flight-level measurements in the middle of the boundary layer in the descending regions. The areas of concentrated subsidence discussed above are potentially significant for many issues related to the dryline. Double drylines, for example, may form in such a manner, similar to that of the observations of Hane et al. (1997, 2001), and that of the modeled "microfronts" of Ziegler et al. (1995).[6] The transport of dry air to the surface in the near-dryline

---

[6] Subsidence would need to be deeper than in the current case for these features to be realized at the surface.

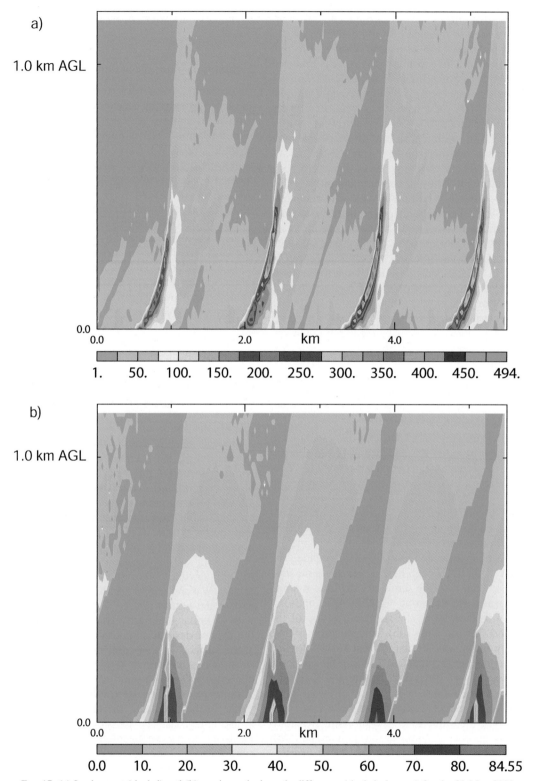

FIG. 17. (a) Look count (shaded) and (b) maximum look angle difference (shaded; degrees) for the 22 May 2002 dryline analyses (Figs. 19 and 20) with a data window cutoff of 60 s imposed (discussed in text).

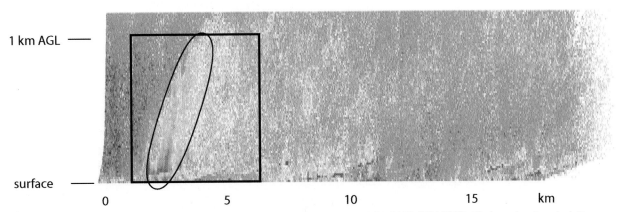

FIG. 18. An east–west display of composite reflectivity from the UMass rolling RHI (0007–0036 UTC). Horizontal and vertical distance scales are indicated. The black box denotes the domain for analysis. The black oval denotes the DCZ.

environment can have substantial effects. In a case presented by Hane et al. (1997), vertical mixing of westerly momentum down to the surface was hypothesized to form a convergence line in the dry air west of the dryline. Severe thunderstorms were later initiated at the intersection point of this convergence line and the dryline.

Data analyzed from the vertical antenna deployment also indicated descending motion immediately to the west of both drylines. This finding agrees with some recent airborne studies of the dryline (e.g., Atkins et al. 1998; Weiss and Bluestein 2002). Because this downward motion transports westerly momentum toward the surface, these observations are consistent with previous observations of westerly component acceleration immediately to the west. More data need to be gathered to assess how ubiquitous this downward motion is, and whether any skill can be realized at predicting convection initiation with these types of observations.

Later during the evening of 22 May 2002, the UMass W-band radar collected RRHI data on the eastern dryline as it retrograded toward the west. The data were analyzed using a variational pseudo-multiple-Doppler radar processing technique, developed specifically for the RRHI collection strategy. This technique was tested using simulated sets of radial velocity derived from fields of known velocity. These tests confirmed the robustness of the technique, with domainwide RMS errors well below the prescribed observational errors. The technique performs the worst when, like any pseudo-Doppler processing algorithm, the observations are close to being collinear. The analysis is biased toward the first guess in these cases.

The assumption of stationarity is central to pseudo-dual- or pseudo-multiple-Doppler data processing. The accuracy of this assumption degrades as the time between observations increases. For airborne pseudo-dual-Doppler techniques (Jorgensen et al. 1995), this elapsed time is a function of range from the aircraft. In the design presented here, areas above and to the near east (in this case) of the vertically pointed radar will have observations separated by the greatest amount of time. However, unlike the airborne dual-Doppler case, there will also be a greater number of observations and a larger separation in angle between each look—both factors that will improve the analysis at these locations. The trade-off between observation collinearity and stationarity is of great significance for practical application of this processing technique. A data cutoff window must be introduced that balances these concerns. For this case, the optimal value of the cutoff window was found to be 60 s. If the dryline were not retreating, a less-restrictive value could have been used.

The OSSEs reveal that the scan strategy used on 22 May 2002 was suboptimal. Faster scan rates would have provided more expansive multiple-Doppler lobes (i.e., regions with sufficient separations in look angle). The introduction of a data cutoff window only exacerbates the negative effects of the slow scans. However, the existing lobes were fortuitously positioned to resolve extrema in vertical velocity near the head of the retreating dryline in this study.

The variational analyses from the 22 May 2002 case depicted the legs of the dryline secondary circulation with great spatial detail. The dryline interface appeared more tilted than seen earlier in the afternoon with the stationary dryline. The rotor circulation and sloped nature of the shear zone mimic the characteristics of a propagating density current (Simpson 1969). Parsons et al. (1991) presented observational evidence that supported the contribution of density current dynamics to the retrograding dryline at the leading edge. The UMass

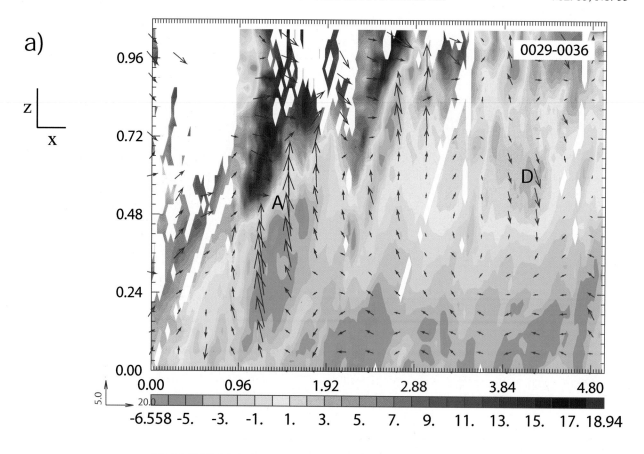

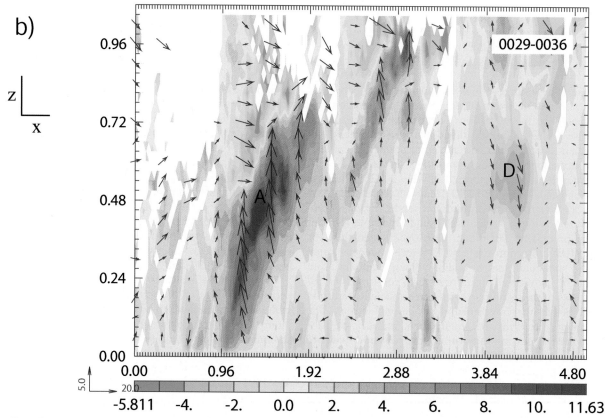

FIG. 19. (a) Ground-relative *u*-component wind (m s^{-1}; contoured) and (b) *w*-component wind (m s^{-1}; contoured) from the variational analysis of the rolling RHI. Cool colors indicate negative component, warm colors indicate positive component. The arrows in (a) and (b) represent *u*/*w* wind vectors. Horizontal and vertical distance scales (km AGL) are indicated. Grid points with fewer than 10 radial velocity observations have been omitted. The areas of strong ascent (A) and descent (D) are identified.

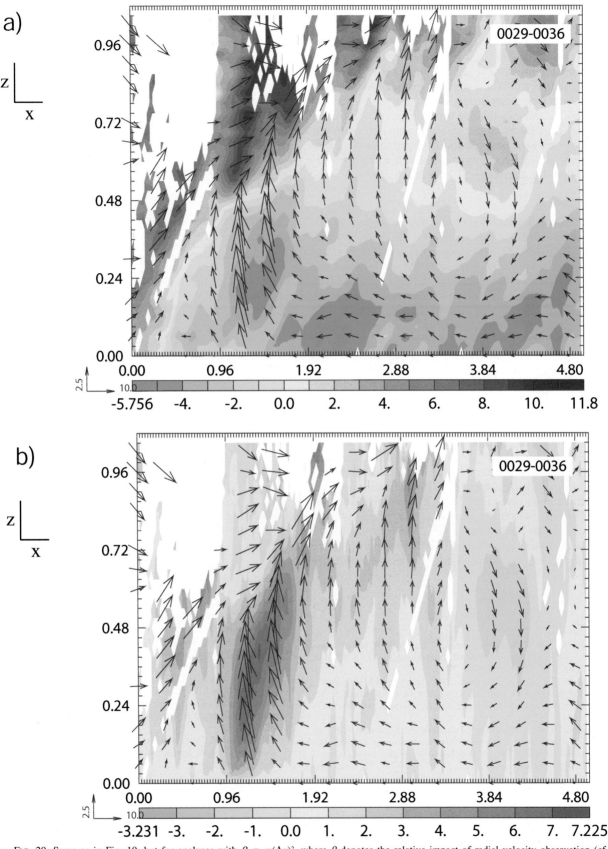

FIG. 20. Same as in Fig. 19, but for analyses with $\beta = m(\Delta x)^2$, where β denotes the relative impact of radial velocity observation (cf. the mass continuity constraint) in the calculation of the cost function, m is the number of radar observations at each analysis point, and Δx is the grid spacing of the analysis.

and UWKA/WCR observations in this study also qualitatively support this hypothesis. However, fine-resolution thermodynamic measurements in the boundary layer are necessary to assess more firmly the validity of density current theory. Attempts to retrieve pressure perturbations from these pseudo-multiple-Doppler analyses were largely unsuccessful because the western extent of the analysis domain was not far enough west of the DCZ to calculate the dryline-normal pressure gradient, nor was it possible to identify accurately a base state ahead of the retreating dryline to use as a boundary condition for the retrieval integration (Parsons et al. 1991).

Acknowledgments. This work was supported by National Science Foundation (NSF) Grants ATM-9912097 and ATM-0241037 (to the University of Oklahoma), and ATM-0129374 (to the University of Wyoming), and was part of the first author's doctoral dissertation at the University of Oklahoma. We are in debt to Evgeni Fedorovich, Carl Hane, Alan Shapiro, David Stensrud, Baxter Vieux, and Conrad Ziegler for useful advice. Dave Leon was instrumental in the WCR intercomparison efforts. Evgeni Fedorovich and Bob Conzemius provided LES data for the testing of the variational analysis scheme. Rick Damiani created the single- and dual-Doppler processing algorithms for the WCR data. Thanks also to David Dowell for his useful insight into variational radar data processing. Curtis Alexander, Nettie Arnott, Mike Buban, Yvette Richardson, and Josh Wurman all assisted in providing mobile radar locations and orientations for the 22 May 2002 case. Paul Markowski provided guidance on GIS applications used in the truck pitch correction. Ming Xue assisted in the use of ZXPLOT, which was used to create the analysis figures. Brendan Fennell drove the UMass vehicle for the 22 May 2002 data collection. We also appreciate the computer support from Mark Laufersweiler at the School of Meteorology.

REFERENCES

Atkins, N. T., R. M. Wakimoto, and C. L. Ziegler, 1998: Observations of the finescale structure of a dryline during VORTEX 95. *Mon. Wea. Rev.,* **126,** 525–550.

Biggerstaff, M. I., and Coauthors, 2005: The Shared Mobile Atmospheric Research and Teaching Radar: A collaboration to enhance research and teaching. *Bull. Amer. Meteor. Soc.,* **86,** 1263–1274.

Bluestein, H. B., 2008: Surface boundaries of the Southern Plains: Their role in the initiation of convective storms. *Synoptic–Dynamic Meteorology and Weather Analysis and Forecasting: A Tribute to Fred Sanders, Meteor. Monogr.,* No. 55, Amer. Meteor. Soc.

——, and A. L. Pazmany, 2000: Observations of tornadoes and other convective phenomena with a mobile, 3-mm wavelength, Doppler radar: The spring 1999 field experiment. *Bull. Amer. Meteor. Soc.,* **81,** 2939–2952.

Cohn, S. A., W. O. J. Brown, C. L. Martin, M. E. Susedik, G. Maclean, and D. B. Parsons, 2001: Clear air boundary layer space antenna wind measurement with the Multiple Antenna Profiler (MAPR). *Ann. Geophys.,* **19,** 845–854.

Crawford, T. M., and H. B. Bluestein, 1997: Characteristics of dryline passage. *Mon. Wea. Rev.,* **125,** 463–477.

Demoz, B., and Coauthors, 2006: The dryline on 22 May 2002 during IHOP_2002: Convective-scale measurements at the profiling site. *Mon. Wea. Rev.,* **134,** 294–310.

Dowell, D. C., and H. B. Bluestein, 2002: The 8 June 1995 McLean, Texas, storm. Part I: Observations of cyclic tornadogenesis. *Mon. Wea. Rev.,* **130,** 2626–2648.

Feltz, W. F., H. B. Howell, R. O. Knuteson, H. M. Woolf, and H. E. Revercomb, 2003: Near continuous profiling of temperature, moisture, and atmospheric stability using the Atmospheric Emitted Radiance Interferometer (AERI). *J. Appl. Meteor.,* **42,** 584–597.

Gao, J., M. Xue, A. Shapiro, and K. K. Droegemeier, 1999: A variational method for the analysis of three-dimensional wind fields from two Doppler radars. *Mon. Wea. Rev.,* **127,** 2128–2142.

Geerts, B., and Q. Miao, 2005: The use of millimeter Doppler radar echoes to estimate vertical air velocities in the fair-weather convective boundary layer. *J. Atmos. Oceanic Technol.,* **22,** 225–246.

Hane, C. E., C. L. Ziegler, and H. B. Bluestein, 1993: Investigation of the dryline and convective storms initiated along the dryline: Field experiments during COPS-91. *Bull. Amer. Meteor. Soc.,* **74,** 2133–2145.

——, H. B. Bluestein, T. M. Crawford, M. E. Baldwin, and R. M. Rabin, 1997: Severe thunderstorm development in relation to along-dryline variability: A case study. *Mon. Wea. Rev.,* **125,** 231–251.

——, M. E. Baldwin, H. B. Bluestein, T. M. Crawford, and R. M. Rabin, 2001: A case study of severe storm development along a dryline within a synoptically active environment. Part I: Dryline motion and an Eta model forecast. *Mon. Wea. Rev.,* **129,** 2183–2204.

Hildebrand, P. H., and Coauthors, 1996: The ELDORA/ASTRAIA airborne Doppler weather radar: High-resolution observations from TOGA COARE. *Bull. Amer. Meteor. Soc.,* **77,** 213–232.

Jorgensen, D. P., T. J. Matejka, and J. D. DuGranrut, 1995: Multibeam techniques for deriving wind fields from airborne Doppler radars. *J. Meteor. Atmos. Phys.,* **58,** 83–104.

LeMone, M. A., 1973: The structure and dynamics of horizontal roll vortices in the planetary boundary layer. *J. Atmos. Sci.,* **30,** 1077–1091.

——, F. Chen, J. G. Alfieri, M. Tewan, B. Geerts, Q. Miao, R. L. Grossman, and R. L. Coulter, 2007: Influence of land cover and soil moisture on the horizontal distribution of sensible and latent heat fluxes in southeast Kansas during IHOP_2002 and CASES-97. *J. Hydrometeor.,* **8,** 68–87.

Miller, D. A., and F. Sanders, 1980: Mesoscale conditions for the severe convection of 3 April 1974 in the east-central United States. *J. Atmos. Sci.,* **37,** 1041–1055.

Parsons, D. B., M. A. Shapiro, R. M. Hardesty, R. J. Zamora, and J. M. Intrieri, 1991: The finescale structure of a west Texas dryline. *Mon. Wea. Rev.,* **119,** 1283–1292.

Russell, R. W., and J. W. Wilson, 1997: Radar-observed "fine lines" in the optically clear boundary layer: Reflectivity contributions from aerial plankton and its predators. *Bound.-Layer Meteor.,* **82,** 235–262.

Sanders, F. S., 1986: Temperatures of air parcels lifted from the surface: Background, application and nomograms. *Wea. Forecasting,* **1,** 190–205.

——, and A. J. Garrett, 1975: Application of a convective plume model to prediction of thunderstorms. *Mon. Wea. Rev.,* **103,** 874–877.

——, and D. O. Blanchard, 1993: The origin of a severe thunderstorm in Kansas on 10 May 1985. *Mon. Wea. Rev.,* **121,** 133–149.

——, and C. A. Doswell, 1995: A case for detailed surface analysis. *Bull. Amer. Meteor. Soc.,* **76,** 505–522.

Sasaki, Y., 1970: Some basic formalisms in numerical variational analysis. *Mon. Wea. Rev.,* **98,** 875–883.

Simpson, J. E., 1969: A comparison between laboratory and atmospheric density currents. *Quart. J. Roy. Meteor. Soc.,* **95,** 758–765.

Weckwerth, T. M., and Coauthors, 2004: An overview of the International H$_2$O Project (IHOP_2002) and some preliminary highlights. *Bull. Amer. Meteor. Soc.,* **85,** 253–277.

Weiss, C. C., and H. B. Bluestein, 2002: Airborne pseudo–dual Doppler analysis of a dryline–outflow boundary intersection. *Mon. Wea. Rev.,* **130,** 1207–1226.

——, ——, and A. L. Pazmany, 2006: Finescale radar observations of the 22 May 2002 dryline during the International H$_2$O Project (IHOP). *Mon. Wea. Rev.,* **134,** 273–293.

Wilson, J. W., and W. E. Schreiber, 1986: Initiation of convective storms by radar-observed boundary layer convergence lines. *Mon. Wea. Rev.,* **114,** 2516–2536.

Ziegler, C. L., W. J. Martin, R. A. Pielke, and R. L. Walko, 1995: A modeling study of the dryline. *J. Atmos. Sci.,* **52,** 263–285.

INTRODUCTION TO PARTS III AND IV

By Lance F. Bosart

Fred Sanders' longstanding interest in the application of fundamental dynamical principles that govern quasigeostrophic (QG) theory to the study of synoptic meteorology motivated his lifelong research investigations of weather analysis and forecasting problems. This interest served him well when it came time to transfer the knowledge gained from weather analysis and forecasting studies to operations. Fred believed that improvements in daily weather forecasts would arise from quantitative research investigations of difficult-to-predict weather events and through a rigorous evaluation of routine forecast skill. These two research themes are common to many of Fred's published papers. Fred was one of the pioneers in the field of probabilistic forecasting and forecast verification even though his work in this area is probably lesser known than his many classical weather analysis and forecast investigations reported throughout this monograph. Fred believed that a quantitative analysis approach to the problem of forecast verification and the assessment of the state-of-the-art skill of routine temperature and precipitation forecasts (probabilistic and categorical) was both necessary and critical to the improvement of the overall skill of weather forecasts. Fred was a true "bilingual" meteorologist. He spoke and appreciated the language of science and operations.

As a part of his commitment to advancing scientific understanding and improving the skill of weather forecasts, Fred and his graduate students (especially the late Robert Burpee), worked with staff members at the National Hurricane Center (NHC) and NOAA's Hurricane Research Division in Miami, Florida, to design and implement an operational objective tropical cyclone (TC) track forecasting scheme in the 1960s and 1970s. This objective TC track forecast scheme was based on the equivalent barotropic model. The Sanders barotropic model, or SANBAR as it became known, produced a 36-hour forecast of the layer-mean tropospheric steering current (assumed to be close to 500 hPa). A vortex was inserted into this layer-mean steering current at the location of the NHC-defined TC position at the time the model was initialized. The vortex was then advected by the layer-mean steering flow in the SANBAR model. SANBAR was used operationally at NHC until the early 1990s and proved to be especially useful in predicting the tracks of Cape Verde–type disturbances in the deep

easterlies to the south of the subtropical ridge axis. A history and details of the SANBAR model are contained in Robert Burpee's contribution to this monograph.

Fred also instilled in his students and colleagues the importance of carefully scrutinizing all of the available data in order to maximize the possibility of uncovering some of the atmosphere's "secrets" when these data were assembled, mapped, analyzed, and interpreted. The paper by John Gyakum on the application of Fred's teaching methods to research on extreme cold-season precipitation events in the St. Lawrence River Valley epitomizes Fred's "the data rules" approach to doing science. Gyakum shows how the signs of an extreme precipitation event are often manifest in planetary- and synoptic-scale precursor signals (anomalies) days in advance of the event. He identifies precursor anomalies relating to transient upper-level disturbances, anomalously large precipitable water values, and the transport of tropical moisture into the region as especially critical to understanding the nature and distribution of extreme precipitation events. In a related study, Steve Tracton poses the question as to why "surprise" snowstorms on occasion still occur in an age of vastly improved numerical weather prediction. He shows that small-scale uncertainties in initialized and forecast fields can sometimes lead to substantial uncertainties in derived model forecast fields such as the track and intensity of cyclones and the amount of moisture entrained in the circulation of these cyclones, all of which can contribute to significant forecast uncertainties and associated forecast "surprises." A motivation for Tracton's work came from Fred's synoptic meteorology class at MIT in which he instilled in his students the idea that the observed large day-to-day variability in forecast skill suggested that forecasting is perhaps best approached in a probabilistic format.

After the conclusion of World War II and before Fred went to MIT, he worked as an operational meteorologist at LaGuardia Airport in Queens, New York. He was the duty forecaster who correctly indicated the potential for a significant snowstorm in the New York City region on 26 December 1947. As usual, Fred based his forecast on a careful analysis of all of the available data, a synthesis of the resulting meteorological analyses, and an understanding of fundamental principles of atmospheric dynamics and thermodynamics. The result was a remarkably successful forecast of that famous New York City snowstorm. Prompted in part by Fred's noteworthy forecast of this event and his seminal research contributions to the study of explosively intensifying marine cyclones, Louis Uccellini, Paul Kocin, Joeseph Sienkiewicz, Robert Kistler, and Michael Baker have conducted a reanalysis and simulation of this famous storm. The simulation was produced from a recent version of the NCEP GFS model that was initialized from a modern reanalysis. Uccellini and colleagues indicate how Fred's modernized quantitative approach to the study of synoptic meteorology based upon QG theory resulted in a greater understanding of the formation and development of explosively deepening oceanic cyclones (which Fred called "bombs") of which the 26 December 1947 event was a prime example. Uccellini and colleagues also show how the availability of satellite-derived scatterometer imagery has enabled marine forecasters to monitor and quantify wind fields in explosively deepening marine cyclones for the first time in a systematic fashion. The resulting improved oceanic analyses based upon the assimilation of these scatterometer winds and overall advances in data assimilation techniques have helped to improve the ability of numerical weather prediction models to forecast explosive oceanic cyclogenesis. To illustrate this ongoing forecast improvement, Uccellini and colleagues produced a modestly successful numerical forecast of the 26 December 1947 snowstorm. The forecast, while far from perfect, was good enough to provide a situational awareness of the event to forecasters that Fred himself demonstrated 60 years earlier from much more primitive but still fundamental means.

During regular map discussions, Fred always talked to his students about the importance of keeping track of the "big picture" in order to maintain an appreciation for how the atmosphere was evolving and why it was evolving as observed. Fred, motivated in part by the research findings of Hurd Willett at MIT and Jerome Namias who had left MIT to take up the challenge of making monthly and seasonal weather forecasts for the Weather Bureau in Washington, D.C., argued that it was important to have an appreciation for how large-scale circulation regimes that could persist for weeks could exert an important influence on the day-to-day weather patterns. Today we know that significant intraseasonal variability within the typical 90-day season contributes importantly to the observed seasonal temperature and precipitation anomalies. Randy Dole, in his contribution on linking weather and climate, has helped to define and quantify the atmospheric circulation features that contribute to temperature and rainfall anomalies on intraseasonal time scales. He makes the point that recent advances in the science of the atmosphere and the oceans coupled with rapid gains in computer power have made it very evident that weather and climate problems can no longer be regarded as mutually exclusive. Randy Dole's contribution to this monograph is important because it suggests "near-term directions for advancing our understanding and capabilities to predict the connections between weather and climate."

An important contributor to intraseasonal temperature and rainfall variability, and one whose importance Fred recognized years ago, is upper-level subtropical anticyclones. When present, these anticyclones may be manifest by extended periods of above normal temperatures and below normal precipitation in subtropical and middle latitudes. Heat waves and severe droughts associated with these anticyclones can have severe adverse impacts on human activities and welfare. Persistent subtropical anticyclones can serve as large-scale circulation "traffic police." The equatorward sides of these subtropical anticyclones can serve as "freeways" for westward-moving tropical disturbances or upper-level mesoscale disturbances that form via fracture from the equatorward ends ends of potential vorticity streamers (aka tails) that originate in midlatitudes. The poleward sides of these subtropical anticyclones can serve as pathways for eastward-moving disturbances that may be associated with severe weather outbreaks in the anomalously strong westerlies that lie poleward of entrenched subtropical anticyclones. Tom Galarneau, Lance Bosart, and Anantha Aiyyer have produced a 54-year (1950–2003) global climatology of 850-, 500-, and 200-hPa closed anticyclones. They find that 500- and 200-hPa closed anticyclones occur preferentially over continental regions in both hemispheres and over the equatorial oceanic warm pool in the western Pacific while 850-hPa closed anticyclones occur preferentially over oceanic basins in both hemispheres. Case studies of the notable but short-lived U.S. heat wave of July 1995, the impressive summer-long western European heat wave of 2003, and the nearly month-long intense Australia heat wave of February 2004 are used to help illustrate the relationship between closed subtropical continental anticyclones and heat waves. Their results indicate the importance of dynamic forcing (e.g., synoptic-scale ridge-induced subsidence) and thermodynamic forcing (e.g., sensible heating over semi-arid elevated terrain) on the production of anomalously hot continental tropical (CT) air masses. Once produced, these anomalously hot CT air masses can be advected eastward from their elevated source regions in the anomalously strong westerly flow on the poleward sides of continental subtropical anticyclones.

Part III
Forecasting

Chapter 11

The Sanders Barotropic Tropical Cyclone Track Prediction Model (SANBAR)

ROBERT W. BURPEE*

Cooperative Institute for Marine and Atmospheric Studies, Miami, Florida

(Manuscript received 3 December 2003, in final form 23 February 2005)

Burpee

ABSTRACT

Sanders designed a barotropic tropical cyclone (TC) track prediction model for the North Atlantic TC basin that became known as the Sanders barotropic (SANBAR) model. It predicted the streamfunction of the deep-layer mean winds (tropical circulation vertically averaged from 1000 to 100 hPa) that represents the vertically averaged tropical circulations. Originally, the wind input for the operational objective analysis (OA) consisted of winds measured by radiosondes and 44 bogus winds provided by analysts at the National Hurricane Center (NHC), which corresponded to the vertically averaged flow over sparsely observed tropical, subtropical, and midlatitude oceanic regions. The model covered a fixed regional area and had a grid size of ∼154 km. It estimated the initial storm motion solely on the basis of the large-scale flow from the OA, not taking into account the observed storm motion.

During 1970, the SANBAR model became the first dynamical TC track model to be run operationally at NHC. Track forecasts of SANBAR were verified from the 1971 TC season when track model verifications began at NHC until its retirement after the 1989 Atlantic TC season. The average annual SANBAR forecast track errors were verified relative to Climatology and Persistence (CLIPER), the standard no-skill track forecast. Comparison with CLIPER determines the skill of track forecast methods. Verifications are presented for two different versions of the SANBAR model system used operationally during 1973–84 and 1985–89. In homogeneous comparisons (i.e., includes only forecasts for the same initial times) for the former period, SANBAR's track forecasts were slightly better than CLIPER at 24–48-h forecast intervals; however, from 1985 to 1989 the average SANBAR track forecast errors from 24–72 h were ∼10% more skillful than homogeneous CLIPER track forecasts.

1. Introduction

The Sanders barotropic (SANBAR) tropical cyclone (TC) track prediction model was developed for the North Atlantic TC basin during the mid- and late 1960s (Sanders and Burpee 1968). At that time, objective op-

erational TC track forecast guidance was largely statistical, but some barotropic track forecasts had been run at the National Meteorological Center [NMC, now the National Centers for Environmental Prediction (NCEP)] from the mid-1950s–60s and were verified in a general manner by Tracy (1966). The purpose in developing SANBAR was to create a barotropic prediction model that would be applied in an appropriate way to operationally available wind data that described the tropical and subtropical circulations.

Sanders' (1970) basic hypothesis was that the motion

---

*Author deceased.

Corresponding author address: Lance F. Bosart, University at Albany, State University of New York, 1400 Washington Ave., Albany, NY 12222.

E-mail: bosart@atmos.albany.edu

of a TC is largely determined by the advection of the mean vorticity by the mean wind where the mean was computed as a mass-weighted average or deep-layer mean (DLM) throughout the depth of the troposphere, assumed to be 1000–100 hPa. Sanders et al. (1980) and others argued that the conservation of absolute vorticity explains the "steering" of a TC by its larger scale environment. Although Charney (1963) had theorized that the lower and upper tropical tropospheric circulations are largely independent, Sanders (1970) maintained that the lack of vertical coupling does not apply in the TC environment where abundant deep convection connects the entire troposphere. Sanders and Burpee (1968) cautioned that SANBAR would primarily be applicable in regions of the Tropics like the Caribbean, the Gulf of Mexico, and western North Atlantic that had adequate wind observations and, even in those areas, only when significant large-scale baroclinic processes were absent.

2. The research SANBAR model

Sanders and Burpee (1968) reasoned that a simple form of the vorticity equation governed the tropical flow:

$$\frac{\partial \zeta}{\partial t} = -\mathbf{V} \cdot \nabla \eta + f \frac{\partial \omega}{\partial p}, \tag{1}$$

where $\mathbf{V}$ is the horizontal velocity, ω is the vertical velocity in p coordinates, ζ is the relative vorticity, f is the Coriolis parameter, and η is the absolute vorticity. They argued that a vertically averaged vorticity equation appropriate for the tropical DLM is one where

$$\frac{\partial \overline{\zeta}}{\partial t} = -\overline{\mathbf{V}} \cdot \nabla \overline{\eta} - \overline{\mathbf{V}' \cdot \nabla \eta'} + f(\omega_{1000} - \omega_{100}). \tag{2}$$

The overbar denotes a vertical average, the prime represents a deviation from this average, and ω is evaluated at 1000 and 100 hPa. Sanders and Burpee justified the reduction of Eq. (2) to

$$\frac{\partial \overline{\zeta}}{\partial t} = -\overline{\mathbf{V}} \cdot \nabla \overline{\eta}. \tag{3}$$

This equation indicates that the time tendency of the DLM relative vorticity is equal to the horizontal advection of the mean absolute vorticity by the DLM wind. They rearranged this form of the vorticity equation in terms of the streamfunction (ψ) of the horizontal wind and obtained

$$\nabla^2 \frac{\partial \psi}{\partial t} = J(\nabla^2 \psi + f, \psi), \tag{4}$$

where $J(\ )$ is the Jacobian operator. This is a Poisson equation that was solved for $\partial \psi / \partial t$ by overrelaxation with ψ constant on the boundary [see King (1966) for more details of the method]. Equation (4) was solved on a fixed grid that covered nearly one-quarter of the

Northern Hemisphere and included most of the tropical North Atlantic, all of the Caribbean and Gulf of Mexico, and part of the eastern Pacific (the area shown in Fig. 1, but with a northern boundary of 55°N). The grid is a Mercator projection with a mesh length of 154 km at the standard latitude of 22.5°N. They did not separate the TC vortex from the large-scale flow. King (1966) demonstrated that the DLM could be represented by the winds at the 10 mandatory pressure levels from 1000 to 100 hPa.

Sanders and Burpee (1968) tested Sanders' hypothesis by computing several experimental track forecasts at the Massachusetts Institute of Technology (MIT) for three TCs of the early 1960s. The results of these forecasts led them to claim that SANBAR could provide substantial improvement over the then operational models for the forecast intervals from 24 to 72 h.

During Sanders' 1967–68 sabbatical at the National Hurricane Research Laboratory [NHRL, now the Hurricane Research Division (HRD) of the National Oceanic and Atmospheric Administration (NOAA)] in Coral Gables, Florida, he conducted additional experiments with SANBAR for two mid-1960s TCs. These TCs had more erratic tracks than the three early 1960s TCs and had accounted for large operational forecast track errors. In these cases, the model errors were similar to those for the earlier experiments, so NHRL's management decided that SANBAR was ready for a real-time test.

This test occurred during Hurricane Gladys (1968). The initial wind analysis and its digitization were completed manually and the data were punched on computer cards to initialize and run the model. Because of the manual procedures and the limited capability of computers at that time, the results of the forecast did not become known until, perhaps, 12–13 h after the observation time. Sanders (1970) described the forecast results as "extremely disappointing." The initial forecast track of Hurricane Gladys was to the northeast. The forecast TC stopped after 24 h and then went toward the southwest, while Gladys only slowed down for a few hours and then continued moving toward the northeast. Sanders et al. (1975) showed that the unrealistic forecast track resulted from northerly flow that developed over most of the model grid and was associated with a type of error that had not occurred in any of the previous research forecasts. They related this error to a spurious retrogression of the long-wave components of the predicted flow. To avoid this problem in the future, they included a high-latitude technique (Cressman 1958) by adding a term to the prediction equation that prevents rapid movement of the longer waves without major effects on the shorter waves. While Sanders et al. (1975) expressed reservations about this type of solution and its possible effects on forecasts of TCs that moved erratically, they included this term in the operational SANBAR.

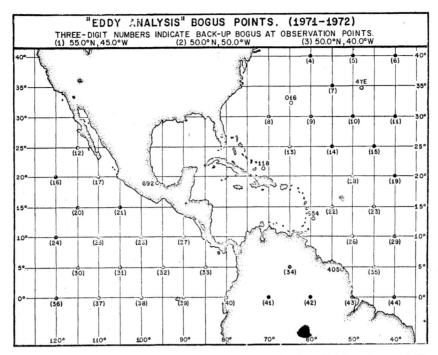

FIG. 1. The base map for the SANBAR TC track forecast model from 0° to 40°N. Locations are indicated for 41 bogus winds used in the initial analysis for SANBAR. The first three bogus points were located outside the area, their positions are listed in the figure heading [reproduced from Pike (1972)].

3. The operational model from the mid-1970s to the early 1980s

When the directors of NHRL and the National Hurricane Center (NHC) decided that SANBAR was ready to become operational, NMC's operational analyses of initial atmospheric conditions did not extend to the equator until 1974 (K. Campana 2003, personal communication). Because these analyses did not include the southern part of the research model's grid, the conversion of SANBAR to an operational model required ei-

ther a change in the domain being forecast (Fig. 1) or an objective analysis (OA) of the DLM winds for SANBAR's existing domain. Sanders decided that SANBAR would have its own OA and selected the Eddy (1967) analysis that employed a statistical technique (Sanders et al. 1975). In this method, the quantity to be analyzed was estimated by a set of multiple regression equations with each grid point as a predictand and the observed values of zonal and meridional wind at nearby radiosonde stations as predictors (see Fig. 2 in Sanders et al. 1975). The OA was structured to determine the departures of the zonal and meridional wind components as deviations from their latitudinal band means.

The radiosonde network was supplemented by 44 bogus winds in data-sparse oceanic regions (Fig. 1) provided by NHC analysts. They determined the bogus winds by subjective interpretation of jet aircraft and surface ship winds, cloud motion vectors from geostationary satellites in those seasons after they became available in 1974 (K. Campana 2003, personal communication), and 12-h prognostic height fields from NMC's previous forecast cycle. The bogus winds were treated in the Eddy OA as if they had been observed. The OA required extensive computing for any wind observed from a platform whose location varied in time (e.g., winds obtained by satellites, aircraft, or moving ships). To minimize computation time, this type of observation was discarded, but used subjectively in estimating the bogus winds. A few months after NMC's operational analyses included the equator, the NHC de-

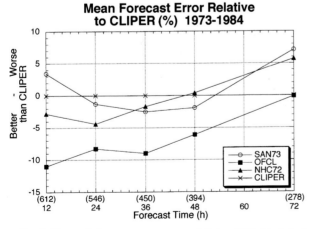

FIG. 2. The track forecast errors averaged during 1973–84 plotted relative to CLIPER at 12-h intervals from 12–72 h for SANBAR, NHC72, and the official NHC forecasts.

veloped objective methods to determine the bogus winds.

Also needed to convert SANBAR to an operational model was a method to remove the effects of the TC winds from any radiosonde winds at stations within the TC's maximum influence radius (MIR) so that the OA could establish the large-scale flow near the TC. Sanders et al. (1975) considered any observed DLM winds within a TC's MIR to be the sum of the TC and large-scale winds. They developed a simple axisymmetric vortex that estimated a TC's tangential wind as a function of radius with three parameters: the TC's maximum surface wind, the radius of the maximum wind (RMAX), and MIR. The profile could be adjusted to be consistent with the observed TC, but in most operational forecasts, RMAX was 37 km, MIR was 555 km, and the maximum surface wind was that reported on the forecast advisory (Sanders et al. 1980). The first operational SANBAR track forecasts were in late 1970 (Pike 1972).

Williams (1972) investigated the performance of the axisymmetric vortex model in operational forecasts and discovered a few forecasts in which the model estimates of the TC wind had resulted in unrealistic large-scale winds, occasionally opposite to the direction of the initial TC motion vector (ITCMV). In these cases, SAN-BAR's initial forecast tracks were not very accurate, so the procedure was modified to avoid these problems from occurring.

Pike (1972) noted that during the first 2 yr of operational SANBAR, the NHC analyst in charge of initiating the SANBAR analysis and track forecast for a forecast cycle did not always compute the DLM from 1000 to 100 hPa. The author recalls that they sometimes chose to average the winds from 1000 hPa to a top level that varied from 400 to 100 hPa with the strength of a TC, 100 hPa being selected for the strongest TCs.

During the computation of SANBAR track forecasts, the predicted TC vorticity maximum and streamfunction minimum tended to drift apart. The TC's location was defined as the average position of the vorticity maximum and the streamfunction minimum.

In the early 1970s, the 12- and 24-h SANBAR forecast tracks were not as accurate as two statistical track models, but the 48- and 72-h SANBAR forecasts were at least as good as other guidance models. Pike (1972) felt that part of the short-term SANBAR track inaccuracy was likely caused by the fact that the statistical models relied on NHC's estimate of the ITCMV to help a forecast start in a reasonable direction whereas SAN-BAR's initial motion depended entirely on the wind field from the Eddy OA. At the early times of the forecast, both the operational and ongoing research SANBAR track forecasts were systematically ~25% slower than the observed TCs. Pike recommended that any observed DLM wind within a TC's MIR be discarded and replaced at each grid point inside the MIR with the sum of the operational estimate of the ITCMV and the appropriate wind from the axisymmetric vortex model. These winds

replaced the winds computed by the Eddy OA in the region within the MIR. The resulting winds were used in the computation of the initial vorticity and the forecast proceeded as before. Pike found that this technique improved the directional bias of the forecasts out to 72 h, but only partially corrected the slow speed bias.

Sanders et al. (1975) determined that most of the slow bias was contributed by TCs that were far from the radiosonde network and deduced that the relaxation procedure for determining the initial streamfunction in these regions, though converging satisfactorily, was consistently underestimating the magnitude of the ITCMV. To remedy the situation, they decided to precalculate the streamfunction in the TC's MIR so that it would agree with the observed ITCMV. Although not mathematically elegant, the procedure to precalculate the streamfunction, part of a method that Sanders et al. (1975) called FAST SANBAR, seemed to converge to a solution and Gaertner's (1973) research forecasts with the method reduced the slow bias to about 3% of the value that had occurred in the operational forecasts. After considerable experimentation, Sanders et al. (1975) frustratingly concluded that they were forced to discard any wind observations within the TC's MIR to obtain a reasonably accurate TC motion. The improvements to the model that Sanders et al. (1975) reported were operationally implemented in 1973 (Goldenberg et al. 1987) and this form of the model is referred to here as SAN73. It includes the Eddy OA, no winds from moving platforms or within the MIR around a TC, the FAST SANBAR technique, a fixed grid, and a TC's position determined by an average of the vorticity maximum and the streamfunction minimum.

In the mid-1970s, the typical NMC database in the lowest 2 km over the oceans included winds from ship observations and low cloud motions estimated from geostationary satellites. In the upper levels (e.g., ~9–12 km), winds from commercial aircraft and satellite-derived high cloud motions were fairly numerous. To use these winds in SANBAR, Sanders et al. (1980) developed regression equations that estimated the DLM wind from a combination of observed winds at low, mid-, or high levels. [They assumed that wind information from midtropospheric levels would soon become available. Satellite winds at these levels became part of the NMC operational database in ~1980 (G. DiMego and C. Velden 2004, personal communications)]. To calculate the equations, Sanders et al. used predictors from tropical and subtropical radiosonde observations during June–October 1971–74. They anticipated that DLM winds determined from the regression equations would help to reduce the dependence in the Eddy OA on isolated oceanic stations. Their technique for estimating the DLM from two levels with winds was implemented in the operational SANBAR without reporting its impact on the accuracy of the forecast tracks.

Sanders et al. (1980) also formulated a new procedure for estimating the ITCMV when two or more radiosonde stations within the MIR reported winds at one observation time. This procedure was tested extensively (Goldenberg et al. 1985) with mixed results and the procedure was never implemented operationally (S. Goldenberg 2003, personal communication).

4. The operational version of the model during the mid- and late 1980s

Goldenberg et al. (1985) felt that confidence in SANBAR's track forecasts was decreasing because the Eddy OA was unable to make use of the newest types of wind observations. These winds, not usually observed at the same location every day, would have significantly increased the computation time of the Eddy OA and, therefore, were withheld from its database. Two types of winds became available in the data-sparse TC environment during the early 1980s. One type of wind observation was derived from the visible infrared spin scan radiometer (VISSR) atmospheric sounder (VAS) on the Geostationary Operational Environment Satellites (Velden et al. 1984). The other was obtained by HRD on a few occasions when NOAA dropped omega dropwindsondes (ODWs) from its WP-3D aircraft during special research flights around TCs (Burpee et al. 1984). The availability of these winds motivated Goldenberg et al. (1985) to replace SANBAR's Eddy OA package with an OA that could assimilate these new data. They tested the Cressman (1959) circular-scan technique with the standard weights and implemented it operationally before the 1985 TC season. The first-guess analysis for the DLM wind in the Cressman OA was determined from a combination of NMC's 10 mandatory-level analyses in the northern portion of the grid and regression equations in the southern portion. These equations estimated the DLM winds from NHC's then operational low-level and 200-hPa tropical wind analyses (Wise and Simpson 1971). The northern and southern regions were joined with a linear transition. At a TC's location, the large-scale DLM was assumed to be approximately equal to the ITCMV rather than everywhere within the MIR. The winds for the bogus points were obtained from the first-guess fields through bilinear interpolation and the number of bogus points was doubled in the east–west direction, but any bogus point closer than 555 km to a radiosonde, ship sounding, or ODW was not included in the database for the Cressman OA.

In addition, Goldenberg et al. (1987) doubled the grid resolution and revised the TC-tracking algorithm to follow only the TC's vorticity center. They noted weaknesses in the FAST SANBAR procedure of Sanders et al. (1975) that occurred because that method produced discontinuities in the vorticity and streamfunction fields that were not allowed to blend fully with the surrounding values. Goldenberg et al. (1987) designed a method that they called NO-FAST SANBAR (because it removed almost all of the FAST SANBAR modifications) that provided better track forecasts. They calculated the vor-

ticity in the MIR as the sum of vorticity from the axisymmetric vortex and the large scale from the Cressman OA. Then they solved for the streamfunction by relaxation so that the discontinuities that resulted from the FAST SANBAR method were removed. The modifications to the analysis by Goldenberg et al. (1985) and to the SAN73 model by Goldenberg et al. (1987) summarized above, led to statistically significant improvements in the developmental sample of track forecasts relative to SAN73 and became operational in 1985 in a version of the model referred to as SAN85.

Velden and Goldenberg (1987) explored the use of special satellite-derived winds (SDWs), in combination with conventional upper-air wind observations, to improve the DLM flow field in the environment of TCs as a possible way to obtain more skillful track forecasts with SAN85. The SDWs consisted of 1) high density cloud track vectors, 2) winds produced from loops of the VAS water vapor channels, and 3) gradient winds derived from VAS sounding height fields (Velden et al. 1984). Working under semioperational constraints, they determined the winds and completed analyses of the DLM around several North Atlantic TCs. To initialize SAN85, Velden and Goldenberg (1987) analyzed these high-quality wind fields in the TC environment with a Barnes (1964) OA scheme in three tropospheric layers (lower, middle, and upper) and estimated DLM winds from the three levels with regression equations (Velden et al. 1984). These DLM winds were incorporated into the Cressman OA and resulted in better TC analyses. The SAN85 forecasts with the SDWs improved the average tracks relative to the corresponding operational SAN73 forecasts by approximately 5%–15% from 24 to 48 h in a homogeneous sample of ~25 cases during 1982–86. These techniques, however, were never implemented operationally, possibly because of the manpower required to estimate the SDWs on a regular basis.

5. SANBAR forecast track verifications

Track forecast verifications for SAN73 and SAN85 are shown in the Climatology and Persistence (CLIPER) diagrams in Figs. 2 and 3, respectively (these samples include only named TCs). The CLIPER model (Neumann and Pelissier 1981) is the standard no-skill model. Skill is defined as

$$\text{skill} = \frac{A - C}{C}, \qquad (5)$$

where A is the track error of any forecast technique for a number of forecasts; C is the track error of the CLIPER model, usually for the same forecasts; and skill is the improvement of technique A relative to and normalized by CLIPER. The skill of CLIPER is represented in a CLIPER diagram by a horizontal line with a value of zero at every forecast time. A model with smaller (larger) average track errors than CLIPER has negative (pos-

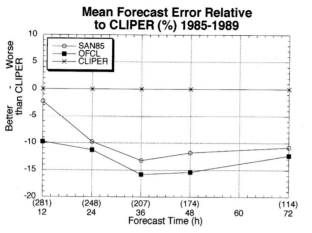

FIG. 3. The track forecast errors averaged during 1985–89 plotted relative to CLIPER at 12-h intervals from 12 to 72 h for SANBAR and the official NHC forecasts.

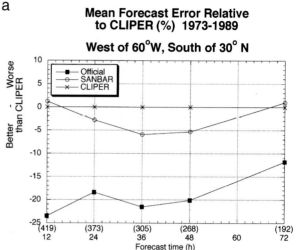

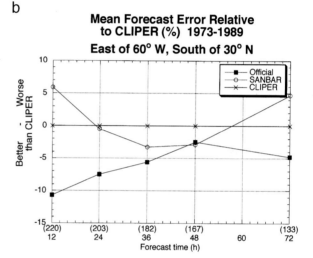

FIG. 4. Track forecast errors south of 30°N for SANBAR and the official forecast averaged during 1973–89 plotted on a CLIPER diagram from 12–72 h: (a) west of 60°W and (b) east of 60°W.

itive) values of skill. The skill of a perfect forecast is −100%.

Both figures also display the corresponding official track forecast. Fig. 2 includes the last (1972 version) purely statistical model to be run operationally at NHC (NHC72; Neumann et al. 1972; Neumann and Pelessier 1981). NHC72 was an amalgamation of an older (1967) version of the NHC statistical model (NMC67; Miller et al. 1968), and CLIPER (Neumann 1972). The model comparisons in the CLIPER diagrams include only those initial times when all of the methods shown in a figure made a track forecast (homogeneous samples). Neumann (1981) pointed out that the degree of difficulty of individual track forecasts varies with a number of factors (e.g., the longitude of a TC at the initial forecast time) so that fair comparisons of different forecast techniques must be homogeneous.

Sanders and Burpee (1968) expected that SANBAR track forecasts would, at least, be competitive with statistical models. The verifications in Fig. 2 indicate that SAN73 was slightly more skillful than NHC72 at 36 and 48 h, but NHC72 was better at 12, 24, and 72 h.

The annual average SAN73 track forecasts (Fig. 2) were only slightly different from those of CLIPER except at 72 h where they were ~7% worse. The modifications to the SANBAR analysis and model that resulted in SAN85 improved the model so that its forecast tracks were ~10% better than CLIPER (Fig. 3) and, at least, in the homogeneous sample, the official forecast tracks were only slightly better than those of SAN85. Inspection of the limited 1985–87 homogeneous sample reveals that SAN85 was also more skillful than NHC72 by ~10%.

Sanders and Burpee (1968) hypothesized that SANBAR would produce better track forecasts in tropical regions with good wind coverage as opposed to regions with few winds. To test this hypothesis, the SANBAR, CLIPER, and official track forecasts during 1973–89

were divided into two samples south of 30°N: one east of 60°W and the other west of 60°W. The western area includes the Caribbean Islands and their radiosonde stations. The annual average SANBAR track forecasts west of 60°W (Fig. 4a) had 2% −3% more skill relative to their CLIPER track forecasts than those east of 60°W (Fig. 4b). The annual average official track forecasts west of 60°W had at least 10% more skill at each forecast time than the corresponding forecasts east of 60°W (Figs. 4a,b).

Before the 1988 TC season, the National Weather Service transferred the official responsibility for TC forecasts in the eastern North Pacific basin from its San Francisco office to NHC. At that time, the SAN85 model was implemented in the eastern Pacific because SAN85 had the capability for flexible placement of its grid (S. Goldenberg 2003, personal communication) and the automated versions of the NHC low-level and 200-hPa tropical wind analyses included that area. Track forecasts for SAN85 were run in 1988, 1989, and part of 1990 until a computer change at NMC ended the possibility of additional SANBAR forecasts. Track forecasts in 1990 were run, until the computer change was completed, as a check on the beta and advection track model (BAM; Marks 1989) that was being implemented that year (J. Gross 2004, personal communication). The idea being tested was that SANBAR and BAM were both based on barotropic steering concepts so NHC and NMC wanted to see how well they compared before terminating SANBAR (M. DeMaria 2005, personal communication). Verifications for SAN85 in the eastern North Pacific TC basin (not shown) revealed that the track forecasts were considerably worse than CLIPER, a result that S. Goldenberg (2004, personal communication) attributed to fewer satellite winds in that basin. The moveable fine mesh model (Hovermale and Livezey 1977) showed little or no skill in its 1988 track forecasts, but the quasi-Lagrangian model (Mathur 1991) performed much better than CLIPER in 1990.

Considerable progress in TC track model forecast guidance has been achieved (DeMaria and Gross 2003) and the length of the TC track forecasts has been extended from 72 to 120 h since the last operational Atlantic basin forecast of SANBAR in 1989. Verifications for the 2003 Atlantic TC season, the most recent season available at the time this manuscript was prepared, indicate that three of the best dynamical models have become much more skillful than those available during the SANBAR era. The forecast skill for the NOAA track models of the Geophysical Fluid Dynamics Laboratory (Kurihara et al. 1998) and the Global Forecast System of NCEP (Kanimitsu 1989) and the Navy Operational Global Atmospheric Prediction System model (Elsberry et al. 1999) of the U.S. Navy, averaged approximately -45% from 24 to 36 h and approximately -55% from 48 to 120 h (these samples include tropical depressions as well as named TCs).

6. Conclusions

The SANBAR model was retired after the 1989 TC season in the Atlantic and in the middle of the 1990 TC season in the eastern North Pacific when a computer change at NMC would have required a large effort to convert the SANBAR program to the new computer. At about the same time, a limited-area barotropic TC track model developed by Vic Ooyama (VICBAR; DeMaria et al. 1992), developed at HRD with superior numerical methods that took information for its boundary conditions from the NMC global forecast model, was ready to be run in real time. A fully operational version of VICBAR, a limited-area sine transformation barotropic model (LBAR; Horsfall et al. 1997) has been running since 1996.

A long-time TC forecaster at NHC (M. Lawrence 2003, personal communication) recalls that the operational availability of SANBAR in 1970 marked the introduction of routine numerical weather prediction in TC track forecasting and the beginning of an era in which TC track forecast errors decreased more rapidly than previously.

Sanders and Burpee (1968) noted that, if the SANBAR technique could be transferred to operations without serious loss of accuracy, it would be a significant advance to the art of TC track forecasting in the western North Atlantic basin. If the SANBAR model was properly transferred from research to operations by 1973, the impact of its track forecasts on the state of the art was more modest. The improvement of the skill of the SAN85 track guidance relative to that of SAN73 suggests that SANBAR may have had more influence on operational TC track forecasting had Sanders selected the Cressman (1959), rather than the Eddy (1967), OA for SANBAR. The SANBAR model, nevertheless, was the first operational dynamical model for TC track prediction in any ocean and was operational in the Atlantic basin for nearly 20 yr. It will be remembered for being at the forefront of the transition from statistical to dynamical operational TC track forecast models that guide TC forecasters.

Acknowledgments. I wrote this SANBAR summary in appreciation of my years spent with "old dad" Fred Sanders, a fantastic teacher and great motivator, who cared about his students and became my Ph.D. advisor and good friend. I have appreciated his enthusiasm in the classroom, keen insights of the weather, and passion for understanding weather events that were forecast incorrectly, particularly those that occurred during his sailboat races. The author is grateful to Mark DeMaria of the Regional and Mesoscale Meteorology Team of the NOAA/National Environmental Satellite, Data, and Information Service (NESDIS) in Fort Collins, Colorado, for supplying the forecast verification statistics in a convenient form and for sharing his knowledge of the history of TC track models. Sim Aberson of HRD produced the CLIPER diagrams and Stan Goldenberg, also of HRD, reviewed an early version of the manuscript, suggested several improvements, and recalled his SAN85 experiences. Goldenberg, Ken Campana of NCEP, Bob Kohler of the Atlantic Oceanographic and Meteorological Laboratory of which HRD is a component, and Jim Gross and Miles Lawrence of NHC provided historical information. Kerry Emanuel of MIT sent information about the M.S. thesis references and a copy of one of the theses. Chris Landsea of HRD, a formal reviewer of the manuscript, suggested several improvements. Chris Velden of the Space Science and Engineering Center, University of Wisconsin, Geoff DiMego of NCEP, and Gary Gray and Gary Ellrod (through Ken Campana) of NESDIS contributed their thoughts on the initial use of midtropospheric winds in the NMC analyses.

The editors gratefully acknowledge the assistance of

Stanley Goldenberg in proofing and editing Bob's paper after his death.

REFERENCES

Barnes, S. L., 1964: A technique for maximizing details in numerical weather map analysis. *J. Appl. Meteor.,* **3,** 396–409.

Burpee, R. W., D. G. Marks, and R. T. Merrill, 1984: An assessment of Omega dropwindsonde data in track forecasts of Hurricane Debby (1982). *Bull. Amer. Meteor. Soc.,* **65,** 1050–1058.

Charney, J. G., 1963: A note on large-scale motions in the tropics. *J. Atmos. Sci.,* **20,** 607–609.

Cressman, G. P., 1958: Barotropic divergence and very long waves. *Mon. Wea. Rev.,* **86,** 293–297.

——, 1959: An operational objective analysis scheme. *Mon. Wea. Rev.,* **87,** 367–374.

DeMaria, M., and J. M. Gross, 2003: Evolution of tropical cyclone forecast models. *Hurricanes: Coping with Disaster, Progress and Challenges since Galveston, 1900,* Robert H. Simpson, Ed., Amer. Geophys. Union, 103–126.

——, S. D. Aberson, K. V. Ooyama, and S. J. Lord, 1992: A nested spectral model for hurricane track forecasting. *Mon. Wea. Rev.,* **120,** 1628–1643.

Eddy, A., 1967: The statistical objective analysis of scalar data fields. *J. Appl. Meteor.,* **6,** 597–609.

Elsberry, R. L., M. A. Boothe, G. A. Ulses, and P. A. Harr, 1999: Statistical postprocessing of NOGAPS tropical cyclone track forecasts. *Mon. Wea. Rev.,* **127,** 1912–1919.

Gaertner, J. P., 1973: Investigation of forecast errors of the SANBAR hurricane track model. M.S. thesis, Dept. of Meteorology, Massachusetts Institute of Technology, 57 pp.

Goldenberg, S. B., S. D. Aberson, and R. E. Kohler, 1985: Incorporation of Omega dropwindsonde data into SANBAR: An operational barotropic hurricane-track forecast model. Preprints, *16th Conf. on Hurricanes and Tropical Meteorology,* Houston, TX, Amer. Meteor. Soc., 44–45.

——, ——, and ——, 1987: An updated fine-grid version of the operational barotropic hurricane-track prediction model. Preprints, *17th Conf. on Hurricanes and Tropical Meteorology,* Miami, FL, Amer. Meteor. Soc., 86–89.

Horsfall, F. M., M. DeMaria, and J. M. Gross, 1997: Optimal use of large-scale boundary and initial fields for limited-area hurricane forecast models. Preprints, *22d Conf. on Hurricanes and Tropical Meteorology,* Fort Collins, CO, Amer. Meteor. Soc., 571–572.

Hovermale, J. B., and R. E. Livezey, 1977: Three-year performance characteristics of the NMC hurricane model. Preprints, *11th Tech. Conf. on Hurricanes and Tropical Meteorology,* Miami Beach, FL, Amer. Meteor. Soc., 367–374.

Kanimitsu, M., 1989: Description of the NMC global data assimilation and forecast system. *Wea. Forecasting,* **4,** 335–342.

King, G. W., 1966: On the numerical prediction of hurricane trajectories with vertically averaged winds. M.S. thesis, Dept. of Meteorology, Massachusetts Institute of Technology, 51 pp.

Kurihara, Y., R. E. Tuleya, and M. A. Bender, 1998: The GFDL hurricane prediction model and its performance during the 1995 hurricane season. *Mon. Wea. Rev.,* **126,** 1306–1322.

Marks, D. G., 1989: The beta and advection model for tropical cyclone track forecasting. *Extended Abstracts, 18th Conf. on Hurricanes and Tropical Meteorology,* San Diego, CA, Amer. Meteor. Soc., 38–39.

Mathur, M., 1991: The National Meteorological Center's quasi-Lagrangian model for hurricane prediction. *Mon. Wea. Rev.,* **119,** 1419–1447.

Miller, B. I., E. C. Hill, and P. P. Chase, 1968: Revised technique for forecasting hurricane motion by statistical methods. *Mon. Wea. Rev.,* **96,** 540–548.

Neumann, C. J., 1972: An alternate to the HURRAN tropical cyclone forecast system. NOAA Tech. Memo. SR-62/, 24 pp.

——, 1981: Trends in forecasting the tracks of Atlantic tropical cyclones. *Bull. Amer. Meteor. Soc.,* **62,** 1473–1485.

——, and J. M. Pelissier, 1981: Models for the prediction of tropical cyclone motion over the North Atlantic: An operational evaluation. *Mon. Wea. Rev.,* **109,** 522–538.

——, J. R. Hope, and B. I. Miller, 1972: A statistical method of combining synoptic and empirical tropical cyclone forecast sustems. NOAA Tech. Memo. NWS SR-63/, 32 pp.

Pike, A. C., 1972: Improved barotropic hurricane track prediction by adjustment of the initial wind field. NOAA Tech. Memo. NWS SR-66, 16 pp.

Sanders, F., 1970: Dynamic forecasting of tropical storms. *Trans. N.Y. Acad. Sci. Ser. II,* **32,** 495–508.

——, and R. W. Burpee, 1968: Experiments in barotropic hurricane track forecasting. *J. Appl. Meteor.,* **7,** 313–323.

——, A. C. Pike, and J. P. Gaertner, 1975: A barotropic model for operational prediction of tracks of tropical storms. *J. Appl. Meteor.,* **14,** 265–280.

——, A. L. Adams, N. J. B. Gordon, and W. D. Jensen, 1980: Further development of a barotropic operational model for predicting paths of tropical storms. *Mon. Wea. Rev.,* **108,** 642–654.

Tracy, J. D., 1966: Accuracy of Atlantic tropical cyclone forecasts. *Mon. Wea. Rev.,* **94,** 407–418.

Velden, C. S., and S. B. Goldenberg, 1987: The inclusion of high density satellite wind information in a barotropic hurricane-track forecast model. Preprints, *17th Conf. on Hurricanes and Tropical Meteorology,* Miami, FL, Amer. Meteor. Soc., 90–93.

——, W. L. Smith, and M. Mayfield, 1984: Applications of VAS and TOVS to tropical cyclones. *Bull. Amer. Meteor. Soc.,* **65,** 1059–1067.

Williams, F. R., 1972: Application of the SANBAR hurricane track forecast model. M.S. thesis, Dept. of Meteorology, Massachusetts Institute of Technology, 39 pp.

Wise, C. W., and R. H. Simpson, 1971: The tropical analysis program of the National Hurricane Center. *Weatherwise,* **24,** 164–173.

Chapter 12

The Application of Fred Sanders' Teaching to Current Research on Extreme Cold-Season Precipitation Events in the Saint Lawrence River Valley Region

JOHN R. GYAKUM

Department of Atmospheric and Oceanic Sciences, McGill University, Montreal, Quebec, Canada

(Manuscript received 28 November 2005, in final form 16 June 2006)

Gyakum

ABSTRACT

Fred Sanders' teaching and research contributions in the area of quasigeostrophic theory are highlighted in this paper. The application of these contributions is made to the topic of extreme cold-season precipitation events in the Saint Lawrence valley in the northeastern United States and southern Quebec.

This research focuses on analyses of Saint Lawrence valley heavy precipitation events. Synoptic- and planetary-scale circulation anomaly precursors are typically identified several days prior to these events. These precursors include transient upper-level troughs, strong moisture transports into the region, and anomalously large precipitable water amounts. The physical insight of Fred Sanders' work is used in the analysis of these composite results. Further details of this insight are provided in analyses of one case of heavy precipitation.

1. Introduction

The purpose of this paper is to highlight the significant teaching and mentoring contributions of Fred Sanders. We relate these contributions to recent research in the understanding of extreme cold-season precipitation events in the Saint Lawrence River valley of the northeastern United States and in the southern region of the Canadian province of Quebec.

We first discuss recent results of precipitation research in the context of concepts that have been elucidated by Fred Sanders in his synoptic meteorology courses. The following sections include analyses of one specific case of extreme precipitation, its large-scale circulation struc-

tures, and discussion of crucial mesoscale details not otherwise provided by composite analyses. These mesoscale details include concepts of frontogenesis, air-mass stratification, and instantaneous precipitation rates articulated by Fred Sanders in his teaching.

2. Extreme cold-season events in the Saint Lawrence and Champlain valleys in the context of quasigeostrophic theory

The recent work of Sisson and Gyakum (2004) has highlighted large-scale circulation structures associated with a 48-case composite of cold-season precipitation events in Burlington, Vermont (BTV), whose amounts ranged from 25 to 50 mm in a 24-h period. Despite the substantial increases in numerical model forecast skill over the years, the problem of quantitative precipitation forecasting remains among the most challenging to both

Corresponding author address: John R. Gyakum, Department of Atmospheric and Oceanic Sciences, McGill University, 805 Sherbrooke Street West, Montreal, QC H3A 2K6, Canada.
E-mail: john.gyakum@mcgill.ca

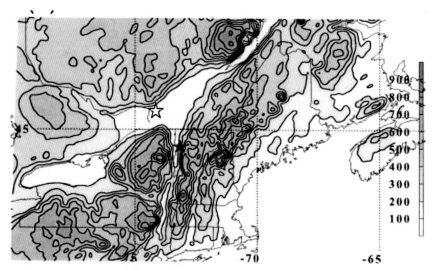

FIG. 1. Terrain elevation (m) for the region surrounding Burlington, Vermont (indicated with a black star), and Montreal (open star). Latitude–longitude lines (dotted) are indicated at intervals of 5°. Adapted from Sisson and Gyakum (2004).

the operational forecaster and to the researcher (Fritsch et al. 1998). Figure 1, adapted from Sisson and Gyakum (2004), shows the terrain surrounding BTV in the north–south-oriented Champlain valley and its location. Montreal is located to the northwest in the southwest–northeast-oriented Saint Lawrence River valley. Both Burlington and Montreal are susceptible to channeling of winds along their respective valleys. Powe (1968, 1969) has documented the preponderance of either southwesterly or northeasterly winds in Montreal. This wind direction bimodality has a significant influence on the weather and climate of Montreal and was crucial in

defining the details of the 1998 ice storm (Gyakum and Roebber 2001; Roebber and Gyakum 2003).

Figure 2 illustrates synoptic-scale sea level pressure composites for the 48-case sample of heavy cold-season precipitation. Composites were chosen for the cases found during the November through March period from 1963 through 1995. Further details concerning the case selection and compositing procedures may be found in the paper by Sisson and Gyakum (2004). The reference time of 0 h refers to the beginning of the maximum 24-h precipitation event.

Two primary conclusions may be made from Fig. 2.

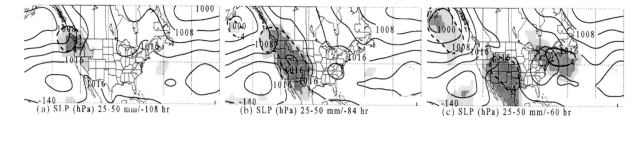

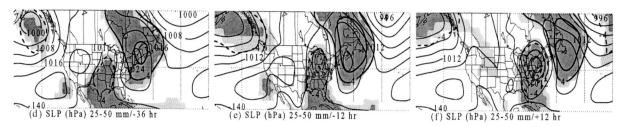

FIG. 2. Sea level pressure (interval of 4 hPa; solid) and anomalies (interval of 4 hPa; heavy dashed for negative and heavy solid for positive) with respect to climatology for the composite of the heavy cases at (a) −108, (b) −84, (c) −60, (d) −36, (e) −12, and (f) +12 h. Light (dark) shading represents statistical significance of the anomalies at the 95% (99%) confidence levels, according to a Student's t test. Latitude–longitude lines (dotted) are shown at intervals of 20°. Adapted from Sisson and Gyakum (2004).

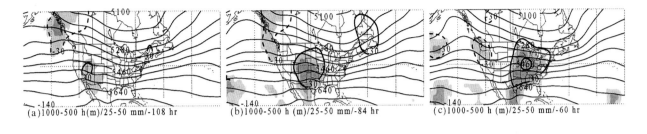

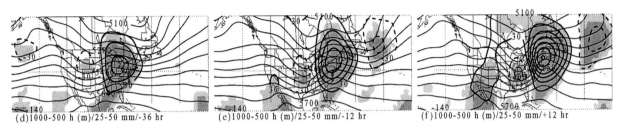

FIG. 3. Thickness of the 1000–500-hPa heights (interval of 60 m; solid) and anomalies (interval of 30 m; heavy dashed for negative and heavy solid for positive) with respect to climatology for the composite of the heavy cases at (a) −108, (b) −84, (c) −60, (d) −36, (e) −12, and (f) +12 h. Light (dark) shading represents statistical significance of the anomalies at the 95% (99%) confidence levels, according to a Student's *t* test. Latitude–longitude lines (dotted) are shown at intervals of 20°. Adapted from Sisson and Gyakum (2004).

The first is that these events are associated with south–southeasterly geostrophic flow with a cyclone to the west and an anticyclone to the east. The second finding is that the synoptic-scale cyclonic disturbance may be detected several days in advance of the event, as it approaches the Washington and British Columbia coastlines. Additionally, these heavy events show stronger cyclone–anticyclone couplets than are found in events with smaller precipitation amounts (not shown).

Figure 3, showing the 1000–500-hPa thicknesses and anomalies for the same composite, and for the same times as for Fig. 2, illustrates the favorable placement of the surface low pressure system downshear of the upper-level thickness trough and upshear of the upper-level ridge. Additionally, the mobile upper-level trough is easily traceable westward to a position west of British Columbia.

The importance of finite-amplitude mobile upper-level troughs was recognized by Fred Sanders throughout his career. Daily charts of 552 dam height contours, compiled by Sanders, showed generations of Massachusetts Institute of Technology (MIT) students that mobile troughs could be traced for many thousands of kilometers. This concept was thoroughly documented by his published work (Sanders 1988). Sanders (1992) showed that such upper-level troughs were crucial to the predictions of rapid east coast cyclogenesis (e.g., Sanders and Gyakum 1980) during the Experiment on Rapidly Intensifying Cyclones over the Atlantic (ERICA; Sanders 1992). More generally, Sanders (1986, 1987) showed that rapid east coast cyclogenesis was nearly always associated with a mobile upper-level trough.

Why is the existence of a finite-amplitude upper

trough so favorable to cyclogenesis and to large-scale precipitation events? The answer may be found in the adiabatic, inviscid form of the quasigeostrophic omega equation (Bluestein 1992, his Eq. 5.6.11):

$$\left(\nabla_p^2 + \frac{f_0^2}{\sigma}\frac{\partial^2}{\partial p^2}\right)\omega = -\frac{f_0}{\sigma}\frac{\partial}{\partial p}[-\mathbf{V}_g \cdot \nabla_p(\zeta_g + f)]$$
$$-\frac{R}{\sigma p}\nabla_p^2(-\mathbf{V}_g \cdot \nabla_p T), \qquad (1)$$

where ∇_p is the horizontal gradient operator on a constant pressure surface, f_0 is a constant value of the Coriolis parameter, σ is the static stability parameter ($g\partial z/\partial p$)($\partial \ln\theta/\partial p$), ω is dp/dt, $\mathbf{V}_g$ is the geostrophic wind vector, ζ_g is the geostrophic relative vorticity, R is the dry air gas constant, and T is the virtual temperature.

Each of the two physical effects on the right side of (1) will be considered independently in the following analysis. Though these two terms are not Galilean invariant, and are therefore not generally independent, the physical insight provided by each of these physical effects is useful.

For the idealized case of an equivalent barotropic atmosphere, we may deduce the effect of differential cyclonic vorticity advection. This effect is shown as the first term on the right side of (1). Either an upward increase in cyclonic vorticity advection or an upward decrease in anticyclonic vorticity advection provides forcing for quasigeostrophic ascent. Either of these situations would decrease the temperature in the troposphere in association with a thickness decrease. The equivalent barotropic assumption precludes any temperature changes from horizontal temperature advection.

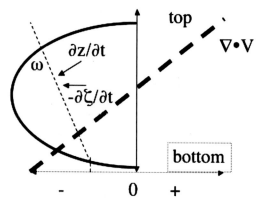 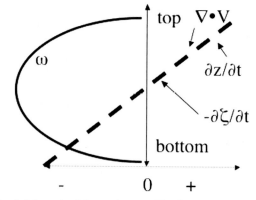

FIG. 4. Schematic of the quasigeostrophic effect of an upper-level increase in cyclonic vorticity advection. The panel shows the vertical tropospheric parabolic profile of omega (ω, solid) and horizontal divergence ($\nabla \cdot \mathbf{V}$, dashed). Also shown on the same vertical scale is one qualitative profile (thin dashed) of the local geopotential height tendency ($\partial z/\partial t$) and the negative of the local tendency of relative vorticity ($-\partial \zeta/\partial t$). "Top" and "bottom" correspond to the top and bottom of the troposphere, respectively.

FIG. 5. Schematic of the quasigeostrophic effect of warm advection. The panel shows the vertical tropospheric parabolic profile of omega (ω, solid). The horizontal divergence ($\nabla \cdot \mathbf{V}$), as well as the local geopotential height tendency ($\partial z/\partial t$) and the negative of the local tendency of relative vorticity ($-\partial \zeta/\partial t$) are all shown qualitatively on the same thick solid line. "Top" and "bottom" correspond to the top and bottom of the troposphere, respectively.

The vertical structure of this height change (and thickness decrease) is shown in Fig. 4, along with a characteristic parabolic profile of omega with "top" indicating the top of the troposphere and "bottom" indicating ground level. Because the only mechanism for a temperature decrease in the troposphere is through vertical motion in a stably stratified column, there must be ascent (Fig. 4), with the accompanying vertical profile of divergence (on level terrain) characterized by lower-tropospheric convergence and upper-tropospheric divergence. The mechanism for surface cyclonic vorticity increase in the lower troposphere is convergence in the presence of cyclonic absolute vorticity. The geostrophic relative vorticity increase occurring aloft is due to the effect of cyclonic vorticity advection overwhelming the opposing effect of divergence. Thus, the vertical profile of relative vorticity change (Fig. 4) is characterized by an upward increase in cyclonic vorticity tendency. This upward increase in geopotential height fall is consistent with the fact that the thickness is decreasing. Near the ground, where there is no vertical motion, there is no temperature change and therefore no near-ground thickness change.

Figures 2 and 3 show that there is a general westward tilt of the cyclonic disturbance, and in fact a very different structure from that of an idealized equivalent barotropic atmosphere. The closed surface pressure systems contrast with the more zonal, wavelike character to the 1000–500-hPa thickness field. Because of this structure, warm advection is also occurring in the region of the heavy precipitation. The quasigeostrophic effect of warm advection is illustrated in Fig. 5 and mathematically in the second term on the right side of (1). The presence of warm advection produces a warm thermal ridge locally, and thus the thermal vorticity decreases locally. Therefore, the tropospheric thickness must be

increasing. This thermodynamic change has the dynamical impact that the local change in relative vorticity aloft must be algebraically less than the relative vorticity change below. Because the only mechanism of vorticity change is through horizontal divergence, and the vertical integral of divergence must be zero in areas of level terrain, the only solution must be convergence in the lower troposphere and divergence aloft, with the accompanying profile of ascent (Fig. 5). The effect of cooling, owing to rising motion in a statically stable air mass, opposes (but typically does not overwhelm) the warm advection. The profile of height falls at lower levels and height rises aloft (Fig. 5) is consistent with the fact that the thicknesses are increasing.

Though the combination of upper-level cyclonic vorticity advection (Figs. 3 and 4) and warm advection (Figs. 2, 3, and 5) forces large-scale ascent, and therefore precipitation in the region of the Saint Lawrence and Champlain valleys, both synoptic-scale and mesoscale details, so crucial in determining the precipitation rate, remain to be documented.

3. An extreme cold-season event in the Saint Lawrence and Champlain valleys in the context of frontogenesis

An extreme precipitation event occurred during January 1998 in the Saint Lawrence and Champlain valleys. As discussed by Gyakum and Roebber (2001), the ice storm of 5–10 January 1998 was associated with a synoptic-scale structure that was persistently favorable for the occurrence of heavy freezing precipitation. The persistence of these structures resulted in 6-day amounts exceeding 100 mm of precipitation (Fig. 6). The banded structure of this accumulated precipitation pattern mimics its instantaneous structure (see, e.g., Fig. 10 of Roebber and Gyakum 2003) as the banded east–west structure

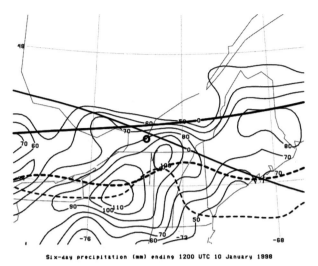

FIG. 6. Precipitation analysis for the 6-day period ending at 1200 UTC 10 Jan 1998. The solid contours show amounts, with an interval of 10 mm. The heavy solid (dashed) contour is that of the 850-hPa (surface) 0° isotherm at 0000 UTC 9 Jan 1998. Montreal is indicated by the open circle. Adapted from Gyakum and Roebber (2001).

persisted for much of this 6-day period. The favorable and persistent synoptic-scale sea level pressure and 1000–500-hPa thickness field is illustrated by Fig. 7. The upstream upper-level thickness trough and the warm advection are apparent, just as has been shown for the cold-season composites (Figs. 2 and 3). Additionally, there is a correspondence of the region of precipitation (Fig. 6) with a zone of geostrophic frontogenesis seen in each of the four panels of Fig. 7.

The importance of geostrophic frontogenesis is seen in Fig. 8, which is adapted from work by Sanders and Hoskins (1990). The idealized geostrophic frontogenetical flow on the left side of the panel is similar to that seen in the region of the banded precipitation occurring in Fig. 7. The instantaneous increase in the thermal gradient is obvious from Fig. 8, even though the geostrophic flow is nondivergent. Nevertheless, we should expect synoptic-scale vertical motions in this pattern of flow. The vertical motions may be understood from the perspective of quasigeostrophic theory, and its omega equation [(1)]. For the illustration in Fig. 8, (1) suggests that we have forcing for ascent in the warm sector and forcing for descent in the cold sector, owing

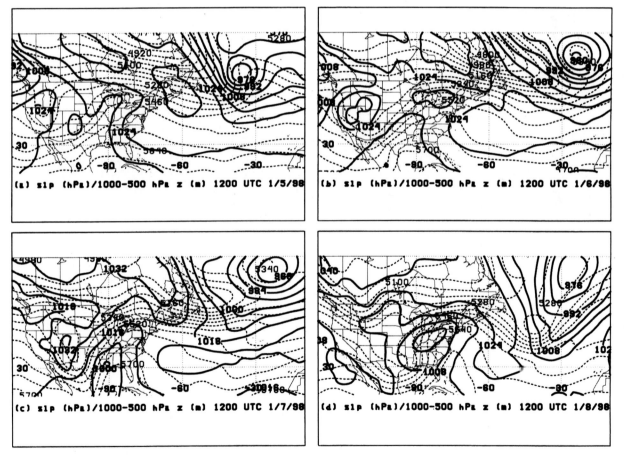

FIG. 7. Sea level pressure (heavy solid; interval of 8 hPa) and 1000–500-hPa thickness (dashed; interval of 60 m) at 1200 UTC (a) 5, (b) 6, (c) 7, and (d) 8 Jan 1998. Adapted from Gyakum and Roebber (2001).

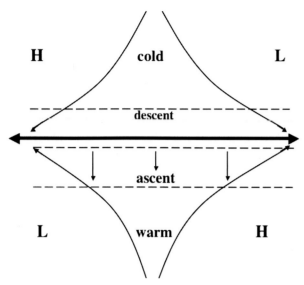

FIG. 8. Heavy double vector indicates the axis of dilatation. An idealized confluent frontogenetical pattern is shown. Light solid lines are sea level pressure isobars with associated arrows indicating the direction of the geostrophic flow, with dashed lines indicating isotherms. The arrows pointing to warmer air are **Q** vectors. Adapted from Sanders and Hoskins (1990).

to the respective warm and cold advections, and to the fact that vorticity advection effects are negligible.

Figure 8 also illustrates the quasigeostrophic motions in perhaps a more elegant form, that of the **Q** vector. The geostrophic frontogenesis function F (Bluestein 1992, his Eq. 5.7.125) may be expressed as

$$F = \frac{D_g}{Dt}|\nabla_p T| = \frac{\sigma p}{R}\frac{1}{|\nabla_p T|}(\nabla_p T \cdot \mathbf{Q}), \qquad (2)$$

where (Bluesteins Eq. 5.7.119)

$$\mathbf{Q} = \frac{R}{\sigma_p}\frac{D_g}{Dt}\nabla_p T. \qquad (3)$$

The **Q** vector is proportional to the rate of change of the horizontal temperature gradient following the geostrophic motion.

Equation (2) also indicates that where **Q** points to warmer air, there is confluent geostrophic frontogenesis (as seen in Fig. 2), and where **Q** points to colder air, there is geostrophic frontolysis.

The thermally direct circulation (Fig. 8), defined by warm air rising and cold air sinking, and associated with geostrophic confluent frontogenesis, may also be understood through the geostrophic adjustment process, so eloquently articulated in Fred Sanders' teaching notes and discussed by Bluestein (1992, his section 5.7.4).

Consider the idealized illustration of geostrophic confluent frontogenetic flow (Fig. 8). The consequence of such a flow, if its effects were felt through a finite depth of the troposphere, is to increase the thickness to the south and to decrease the thickness between two pressure surfaces to the north (Fig. 9), and simultaneously, vertical variations in the geostrophic advection of momentum act to reduce the vertical geostrophic shear. The associated change in pressure gradient accompanying this increased thickness gradient upsets the geostrophically balanced flow, and the advective change in shear is in the opposite direction to what would be needed to maintain balance. Thus, the atmosphere requires an ageostrophic adjustment to accommodate the new horizontal pressure gradients. This adjustment is illustrated in Fig. 10. The westerly winds increase aloft, while the easterlies increase at lower levels, thereby increasing the westerly shear to adjust to the new stronger tropospheric north–south temperature gradient. The ageostrophic thermally direct secondary circulation is the means by which this increased vertical shear is produced (Fig. 10). This process may be understood by utilizing the combination of the geostrophic wind relationship and the frictionless form of the momentum equation to get the relationship between the ageostrophic wind, $\mathbf{V} - \mathbf{V}_g$, and the horizontal acceleration, $d\mathbf{V}/dt$:

$$\mathbf{V} - \mathbf{V}_g = (1/f)\mathbf{k} \times (d\mathbf{V}/dt), \qquad (4)$$

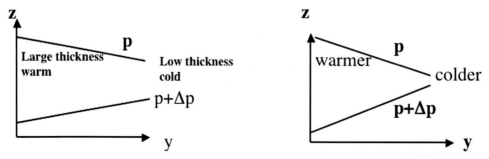

FIG. 9. Vertical cross sections illustrating thickness gradient increase associated with the geostrophic confluent frontogenetic effect of Fig. 8. Adapted from Bluestein (1992).

The upper-level acceleration from the west:

$$\mathbf{V}\text{-}\mathbf{V}_g = (1/f)\mathbf{k} \times (d\mathbf{V}/dt)$$

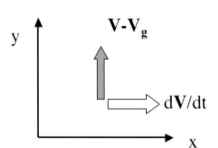

Later, with stronger thickness gradient:

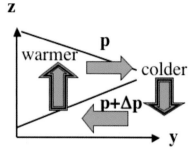

FIG. 10. Illustration of the upper-level horizontal ageostrophic wind, $\mathbf{V} - \mathbf{V}_g$ [shown as (left) the shaded arrow and as (right) the northward-directed arrow]. The horizontal acceleration vector $d\mathbf{V}/dt$ (open arrow) points to the right of $\mathbf{V} - \mathbf{V}_g$ in the left panel, thereby increasing the westerly component of the wind aloft. The upward and downward motions in the right panel are indicated as shaded double arrows.

where f is the Coriolis parameter and $\mathbf{k}$ is the vertical unit vector. Equation (4) states that, in the absence of friction, the acceleration points 90° to the right of the ageostrophic wind in the Northern Hemisphere. The new enhanced pressure decrease to the north aloft requires an ageostrophic wind component to the north, with its accompanying acceleration to the east (left panel of Fig. 10). The upper-level acceleration is from the west, thereby increasing the westerlies. The lower levels are characterized by a new enhanced pressure decrease to the south, with its accompanying ageostrophic flow to the south, and an acceleration to the west, with an increase in the easterlies. The vertical motions of descent in the colder air and ascent in the warmer air close this ageostrophic circulation. The ageostrophic thermally direct circulation is therefore an adjustment to the new stronger thermal gradient, which in turn acts to weaken the gradient, and thereby works in the sense needed to restore thermal wind balance. As pointed out by Hoskins et al. (1978), the role of the ageostrophic motion is to restore the thermal wind balance that the geostrophic flow tends to destroy.

Though the geostrophic frontogenesis associated with the 1998 ice storm persisted and forced the synoptic-scale vertical motions and precipitation for this extended period (Figs. 6 and 7), the mesoscale details of this precipitation event were crucial to the large amounts of freezing rain recorded. The orographic channeling of winds was prevalent through much of the January 1998 ice storm. A representative example of this channeling is shown in Fig. 11, in which the winds along the Saint Lawrence valley are northeasterly and are oriented approximately 90° to the sea level isobars. This nongeostrophic cold advection contributed substantially to the warm frontogenesis during the ice storm (Fig. 11). The surface observations also show freezing rain occurring north of the baroclinic zone in the Saint Lawrence val-

ley, with rain occurring in the southerly flow to the south. Roebber and Gyakum (2003) have also shown that the orographic channeling of surface winds was responsible for the strong surface baroclinic zone and the strong vertical circulations. Composite radar reflectivities throughout the ice storm indicate an east–west-oriented precipitation structure (not shown).

A particularly crucial detail concerning the ice storm, and many other cases of heavy precipitation in the middle latitudes, is that of the airmass stratification. Figure 12 (Gyakum and Roebber 2001) shows representative air parcel trajectories that have been traced from the region of precipitation to their positions several days earlier. The analysis of Gyakum and Roebber (2001) shows that the air associated with the precipitation in the Saint Lawrence valley had a history of being warmed and moistened in the boundary layer of the subtropical oceans. This particular finding is similar to results found for other cases of middle-latitude heavy precipitation, particularly that of Lackmann and Gyakum (1999), in their study of heavy precipitation along the west coastal region of North America, and the commonly used term *pineapple express*. Such a term is applied to meteorological conditions in which subtropical moisture is transported from the region of the Hawaiian Islands poleward to the states of Washington and Oregon and to the Canadian province of British Columbia.

An important consequence of the subtropical air mass being associated with any precipitation event is its potent combination of weak static stability and large amounts of precipitable water. As seen in (1), we expect more synoptic-scale ascent to occur as the static stability decreases. Though precipitation occurring during the 1998 ice storm was associated with a strong, lower-tropospheric inversion (Fig. 13), as is the case in many northeasterly flow situations, the stratification above this inversion is considerably weaker. During the later phas-

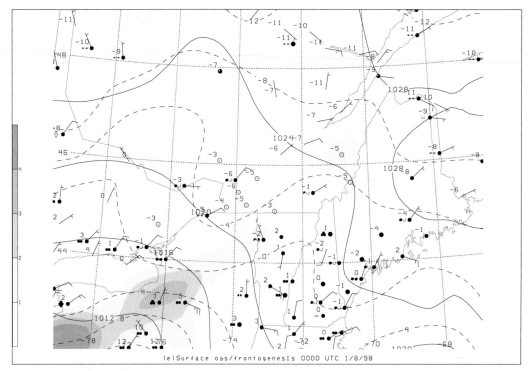

FIG. 11. Surface map for 0000 UTC 8 Jan 1998 with observed wind speed (half barb, 2.5 m s^{-1}; full barb, 5 m s^{-1}) and direction, sea level pressure (solid) at 4-hPa intervals, temperatures (dashed) at 4°C intervals, and total wind frontogenesis (shading convention for the range from 1° to 5° (100 km)$^{-1}$ (3 h)$^{-1}$. Station model includes the temperature (°C), weather, and cloud cover. Adapted from Roebber and Gyakum (2003).

es of the ice storm, anecdotal reports of thunder by residents of Montreal were given, suggesting the presence of moist convection. The time section of soundings shown in Fig. 13 illustrates the remarkable warmth aloft with temperatures exceeding 6°C at 900 hPa during the early phases of the ice storm and similar temperatures of nearly 6°C at 800 hPa during the latter part of the case on 9 January 1998. The atmosphere was nearly saturated during this period.

Aside from the positive effect of weak stratification on ascent, the additional effect of the air mass on pre-

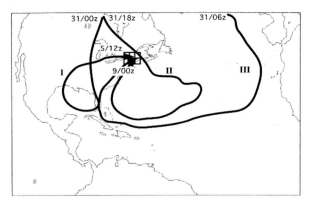

FIG. 12. Paths of three representative trajectories are shown for the ice storm of 1998. Initial and final times are indicated for each trajectory path (date/time). Adapted from Gyakum and Roebber (2001).

cipitation rate has been eloquently illustrated in the teachings of Fred Sanders. Assuming that the condensation rate is the same as the precipitation rate, and assuming that saturated air parcels are ascending moist adiabatically, the expression for precipitation rate is

$$P = -(1/g) \int \omega(dr_s/dp)_{ma} \, dp, \qquad (5)$$

where g is gravity, the vertical integral extends from 1000 to 200 hPa, r_s is the saturation mixing ratio, and the subscript ma represents the appropriate moist adiabat. It is further assumed that omega vanishes at both the 1000- and 200-hPa levels. A particularly insightful calculation results when one assumes a characteristic parabolic ascent profile between 1000 and 200 hPa, in which the maximum ascent is -30×10^{-4} hPa s^{-1}. This results in an instantaneous precipitation rate of 1.1, 0.6, and 0.3 mm of precipitation during a 1-h period for respective air masses of maritime tropical, maritime polar, and continental polar. Therefore, the transports of tropical airmass characteristics are crucial to the production of heavy precipitation in the extratropical latitudes. The additional impact of generally weaker static stability, namely, the effective static stability as the lapse rate is referenced to the moist adiabat (Sanders and Olson 1967), associated with maritime tropical air masses suggests stronger ascent than would be expected for the

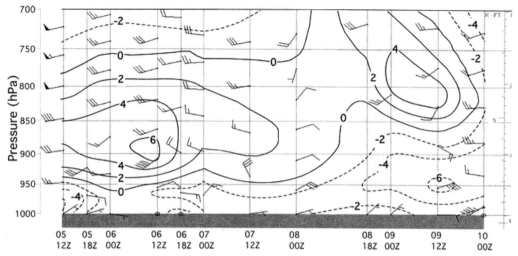

FIG. 13. Time–height cross section at Ottawa (YOW) for the period from 1200 UTC 5 Jan to 0000 UTC 10 Jan 1998. Isotherms are shown with a contour interval of 2°C (solid for greater than or equal to 0°C and dashed for colder than 0°C). Wind barbs are shown with the same convention as for Fig. 11. Adapted from Roebber and Gyakum (2003).

same amount of quasigeostrophic forcing in other air masses. Sanders and Olson (1967) also found that the tropospheric temperature lapse rates were only slightly more stable than that of the moist adiabat for a sample of cold-season rainstorms in the central and eastern United States. Thus, the combined effects of stronger ascent, coupled with the larger values of incremental changes in saturation mixing ratios associated with the maritime tropical air masses [(5)], suggests an even greater ratio of precipitation rates in maritime tropical air masses to those precipitation rates in maritime polar air masses, and to those precipitation rates in continental polar air masses, than those discussed above.

4. Summary

This paper has discussed concepts, elucidated by Frederick Sanders in his teaching and mentoring of MIT students, as they apply to the continuing research problem of understanding heavy precipitation events during the cold season in the extratropical latitudes. We find that Fred Sanders' teaching has provided illuminating physical insight into this research problem. As has been pointed out by Professor Sanders, the mesoscale detail and structures of precipitation, including a better understanding of moist convection, remain outstanding research issues. Nevertheless, the concepts discussed in this paper provide a conceptual background to this problem of cold-season precipitation forecasting in extratropical latitudes. Such physically based conceptual understanding may assist the researcher who may otherwise be overwhelmed with a plethora of numerical model output.

Acknowledgments. This work has been sponsored by research grants from the Natural Sciences and Engineering Research Council of Canada and the Canadian Foundation for Climate and Atmospheric Sciences. The author acknowledges the constructive comments of Dr. Gary Lackmann and the other, anonymous, reviewer. The author is grateful for the education and guidance provided by his graduate supervisor at MIT, Professor Fred Sanders.

REFERENCES

Bluestein, H. B., 1992: *Synoptic-Dynamic Meteorology in Midlatitudes.* Vol. I. *Principles of Kinematics and Dynamics,* Oxford University Press, 431 pp.

Fritsch, J. M., and Coauthors, 1998: Quantitative precipitation forecasting: Report of the Eighth Prospectus Development Team, U. S. Weather Research Program. *Bull. Amer. Meteor. Soc.,* **79,** 285–299.

Gyakum, J. R., and P. J. Roebber, 2001: The 1998 Ice Storm—Analysis of a planetary-scale event. *Mon. Wea. Rev.,* **129,** 2983–2997.

Hoskins, B. J., I. Draghici, and H. C. Davies, 1978: A new look at the ω-equation. *Quart. J. Roy. Meteor. Soc.,* **104,** 31–38.

Lackmann, G. M., and J. R. Gyakum, 1999: Heavy cold-season precipitation in the northwestern United States: Synoptic climatology and an analysis of the flood of 17–18 January 1986. *Wea. Forecasting,* **14,** 687–700.

Powe, N. N., 1968: The influence of a broad valley on the surface winds within the valley. Canadian Department of Transport, Meteorology Bureau, Rep. TEC-668, 9 pp. [Available from the Canadian Government Publishing Centre, Supply and Services Canada; Ottawa, ON K1A 0S9, Canada.]

——, 1969: The climate of Montreal. Canadian Department of Transport, Meteorology Bureau, Climatology Study 15, 51 pp. [Available from the Canadian Government Publishing Centre, Supply and Services Canada; Ottawa, ON K1A 0S9, Canada.]

Roebber, P. J., and J. R. Gyakum, 2003: Orographic influences on the mesoscale structure of the 1998 ice storm. *Mon. Wea. Rev.,* **131,** 27–50.

Sanders, F., 1986: Explosive cyclogenesis in the west-central North Atlantic Ocean, 1981−84. Part I: Composite structure and mean behavior. *Mon. Wea. Rev.,* **114,** 1781–1794.

——, 1987: A study of 500 mb vorticity maxima crossing the east coast of North America and associated surface cyclogenesis. *Wea. Forecasting,* **2,** 70–83.

——, 1988: Life history of mobile troughs in the upper westerlies. *Mon. Wea. Rev.,* **116,** 2629–2648.

——, 1992: Skill of operational dynamical models in cyclone prediction out to five-days range during ERICA. *Wea. Forecasting,* **7,** 3–25.

——, and D. A. Olson, 1967: The release of latent heat of condensation in a simple precipitation forecast model. *J. Appl. Meteor.,* **6,** 229–236.

——, and J. R. Gyakum, 1980: Synoptic-dynamic climatology of the "bomb." *Mon. Wea. Rev.,* **108,** 1589–1606.

——, and B. J. Hoskins, 1990: An easy method for estimation of Q-vectors from weather maps. *Wea. Forecasting,* **5,** 346–353.

Sisson, P. A., and J. R. Gyakum, 2004: Synoptic and planetary-scale precursors to significant cold-season precipitation events in Burlington, Vermont. *Wea. Forecasting,* **19,** 841–854.

Chapter 13

Must Surprise Snowstorms be a Surprise?

M. STEVEN TRACTON

Office of Naval Research, Arlington, Virginia

(Manuscript received 7 December 2005, in final form 2 June 2006)

Tracton

ABSTRACT

Today, even with state-of-the-art observational, data assimilation, and modeling systems run routinely on supercomputers, there are often surprises in the prediction of snowstorms, especially the "big ones," affecting coastal regions of the mid-Atlantic and northeastern United States. Little did the author know that lessons from Fred Sanders' synoptic meteorology class at the Massachusetts Institute of Technology (1967) would later (late 1980s) inspire him to pursue practical issues of predictability in the context of the development of ensemble prediction systems, strategies, and applications for providing information on the inevitable case-dependent uncertainties in forecasts. This paper is a brief qualitative and somewhat colloquial overview, based upon this author's personal involvement and experiences, intended to highlight some basic aspects of the source and nature of uncertainties in forecasts and to illustrate the sort of value added information ensembles can provide in dealing with uncertainties in predictions of East Coast snowstorms.

1. Introduction

Ever since I can remember (let us say ~50 yr ago), I have been awed, fascinated, inspired, passionate, delighted, infatuated, maniacal, and otherwise stir-crazy about East Coast snowstorms, especially the really "big ones" (usually "Nor'easters") with snow and wind combining into blizzard conditions producing mountainous drifts. There was always the exhilaration of experiencing a big one at my home in Brockton, Massachusetts, whether it was expected from monitoring media broadcasts or arrived as a complete surprise. And, of course, there was the profound disappointment when

Corresponding author address: M. Steven Tracton, 379 N St., SW, Washington, DC 20024.
E-mail: s.tracton@hotmail.com

a promised storm failed to materialize. I vividly recall wondering about the nature of these storms and why forecasts even a day or less ahead were not totally reliable "yes" or "no" statements, rather than some outlandish scheme for manipulating my emotional well-being.

Of course, those were the days (1950s) before operational numerical weather prediction (NWP). However, even today with state of the art observational data assimilation and modeling systems run on supercomputers, there still can be surprises associated with major snowstorms affecting the East Coast, with red-faced forecasters and media outlets eating "humble pie" and blaming the busts on "the models."

One of the first things I learned in Fred Sanders' synoptic meteorology class at the Massachusetts Institute of Technology (1967) was that forecast skill varied

251

from one day to the next, and forecasting is best approached in a probabilistic framework. This was before either he or I knew anything about Ed Lorenz's early work on predictability and the "Butterfly Effect" that later was recognized as the scientific rationale of what I am sure to Fred was intuitive. Additionally, it became evident in that same course while studying Fred's analytic model of cyclogenesis (Sanders 1971) how sensitive the development process was to small changes in the adjustable model parameters (e.g., wavelength, phase relationships, vertical stability, and diabatic heating).

Little did I know then how these influences many years later (late 1980s) would inspire me to pursue practical issues of predictability (i.e., the nature and consequence of small changes in initial conditions and/or model physics in modern data assimilation and forecast systems). This in turn led to being one among only a relatively few "true believers" at the forefront then in the development and application of operational ensemble prediction systems (EPSs). Today, it is generally accepted that EPSs are "the wave of the future" in NWP, a practical and operationally feasible capability that can provide quantitative estimates of the inevitable uncertainties in forecasts. Ensemble forecasting now is the mainstay of operational medium-range NWP based upon global models at virtually all major numerical weather prediction centers internationally and rapidly becoming so in the context of regional modeling systems for shorter-range, mesoscale prediction.

In the following I shall highlight some basic aspects of the source and nature of uncertainties in forecasts and of dealing with them in the context of East Coast snowstorms. This paper is not a comprehensive review of EPS systems, strategies, and applications for which there is an increasingly voluminous body of Web-based information, books, preprint volumes, and journal publications (e.g., Eckel and Mass 2005; Lewis 2005; see also information and references at http://wwwt.emc.ncep.noaa.gov/gmb/ens/ens_info.html). Nor is it a meaningful evaluation of the capabilities and limitations of EPS systems generally or those specifically used in the illustrative case studies below. Rather, it is a brief qualitative and somewhat colloquial overview based upon this author's personal involvement and experiences, intended to illustrate the sort of information EPS can provide on forecast uncertainties and to demonstrate the potential value added by incorporating this information into the forecast products.

2. Basic principles

The basic premise of ensemble prediction is that weather forecasting is stochastic, not deterministic in nature. Because the evolution of atmospheric states is fundamentally chaotic, uncertainties in initial conditions means there is no single solution, but only an array of possibilities—even if models were perfect. Of course models are not and never will be perfect, so uncertainties in model formulation (e.g., physics parameterization) contribute to the envelope of possible outcomes. Ensemble prediction systems address this through the divergence ("spread") among a set of multiple runs of one or more models, which differ as functions of perturbations to initial conditions and/or model details. The spread and its evolution through the model integrations are dependent upon factors such as season, circulation regime, and scale of weather system, parameter, level, and location. In contrast to a single deterministic model run, the spread provides explicit situation-contingent information on the array of possibilities as, for example, in the intensity and motion of a possible storm and associated sensible weather. The prospective value of EPS is the degree to which it can reliably discriminate between varying levels of confidence as inversely related to the spread, which can range from sufficiently small to be inconsequential (complete confidence) to so large that confidence on the outcome is effectively nil.

How one expresses and exploits this information is highly user-dependent. It can be simply assigning some qualitative level of confidence (e.g., very high to extremely low) to the expected most likely scenario. Additional information might be highlighting alternative possibilities, such as shifts in the track of a developing storm. The most comprehensive approach is providing the full probability distribution of selected parameters derived from the ensemble spread as, for example, the probability of snow greater than 15.2 cm (6 in.). The latter is especially applicable in decision making and risk analysis and for quantitative estimates of forecast value in contexts such as cost–benefit analyses (e.g., Richardson 2000; Mylne 2002).

3. Some historical perspective

One of the early challenges in the development of operational EPS was how to extract, condense, and ultimately convey the seemingly overwhelming amount of information generated by the multiple model runs composing the ensemble, and to do so in a "user-friendly" manner tailored to suit particular applications. The first formal attention addressing the realistic possibility of operational EPS—in the context of global models—was at the 1986 *Workshop on Predictability in the Medium and Extended Range* at the European Centre for Medium-Range Weather Forecasts (ECMWF). However, the problem of postprocessing and presenting ensemble forecast results was not explicitly addressed until 1991 at the ECMWF *Workshop on New Developments in Predictability*, where a listing of generic products was provided. To a fairly considerable degree the list reflected the recommendations of this author[1] based on recent experience "playing around" with output from

---

[1] In my position at the Climate Prediction Center (CPC) of the National Centers for Environmental Prediction (NCEP).

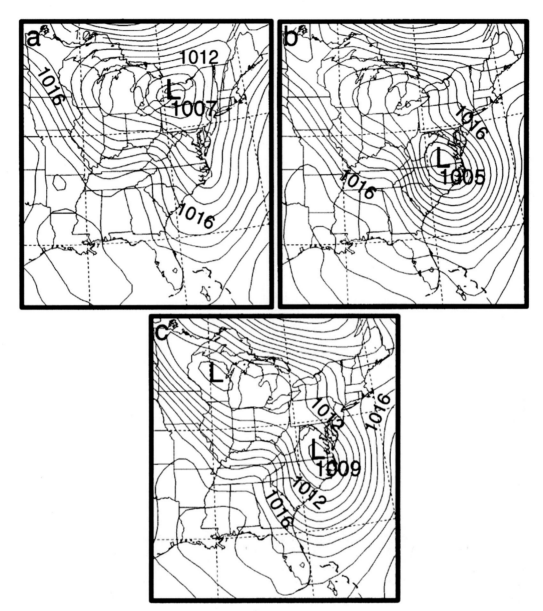

FIG. 1. Contours of sea level pressure (hPa) at 48 h, valid at 1200 UTC 22 Sep 1996, from an ensemble member with (a) no coastal cyclone, (b) a strong coastal cyclone and a weak parent cyclone, and (c) a combination of a coastal cyclone and a stronger parent cyclone. Sea level pressure contoured every 1 hPa. From Stensrud et al. (1999).

a concept demonstration global model [Medium-Range Forecast model (MRF)] ensemble and utilizing a crude (by today's standards) paper-based graphics system. It was in this construct that the "spaghetti chart" emerged as arguably the most recognized and symbolic product of ensemble forecasting. Spaghetti charts, where a single selected contour of a variable (e.g., 558-dam 500-hPa height contour) is plotted for each ensemble member, are a disarmingly simple way to visually display the array of possible solutions and to permit appraising the relative likelihood of various outcomes (Tracton and Kalnay 1993). Of special relevance for this monograph is that the inspiration for adopting spaghetti charts in

the EPS context was engendered inadvertently by Fred in his synoptic laboratory course where "continuity charts" were used to describe the evolution of upper-level mobile troughs (Sanders 1988, his Fig. 2). Not coincidentally, a prime example of the potential value of EPS in Tracton and Kalnay (1993, their Figs. 11, 12) was a sequence of spaghetti charts showing the possibility of a major winter storm developing over the northern Gulf of Mexico and its subsequent movement to the northeast. While several other generic EPS products come into play (e.g., ensemble mean, spread, clusters, and probability charts), it is reasonable to assert that spaghetti charts were critical in the early 1990s for dem-

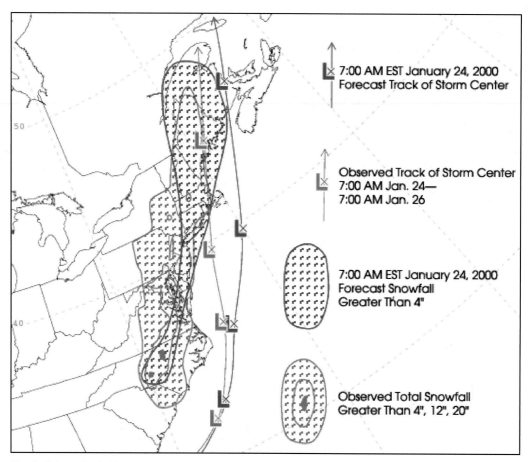

FIG. 2. Schematic of difference between observed and forecast behavior of the 24–25 Jan 2000 surprise snowstorm. The actual storm tracked closer to the coastline than predicted, depositing unexpected heavy snow.

onstrating and "selling" the concepts and operational applications of EPS.

At that time, as global model ensemble systems were coming to the fore at operational centers, there was much skepticism that ensembles had any role in short-range, regional model–based predictions. This appeared mostly to reflect the perception, from popular reference to the "two week" limit, that predictability considerations were significant only for larger scales at medium and longer ranges. The barriers on the predictability of smaller scales at short ranges were rationalized as being only the central processing unit (CPU) constraints on increasing model resolution. As highlighted at a workshop held at the National Centers for Environmental Prediction in 1994 (Brooks et al. 1995), however, it was not long before the concepts of ensemble prediction were acknowledged widely as equally relevant to mesoscale systems at short ranges. With this came the recognition that issues entailed with developing regional model–based short-range ensemble forecasting (SREF) systems and strategies (such as methodology for generating initial and/or model physics perturbations, dealing with boundary conditions, and competition for peo-

ple and CPU resources) were considerably more challenging than for the global models.

To begin exploring the challenges and prospects for SREF, the National Centers for Environmental Prediction (NCEP) launched a pilot short-range ensemble project in 1996 along the lines recommended in the aforementioned workshop. The details and encouraging results from this study were documented by Stensrud et al. (1999) and Hamill and Colucci (1997). Stensrud et al. show an especially relevant example of the significant diversity that can occur between ensemble members in a case of cyclogenesis along the East Coast (Fig. 1). More generally, the key findings from this pilot study and subsequent investigations at NCEP showed that enhanced diversity of solutions (spread) was obtained with a multimodel ensemble [Eta Model and Regional Spectral Model (RSM)],[2] higher resolution (48 versus 80 km), and regional model–based rather than interpolated global model "bred" perturbations (Toth and Kalnay 1993) to initial conditions. Most importantly, these studies, together with results from investigations such as

---

[2] Eta Model (Black 1994) and RSM (Juang and Kanamitsu 1994).

FIG. 3. Examples of media reaction to the surprise snowstorm. Cartoon by Tom Oliphant has an NWS person digging out of a snowdrift saying, "... with our new equipment, our supercomputer models and enhanced programming, we will now be able to be wrong much quicker."

that of Du et al. (1997) and Mullen et al. (1999), illustrated the practical significance of uncertainties in short-range regional model–based predictions, demonstrated the potential of SREF to provide operationally useful information, and provided a basis for a prototype operational SREF system at NCEP.

It was not until June 2001, however, that a prototype system intended for operational acceptance began running in a "real time test and evaluation" mode (RTT&E) at NCEP. It consisted of 10 members running from both 0900 and 2100 UTC for 63 h.[3] It was composed of a control (unperturbed) Eta Model forecast plus four Eta perturbed runs and a control RSM run plus four perturbed RSM forecasts. All forecasts were with 48-km resolution versions of the models over the operational Eta domain. Lateral boundary conditions were provided by the 9-h forecasts from the respective members of the NCEP global model ensemble. Initial-state perturbations

were generated by independent "breeding" cycles, as for the global ensembles, but in the context of the respective regional models. Each member of the Eta ensemble was with the same version of the model, and likewise for the RSM (i.e., no physics diversity within each model set).

Prior to this, essentially the same SREF system began running routinely at NCEP in an experimental, low-profile, resource-deprived mode and offline from the main thrust driving model development efforts (increasing model resolution). The issue alluded to here is whether an ensemble, which necessarily must employ lower-resolution versions of models, can provide more useful information than a single high-resolution deterministic run.[4] The "tipping point" leading to a consensus that this might very well be true was a set of winter storm case studies described below. The cases were largely responsible for the acceleration of SREF development that led to the RTT&E and ultimately to the

---

[3] In the context of the overall NCEP computer suite the initial times of 0900 and 2100 UTC were (and remain) the only approach that enable completion of runs about the same time as the operational Eta model from 1200 and 0000 UTC becomes available.

[4] See Roebber et al. (2004) for an informative exposition of the tradeoffs involved.

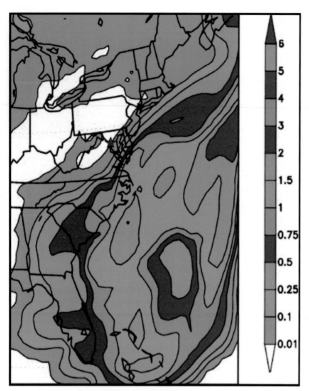

FIG. 4. Operational Eta from 1200 UTC 24 Jan, 24-h accumulated QPF (in.) for the period ending 1200 UTC 25 Jan.

inclusion of SREF as an integral component of the NCEP production suite.

4. Snowstorm case studies

Two significant snowstorm cases along the East Coast are described with special attention focused parochially on events in the Washington, D.C., vicinity. The first is the "surprise" heavy snowfall associated with the cyclogenesis along the East Coast on 24–26 January 2000. Second is the so-called end of millennium storm that deposited copious amounts of snow over the Northeast on 29–30 December 2000, but not at Washington, D.C.—a surprise no big one. There are several references dealing with various aspects of the development and predictability of these storms (e.g., Zhang et al. 2002; Wasula and Bosart 2004). Except where especially relevant, the intent here is not to review or relate these investigations to what follows. Rather, it is to illustrate with these cases, which are not necessarily exceptional, the sort of uncertainties that can arise in forecasting major snow events. At issue is whether in retrospect these events need have been a complete surprise, at least in terms of model guidance. In addressing this question—to which the answer will come as no surprise—what follows demonstrates the prospective value of SREF in identifying uncertainties and the importance of accounting for them in the forecast process.

a. January 2000 surprise snowstorm

The storm of 25–26 January 2000 deposited a blanket of heavy snow from North Carolina northward to New England and eastern New York. As much as 50.8 cm (20 in.) were recorded in sections of North Carolina, and up to 30.5–35.6 cm (12–14 in.) in the metropolitan Washington, D.C. area, between the early morning hours and the evening of Tuesday 25 January. The storm's notoriety arises from the fact that heavy snow was unpredicted in D.C. until about 6 h before beginning to accumulate. This mostly reflected the fact that *all* operational forecast models routinely available at the time gave little, if any, clue to the imminence of this major weather event as late as the runs from Monday, 1200 UTC 24 January. (An overview of the storm and discussion on the short- and medium-range model guidance are available at http://www.emc.ncep.noaa.gov/mmb/research/blizz2000/.) For the most part the models underpredicted storm intensity and forecasted the track and accompanying precipitation shield too far east (Fig. 2). The official National Weather Service (NWS) forecast Monday morning (10 A.M.) for Tuesday was "mostly cloudy and cold." By the afternoon of the 24th, later model guidance becoming available reinforced an earlier trend to a somewhat farther westerly storm track. Together with potentially "ominous" signals seen on radar and satellite imagery, it was clear that D.C. was in for at least a close call. Nevertheless, since all models insisted on keeping the main area of precipitation to the east, the late afternoon (4 P.M.) NWS forecast for Tuesday called only for "cloudy and cold. Chance of snow . . . accumulation of an inch possible." Likewise, the local early evening broadcast meteorologists unanimously and categorically dismissed the possibility of a big snowstorm for the Washington metropolitan area. It was not until 10 P.M. that an official "winter storm warning" was issued with the forecast now calling for 4–8-inch accumulations. Unfortunately many folks, including emergency managers and public school officials, retired before then and missed the media beginning to disseminate the warning. Suffice it to say when these same folks arose the following morning to unexpected near-blizzard conditions, consternation was in abundance (Fig. 3), to say nothing of the dismay and embarrassment on the part of the NWS and local television meteorologists.[5]

To assess whether SREF might have provided useful information in the context of the operational scenario just described, an experimental 10-member Eta Model ensemble was run retrospectively from 1200 UTC 24 January. The version of the Eta used was essentially the

---

[5] Especially given the recent announcement by NWS Headquarters of the transition to the latest supercomputer with pronouncement of the new "no surprise" weather service. Washington Post columnist Tony Kornheiser wrote, "Models? Next time, read pig entrails."

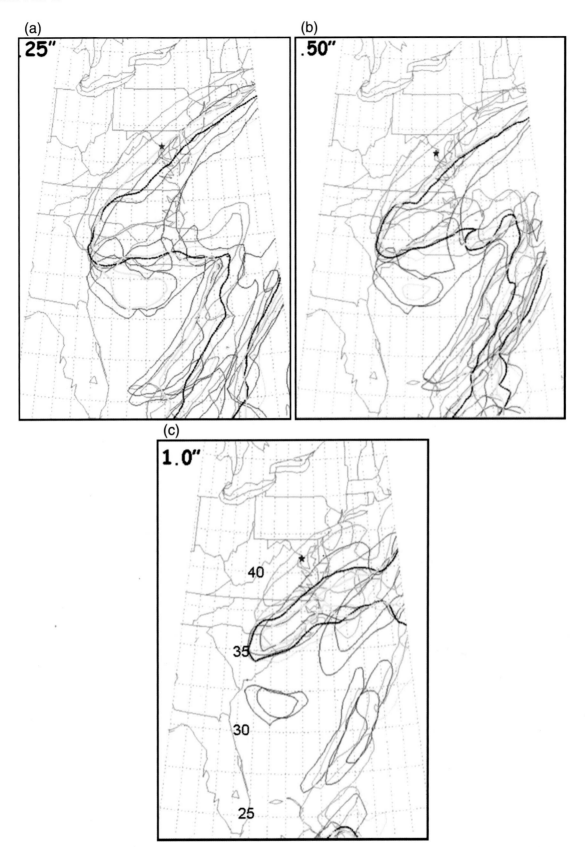

FIG. 5. The 24-h Eta ensemble from 1200 UTC 24 Jan 2000: spaghetti diagrams of 12-h (a) 0.6-, (b) 1.3-, and (c) 2.5-cm (0.25, 0.50, and 1.0 in., respectively) QPF for period ending 1200 UTC 25 Jan. The location of Washington, D.C., identified by red symbol.

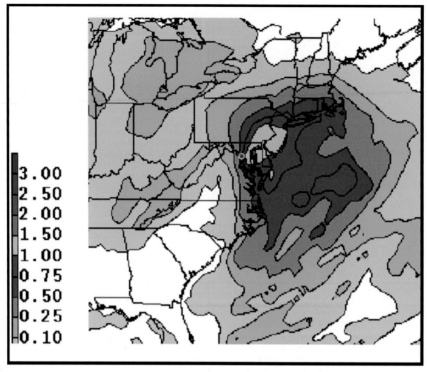

FIG. 6. Operational Eta from 1200 UTC 29 Dec 2000, 24-h accumulated QPF (in.) for the period ending 1200 UTC 30 Dec.

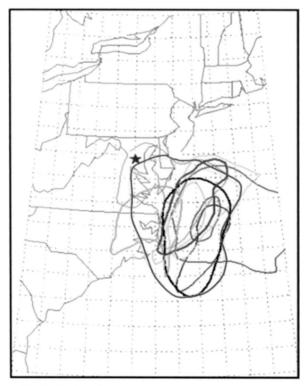

FIG. 7. The 24-h Eta ensemble from 1200 UTC 29 Dec 2000; spaghetti diagrams of 24-h accumulated QPF (in.).

same as the operational model at the time, except with 48- rather than 32-km horizontal resolution. A 10-member, Eta–RSM multimodel ensemble was also run. Qualitatively, results from the two versions of SREF produced the same "screaming message," namely, a clear "heads up" on Monday morning of the distinct possibility of a major snow event for D.C. the following day. For clarity in comparisons with the more familiar Eta Model, the SREF output discussed below is only from the Eta member ensemble. In this single model ensemble only perturbations to initial conditions are pertinent.

Figure 4 displays the operational 24-h accumulated Eta quantitative precipitation forecast (QPF) from 1200 UTC 24 January valid at 1200 UTC 25 January. Figure 5 shows the corresponding SREF spaghetti diagrams for 0.6, 1.3, and 2.5 cm (0.25, 0.50, and 1.0 in.). For this storm the thermodynamic structure in all predictions was sufficiently cold such that the possibility of the precipitation being anything but snow in the D.C. vicinity was nil.

It can be seen in Fig. 4 that the operational Eta, as true of *all* models nominally available at NCEP Monday morning, kept significant precipitation to the south and east of D.C. as the forecast low tracked farther east than verified—admittedly a close call, but as a famous coach of the Boston Celtics once said, "close, but no cigar." As noted above, this was the universal "buy-in" until

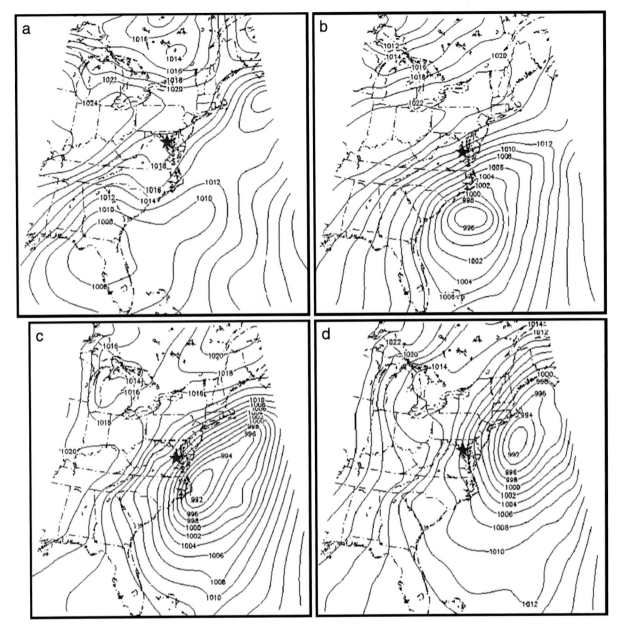

FIG. 8. Sequence of (a) 0-, (b) 12-, (c) 24-, and (d) 36-h MSLP charts for control of Eta ensemble from 1200 UTC 24 Jan.

late in the day by NWS forecasts and local media outlets. On the other hand the SREF (Fig. 5) indicates four members with greater than or close to 0.25 in. melted over D.C., two members with greater than 0.5 in., and those same two members indicating 1.0 in. or more. Given the extreme cold, the snow to melted ratio was likely greater than the nominal 10 to 1 standard. Hence, the SREF clearly indicated a demonstrable *possibility* of a heavy snow event in the Washington area. Moreover, unlike the case with the operational models available at the time, this possibility was consistent with surface data, radar, and satellite imagery, which all suggested by Monday morning, the 24th, that trouble was

brewing (Bosart 2003).[6] If this information had been available in real time, along with sufficient forecaster experience to establish reasonable assurance of SREF reliability, it is easy to speculate that a heads up of some

----

[6] In an interview with "Earth Observatory," the Director of NCEP, Louis Uccellini, is quoted as follows: "Uccellini admits that there was some uncertainty in the minds of meteorologists as to what path the storm would take. He says the NWS could have emphasized that uncertainty more in its forecasts. Forecasters know the models aren't perfect. There should be some way to present alternate scenarios without following the mainstream (models) and creating a panic (i.e., 'crying wolf')." Sentiments of this sort relate to the impetus to develop SREF at NCEP, as mentioned in section 3.

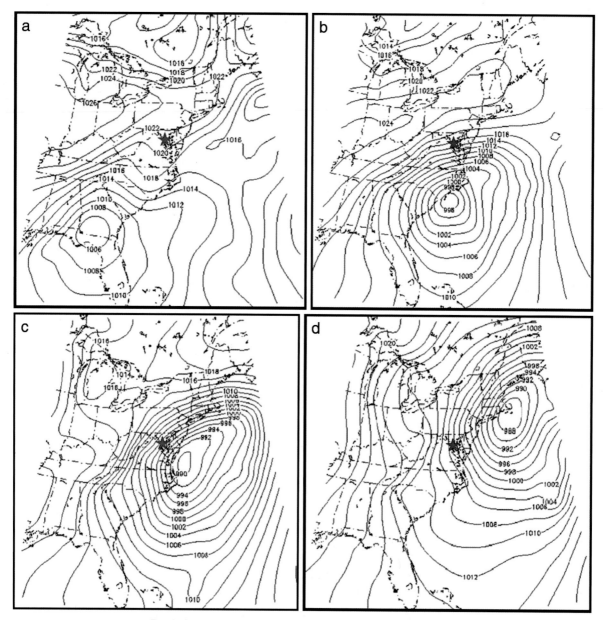

FIG. 9. Same as Fig. 8, but for the best member of the Eta ensemble.

sort on this possibility might have been issued Monday morning rather than what amounted simply to a categorical "no" to that chance.

b. 29–30 December 2000 "end of millennium" snowstorm

Very heavy snow fell over much of the Northeast with over 2 feet dumped on northern New Jersey associated with a surface low that developed off of the Carolina coast the evening of 29 December. But amounts dropped off dramatically just to the west of the areas that received the heaviest accumulations. As late as the 1200 UTC run from 29 Friday December, the Eta pre-

dicted heavy snowfall (~1.0 in. melted) to occur in the Washington–Baltimore corridor by early Saturday morning (Fig. 6). As it turned out, the predicted 24-h forecast precipitation shield was too large and too far west and south. The later arriving MRF run predicted a lighter accumulation (not shown), and the NWS official forecast was a compromise—a winter storm warning across the D.C. area, with predictions of 7.6–15.2 cm (3–6 in.) in Washington and 12.7–25.4 cm (5–10 in.) in Baltimore. That warning was kept through the nominal 10 P.M. Friday evening update. In reality, Washington and Baltimore woke up on Saturday morning with sunny skies and news reports of the storm that missed the local area but was burying areas to the north-

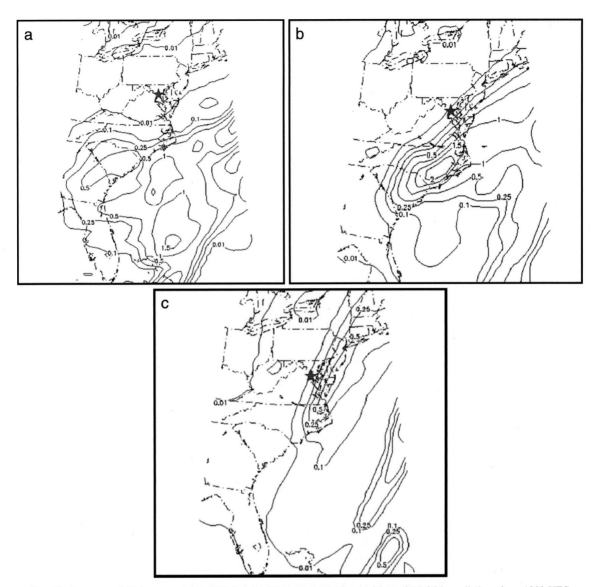

FIG. 10. Sequence of 12-h accumulated QPF (in.) for Eta control (a) 12-, (b) 24-, and (c) 36-h predictions from 1200 UTC 24 Jan.

east. Suffice it to say it was another embarrassing "bad hair day" for the large cadre of local meteorologists, professional and otherwise—to say nothing about the extreme disappointment of the "snow crows" like myself. The surprise here was like that of the January case, except in reverse—a surprise no big one.

The same 10-member Eta ensemble used in the January case was run retrospectively from 1200 UTC 29 December. Figure 7 displays the SREF 24-h QPF 0.50-in. spaghetti diagrams of 24-h accumulated precipitation valid at 1200 UTC 30 December. The figure serves to illustrate that the ensemble at most indicated a 10%–20% chance of significant snowfall overnight or, equivalently, indicated an 80%–90% chance of a much less substantial event. With the same caveats of forecaster experience mentioned above, it is reasonable to spec-

ulate that had this SREF been available operationally it might have prompted backing off much earlier from the parlayed certainty of a major snow event in the Washington area, and avoided having to suffer pangs of a high-visibility false alarm.[7]

To summarize, in the face of the respective deterministic operational model predictions to the contrary, the ensemble runs sent a clear signal for the possibility of heavy snow in the Washington region in the January case and for the possibility of something other than a substantial snowstorm in the December case. As noted by the duly impressed director of NCEP (Louis Uccel-

_____

[7] The CNN headline story on Saturday morning proclaimed, "The storm unexpectedly spared Washington . . . forecasters said."

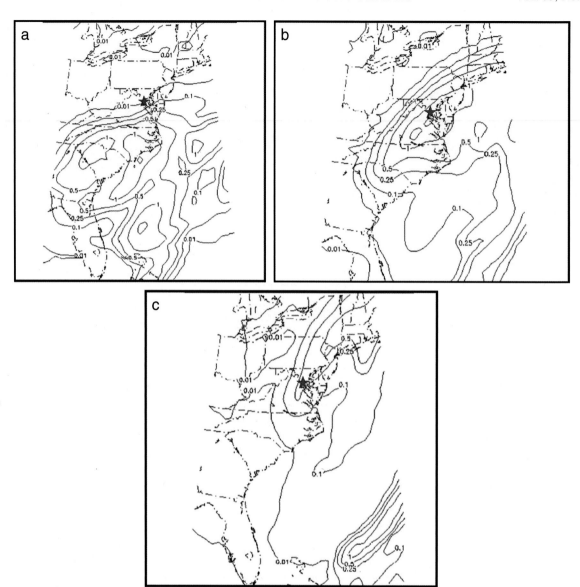

FIG. 11. Same as Fig. 10, but for best member of Eta ensemble.

lini) on several occasions, the bottom line is "deterministic forecasting is not healthy!" [8]

5. So, why the uncertainty?

It is well known that small errors (i.e., uncertainties) in the initial conditions of *any* NWP model can amplify significantly over time, thereby leading to the divergence of solutions. In the cases above this is reflected by the array of possibilities shown with the QPF spaghetti diagrams. This is explored further by examining differences for the January 2000 case between the un-

_____

[8] As per slide presentations at several meetings and workshops, but for which there apparently is not any citable reference or Web site.

perturbed "control" (C), which is most comparable to the operational Eta, and the "best" (B) member of the Eta ensemble run from 1200 UTC 24 January. Best here refers to the member giving the strongest signal of the impending snowstorm. Figures 8 and 9, respectively, display C and B for the 0-, 12-, 24-, and 36-h mean sea level pressure (MSLP) charts. The corresponding 12-h accumulated 12-, 24-, and 36-h QPF appears in Figs. 10 and 11. At the initial time, the low center in the control is somewhat deeper and farther north over the Florida and Georgia border region, but otherwise differences in MSLP are rather innocuous. The 500-mb height fields are likewise nearly indistinguishable (not shown). It is doubtful that, all else being equal, one would conclude from visual inspection of these charts

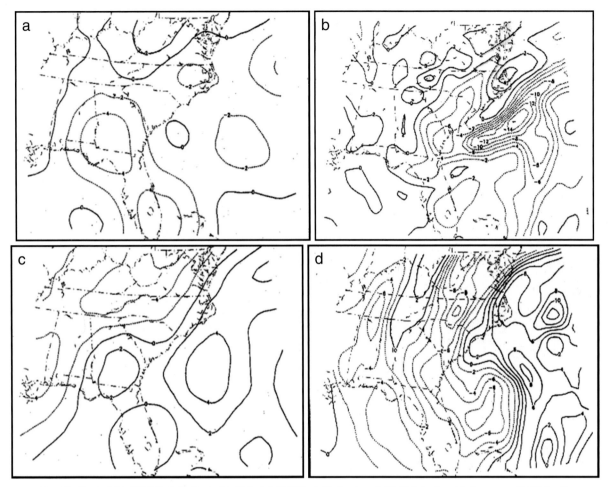

FIG. 12. Control (a) 850-hPa initialized, (b) 12-h, (c) 250-hPa initialized, and (d) 12-h horizontal divergence ($\times 10^{-5}$ s^{-1}).

that the C and B forecasts would be so significantly different. Of most consequence is that B intensifies as it tracks closer to the coastline with its shield of heavy precipitation clearly overspreading D.C. and its environs.

Particularly interesting is that further evaluation of the initial-state perturbations (B minus C) and differences between forecasts thereafter are consistent with dynamically reasonable expectations. Thus, for example, the surface system in B was associated with a more developmental vertical wind profile and low-level thermodynamic structure. To illustrate, Figs. 12 and 13, respectively, show the C and B 0- and 12-h fields of horizontal divergence of the 850- and 250-hPa winds. Of note is the greater intensification of the low-level convergence over the Southeast in B and, especially, the more concentrated band of convergence along the Georgia coastline. The 250-mb upper-level divergence is likewise more intense, but with less structural detail. The low-level convergence, in turn, is associated with increased instability and precipitation (largely convective) along the coast (Fig. 14) and is presumably coupled

to the intensifying storm in B and its track closer to the coast. This result parallels the Zhang et al. (2003) investigation of this storm, which found that moist processes were integral in the strong initial condition sensitivity of 1–2-day forecasts. Also, it can be noted that the results here illustrate that perturbations to initial conditions generated with the breeding can appropriately represent dynamically significant and rapidly growing modes of analysis uncertainties in the context of regional as well as global model ensemble systems.

6. Discussion

The nature and degree of uncertainties in the two illustrative cases are undeniably model- and ensemble system–dependent. Of import is that, given the multitude of possible nonlinear interactive atmospheric processes and interactions between the atmosphere and lower boundary, uncertainties are inevitable and unavoidable regardless of system. Advanced observational platforms and strategies, improved data assimilation methodologies, higher-resolution models, and more so-

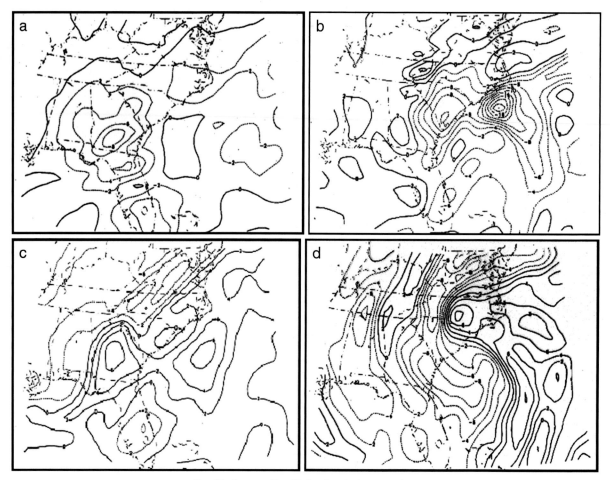

FIG. 13. Same as Fig. 12, but for the best member.

phisticated physical parameterizations will contribute to reducing analysis and forecast model errors, but can never eliminate them. With this fundamental fact of life—uncertainty is the only thing that is certain—ensemble systems offer the only realistic approach for providing case-dependent information on uncertainty levels, as for example in the timing, intensity, and track of storms and related precipitation amounts and type.

However, it is important to recognize that just as forecast models can never be perfect, ensemble systems and strategies will inevitably be less than totally reliable in providing uncertainty information. This brings us finally to an answer to the question posed by the title of this paper, "Must surprise snowstorms be a surprise?" Clearly the answer is "no," but some surprises, even at short ranges (<2 days), are unavoidable! In the extreme, for example, most (if not all) ensemble members might point to a given scenario or set of scenarios (e.g., snowstorm) that proved ultimately not to include the actual outcome (no snowstorm).

It is undeniable that the prototype SREF used in these experiments essentially was not much beyond a "starter system" phase relative to the many outstanding scien-

tific, technical, and practical challenges of developing SREF systems, strategies, and applications. Nevertheless, the case studies described above, as well as additional cases and support from available verification statistics, provided a high-visibility demonstration of the potential value of SREF operationally in winter storm scenarios. This provided the impetus for accelerating SREF development and evaluation, including it becoming a major component in the NWS Winter Weather Experiment (see http://www.hpc.ncep.noaa.gov/html/ WWE_reports/wwe2hpcfinalreport.pdf) aimed at improving public winter weather services. These activities led first to the RTT&E mentioned above and ultimately to the first (and, at this writing, still only) operational regional model–based ensemble prediction system running at a major NWP center. Information on the latest version of SREF along with an impressive array of general and specialized graphical products, including guidance on winter weather forecast scenarios, can be found online (see http://www.emc.ncep.noaa.gov/mmb/SREF/ SREF.html). At NCEP, these and some additional products are available for operational use on the NCEP Advanced Weather Interactive Processing System (NA-

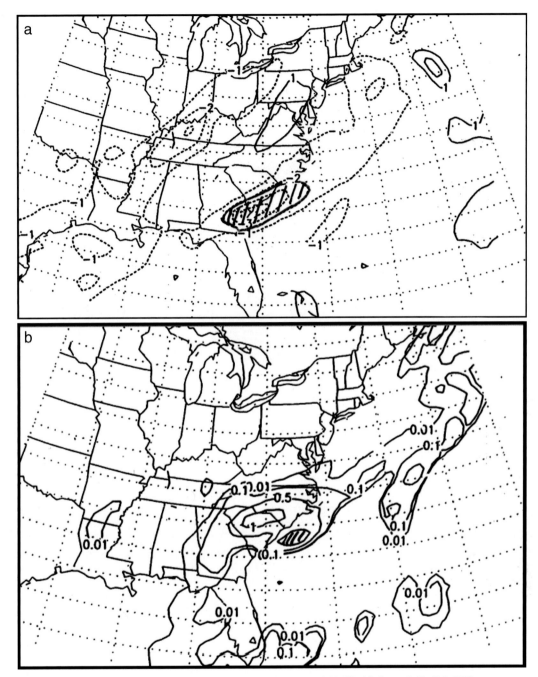

FIG. 14. Difference (best minus control) 12-h forecasts of (a) lifted index and (b) 12-h QPF.

WIPS). Elsewhere, other NWP centers are striving to get on the "wave" of regional model ensemble prediction systems. Perhaps most notable are the efforts underway at the Met Office and development of the U.S. Air Force and Navy Joint Ensemble Forecast System (JEFS). To be noted also is the University of Washington Mesoscale Ensemble (Grimit and Mass 2002) now running routinely and available for use by the Seattle NWS forecast office.

A critical issue was and remains how best to convey information on the nature and degree of uncertainties to prospective users. That in turn depends upon the specific needs and requirements of users, which varies from the general interest of the "person on the street" to more sophisticated customers who can benefit from cost versus loss type considerations in regard to application specific thresholds. In the former case a simple qualitative estimate in the degree of confidence (e.g., very high to very low) applied to the most likely outcome might suffice. To this might be added information on alter-

FIG. 15. Tom Toles cartoon from the 2 Mar 2005 edition of the Washington Post.

native scenarios, for example, the ramifications of a possible shift in storm track. For more sophisticated users and applications, providing quantitative estimates of the full probability distribution of some parameter is likely the most useful approach.

In the two snowstorm cases, whether it does or does not snow might be irrelevant to some individuals, perhaps dedicated "couch potatoes" who are going to stay put no matter what. For others, simply recognizing qualitatively that the snow forecast is "in the bag" versus just an outside but nonzero chance could enable a more informed decision, for example, of whether to stock up on supplies of bread, milk, and toilet paper. Not that many years ago (~25 yr) it is probably safe to say that these folks paid little attention to snow or no-snow forecasts or how the forecasts verified because it was "common knowledge" that the forecasts would most likely be wrong. Over the past several years, however, public expectations have been raised (justifiably) by generally improved and more reliable forecasts, so much so that when a strictly yes or no snow forecast does not verify one can almost hear a collective sigh of near or actual embitterment—too little bread, milk, and toilet paper (e.g., January 2000 case) or unneeded supplies of the same (e.g., December case).

In a more serious vein, a quantitative probability forecast of, let us say more than 6 in. of accumulating snow, is potentially useful information that emergency managers could use in deciding whether to stage snowplows in anticipation of a storm. Ultimately, a decision of this sort depends on the "threshold of pain" in the willingness to absorb the cost of staging plows unnecessarily, even in high-probability scenarios, or suffering the consequences of not being prepared should the snow occur, even when the chance is relatively small.[9]

7. Concluding remarks

There can be little doubt that ensemble prediction will continue to become a dominant theme in operational NWP. Its promise and potential has been firmly established with prospective applications wide and varied. This was true first in the context of medium–extended range global models and more recently with regard to regional models for shorter-range, mesoscale prediction. In the construct of NCEP's "Seamless Suite of Products" (see http://www.ncep.noaa.gov/director/), global and regional model ensembles permit estimates of forecast confidence/possibilities of specific weather events, first in the context of the requisite larger-scale circulation pattern at longer ranges and then in the details of the relevant weather system in the short range.

Perhaps most importantly, ensembles arguably have or are becoming largely recognized as indispensable by

---

[9] In Washington, D.C., it might not have made a difference in light of a former mayor's response to a question of the city's snow removal plan, namely, "spring!" Or, as in a cartoon by Tom Toles (reproduced in Fig. 15) depicting consternation of what little snow, or even just the threat of snow, might have on Washington.

operational forecasters. This coming of age, as it justifiably might be referred to, reflects a significant break with the long-held paradigm of "deterministic thinking" where the focus is the "model of the day" and single best forecast. In contrast, ensembles enable forecasters, *in principle,* to provide science-based, case-dependent and user-specific information on uncertainties extant in *all* forecasts to some degree. The emphasis on "in principle" refers to the fact that many distributed products by the NWS remain deterministic, single valued in nature, regardless of forecaster insight on distinguishing between, for example, high versus low confidence in a potential big one. Nevertheless, there is reason to believe this will change over time as ensemble systems and strategies continue to improve and users of weather information become increasingly more cognizant of the value of having uncertainty information as a fundamental component of the products they rely upon. The only question likely is the pace at which this occurs.

Viewed narrowly and with some stretch of imagination, it might be argued that the provision and use of ensembles emerged largely from the "butterfly" of intellectual seeds planted by Fred Sanders in a former student and that student's epiphany of adopting Fred's "continuity chart" concept (a.k.a. spaghetti diagram), which in turn proved integral to the initial success in "selling" operational ensemble prediction.

Acknowledgments. The development, testing, and operational implementation of SREF at NCEP would not have been possible without the dedication, perseverance, and know-how of the relevant S&T by Dr. Jun Du.

REFERENCES

Black, T., 1994: The new NMC mesoscale Eta model: Description and forecast examples. *Wea. Forecasting,* **9,** 265–278.

Bosart, L., 2003: Whither the weather analysis and forecasting process? *Wea. Forecasting,* **18,** 520–529.

Brooks, H. E., M. S. Tracton, D. J. Stensrud, G. DiMego, and Z. Toth, 1995: Short-range ensemble forecasting: Report from a workshop, 25–27 July 1994. *Bull. Amer. Meteor. Soc.,* **76,** 1617–1624.

Du, J., S. Mullen, and F. Sanders, 1997: Short-range ensemble forecasting of quantitative precipitation. *Mon. Wea. Rev.,* **125,** 2427–2459.

Eckel, F., and C. Mass, 2005: Aspects of effective mesoscale, short-range ensemble forecasting. *Wea. Forecasting,* **20,** 328–350.

Grimit, E. P., and C. F. Mass, 2002: Initial results of a mesoscale short-range ensemble forecasting system over the Pacific Northwest. *Wea. Forecasting,* **17,** 192–205.

Hamill, T. M., and S. J. Colucci, 1997: Verification of Eta–RSM short-range ensemble forecasts. *Mon. Wea. Rev.,* **125,** 1312–1327.

Juang, H.-M. H., and M. Kanamitsu, 1994: The NMC nested regional spectral model. *Mon. Wea. Rev.,* **122,** 3–26.

Lewis, J. M., 2005: Roots of ensemble forecasting. *Mon. Wea. Rev.,* **133,** 1865–1885.

Mullen, S. L., J. Du, and F. Sanders, 1999: The dependence of ensemble dispersion on analysis–forecast systems: Implications to short-range ensemble forecasting of precipitation. *Mon. Wea. Rev.,* **127,** 1674–1686.

Mylne, K. R., 2002: Decision-making from probability forecasts based on forecast value. *Meteor. Appl.,* **9,** 307–315.

Richardson, D. S., 2000: Skill and relative economic value of the ECMWF ensemble prediction system. *Quart. J. Roy. Meteor. Soc.,* **126,** 649–667.

Roebber, P. J., D. M. Schultz, B. A. Colle, and D. J. Stensrud, 2004: The risks and rewards of high resolution and ensemble numerical weather prediction. *Wea. Forecasting,* **19,** 936–949.

Sanders, F., 1971: Analytic solutions of the non-linear omega and vorticity equation for a structurally simple model of disturbances in the baroclinic westerlies. *Mon. Wea. Rev.,* **99,** 393–407.

——, 1988: Life history of mobile troughs in the upper westerlies. *Mon. Wea. Rev.,* **116,** 2629–2648.

Stensrud, D. J., H. E. Brooks, J. Du, M. S. Tracton, and E. Rogers, 1999: Using ensembles for short-range forecasting. *Mon. Wea. Rev.,* **127,** 433–446.

Toth, Z., and E. Kalnay, 1993: Ensemble forecasting at NMC: The generation of perturbations. *Bull. Amer. Meteor. Soc.,* **74,** 2317–2330.

Tracton, M. S., and E. Kalnay, 1993: Ensemble forecasting at NMC: Practical aspects. *Wea. Forecasting,* **8,** 379–398.

Wasula, T. A., A. C. Wasula, and F. L. Bosart, 2004: A multi-scale analysis of the end of the millennium snowstorm. Preprints, *20th Conf. on Weather Analysis and Forecasting,* Seattle, WA, Amer. Meteor. Soc., P1.30.

Zhang, F., C. Snyder, and R. Rotunno, 2002: Mesoscale predictability of the "surprise" snowstorm of 24–25 January 2000. *Mon. Wea. Rev.,* **130,** 1617–1632.

——, ——, ——, 2003: Effects of moist convection on mesoscale predictability. *J. Atmos. Sci.,* **60,** 1173–1185.

Chapter 14

Fred Sanders' Roles in the Transformation of Synoptic Meteorology, the Study of Rapid Cyclogenesis, the Prediction of Marine Cyclones, and the Forecast of New York City's "Big Snow" of December 1947

LOUIS W. UCCELLINI

National Centers for Environmental Prediction, Camp Springs, Maryland

PAUL J. KOCIN

Hydrometeorological Prediction Center, National Centers for Environmental Prediction, Camp Springs, Maryland

JOSEPH SIENKIEWICZ AND ROBERT KISTLER

Ocean Prediction Center, National Centers for Environmental Prediction, Camp Springs, Maryland

MICHAEL BAKER

Meteorological Development Laboratory, Silver Spring, Maryland

(Manuscript received 3 November 2005, in final form 1 November 2006)

Uccellini   **Kocin**

Sienkiewicz   **Kistler**   **Baker**

Corresponding author address: Dr. Louis Uccellini, NOAA/NWS/NCEP, 5200 Auth Road, Camp Springs, MD 20726.
E-mail: louis.uccellini@noaa.gov

ABSTRACT

Fred Sanders' career extended over 55 yr, touching upon many of the revolutionary transformations in the field of meteorology during that period. In this paper, his contributions to the transformation of synoptic meteorology, his research into the nature of explosive cyclogenesis, and related advances in the ability to predict these storms are reviewed. In addition to this review, the current status of forecasting oceanic cyclones 4.5 days in advance is presented, illustrating the progress that has been made and the challenges that persist, especially for forecasting those extreme extratropical cyclones that are marked by surface wind speeds exceeding hurricane force. Last, Fred Sanders' participation in a forecast for the historic 1947 snowstorm (that produced snowfall amounts in the New York City area that set records at that time) is reviewed along with an attempt to use today's operational global model to simulate this storm using data that were available at the time. The study reveals the predictive limitations involved with this case based on the scarcity of upper-air data in 1947, while confirming Fred Sanders' forecasting skills when dealing with these types of major storm events, even as a young aviation forecaster at New York's LaGuardia Airport.

1. Introduction

Fred Sanders' career as a synoptic meteorologist spanned a remarkable period in meteorological history. He entered the field in the 1940s when observing the weather was based on surface observations, a pilot balloon (pibal) network, and an expanding post–World War II radiosonde network. Forecasting the weather was based mostly on these limited observations and was considered more of a subjective art that could only be practiced 24–36 h in advance, at best. Perhaps the spirit of forecasting in the 1940s was best captured by Petterssen who wrote the following: "Forecasting is exciting. The competitive instinct . . . swells and drives the forecaster into action . . . The challenge and the elation are largely lost in general-purpose forecasting . . . But the scientific challenge and the elation that stem from an exalted sense and purpose and usefulness are enormously enriched when you are called on the forecast for a specific and important operation." (Petterssen 2001, 132–133). Fred Sanders always made synoptic meteorology and forecasting "exciting" even as the profession was transformed into a mathematically based, applied science. Over a period of little more than half a century, he participated in the revolutionary changes in the forecast process that transformed meteorology into a science securely rooted in mathematics and physics.

Modern weather forecasts are now based on advanced technologies that utilize global observation systems, powerful computers, and sophisticated numerical models of the atmosphere, ocean, land, and hydrological processes. Furthermore, today's forecasters require a rigorous technical background in math and physics that can be applied to interpret and improve upon numerical forecasts. Today's weather services include those related to aviation, maritime operations, tropical cyclones, winter storms, flood situations, medium-range predictions, and fire weather. The net result is increasingly accurate, high-resolution weather forecasts for routine and extreme weather events that are now made operationally up to seven days in advance with documented skill (see e.g., Lalaurette et al. 2006).

With these advances, the role of the "synoptician"

has also changed. From the 1950s into the 1980s, a synoptic meteorologist was known as an analyst who strived to find "possible" relationships between dynamic and thermodynamic processes and the "associated" weather phenomena related to storm systems. The synoptic analysis process, through the careful and tedious study of many case studies, did more to establish a hypothesis than to study cause-and-effect relationships. These hypotheses were then used by more theoretical meteorologists to solve problems using simplified linear forms of the primitive equations based on quasigeostrophic assumptions (or more general balance constraints) that filtered out important mesoscale details related to the baroclinic nature of cyclones, fronts, and related weather events.

Beginning in the 1980s[1] and continuing today, synopticians have increasingly relied on sophisticated numerical models as a laboratory of the earth's system, to 1) dissect the sequential chain of events from the global scale to the mesoscale that mark notable weather events, 2) address cause-and-effect relationships of important dynamical and physical interactions that lead to the generation, evolution, and decay of cyclones, fronts, and their associated weather, and 3) provide feedback to the modeling community in order to continue improving the forecast skills of these models.

Fred Sanders played a key role in this dramatic transformation by placing himself in the synoptic world of analysis and forecasting in the late 1940s and 1950s, providing detailed analysis of fronts and cyclones throughout his career. He joined Dick Reed and Chester Newton (among others), who followed pioneers like R.C. Sutcliffe (see Hoskins 1999) and linked the theories and the new upper-air observation network to address synoptic-scale issues associated with cyclogenesis, fronts, and jet streams. He also challenged the assumptions employed by the more theoretical component of the meteorological community in the 1950s and 1960s in their research on baroclinic instability, while appearing to enjoy the conflicts that arose among those

---

[1] The use of numerical models to perform synoptic, diagnostic studies was first reviewed by Keyser and Uccellini (1987).

that promoted the theoretical applications and those that provided the real-world analyses of fronts and cyclones. Early on, Fred Sanders recognized the important contribution of numerical models to understand cyclone and frontal structures and how these advances contributed to improved weather forecasts. He also began a long-term effort to assess the strengths and weaknesses of these models, especially for improving forecasts of rapid developing cyclones and extreme weather events. In short, Fred Sanders acted as a referee for more than 50 yr between those in the research and modeling communities that devised the necessary assumptions, simplifications, and balance constraints to describe the basic workings of the atmosphere and those in the forecast community who dealt with the realities of synoptic and mesoscale weather processes that are so difficult to forecast on a day-to-day basis.

In section 2, a review of Fred Sanders' contributions to the study of rapid cyclogenesis is presented, with emphasis placed on "explosive" cyclogenesis over warm ocean currents and the inability of early models to predict these so-called "bombs." Recent advances in predicting rapid cyclogenesis are discussed in section 3, emphasizing statistics of four-day forecasts of cyclones over the Atlantic and Pacific Ocean basins and illustrating the remarkable progress made in predicting explosive oceanic cyclogenesis while also noting the remaining forecast challenges. Section 4 is devoted to a review and simulation of one such rapidly developing cyclone, the 26 December 1947 Northeast snowstorm that produced the heaviest 24-h snowfall in New York City's (NYC's) recorded history at the time. This storm was poorly forecast in 1947 but was not a complete surprise to one young meteorologist working as an aviation forecaster at New York's LaGuardia Airport. That forecaster's name was Fred Sanders.

2. Fred Sanders' contributions to the study of rapid cyclogenesis

The post–World War II period of the 1940s and 1950s was marked by an expanding global upper-air network that provided 12-hourly synoptic snapshots of the atmosphere. The principal goal of this worldwide effort was to improve weather forecasting, especially for the expanding commercial aviation industry (see, e.g., Petterssen 2001). The expanding network was also a boon for research and synoptic meteorologists who would use this data to revolutionize weather forecasting by transforming the forecast process from one based on subjective evaluations of analyses to a mathematical, objectively based process using numerical models initialized in real-time using the upper-air data provided by the radiosonde network (see, e.g., Uccellini et al. 1999).

The data were also a gold mine for the so-called "Chicago school" of meteorology directed by Erik Palmén in the early 1950s, which made significant contributions to the understanding of how dynamic, upper-level pro-

cesses influenced surface cyclogenesis, culminating in the classic book, *Atmospheric Circulation Systems,* written by Palmén and Newton (1969). The immediate impact of the global upper-air network was to provide supporting evidence for the importance of troughs and ridges aloft in the development and evolution of surface low and high pressure systems, as first proposed by Bjerknes and Holmboe (1944), and relate them to day-to-day forecast challenges. This period was also marked by the increasing use of "development equations" that applied vorticity and thermal advections to the surface pressure tendency and development rates of surface cyclones. The most recognized of these development equations was derived by Sutcliffe (see Hoskins 1999; Bosart 1999) and Petterssen (1956) and was used extensively by synoptic meteorologists from the 1950s through the 1970s.

A different approach was taken by theoretical meteorologists during the same period, with Charney (1947) and Eady (1949) developing baroclinic instability theories to differentiate flow regimes from which cyclones would either amplify or decay over time. More importantly, advances in dynamic meteorology, the emergence of new "high-speed" computers, combined with the expanding upper-air network, all provided an opportunity to apply mathematical concepts and initial-value numerical models to weather forecasting. Charney et al. (1950) derived a set of equations that ultimately led to the first successful application of a numerical model to the predictions of cyclogenesis (Uccellini et al. 1999; Phillips 2000).

It was during this exciting period that Fred Sanders launched his career applying development equations and two- and three-layer models to the study of cyclogenesis (see, e.g., Sanders et al. 1960). While Sanders and others used analytic approaches to document the importance of trough–ridge systems and the interconnection of dynamic and diabatic processes to surface cyclogenesis, his work also highlighted critical deficiencies in the baroclinic theories, and he increasingly focused on the deficiencies of the early barotropic (and then baroclinic) numerical models that were quickly being introduced into the day-to-day forecasts in the United States during the middle to late 1950s and into the 1960s (see Shuman 1989).

While early numerical models and applications based on baroclinic instability theories produced large-scale features associated with surface cyclones and upper-level troughs and ridges, these early applications were deficient in capturing mesoscale, weather-related details related to fronts and rapidly deepening cyclones, leading Fred Sanders to document the three-dimensional structure of the temperature gradients and related fronts that marked these extratropical storm systems, as described in other papers throughout this monograph. In this paper, we will focus on studies of cyclones and their deepening rates, especially those that developed explosively over the ocean.

a. Characteristics of rapid cyclogenesis

Early applications of baroclinic instability theory in the 1950s, use of a simplified system of "balanced" equations, and the application of two- and three-layer numerical models to day-to-day forecasts (Charney 1950, 1955; Eliassen 1956; among others) led to a realization that 1) the interplay between upper- and lower-level "baroclinic" systems, 2) their related temperature gradients and vertical wind shears, and 3) the amplification of trough–ridge systems and development of surface cyclones were critical factors for accurate model simulations and forecasts. Nevertheless, a major deficiency in the early theories also became readily apparent: the development rates for cyclones derived from the theoretical approach and model applications were too slow. The theory yielded slowly deepening cyclones over periods of greater than 60 h, while actual cyclogenesis occurred much more rapidly, usually within a 24-h period. Furthermore, cyclones derived from the theoretical and early models were marked by central pressures that were not deep enough when compared to actual cyclones. This deficiency in the baroclinic instability theory was highlighted by Eliassen (1956) and many others since.

The inability of these early operational numerical models to predict rapid cyclogenesis, and to overpredict the number of storms, was a major concern among model pioneers. Shuman (1989, p. 287) noted that forecasters could easily outperform the operational models introduced in the 1960s. The inability to predict rapid cyclogenesis over the oceans and "secondary" cyclogenesis off the east coast of the United States continued to frustrate many forecasters and modelers well into the 1970s and 1980s. This situation led Ramage (1976) to call for an end to the application of numerical models for cyclone prediction and urged the meteorological community to "abandon research that uses weather sequences in a computer as bases for deduction about the real atmosphere" (p. 9).

While others were frustrated, Sanders was intrigued by the problems associated with rapid cyclogenesis. Therefore, he began studying the atmospheric processes that contributed to the rapid development rates, and along with an increasing number of research meteorologists, uncovered those deficiencies that plagued the operational models, with the intent of enabling numerical models to produce reliable operational forecasts of extreme events. The related debates of the relative importance of diabatic and dynamic processes documented in the literature focused on cyclones, cyclogenesis, and our inability to predict these storms (see Uccellini 1990).

Sanders' and his students' efforts to document and explain rapid oceanic cyclogenesis are best contained in four papers: Sanders and Gyakum (1980), Roebber (1984), and Sanders (1986a,b). The first paper (Sanders and Gyakum 1980) is a synoptic–dynamic climatology of storms characterized as explosive cyclogenesis, or bombs. The bomb criterion was derived from a 3-yr sample of oceanic cyclogenesis as an average 1 hPa h^{-1} decrease in the central pressure of a surface low pressure system for a 24-h period, normalized to 60° latitude. As shown in Figs. 1 and 2, these storms are most often found over the ocean, especially along the sea surface temperature (SST) gradients associated with the Kuroshio and Gulf Stream. Furthermore, the four papers clearly show that development rates based on the quasigeostrophic approach were far less than observed and noted the almost total inability of the operational models to predict these storms during the 1980s.

Roebber (1984) provided the distribution of the 24-h deepening rates for rapidly developing marine cyclones (Fig. 3) and illustrated a class of cyclogenesis with extreme development rates exceeding 20 to 40 hPa (24 h)$^{-1}$. A classic illustration of an explosively developing oceanic storm is provided by Reed and Albright (1986) in which a cyclone on 12–14 November 1981 underwent explosive deepening in the eastern Pacific Ocean, deepening 38 hPa in only 12 h (Fig. 4). This storm was completely misforecast by the operational numerical models of that time. Other famous cases of explosive cyclogenesis that engaged the research communities in the 1980s and eluded any accurate forecasts, even 12 h in advance, included the February 1979 "Presidents' Day" storm (Bosart 1981; Bosart and Lin 1984; Uccellini et al. 1984, 1985; Whitaker et al. 1988) and the 9–11 September 1978 "QE II" storm (Anthes et al. 1983; Gyakum 1983a,b, 1991; Uccellini 1986; Manobianco et al. 1992).

The synoptic-climatological studies culminated with Sanders' (1986a,b) papers, which classified "strong," "moderate," and "weak" bombs. These rapidly developing cyclones displayed more of a northeastward motion (rather than eastward) than slower-deepening cyclones, and remained over warmer water for a longer period of time than their slower-developing counterparts. The rapid deepening occurs over 24–36 h, with the most rapid deepening exceeding 24 hPa (12 h)$^{-1}$, occurring generally as the surface low crosses the strong sea surface temperature gradients on the northern flank of the Gulf Stream. The analysis also pointed to a complex interaction among upper- and lower-tropospheric dynamic and thermodynamic processes by highlighting 1) the prominent absolute vorticity advection at 500 hPa related to preexisting troughs approaching from the northwest and located within 400 km of the developing surface low, 2) the existence of low-level warm-air advection in the planetary boundary layer influenced by the warm sea surface temperature and large heat fluxes immediately above the Kuroshio and Gulf Stream in the Pacific and Atlantic Oceans, respectively, and 3) the release of latent heat and possible role of deep convection often observed during explosive cyclogenesis (also emphasized by Tracton 1973). The net result was that Sanders and collaborators provided a description of the bomb as a fundamentally baroclinic system whose ini-

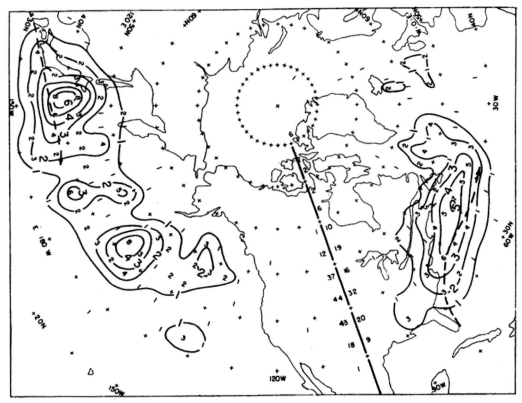

FIG. 1. Distribution of bomb events during three cold seasons. Raw nonzero frequencies appear in each 5° × 5° quadrilateral of latitude and longitude. Isopleths represent smoothed frequencies, obtained as one-eighth of the sum of 4 times the raw central frequency plus the sum of the surrounding raw frequencies. The column of numbers to the left and right of the heavy line along longitude 90°W represent, respectively, the normalized frequencies for each 5° latitude belt in the Pacific and Atlantic regions, using a normalization factor of [cos(42.5°)/cosφ]. Heavy dashed lines represent the mean winter position of the Kuroshio and the Gulf Stream. From Sanders and Gyakum (1980).

tiation is linked to preexisting upper-level troughs, related horizontal temperature gradients and vertical wind shear, and whose development rate is linked to the small effective static stability (especially in the middle to lower troposphere) and diabatic processes related to boundary layer heating and latent heat release. The debate(s) surrounding the relative importance of the various physical processes noted above in the development rates of marine cyclones are reviewed by Uccellini (1990).

b. Fred Sanders' contributions to unraveling the prediction problem

As noted earlier, Sanders worked as a referee between the theoretical, modeling, and synoptic forecast meteorological communities studying the interplay between dynamic and diabatic processes and their contribution to the explosive cyclogenesis observed over or near warm ocean currents. A series of cyclone workshops in the 1980s and 1990s, and the Genesis of Atlantic Lows Experiment (GALE; Dirks et al. 1988) and Experiment on Rapidly Intensifying Cyclones in the Atlantic (ERICA; Hadlock and Kreitzberg 1988) field programs all worked toward increasing the knowledge of rapidly de-

veloping cyclones, the improvement of the operational models, and accelerating advancements in the ability to predict rapid cyclogenesis. A unique aspect of Sanders' work was to relate research results directly to the deficiencies of operational numerical models and their subsequent inability to predict the explosive oceanic cyclogenesis. A series of "report cards" was published by Sanders (1986a,b, 1987, 1992). The papers noted the flaws of the operational models, emphasized the deficiencies, and captured the breakthroughs that marked the 1980s, and are particularly noteworthy for the attention that was brought first to the deficiencies of the models and eventually to the breakthroughs and advances in predicting the explosive nature of these storms (Uccellini et al. 1999).

Important advances in the operational numerical models did occur in the 1980s, advances that were spurred on by the results of the First Global Atmospheric Research Program (GARP) Global Experiment (FGGE) in 1978 and 1979 (see, e.g., Johnson 1986), improved computer capacity and speed, and the improvements in the global observing network, including satellite observations. At the National Meteorological Center (NMC), a Regional Analysis and Forecast System (RAFS) based

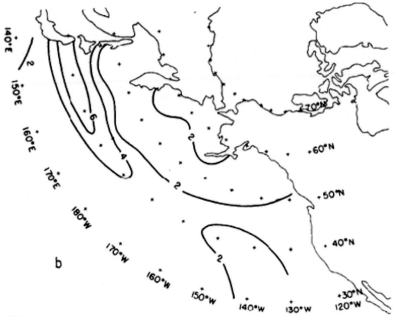

Pacific

Atlantic

FIG. 2. Analysis of the magnitudes of the local SST gradient [isopleths, °C (180 n mi)$^{-1}$] evaluated over a 180-n mi distance interval for 15 Jan 1979 in the Atlantic and Pacific basins. From Sanders and Gyakum (1980).

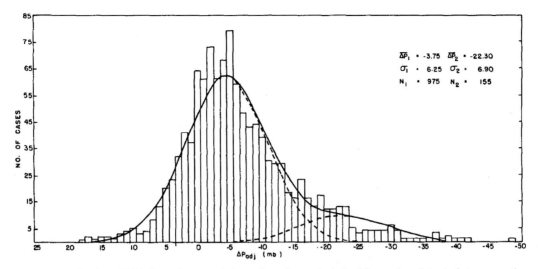

FIG. 3. Distribution of 24-h deepening rates for oceanic cyclogenesis derived from a 1-yr dataset of storms meeting the bomb criteria. The solid line indicates the sum of two normal curves, and the dashed lines indicate the two separate curves. From Roebber (1984).

on the Nested Grid Model (NGM; Hoke et al. 1989) and the global Medium-Range Forecast (MRF) model (Kanamitsu 1989) were introduced in the mid-1980s. These models included increased horizontal and vertical resolution, improved physics and the representation of diabatic processes, direct access and real-time analysis of an expanding global observing network, and a focused community-wide effort to understand rapid cyclogenesis (related to the cyclone workshops noted earlier).

The impact on the prediction of cyclogenesis was immediate. In 1986, Sanders published a paper docu-

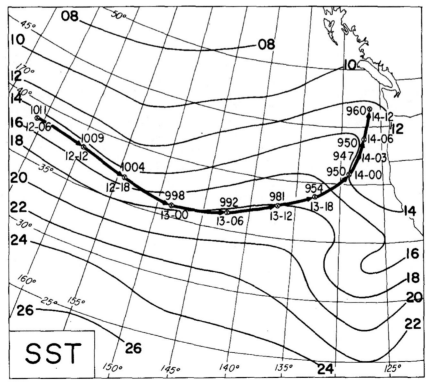

FIG. 4. Storm track for an intense oceanic cyclone. The 6-h positions and corresponding central pressures are marked for the period from 0600 UTC 1200 Nov to 12 UTC 14 Nov. Light solid lines are SSTs (°C). From Reed and Albright (1986).

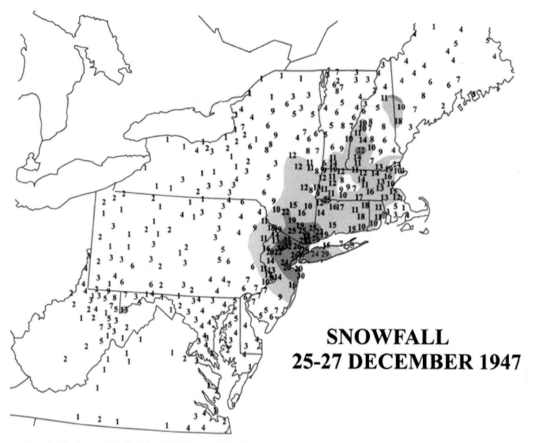

FIG. 5. Total snowfall (in.) for 25–27 Dec 1947. Colored shading represents increments of 10 in. (25 cm) for amounts of 10 in. (25 cm) and greater. From Kocin and Uccellini (2004b).

menting the poor performance of the Limited Fine Mesh (LFM) model (the predecessor to the NGM noted above) in predicting rapid or secondary cyclogenesis. He noted that while the operational LFM essentially captured the "baroclinic nature of cyclogenesis, the intensity of the response to the baroclinic forcing remains intractable" (Sanders 1986b, p. 2207). With the introduction of the NGM and MRF in 1985, models marked by improved horizontal and vertical resolutions and physical parameterizations, major improvements in cyclone forecasting were noted but not immediately reported. One year after his 1986 paper, Sanders (1987) emphasizes that the NGM and MRF demonstrated consistent improvements and "useful skill" in predicting rapid and secondary cyclogenesis, noting that this skill extended out to 48 h in the NGM and 60 h in the MRF. His analysis also highlighted more useful skill in the Atlantic Ocean basin than in the Pacific, suggesting that the more dense, upstream upper-air observations over North America were a probable factor while also emphasizing the impact of improved latent heat, boundary layer physics, and horizontal and vertical resolution.

Sanders' later work documented the continued improving capabilities of the newer operational models to predict rapid cyclogenesis. Furthermore, it spurred

many others (see Uccellini et al. 1999) to continually document the operational forecast models documenting the steady improvements related to the prediction of explosive cyclogenesis. Sanders (1992) and Grumm (1993) show just how rapid the progress was. By the early 1990s useful skill was documented for the operational MRF to predict rapid cyclogenesis three to five days in advance, with Sanders noting more predictive skill over the western Atlantic than over the Pacific Ocean. Grumm also documented a decreasing false alarm rate and noted that if the global model predicted a cyclone at day 3, it was likely that it would happen. This consistent capability of numerical models to predict major storms and rapid cyclogenesis five days in advance was quickly grasped by the forecast communities in the 1990s, who began predicting extreme events in the three-, four-, and five-day range with a level of confidence that would not have been possible even five years before, with the defining moment being the successful five-day prediction of the March 1993 "Superstorm" along the east coast of the United States (Uccellini et al. 1995).

The report cards that Sanders provided for the operational models in the 1980s to mid-1990s document how oceanic bombs were now being predicted with

TABLE 1. The number of verifying forecasts, MSLP errors (hPa), and standard deviation of MSLP error of 96-h forecasts for three classification of cyclones: 980 hPa or less, 965 hPa or less, and cyclones containing hurricane force winds for the North Pacific (PAC) and North Atlantic (ATL) Oceans. The categories of 965 hPa or less and hurricane force are subsets of the 980-hPa or less category. Positive values indicate an underforecast. The dataset used for computing the MSLP errors for the Pacific Ocean in 1992/93 no longer exists so the varying parameters marked with a dash could not be computed. The MSLP errors shown are taken directly from Uccellini et al. (1999).

		≤980 hPa			≤965 hPa			Hurricane force		
	Season	No.	MSLP error (hPa)	Std dev	No.	MSLP error (hPa)	Std dev	No.	MSLP error (hPa)	Std dev
PAC	1992/93	—	3.6	—	—	4.2	—	—	—	—
	2002/03	204	2.8	8.8	42	5.8	9.5	15	5.5	12.9
	2003/04	187	3.0	8.9	35	3.0	8.8	18	7.9	11.7
	2004/05	197	3.4	8.7	52	7.4	10.2	29	12.4	10.9
ATL	2002/03	153	4.8	8.7	43	7.1	8.2	24	7.9	10.5
	2003/04	108	3.4	7.7	9	7.3	10.4	10	8.4	7.0
	2004/05	102	4.8	8.9	24	7.9	10.3	30	7.2	10.6

some regularity with a noticeably decreasing false alarm rate. The focused efforts of the theoretical, modeling, and synoptic communities were finally paying off in better prediction for rapid cyclogenesis, especially for the marine environment and provided a basis for the National Weather Service's marine forecasters to extend and improve their forecasts and warnings of extreme oceanic cyclones.

3. Current capabilities: Predicting oceanic cyclogenesis

As Fred Sanders documented the progress in predicting rapid oceanic cyclogenesis for the Atlantic and Pacific Oceans, the National Meteorological Center created a Marine Prediction unit that provided increasingly accurate analyses, forecasts and warnings for what was traditionally considered a data-sparse area (Uccellini et al. 1999). Uccellini et al. (1999) found that in the 1990s four-day forecasts of the initiation, maturation, and decay of cyclones was remarkably accurate for storms over the Atlantic and also the Pacific Oceans. Clearly, extraordinary progress was being made by operational models and forecasters in predicting explosive oceanic cyclogenesis two, three, and even four days in advance, and the dream of predicting oceanic cyclones that provided the catalyst for the Bergen school and motivated many meteorologists was being realized.

In this section, the current skill of forecasting ocean cyclones at day 4 at the National Oceanic and Atmospheric Administration/National Centers for Environmental Prediction (NOAA/NCEP) Ocean Prediction Center (OPC) is assessed. Forecaster-generated day 4 forecasts were compared against the verifying operational OPC 1200 UTC surface analyses. OPC forecasters have available global models from a variety of national centers including the European Centre for Medium-Range Weather Forecasts (ECMWF), Environment Canada, the Met Office, the U.S. Navy Fleet Numerical Meteorology and Oceanography Command, and NCEP. Specific forecasts can be based on a single model or ensemble of national models. This study examines sig-

nificant cyclones from both the North Pacific and Atlantic, covering three separate winter seasons from the period of 1 October 2002–31 March 2005. The first 6-month period from 1 October 2002 to 31 March 2003 extends the study from Uccellini et al. (1999) to gain a 10-yr perspective of North Pacific cyclone forecast skill. The additional two winter seasons allow a comparison of forecast skill between the winter North Atlantic and North Pacific Oceans and an assessment of the variability of skill from year to year.

a. Mean pressure errors: 1 October 2002–31 March 2005

Mean sea level pressure (MSLP) errors were calculated for all cyclones of 980 hPa or less, a subset of these storms of 965 hPa or less, and a smaller subset of storms that were marked by hurricane force winds. The National Aeronautics and Space Administration (NASA) Quick Scatterometer (QuikSCAT; Atlas et al. 2001) has given OPC forecasters the ability to detect and verify hurricane force conditions within intense extratropical cyclones. For this evaluation, cyclones that were observed to possess a central pressure of 965 hPa or less are also included in the "980 hPa or less" category. Likewise, cyclones that contained "hurricane force winds" are included for all samples covering both the 980 and 965 hPa or less categories. Verifications of intensity and position for all cyclones were taken from the operational OPC 1200 UTC North Atlantic and North Pacific surface analyses. No reanalysis was done.

As shown in Table 1, MSLP forecast error for the entire sample of storms (980 hPa or less) in the North Pacific has been reduced by over 0.8 hPa over the 10-yr period from 1992/93 to 2002/03. For the subsequent two years, MSLP error trends show continued improvement over the 1992/93 errors, indicating that the positive trend is sustained over a larger sample. For the category of 965 hPa or less, only the 2003/04 value showed improvement over the 1992/93 MSLP error. In essence no improvement in the intensity forecast has been observed over the past three winter seasons for the subsample of

TABLE 2. The number of verifying forecasts, MPE (km), and standard deviation of MPE of 96-h forecasts for three classification of cyclones: 980 hPa or less, 965 hPa or less, and hurricane force conditions on the North Pacific and North Atlantic Oceans.

	Season	≤980 hPa			≤965 hPa			Hurricane force		
		No.	MPE (km)	Std dev	No.	MPE (km)	Std dev	No.	MPE (km)	Std dev
PAC	1992/93	—	527.0	—	—	461	—	—	—	—
	2002/03	204	400.3	246.2	42	381.7	280.7	15	405.5	282.6
	2003/04	187	481.3	309	35	386.8	207.4	18	516.7	266.5
	2004/05	197	439.7	329.7	52	350.6	196.3	29	387.3	212.1
ATL	2002/03	—	379.1	253.3	42	371.6	279.9	24	451.6	338.1
	2003/04	108	385.2	230.8	9	342.6	162.0	10	448.1	262.3
	2004/05	102	473.4	438.2	24	380.5	344.3	30	594.3	457.7

the strongest cyclones over the North Pacific. The comparison of both ocean basins also shows that, for nearly all cyclone categories of the North Atlantic, MSLP errors were significantly larger in the Atlantic Ocean basin than for the North Pacific. Although these comparisons are for only three winter seasons, it does appear that the operational models and forecasters have reversed their fortunes, now making as good or better forecast for the North Pacific Ocean cyclones than those that formed in the North Atlantic. Similar to 1992/93, MSLP errors in all categories were positive, indicating that the tendency continues for forecasters (and the numerical models used as guidance to produce these forecasts) to underforecast the depth of intense oceanic cyclones.

MSLP errors for the smallest subset of the most intense cyclones containing hurricane force winds tend to be significantly larger in both ocean basins than for the other two categories of cyclones. In Table 1, MSLP errors for the 980 and 965 hPa or less categories, excluding the hurricane force cyclones, are shown in parentheses. It appears that significant MSLP error for the total sample is due to the extreme cyclones with hurricane force winds being underforecast four days in advance. Despite this deficiency, we are encouraged that, unlike the era prior to the 1990s, these storms are not being missed completely, but rather underforecast with respect to the extreme intensities they attain.

b. Position errors

Mean position errors (MPE; km) were also calculated for all observed cyclones (Table 2). The MPE for all the categories of cyclones show a significant improvement over the Pacific 1992/93 values. In fact, the 400.3-km MPE for the North Pacific of 2002/03 is a 25% improvement over the 527-km MPE of 1992/93. In addition, MPE was less for the stronger category of 965 hPa or less cyclones as compared with the 980 hPa or less cyclones. This is consistent with the results of Uccellini et al. (1999) and suggests that the position and related track associated with the stronger category of cyclones are indeed more predictable. This is the good news. However, the MPE for the subset of most intense cyclones that were observed to have winds of hurricane force were higher than all cyclones observed at both

980 and 965 hPa or less, indicating that the most intense storms still pose bigger challenges to the OPC forecasters. In a separate study it has been found that hurricane force cyclones are explosive deepeners, with extreme winds developing within 24 h of the cyclone reaching minimum central pressure. Therefore the hurricane force cyclones were indeed undergoing rapid intensification. Many of the 965- or 980-hPa cyclones may have been filling cyclones. These results suggest that there is a need for increased upstream observations and improvements in the model initialization, increased resolution, and model physics to more accurately predict the intensity and tracks of the rapidly intensifying storms.

The overall results from this review are positive. North Pacific MSLP error and mean position error for all cyclones with MSLP < 980 hPa have been reduced over the past 10 yr. We continue to see smaller MPE for the stronger cyclone category of 965 hPa or less, which is again encouraging news. All MSLP errors or bias for the three categories of cyclone were positive, with the largest for the hurricane force category. For two of the three years, North Atlantic MPE were less than Pacific MPE, however, MSLP errors tend to be larger for the Atlantic cyclones for all three categories.

Table 3 shows a comparison of track or position errors in nautical miles (n mi) for 48- and 96-h forecast periods for 2004/05 winter season Atlantic extratropical cyclones, 2003 Atlantic basin tropical cyclones, and the 1993–2002 Atlantic mean track error from the Tropical Prediction Center (TPC) as given by Lawrence et al. (2005). OPC position errors for all extratropical cyclones, those reaching a central pressure of 980 hPa or less and 965 hPa or less, and those containing hurricane force winds, are shown. Certainly remarkable progress has been made in recent years in predicting the position of tropical cyclones. This progress can been seen by comparing the 2003 and 1993–2002 10-yr mean error for 48- and 72-h forecast periods. The OPC 48-h mean forecast error for all Atlantic basin cyclones equals the very impressive 123 n mi error of the 2003 tropical season. The 96-h forecast error for cyclones reaching 965 or less hPa approached the 96-h TPC 2003 error and less than the 10-yr 72-h mean error.

TABLE 3. A comparison of track error for 48-, 72-, and 96-h forecast periods for 2003 Atlantic basin tropical cyclones, the 1993–2002 Atlantic basin tropical cyclone average track error, and OPC Atlantic extratropical cyclone position error (n mi) for the 2004/05 winter season for all cyclones, those obtaining central pressures of 980 or 965 hPa or less, and those containing hurricane force winds. Shown in parentheses for each category is the number of verifying forecasts. OPC forecasts are verified for the 1200 UTC forecast cycle only, thus the limited number of forecasts.

	Forecast period (h)		
	48	72	96
2003 TPC Atlantic track error (n mi)	123 (256)	161 (197)	191 (125)
1993–2002 TPC track error (n mi)	150 (2229)	224 (1818)	—
2004/05 OPC Atlantic track error (n mi) all cyclones	123 (329)	—	247 (300)
2004/05 OPC Atlantic track error (n mi) ≤980	111 (107)	—	256 (102)
2004/05 OPC Atlantic track error (n mi) ≤965	88 (26)	—	205 (24)
2004/05 OPC Atlantic track error (n mi) hurricane force	127 (30)	—	321 (30)

This comparison indicates equivalent skill in the track predictions for hurricanes and intense extratropical cyclones at 48 h. Comparative skill does not exist at 96 h. This is likely because of two factors. The average life cycle for an oceanic extratropical cyclone is approximately five days. Therefore, many of the extratropical cyclones whose positions were verified in this study at the 96-h forecast period did not yet exist in the real atmosphere at the time the day 4 forecast was made. Tropical cyclone forecasts (at all forecast hours) are made for only (and understandably) *existing* cyclones. Second, extratropical cyclones embedded in the midlatitude westerlies move significantly faster than tropical cyclones in the easterlies. Thus, the fact that 48- and 96-h forecast errors for extratropical and tropical cyclones are comparable is a tribute to the forecasters responsible for issuing the forecasts and the quality of the numerical guidance available to the forecasters.

Although the results are not shown here, wind warning categories were also verified for the day 4 forecasts. Wind warning categories are as follows: gale (17.2–24.4 m s^{-1}), storm (24.5–32.6 m s^{-1}), and hurricane (≥32.7 m s^{-1}) force. While position and MSLP errors have shown some improvement for the 980 hPa or less category, there is a clear inability to be able to confidently forecast the highest impact, extreme cyclone (containing hurricane force winds) four days in advance. MPE and MSLP error were the largest for this category of cyclones. Warning category forecasts for hurricane force conditions continue to fall short, with an average between 7% and 31% correctly forecast for the North Pacific and 12% and 43% for the North Atlantic. However, for both basins, 75% of the incorrect forecasts that missed the hurricane force conditions fell just one warning category low (storm force with winds 24.5–32.6 m s^{-1}).

There are several possible reasons for the underforecast of the intensity of these storms. The positive bias in MSLP forecast of between 5 and 12 hPa for the hurricane force category, and the highest MPEs for the same category of storms, point to contributions related to timing the onset of these storms, the related inability to predict the most rapid period of intensification, and the underlying lack of high-resolution observations to best define the initial state of the atmosphere. Using the cyclone phase diagrams developed by Hart (2003), we found that 70% of the hurricane force cyclones for the 2003/04 season had shallow asymmetric warm cores of a warm seclusion. This structure strongly suggests the presence of active convection near the shallow warm core and supports the role of convection in rapid development discussed by Tracton (1973). Last, there may also be some hesitancy on the part of the forecasters to "go for" the extreme event 96 h in advance. Just like the tropical cyclone problem, remarkable progress has been made with track forecasts for strong extratropical cyclones, however rapid intensification and de-intensification still possess serious challenges for the ocean forecaster and science community.

Regarding the high-impact cyclone, there are still predictability issues to be resolved, especially for the most rapid intensification phase, which is still underforecast by the operational numerical models. It appears that the same deficiencies related to model resolution, model physics, and the interplay between dynamics and physical processes still inhibit the ability of these global *operational models* to predict the most intense portion of the rapid development phase of oceanic cyclogenesis. This result is similar to those derived from the high-resolution, limited-area *research models* in the 1980s and early 1990s when applied to storms such as the QEII (Anthes et al. 1983; Manobianco et al. 1992), the 12–14 November 1981 Pacific Storm (Fig. 5; Kuo and Reed 1988) and the Presidents' Day cyclone (Uccellini et al. 1987; Whitaker et al. 1988). The OPC has now instituted a formal verification for 48- and 96-h forecasts of oceanic cyclones, designed to follow Sanders' lead. The goal of this effort is to verify intensity and track forecasts for the high-impact oceanic cyclone and provide feedback to the modelers to continue improving the ability of the operational numerical models to forecast these storms.

4. The December 1947 New York City snowstorm

As mentioned, Fred Sanders' career spanned an era of revolutionary changes in the process of forecasting, which evolved from a reliance on surface weather maps and subjective interpretations to one using highly complex numerical weather prediction solutions with skill out to seven days in advance. Questions are often asked along the lines of how today's forecast models would have predicted historic storms that are well known for their impact and especially those that were forecast failures at the time. One example is a storm that occurred on 26 December 1947, which produced the heaviest 24-h snowfall in New York City's recorded history up to that time,[2] much of which was not forecast, even 12 h in advance.

This storm was a huge forecast challenge to forecasters who had to deal with it, including Fred Sanders, who, at the time, was an aviation forecaster at LaGuardia Airport in New York City. Thus, simulating this storm with the current operational global numerical model run at NCEP serves to highlight the challenges confronting the forecasters in 1947 and to provide a measure of progress in the observation–model systems during Fred Sanders' professional career.

a. Background

A description of the 26 December 1947 can be found in Kocin and Uccellini (2004b). The storm was notable given 26.4 in. (67.1 cm) of snow fell in New York City in less than 24 h, with more than 30 in. (75 cm) accumulating in some of the surrounding suburbs (Fig. 5).

Several photographs illustrate the impact of the storm (Fig. 6), which brought NYC to a standstill. In a "Weather Talk" contribution to *Weatherwise*, Zucker (2003) describes the day from the viewpoint of his uncle, who owned a bakery in NYC, and his effort to keep it open on the day after Christmas. The description paints the picture of a store owner severely impacted (no customers and no deliveries until the end of the day when a Mr. Cohn finally arrives with a delivery of cakes and bread), yet Zucker's uncle marvels at the incredible beauty of the prolonged snowburst that was not supposed to have happened in NYC. The article also provides a description of the sense of adventure and danger confronting the citizens of the New York metropolitan area that so often makes these storms a part of many family histories for generations to come.[3]

The review of the "Effects of the Storm" provided by the Weather Bureau (1948, 4–5) is more stark and speaks to the unprecedented impact of this storm:

> The unprecedented snowfall which New York did receive disrupted almost all forms of transportation and interfered critically with the normal activities of the great metropolitan district. Streets and highways were rendered impassable to vehicular traffic; at least 10,000 automobiles were abandoned in the snow, and bus, trolley, elevated, and subway lines were almost totally paralyzed. All airplane flights were suspended, harbor shipping was brought to a virtual halt, and railroad services were can-

celled or subjected to long delays. Many business establishments dismissed their employees early on the 26th, and scenes of indescribable confusion were enacted at transportation terminals, as passengers vainly sought to reach their destinations. Thousands of commuters were stranded, or reached home only after hours of exertion. Since hotel accommodations were quickly exhausted, movie houses, armories, and other public places provided all-night shelter for those stranded by the storm. Charitable organizations and military units responded to the emergency. Fortunately, loss of life was comparatively small; incomplete reports attributed one death to the storm in New York City, three in Westchester County and twenty-three in New Jersey.

> New York City appropriated almost $7,000,000 in addition to the $1,287,000 regularly allotted for snow removal in order to rid its 5,000 miles of streets of the record snowfall. The Department of Sanitation hired as many as 24,000 extra workers to serve on snow gangs, and threw over 3,200 pieces of equipment into the struggle. The task of snow clearance was greatly hampered both by the vehicles parked or abandoned on the streets, and by the commercial and private traffic which attempted to resume usual movement. This fact led Mayor O'Dwyer to proclaim an embargo on the movement of all but essential supplies by truck in the city, and to issue a ban on private cars using the streets, effective midnight December 30th. Thousands of "snow parking" violators were arrested and fined until the Mayor lifted the ban on private driving at noon, January 3rd, two days after trucks had been allowed to operate again. The city's bridges and tunnels suffered a loss in revenue of about $625,000 on account of the drop in traffic during the storm and its aftermath. In the Great Port of New York, harbor traffic fell to 10% of normal on the 26th and to 50% on the 27th, but thereafter recovery was swifter than on land. On January 4th, all train, bus, subway, elevated, street car, and airplane services were reported functioning fully and for the most part on schedule for the first time since the storm began, and over half the city's streets were reported cleared for traffic.

> The economic loss, direct and indirect, which the New York area's innumerable commercial enterprises suffered during the period December 26–January 4 must be measured in the millions of dollars.

b. The synoptic setting

While some East Coast snowstorms develop in simple synoptic weather patterns that can be forecast from surface weather maps a couple of days in advance, such as strong low pressure systems that develop in the Gulf of Mexico, in combination with large cold anticyclones located to their north (such as the Blizzard of 1899; Kocin et al. 1988; Kocin and Uccellini 2004a,b), other storms do not appear to be much of a threat even the day before the storm actually strikes (Kocin and Uccellini 2004a). Such was the case in December 1947.

---

[2] The record was broken on 11–12 February 2006 when 26.9 in (68.3 cm) was measured in a 24-h period.

[3] During this storm, NYC Fire Commissioner Quayle announced that "life and property from fire hazards have never been in such jeopardy."

Fig. 6. Scenes from New York City during the 25–27 Dec 1947 snowstorm. From Kocin and Uccellini (2004b).

The synoptic setting for the December 1947 snowstorm featured some elements typical for a classic northeast U.S. heavy snow scenario (Kocin and Uccellini 2004a,b) with a low pressure system located along the southeast coast, while an anticyclone poised over the Northeast provided cold temperatures for snow (Fig. 7). However, the surface low was relatively weak, precipitation was generally light, and the surface anticyclone was not particularly impressive, characterized by pressures no higher than 1024 hPa. By 1830 UTC 25 December, the low pressure center was located east of the South Carolina coast downwind of an upper trough and

associated upper-level jet streak over the southeastern United States (Fig. 9; see 1500 UTC 25 December). Meanwhile, the 1022-hPa surface anticyclone is found within the confluent entrance region of a separate jet streak rounding the base of a trough over southeastern Canada (Fig. 8). These upper-level maps are from a reanalysis of the event (see section 4d).

While the surface maps showed relatively modest surface low and high pressure systems, a separate low pressure area and associated fronts were also observed propagating across the upper Midwest (Fig. 7). These surface features are associated with an upper-level trough/jet

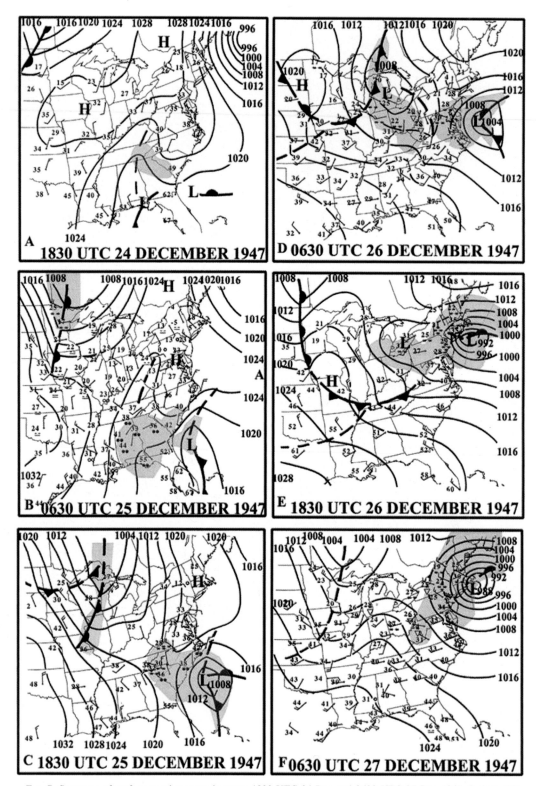

FIG. 7. Sequence of surface weather maps between 1830 UTC 24 Dec and 0630 UTC 27 Dec 1947. Isobars (hPa; solid lines; 4-hPa interval), fronts, and station symbols [temperatures (°F), winds, and current weather] are also shown. Areas of snow are shaded blue, and areas of rain are shaded green. Surface wind for this analysis sequence is from the NCDC. From Kocin and Uccellini (2004b).

FIG. 8. Twelve-hourly analyses of 500-hPa geopotential height and upper-level wind fields for 0300 UTC 25 Dec–1500 UTC 27 Dec 1947 based on a reanalysis approach discussed by Kocin and Uccellini (2004a,b). Analyses include locations of geopotential height maxima (H) or minima (L), contours of geopotential height (solid; 60-m interval; 522 = 5220 m), locations of 500-hPa absolute vorticity maxima (yellow, orange, and brown areas beginning at 16×10^{-5} s^{-1}; interval of 4×10^{-5} s^{-1}), and 400-hPa wind speeds exceeding 30 ms^{-1} (10 ms^{-1} interval; alternate blue/white shading).

system (Fig. 8) that played an important role in the "merging" or "phasing" of two or more upper-level trough/jet systems similar to that shown for the March 1993 Superstorm (Bosart et al. 1996; Dean and Bosart 1996; Dickinson et al. 1997) and the so-called "Cleveland Superbomb" of January 1978 (Hakim et al. 1995, 1996) that acted to produce the intensifying cyclone over the northeast United States during the following 24 to 30 h. This trough and its associated upper-level

jet streak are readily apparent over the Great Lakes and upper Midwest at 1500 UTC 25 December and 0300 UTC 26 December.

By 0630 UTC 26 December (Fig. 7), the surface low moved northward to a position east of Virginia, deepening a modest 4 hPa during the previous 12 h, with snow spreading from northern Virginia to New Jersey. In the 12 h following 0630 UTC, the low pressure center intensified more rapidly as it continued to move north-

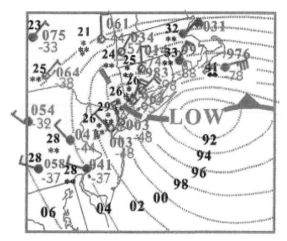

FIG. 9. Surface analysis at 1830 UTC 26 Dec 1947, including surface temperature (°F), winds (conventional notation), pressure (064 = 1006.4 hPa), and pressure tendency [−41 = −4.1 hPa (3 h)$^{-1}$] depicting the inverted trough and wind shift with dashed line. From Kocin and Uccellini (2004a).

ward to a position southeast of Long Island, New York, deepening to near 990 hPa by 1830 UTC 26 December as heavy snow developed and enveloped the NYC metropolitan area.

During this period, a mesoscale surface boundary also formed in the northwest quadrant of the surface low (Fig. 7e). A detailed surface analysis at 1830 UTC 26 December (Fig. 9) shows the boundary as a distinct wind shift line embedded within an inverted pressure trough (and associated rapid pressure falls) extending from Long Island into extreme southwestern Connecticut and southeastern New York State. The wind shift line separated mostly northeasterly flow to the east of the boundary from northwesterly flow to the west. The heaviest snow accumulations (Fig. 5) occur roughly along and to the southwest of the boundary (Fig. 9), indicating that the record-breaking aspects of the snowfall appear linked to the enhanced ascent and moisture convergence associated with the development of the inverted trough and associated mesoscale wind and temperature boundaries [see Kocin and Uccellini (2004a) for discussions of similar features that locally enhance heavy snowfall, including one on 13 December 1988 across Long Island; see Fig. 6-11, p. 191]. This feature was not accounted for in the official Weather Bureau report (Weather Bureau 2003), which instead pointed to the northward progression of the intensifying surface low across Long Island. The surface low continued to deepen to 983 hPa by 0630 UTC 27 December as it drifted slowly eastward from a position near Block Island, Rhode Island to near Nantucket, Massachusetts.

While the initial development of the low pressure system near the Southeast coast appeared to be associated with the upper-level trough/jet system over the southeastern United States, the storm's rapid intensification on 26 December appears to be influenced by the evolution of the separate, amplifying upper-level trough/jet system diving southeastward over the Canadian prairies at 0300 UTC 25 December, to a position over the Midwest by 0300 26 December 1947, and to the mid-Atlantic coast by 1500 UTC 26 December (Fig. 8). The rapid development and northward movement of the surface low on 26 December occurred within the diffluent exit region of the amplifying Midwest jet system as it reached the Atlantic coast by 1500 UTC 26 December. Therefore, it appears that the northward movement and subsequent intensification of the surface low near Long Island was associated with an interaction of the southeastern upper trough/jet system with an amplifying, second upper trough/jet system diving southeastward from Canada to the mid-Atlantic coast on 25–26 December.

c. Forecasting the 26–27 December snowstorm

One of the most enduring legacies of this storm is that its onset was poorly forecast, a factor that played a major role in the immediate impacts on NYC described earlier in the U.S. Weather Bureau report (Weather Bureau 2003; see extracts below). Twelve hours before heavy snow was accumulating in NYC, the public forecast read as follows:

> December 25, 9:30 pm. Tonight cloudy with some snow possible toward morning. Friday—cloudy with occasional snow ending during the afternoon followed by partial clearing.

As the storm gathered strength during the morning of 26 December, the forecast was amended as follows:

> December 26, 10:30 am. About seven inches of snow has fallen up to this time. Indications are for snow to continue to late afternoon or evening and accumulate 8 to 10 inches.

These forecasts point to the obvious difficulties the public forecasters had with 1) predicting the storm even as late as Christmas night (6 h before light snow commenced in NYC), and 2) then keeping up with the unprecedented and prolonged snowfall rates once the storm took off in earnest. In the official review of the storm, the Weather Bureau notes the difficulty of the forecast situation and expectations that a weak storm off the Carolina coast would move northeastward and spare the city. At 9:30 P.M. 25 December, forecasters logged the following:

> This is a borderline situation and amount of snow and time of precipitation depend on movement of storm centered at 7:30 p.m. off the Carolina coast. There is one chance in four that snow may be heavy and last most of the day and one chance in four that very little snow may fall.

Clearly, as New York City slept, none of these uncertainties were conveyed in the official forecast released that night. Thus, the light snow at sunrise did not ring any alarms for people venturing into the city and going about their business. Furthermore, even as the heavy snow developed during the morning, the updates from the Weather Bureau literally could not keep up with a prolonged snowburst with snowfall rates exceeding 3 in. h^{-1} (7.5 cm h^{-1}). Only later did the Weather Bureau recognize that the storm intensified in concert with an upper-level deepening trough that forced it to move further north and deepen rapidly. But as noted earlier, other mesoscale factors associated with the inverted trough extending northward across Long Island into Connecticut also played an important role.

In the late 1940s, the public forecast was not the only forecast issued by the U.S. Weather Bureau. Aviation forecasters made separate forecasts to support the expanding commercial aviation industry. As fate would have it, young Fred Sanders was an aviation forecaster on duty at New York City's LaGuardia Airport on 25 December 1947. During this period, upper-level analyses had only recently become operational and data coverage was still limited to the eastern half of the United States, as the raob network was being expanded westward during this period. There were no numerical weather forecasts, and forecasting the weather was often little more than trying to move weather systems as deduced from surface weather analyses.

As Fred Sanders remembered in an e-mail sent to the primary author (F. Sanders 2003, personal communication), "we forecast only one day in advance," "there was no model guidance," "upper-level observations were greatly expanded during the war. No hemispheric analysis, though." Finally, "we were aware of the interaction between surface and upper-level features, but mostly steering of surface cyclones by the upper flow, with little idea of development beyond simple extrapolation."

In other words, weather systems that were neither developing nor decaying were probably forecast with acceptable reliability, while developing (or decaying) weather systems that exhibited rapid changes were not going to be predicted with any degree of accuracy. Cressman's (1970) analysis of predictability for heavy snowstorms does indeed show that predictive skill was limited to 12 h or less, even into the early 1970s (Fig. 10).

Because of the complexity of the factors that played into the development of the storm, it is not hard to see why the public forecasts played "catch up" with the forecast of the snowfall in December 1947. Despite these complexities, the storm was not an entire surprise to the aviation forecasters at LaGuardia Airport, who did a pretty good job, in retrospect. According to a letter sent to *Weatherwise* by Tom Morgan (a retired aviation forecaster), the LaGuardia forecast staff felt that there was the potential for a significant snowstorm in spite of the public forecast. Tom Morgan recounts how lead forecaster Frank Zucker, one of the "best terminal forecasters," predicted "moderate to occasionally heavy

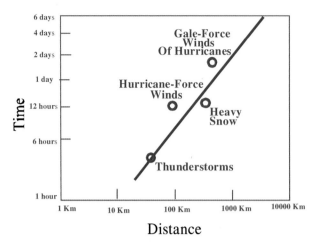

FIG. 10. Limits of predictability of public weather forecasts. From Cressman (1970).

snow though a good part of the day" (with a 12–24-h lead time). The letter also quoted a "young forecaster" from the International Aviation Forecast Center, Fred Sanders, who stated, "Get out your snow shovels boys—we're going to have a foot" (Morgan 2003). There is no doubt that Fred Sanders was already bringing a sense of excitement to the forecast challenge confronting the meteorological community on 25 and 26 December 1947.

d. Simulating the 26–27 December 1947 snowstorm: Background

The 26–27 December 1947 snowstorm posed an enormous challenge to the forecast community, yielding noticeably different forecasts from distinct forecast units operating within the Weather Bureau in the same metropolitan area. An interesting question confronting us is the following: Would today's numerical forecast models, working off the growing radiosonde network of late 1947, have provided any useful guidance for the forecasters? And if so, how far in advance could this event have been predicted?

The approach to answering these questions is based on the Reanalysis Project described by Kalnay et al. (1996), which provides 50 yr of global analyses extending back to January 1948. This analysis approach has been used in numerical experiments of famous cases, including the 1950 "Appalachian Storm" to show that with the data distribution at that time, an intense cyclone in the eastern United States was still predictable four to five days in advance using today's numerical models (Phillips 2000; also discussed by Kocin and Uccellini 2004a). More recent model simulations, using a 55-km version of the NCEP Global Forecast System (GFS), was applied to the November 1950 storm by Robert Kistler (not shown). These simulations predicted heavy snowfall in northwestern Pennsylvania and West Virginia (where up to 60 in. or 150 cm was reported) as much as 2.5 days in advance. The challenge of assessing the predictability of the December 1947 storm is modeled on the success of simulating the 1950 storm, as well as the application of the reanalysis for all of the analyses provided in Kocin and Uccellini's (2004b) *Northeast Snowstorms* monograph.

Ironically, the original Reanalysis Project provides analyses back through 1 January 1948, just days *after* New York City's snowstorm. This cutoff was decided because the number of upper observations decreased significantly prior to 1 January 1948, prohibiting the reanalysis from extending from January 1948 much further backward in time. Nevertheless, working with the National Center for Atmospheric Research (NCAR) and the National Climatic Data Center (NCDC), we extended the reanalysis back through December 1947 using this twice-daily pibal data that extended throughout the United States.[4] Approaching this study we were concerned whether this network contained sufficient information to allow NCEP's 2003 version of the GFS to predict even the synoptic-scale features of the December 1947 snowstorm, much less address the issue of how far in advance such predictions could be made.

Extending the reanalysis further back in time involved other challenges. We needed to determine how far back to extend the analysis in order to have a sufficient period of time to perform the simulations, which was an issue in addition to assimilating a sparse network of observations. Last, the GFS normally executes a 384-h prediction every six hours, a forecast length that is based on data quality and coverage in 2003, not 1947. These issues were resolved based on the experience gained in simulating the 1950 Appalachian Storm and Europe's infamous 1953 Gale,[5] where useful skill was realized in the 3- to 4-day time frame. In these experiments, the models responded quickly to the data assimilation process starting from climatology, and initial conditions for the start of the experiment were created over a multiday period of assimilating all available observations. Based on these other case studies, a starting time of 2100 UTC 15 December 1947 was selected to begin the forecast initialization, and 0300 UTC 21 December 1947 was selected as the time of the first 5-day prediction.

To start the data assimilation process, a climatology was created by averaging 1800 UTC 15 December and 0000 UTC reanalysis data files every five years from 1970 to 1995. These separate 1800 and 0000 UTC files were then averaged to obtain a 2100 UTC 15 December analysis to be used as an initial climatological background, which was then interpolated from the T62L28 configuration of the NCEP–NCAR reanalysis to the

T254L64 resolution of the GFS to become the initial background file for the GFS data assimilation cycle. To fully utilize the sparse data during the first two days of the data assimilation cycle, the quality control programs were disabled and the relative weighting of the observations to the analysis background was increased in order to accept observations normally rejected as having excessive departures from the background. Last, the ingest of surface boundary fields such as SSTs, sea ice, and snow depth was adjusted for their availability in 1947. Weekly SSTs were available and interpolated to the daily values that the GFS expects. Sea ice was set to climatology. Snow depth evolved according to the assimilation model precipitation and surface energy budget since analyses were not available at that time.

An example of initial conditions at 1500 UTC 25 December 1947, related pibal observations, and gridded model data are shown in Fig. 11. Note that the data are sparse, with few observations across south-central and western southwest Canada, and the southeast and western United States. Also note that the model-gridded wind data sometimes differ quite dramatically from the nearby observations, pibal, and radiosonde wind. The apparent erroneous wind observations, as compared with the analysis, are likely related to a low observation angle of the balloon measurement systems, which are known to yield large errors in wind measurements for the middle and upper troposphere marked by strong wind regions. Despite the obvious issues related to the quality of the upper-air data, the initial analysis at 1500 UTC 25 December shows evidence of the weak trough over the southeastern United States extending vertically from 850 hPa through at least 500 hPa, upper-level confluence over the northeastern United States upstream of a closed low center over Newfoundland, and an amplifying trough over south-central Canada associated with a strong jet streak evident in both the observed and gridded winds.

e. Numerical prediction for 26–27 December 1947: Results

Forecasts of 24-h precipitation amounts valid for the period between 1500 UTC 26 December and 1500 UTC 27 December from the 72-, 48-, and 36-h GFS simulations are provided in Fig. 12. Forecasts initialized from 0300 21 December 1947 through 1500 UTC 24 December 1947 showed little or no skill in predicting cyclogenesis and snowfall across the northeast United States verifying on 26–27 December. Only a broad area of light to moderate precipitation is forecast from the Great Lakes eastward to New England in the 72-h forecast (Fig. 12), with little or no precipitation forecast from Virginia to New York City, indicating little evidence of east coast cyclogenesis. Starting with the forecast initialized at 1500 UTC 25 December, forecasts began to indicate the potential for snowfall in New York City,

---

[4] See the acknowledgments.

[5] The North Sea Gale of 31 January–2 February 1953 produced record floods in the Netherlands and Great Britain, killing thousands of people. Numerical weather forecasts of the onset of this event are being examined by Huug Van den Dool in the Climate Prediction Center.

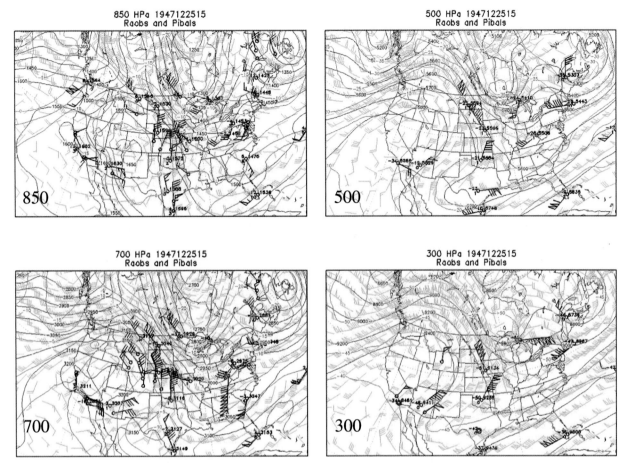

FIG. 11. Raob and pibal observations available at 1500 UTC 25 Dec 1947 at 850, 700, 500, and 300 hPa [site station circles, temperature (°C), height (m), and winds (kt)]. Corresponding GFS model analyses (overlaid on observations) for the same time, showing geopotential height (m; solid black lines; 50-m interval for 850 and 700 hPa, 100-m interval for 500 hPa, and 200-m interval for 300 hPa), temperature (°C; colored lines; 5°C interval, red > 0°C, blue ≤ 0°C), and wind analyses (kt).

although none of the forecasts verified the excessive amount of snow that fell over New York City.

However, the forecast initialized at 1500 UTC 25 December (Fig. 12b) shows a north–south band of precipitation extending from east of the mid-Atlantic coast to New Jersey, New York City, and much of New England. Forecast precipitation amounts range from 0.5 to 0.75 in. (1.2 to 1.8 cm) in New York City to more than 1 in. (2.5 cm) across Long Island. The forecast initialized at 0300 UTC 26 December (Fig. 12c) shows a similar precipitation distribution, but this forecast increased the precipitation amounts across Long Island to greater than 1.5-in. (3.7 cm) liquid equivalent, while still showing about 0.75 in. (1.8 cm) over New York City. Therefore, it appears that the potential for significant snow in New York City first appears on the GFS forecast initialized at 1500 UTC 25 December (24 h before the event), with the forecast of heaviest precipitation focused in the 24–36-h range that covers this period of heaviest snowfall observed on 26 December.

The GFS forecasts initialized at 0300 UTC 25 December (not shown) fail to generate significant precip-

itation and cyclogenesis even 36 h prior to the event. A cyclone is predicted to develop off the southeast U.S. coast but then moves offshore. The GFS also forecasts a weaker cyclone/frontal system that develops over south-central Canada that moves eastward and weakens as an inverted trough passes across the northeast United States and then evolves into a rapidly developing cyclone off the coast of Nova Scotia. The corresponding temporal sequence of 500-hPa heights, absolute vorticity, and winds shows the two distinct upper-level troughs, one over the southeast United States and the other over south-central Canada associated with the two separate surface low pressure systems. However, the GFS forecasts the southeast U.S trough to move rapidly eastward while the upper trough just north of the Great Lakes amplifies southeastward across the northeast United States by 36 h and then continue to move it off the New England coast by 48 h. Both the southeast U.S. and southern Canadian troughs appear to progress rapidly eastward, rather than deepen and interact with each other closer to the coast. Therefore, two separate storm systems move eastward off the East Coast, rather than

a

72 hr GFS FCST FROM 1947122415 valid 1947:12:27:15
72 hr TOTAL PRECIP(IN)

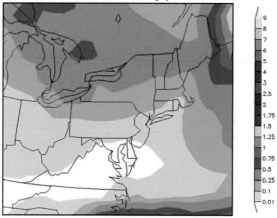

b

48 hr GFS FCST FROM 1947122515 valid 1947:12:27:15
48 hr TOTAL PRECIP(IN)

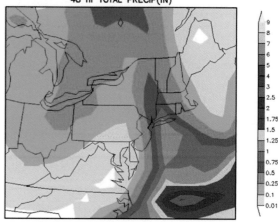

c

36 hr GFS FCST FROM 1947122603 valid 1947:12:27:15
36 hr TOTAL PRECIP(IN)

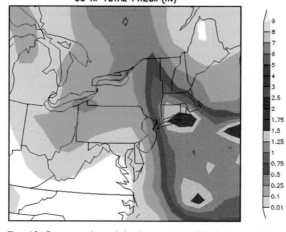

FIG. 12. Storm total precipitation amounts (in.) from the GFS model at forecast lead times of 72, 48, and 36 h, all ending at 1500 UTC 27 Dec 1947.

interacting and developing into the intense cyclone that was observed.

The GFS forecast initialized at 1500 UTC 25 December 1947, 24 h prior to the onset of heavy snow in New York City (Figs. 13 and 14), captures some of the features that appear to play a role in the development of heavy snowfall, although the forecast is far from perfect. The 6-hourly sequence of forecasts of sea level pressure and 12-hourly precipitation amounts (Fig. 13) show that the two cyclones appear to exist relatively close to each other during the simulation initialized 12 h earlier. The southeast U.S. cyclone is located closer to the North Carolina coastline by 0300 26 December, while the low pressure system over south-central Canada is a little slower and located further southwestward than the ear-

lier forecast. In the following 12 to 24 h, the southeast U.S. low moves north and then east over the western Atlantic while the more northern system weakens but leaves a distinct inverted surface trough extending from the Atlantic low northwestward toward New York City. This trough remains over New England by 36 h, even as the main oceanic storm system continues moving away from the coast. This inverted trough resembles the mesoscale trough in the analyses on 26 December, which played a crucial role in the development of the record snowfall in the New York City area. Although the simulation does not predict cyclogenesis to occur near the southern New England coast, the forecast does predict more significant precipitation with the inverted surface trough.

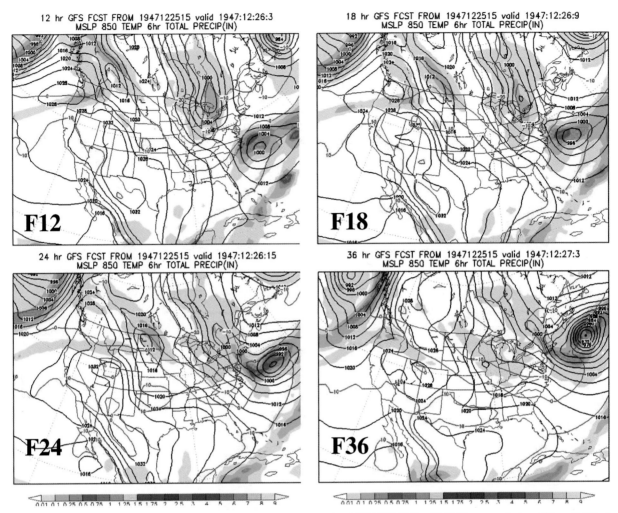

FIG. 13. Sequence of 12-hourly forecasts from the GFS model initialized at 1500 UTC 25 Dec 1947. Shown are isobars (hPa; solid black lines; 4-hPa interval), 850-hPa temperature (°C; colored lines; 10°C interval, red > 0°C, blue ≤ 0°C), and 6-h total precipitation (in.; shaded areas; scale at bottom of figure).

The corresponding 500-hPa forecasts (Fig. 14) also appear to improve upon the previous forecast cycle in that the upper trough over the southeast United States progresses eastward more slowly, while the southern Canadian trough amplifies more significantly. This results in a more pronounced trough along the East Coast, and a weak upper trough over the western Atlantic that still results in two separate cyclonic systems along the East Coast, rather than the one explosive cyclone near the New England coast observed by 0300 UTC 27 December.

While this forecast still misrepresents the interaction of the two separate systems, it at least hints at their merger and, as a result, predicts some of the elements that led to the heavy snowfall over New York City. Most notably, this numerical simulation forecast the pronounced amplification of the upper-level trough from south-central Canada to the Atlantic coast, the slower eastward movement of the southeast U.S. storm system,

the development of an inverted sea level pressure trough between the Canadian low and the southeast U.S. storm system, and the development of significant precipitation across the New York City metropolitan area and southern New England.

Six-hourly forecasts initialized 12 h later at 0300 UTC 26 December 1947 (12 h prior to the onset of the heaviest snowfall over New York City) continue to show that the forecast of this storm remains imperfect, with the main low predicted several hundred kilometers east of its actual position near New England early on 27 December (Figs. 15 and 16). However, the impressive inverted sea level trough was predicted to develop and extend across western Long Island into southeast New York and to focus heavy precipitation from New Jersey through New England, producing 1 to 1.5 in. (2.50 to 3.50 cm) of precipitation over the NYC metropolitan area. Furthermore, the model prediction captures a more distinct phasing of the two upper-level trough systems

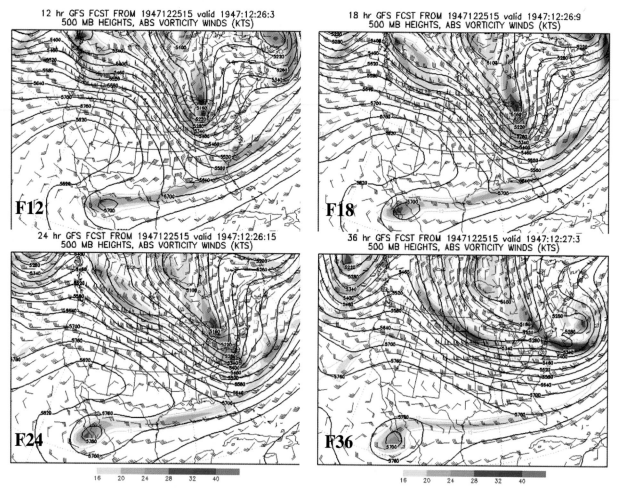

FIG. 14. Same as in Fig. 13, but showing 500-hPa geopotential heights (m; solid lines; 60-m interval), absolute vorticity (s⁻¹; shaded areas; scale at bottom of figure), and wind (kt).

supporting the intense surface cyclogenesis east of New England (Fig. 16).

These experiments illustrate that the predictability of the 26–27 December 1947 snowstorm was limited by the restricted areal extent of the upper-air observation network at the time (where the expanding radiosonde network only extended halfway across the nation). Thus, unlike the 1950 Thanksgiving storm, where the intense cyclone was predicted three, four, and five days in advance using a similar model experiment (Phillips 2000), the potential impact of the intense December 1947 East Coast snowstorm was only predictable 12–18 h in advance even with the global operational prediction system employed at NCEP in 2003. Nevertheless, the numerical simulations locked onto a mesoscale inverted trough extending west to northwest of the deepening cyclone that acted to localize the heaviest snowfall in NYC, with the model forecast initialized 12–24 h prior to this event predicting well over a foot of snow. The excitement conveyed by Fred Sanders' forecast the night before the big storm enveloped New York City; "Get

out your snow shovels boys—we're going to have a foot" provides supporting evidence that even with the restricted amount of observations available in December 1947, Fred Sanders' forecasting skill was 56 yr ahead of his time!

5. Summary

Fred Sanders' career as a synoptic meteorologist from the 1940s through the beginning of the twenty-first century spanned a truly remarkable period, during which Fred contributed toward

utilizing model-based diagnostic approaches by synoptic meteorologists to increase our understanding of atmospheric processes that contribute to significant weather events;

a fundamental increase in our understanding of the dynamical and physical processes contributing to explosive cyclogenesis; and

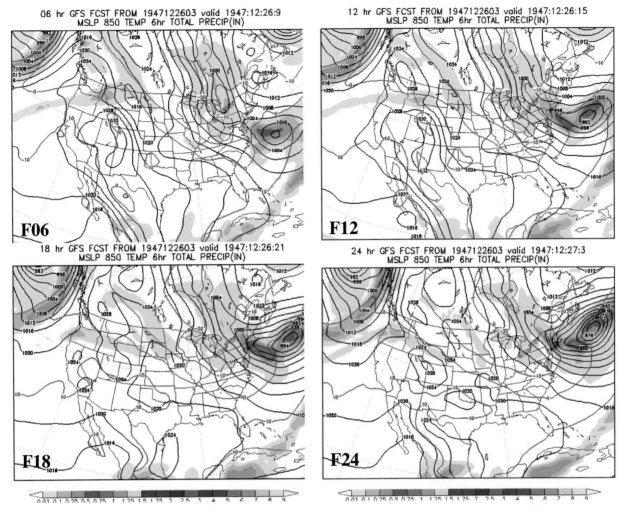

06 hr GFS FCST FROM 1947122603 valid 1947:12:26:9
MSLP 850 TEMP 6hr TOTAL PRECIP(IN)

12 hr GFS FCST FROM 1947122603 valid 1947:12:26:15
MSLP 850 TEMP 6hr TOTAL PRECIP(IN)

18 hr GFS FCST FROM 1947122603 valid 1947:12:26:21
MSLP 850 TEMP 6hr TOTAL PRECIP(IN)

24 hr GFS FCST FROM 1947122603 valid 1947:12:27:3
MSLP 850 TEMP 6hr TOTAL PRECIP(IN)

FIG. 15. Same as in Fig. 13, but for the GFS initialized at 0300 UTC 26 Dec 1947, and showing 6-, 12-, 18-, and 24-h forecasts.

revolutionary changes in the forecast methodology from a subjective process, based primarily on surface data and a sparse network of upper-air observations, to a more objective approach, based primarily on a global- to regional-scale suite of numerical weather prediction models operating off a global observing network.

His leadership role in all aspects of the advancements in understanding and predicting rapid cyclogenesis is reflected by his 1) instinctive forecast skills as demonstrated for the famous December 1947 New York City snowstorm, 2) careful climatological studies that identified the nature and characteristics of explosive oceanic cyclogenesis, which came to be called the bomb, 3) advancing the understanding of how larger-scale dynamic features associated with upper-level trough and ridge patterns interact with physical processes such as latent and sensible heat release and contribute to the period of most rapid development of these storms, and 4) evaluation of the operational numerical models that were introduced in the 1960s through the 1990s and the

related insights to the deficiencies of these models that needed to be addressed (resolution, data, physical parameterization, and air–sea interaction) in order to improve the prediction of rapidly developing oceanic cyclones.

The ability of today's operational models to predict these storms three, four, and five days in advance (and for some cases even further in advance) and the ability of today's forecasters to issue accurate 4- to 5-day forecasts for explosive oceanic cyclogenesis and provide warnings for dangerous winds and waves to ships at sea have their roots in Fred Sanders' career accomplishments.

While remarkable progress in forecasting rapid oceanic cyclogenesis has been made, some of the same issues that were identified through Sanders' work still exist. For example, the most rapid development phase of oceanic cyclogenesis, and related deepening of the central pressure, is still underpredicted by today's operational numerical models. These deficiencies are re-

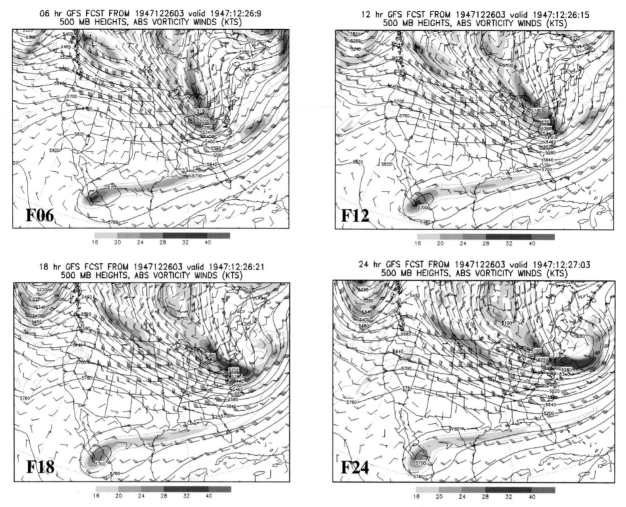

FIG. 16. Same as in Fig. 14, but for the GFS initialized at 0300 UTC 26 Dec 1947, and showing 6-, 12-, 18-, and 24-h forecasts.

flected by relatively lower skill levels exhibited by forecasters when dealing with the deepening rates for those storms marked by hurricane force surface winds. Thus, like current hurricane predictions, numerical models still do not capture the most rapid development phase of these very intense storms. This situation points to the continuing need for the ongoing research to determine 1) exactly how the dynamical and physical processes interact on the meso- to microscale to produce the explosive deepening rates of these storms, and 2) the related requirements of the operational numerical models to predict these dangerous storm systems on a reliable basis while continuing to extend the range of predictability four–seven days in advance. Last but not least, the limited ability to predict the famous December 1947 New York City snowstorm with today's sophisticated global numerical models running off the sparse observational network that forecasters (including a young Fred Sanders) had to use in 1947 illustrates the important role of today's extensive global observing network based primarily on satellite data in making accurate forecasts of these rapidly developing cyclones three to five days in advance. Today, we continue with the issues Fred Sanders raised throughout his career (defining observational requirements, increasing the fundamental understanding of cyclones, and documenting the attributes of operational model strengths and weaknesses) to continue our move forward and improve the prediction of the most rapid development phase of the explosive extratropical cyclone with increased reliability three, five, and even seven days in advance.

Acknowledgments. We kindly acknowledge the work of Roy Jenne and his colleagues at NCAR, NCDC Director Tom Karl and colleague Joe Elms, and our NCEP historical observation expert, Jack Woollen, who recovered a sparse, coast-to-coast coverage of twice-daily upper-air pibal observations used in the model study for the December 1947 snowstorm. We also thank Commander Mark Moran, NOAA, and Lauren Morone for their assistance in the final preparation of the figures and manuscript, and the reviewers, whose suggestions have improved the final version of this paper.

REFERENCES

Anthes, R. A., Y.-H. Kuo, and J. R. Gyakum, 1983: Numerical simulations of a case of explosive marine cyclogenesis. *Mon. Wea. Rev.,* **111,** 1174–1188.

Atlas, R., and Coauthors, 2001: The effects of marine winds from scatterometer data on weather analysis and forecasting. *Bull. Amer. Meteor. Soc.,* **82,** 1965–1990.

Bjerknes, J., and J. Holmboe, 1944: On the theory of cyclones. *J. Meteor.,* **1,** 1–22.

Bosart, L. F., 1981: The Presidents' Day snowstorm of 18–19 February 1979: A subsynoptic-scale event. *Mon. Wea. Rev.,* **109,** 1542–1566.

——, 1999: Observed cyclone life cycles. *The Life Cycles of Extratropical Cyclones,* M. A. Shapiro and S. Grønås, Eds., Amer. Meteor. Soc., 187–213.

——, and S. L. Lin, 1984: A diagnostic analysis of the Presidents' Day storm of February 1979. *Mon. Wea. Rev.,* **112,** 2148–2177.

——, G. J. Hakim, K. R. Tyle, M. A. Bedrick, W. E. Bracken, M. J. Dickinson, and D. M. Schultz, 1996: Large-scale antecedent conditions associated with the 12–14 March 1993 cyclone ("Superstorm '93") over eastern North America. *Mon. Wea. Rev.,* **124,** 1865–1891.

Charney, J. G., 1947: The dynamics of long waves in a baroclinic westerly current. *J. Meteor.,* **4,** 135–162.

——, 1950: Progress in dynamic meteorology. *Bull. Amer. Meteor. Soc.,* **31,** 231–236.

——, 1955: The use of primitive equations of motion in numerical weather prediction. *Tellus,* **7,** 22–26.

——, R. Fjortoft, and J. Von Neuman, 1950: Numerical integration of the barotropic equation. *Tellus,* **2,** 237–254.

Cressman, G., 1970: Public forecasting: Present and future. *A Century of Weather Progress,* J. E. Caskey Jr., Ed., Amer. Meteor. Soc., 71–77.

Dean, D. B., and L. F. Bosart, 1996: Northern Hemisphere 500-hPa trough merger and fracture: A climatology and case study. *Mon. Wea. Rev.,* **124,** 2644–2671.

Dickinson, M. J., L. F. Bosart, W. E. Bracken, G. J. Hakim, D. M. Schultz, M. A. Bedrick, and K. R. Tyle, 1997: The March 1993 superstorm cyclogenesis: Incipient phase synoptic- and convective-scale flow interaction and model performance. *Mon. Wea. Rev.,* **125,** 3041–3072.

Dirks, R. A., J. P. Kuettner, and J. A. Moore, 1988: Genesis of Atlantic Lows Experiment (GALE): An overview. *Bull. Amer. Meteor. Soc.,* **69,** 148–160.

Eady, E. T., 1949: Long waves and cyclone waves. *Tellus,* **1,** 33–52.

Eliassen, A., 1956: Instability theories of cyclone formation. *Weather Analysis and Forecasting,* 2d. ed. S. Petterssen, Ed., McGraw-Hill, 305–319.

Grumm, R. H., 1993: Characteristics of surface cyclone forecasts in the aviation run of the global spectral model. *Wea. Forecasting,* **8,** 87–112.

Gyakum, J. R., 1983a: On the evolution of the *QE II* storm. I: Synoptic aspects. *Mon. Wea. Rev.,* **111,** 1137–1155.

——, 1983b: On the evolution of the *QE II* storm. II: Dynamic and thermodynamic structure. *Mon. Wea. Rev.,* **111,** 1156–1173.

——, 1991: Meteorological precursors to the explosive intensification of the *QE II* storm. *Mon. Wea. Rev.,* **119,** 1105–1131.

Hadlock, R., and C. W. Kreitzberg, 1988: The Experiment on Rapidly Intensifying Cyclones over the Atlantic (ERICA) field study: Objectives and plans. *Bull. Amer. Meteor. Soc.,* **69,** 1309–1320.

Hakim, G. J., L. F. Bosart, and D. Keyser, 1995: The Ohio Valley wave-merger cyclogenesis event of 25–26 January 1978. Part I: Multiscale case study. *Mon. Wea. Rev.,* **123,** 2663–2692.

——, D. Keyser, and L. F. Bosart, 1996: The Ohio Valley wave-merger cyclogenesis event of 25–26 January 1978. Part II: Diagnosis using quasigeostrophic potential vorticity inversion. *Mon. Wea. Rev.,* **124,** 2176–2205.

Hart, R. E., 2003: A cyclone phase space derived from thermal wind and thermal asymmetry. *Mon. Wea. Rev.,* **131,** 585–616.

Hoke, J. E., N. A. Phillips, G. J. DiMego, J. J. Tuccillo, and J. G. Sela, 1989: The regional analysis and forecast system of the National Meteorological Center. *Wea. Forecasting,* **4,** 323–334.

Hoskins, B. J., 1999: Sutcliffe and his development theory. *The Life Cycles of Extratropical Cyclones,* M. A. Shapiro and S. Grønås, Eds., Amer. Meteor. Soc., 81–86.

Johnson, D. R., 1986: Summary of the Proceedings of the First National Workshop on the Global Weather Experiment. *Bull. Amer. Meteor. Soc.,* **67,** 1135–1143.

Kalnay, E., and Coauthors, 1996: The NCEP/NCAR 40-Year Reanalysis Project. *Bull. Amer. Meteor. Soc.,* **77,** 437–471.

Kanamitsu, M., 1989: Description of the NMC global data assimilation and forecast system. *Wea. Forecasting,* **4,** 335–342.

Keyser, D., and L. W. Uccellini, 1987: Regional models: Emerging research tools for synoptic meteorologists. *Bull. Amer. Meteor. Soc.,* **68,** 306–320.

Kocin, P. J., and L. W. Uccellini, 2004a: *Overview.* Vol. 1, *Northeast Snowstorms, Meteor. Monogr.,* No. 54, Amer. Meteor. Soc., 296 pp.

——, and ——, 2004b: *The Cases.* Vol. 2, *Northeast Snowstorms, Meteor. Monogr.,* No. 54, Amer. Meteor. Soc., 818 pp.

——, A. D. Weiss, and J. J. Wagner, 1988: The great Arctic outbreak and East Coast blizzard of February 1899. *Wea. Forecasting,* **3,** 305–318.

Kuo, Y.-H., and R. J. Reed, 1988: Numerical simulation of an explosively deepening cyclone in the eastern Pacific. *Mon. Wea. Rev.,* **116,** 2081–2105.

Lalaurette, F., J. Bidlot, L. Ferranti, A. Ghelli, F. Grazzini, M. Leutbecher, D. Richardson, and G. van der Grijn, 2006: Verification statistics and evaluations of ECMWF forecasts in 2004–2005. ECMWF Tech. Memo. 501, 16 pp.

Lawrence, M. B., L. A. Avila, J. L. Beven, J. L. Franklin, R. J. Pasch, and S. R. Stewart, 2005: Atlantic hurricane season of 2003. *Mon. Wea. Rev.,* **133,** 1744–1773.

Manobianco, J., L. W. Uccellini, K. F. Brill, and Y.-H. Kuo, 1992: The impact of dynamic data assimilation on the numerical simulations of the *QE II* cyclone and an analysis of the jet streak influencing the precyclogenetic environment. *Mon. Wea. Rev.,* **120,** 1973–1996.

Morgan, T., 2003: Letter to the editor. *Weatherwise,* **56** (May–June), 6.

Palmén, E., and C. W. Newton, 1969: *Atmospheric Circulation Systems.* Academic Press, 603 pp.

Petterssen, S., 1956: *Weather Analysis and Forecasting.* Vol. 1, McGraw-Hill, 428 pp.

——, 2001: *Weathering the Storm: Sverre Petterssen, the D-Day Forecast, and the Rise of Modern Meteorology,* J. R. Fleming, Ed., Amer. Meteor. Soc., 329 pp.

Phillips, N. A., 2000: A review of theoretical questions in the early days of NWP. *50th Anniversary of Numerical Weather Prediction Commemorative Symposium,* A. Spekat, Ed., Deutsche Meteorologische Gesellschaft, 13–28.

Ramage, C. S., 1976: Prognosis for weather forecasting. *Bull. Amer. Meteor. Soc.,* **57,** 4–10.

Reed, R. J., and M. D. Albright, 1986: A case study of explosive cyclogenesis in the eastern Pacific. *Mon. Wea. Rev.,* **114,** 2297–2319.

Roebber, P. J., 1984: Statistical analysis and updated climatology of explosive cyclones. *Mon. Wea. Rev.,* **112,** 1577–1589.

Sanders, F., 1986a: Explosive cyclogenesis in the west-central North Atlantic Ocean, 1981–1984. Part I: Composite structure and mean behavior. *Mon. Wea. Rev.,* **114,** 1781–1794.

——, 1986b: Explosive cyclogenesis over the west-central North Atlantic Ocean, 1981–1984. Part II: Evaluation of LFM model performance. *Mon. Wea. Rev.,* **114,** 2207–2218.

——, 1987: Skill of NMC operational dynamical models in prediction of explosive cyclogenesis. *Wea. Forecasting,* **2,** 322–336.

——, 1992: Skill of operational dynamical models in cyclone prediction out to five-days range during ERICA. *Wea. Forecasting,* **7,** 3–25.

——, and J. R. Gyakum, 1980: Synoptic-dynamic climatology of the "bomb." *Mon. Wea. Rev.,* **108,** 1589–1606.

——, A. J. Wagner, and T. N. Carlson, 1960: Specification of cloudiness and precipitation by multilevel dynamical models. Scientific Rep. 1, AFF 19(604)-5491, Air Force Cambridge Research Laboratories, 111 pp.

Shuman, F. G., 1989: History of numerical weather prediction at the National Meteorological Center. *Wea. Forecasting,* **4,** 286–296.

Tracton, M. S., 1973: The role of cumulus convection in the development of extratropical cyclones. *Mon. Wea. Rev.,* **101,** 573–592.

Uccellini, L. W., 1986: The possible influence of upstream upper-level baroclinic processes on the development of the *QE II* storm. *Mon. Wea. Rev.,* **114,** 1019–1027.

——, 1990: Processes contributing to the rapid development of Extratropical Cyclones. *Extratropical Cyclones: The Erik Palmén Memorial Volume,* C. W. Newton and E. O. Holopainen, Eds., Amer. Meteor. Soc., 81–105.

——, P. J. Kocin, R. A. Petersen, C. H. Wash, and K. F. Brill, 1984: The Presidents' Day cyclone of 18–19 February 1979: Synoptic overview and analysis of the subtropical jet streak influencing the pre-cyclogenetic period. *Mon. Wea. Rev.,* **112,** 31–55.

——, D. Keyser, K. F. Brill, and C. W. Wash, 1985: Presidents' Day cyclone of 18–19 February 1979: Influence of upstream trough amplification and associated tropopause folding on rapid cyclogenesis. *Mon. Wea. Rev.,* **113,** 962–988.

——, R. A. Petersen, K. F. Brill, P. J. Kocin, and J. J. Tuccillo, 1987: Synergistic interactions between an upper-level jet streak and diabatic processes that influence the development of a low-level jet and secondary coastal cyclone. *Mon. Wea. Rev.,* **115,** 2227–2261.

——, P. J. Kocin, R. S. Schneider, P. M. Stokols, and R. A. Dorr, 1995: Forecasting the 12–14 March 1993 superstorm. *Bull. Amer. Meteor. Soc.,* **76,** 183–199.

——, J. M. Sienkiewicz, and P. J. Kocin, 1999: Advances in forecasting extratropical cyclogenesis at the National Meteorological Center. *The Life Cycles of Extratropical Cyclones,* M. A. Shapiro and S. Grønås, Eds., Amer. Meteor. Soc., 317–336.

Weather Bureau, 1948: The snowstorm of December 26–27, 1947 with comparative data of earlier heavier snowstorms. U.S. Department of Commerce.

Whitaker, J. S., L. W. Uccellini, and K. F. Brill, 1988: A model-based diagnostic study of the rapid development phase of the Presidents' Day cyclone. *Mon. Wea. Rev.,* **116,** 2337–2365.

Zachter, M, 2003: Waiting for Cohn. *Weatherwise,*(January–February), 36–37.

Part IV
Climate and Climatology

Chapter 15

Linking Weather and Climate

RANDALL M. DOLE

NOAA/Earth System Research Laboratory, Boulder, Colorado

(Manuscript received 1 March 2006, in final form 14 September 2006)

Dole

ABSTRACT

Historically, the atmospheric sciences have tended to treat problems of weather and climate separately. The real physical system, however, is a continuum, with short-term (minutes to days) "weather" fluctuations influencing climate variations and change, and, conversely, more slowly varying aspects of the system (typical time scales of a season or longer) affecting the weather that is experienced. While this past approach has served important purposes, it is becoming increasingly apparent that in order to make progress in addressing many socially important problems, an improved understanding of the connections between weather and climate is required.

This overview summarizes the progress over the last few decades in the understanding of the phenomena and mechanisms linking weather and climate variations. The principal emphasis is on developments in understanding key phenomena and processes that bridge the time scales between synoptic-scale weather variability (periods of approximately 1 week) and climate variations of a season or longer. Advances in the ability to identify synoptic features, improve physical understanding, and develop forecast skill within this time range are reviewed, focusing on a subset of major, recurrent phenomena that impact extratropical wintertime weather and climate variations over the Pacific–North American region. While progress has been impressive, research has also illuminated areas where future gains are possible. This article concludes with suggestions on near-term directions for advancing the understanding and capabilities to predict the connections between weather and climate variations.

1. Introduction

Through his research and teaching, Professor Fred Sanders has contributed greatly to advancing our understanding and predictions of a broad range of atmospheric phenomena. Although most noted for his work on meso- and synoptic-scale phenomena, Professor Sanders also helped to stimulate new research linking slowly varying features of the planetary circulation with changes in synoptic-scale storm activity. This overview summarizes major developments in this area from early work (some of which was guided by Professor Sanders) through to our present understanding. A major underlying theme is the value of both synoptic and general circulation research methods in developing an understanding of how weather and climate phenomena are related, and how this knowledge can be used to improve forecasts.

In approaching this broad problem, it is useful to consider three related subquestions. First, how do climate variations affect weather phenomena? Second, how do weather phenomena affect climate variations?

Corresponding author address: Randall M. Dole, NOAA/Earth System Research Laboratory, Boulder, CO 80305.
E-mail: randall.m.dole@noaa.gov

And third, what are key phenomena and processes that bridge the time scales between synoptic-scale weather variability and climate variations of a season or longer? As one example of the first question, we might ask: How do climate variations influence hurricane behavior? Individual hurricanes evolve on weather time scales and cannot be predicted, say, a month or season in advance. However, climate phenomena such as El Niño–Southern Oscillation (ENSO), the quasi-biennial oscillation (QBO), and slowly evolving tropical ocean conditions have been shown to alter the probability that a given hurricane season will be more or less active (e.g., Gray 1984a,b; Bove et al. 1998; Vitart and Stockdale 2001). Because of such relationships, skillful seasonal hurricane activity outlooks can be made months and even seasons in advance (e.g., Owens and Landsea 2003; Klotzbach and Gray 2003, 2004).

As an example of the second question, we might ask: What are the effects of hurricanes on the climate system? Hurricanes clearly impact the atmospheric heat budget through the transfer of heat from the tropical oceans into the tropical upper atmosphere and to higher latitudes. More subtly, by altering the upper-ocean heat balance hurricanes can change the ocean thermohaline circulation, and thereby induce longer-term climate variations (Emanuel 2001, 2002). As another example, modeling studies show that transient eddy fluxes associated with synoptic-scale disturbances play a major role in determining the extratropical response to ENSO (Kok and Opsteegh 1985; Held et al. 1989; Hoerling and Ting 1994). More generally, many hydrological processes, such as clouds, moist convection, and water vapor transports operate on fast "weather" time scales, and yet also profoundly impact climate variations.

While we will touch on these first two questions, our main emphasis will be on the third question. More specifically, we will focus on key phenomena and processes that operate on time scales longer than typical synoptic-scale periods but less than a season, with a primary emphasis on understanding extratropical flow variations between approximately 1 week and a few months. On these time scales, initial conditions and boundary variations can each play important roles, and hence both need to be considered in attempts to advance predictive skill. Because of the vast body of related literature, we cannot hope to cover all relevant topics. Rather, our primary intent is to expose future researchers, and especially current students, to some key concepts and scientific steps that have led to progress in advancing understanding and predictions at the interface between weather and climate, and to suggest promising directions for future progress.

Some basic questions can be motivated by examining Fig. 1. This figure shows daily time series of precipitation in Los Angeles for two winters in which sea surface temperatures (SSTs) in the central and eastern equatorial Pacific were anomalously warm (El Niño conditions), and a third winter in which SSTs in the same

region were anomalously cold (La Niña conditions). It is immediately apparent that the seasonal-average rainfall is greater in the two El Niño years than in the La Niña year. More comprehensive statistical analyses show a significantly increased probability of above-normal precipitation during El Niño years over much of southern California and the southwestern United States (Schonher and Nicholson 1989; Cayan et al. 1999), although it should be noted that El Niño conditions are neither necessary nor sufficient for wet conditions to occur in this region (Schonher and Nicholson 1989). The seasonal averages themselves, however, provide only bland characterizations of the season as a whole. In particular, they do not reveal important differences in temporal behavior that occur within the seasons, the *intraseasonal variability* that is our primary focus here.

For example, even during the El Niño years there are extended periods of up to a month in which no rainfall occurs. While this behavior is well known to meteorologists, this knowledge has not always been conveyed adequately to the public, as evidenced by frequent comic references to "El No-Show" when midwinter 1997/98 was relatively dry in southern California despite official seasonal forecasts for abnormally wet wintertime conditions (which subsequently materialized in late January). Interspersed with these extended dry regimes are periods of several days to a few weeks in which the rainfall shows multiple peaks separated by a few days, the peaks being reflective of individual synoptic storms, which would be a primary focus for short-range forecasters. The longer-period organization of storm tracks is a crucial problem that we need to consider.

There are also indications of other differences that are highly relevant to emergency planning and resource management; for example, more frequent rainy days and more extreme rainfall events in the El Niño years, relationships that again appear in more detailed studies (Gershunov and Barnett 1998; Cayan et al. 1999; Gershunov and Cayan 2003; Andrews et al. 2004). While the extreme events themselves are clearly manifestations of individual weather events, how climate variability and change affect the likelihood of their occurrence is clearly a question of major societal significance. From a forecast perspective, we need to consider to what extent these probability shifts might be predictable, how they might be forecasted, how to identify the maximal time limits of predictability, and how to determine the factors that presently limit forecast skill.

This overview will begin with an informal, personal synopsis of the situation circa 1980, a time when the present author was a graduate student under Professor Sanders' guidance. Several major developments that occurred around and shortly after this time were vital to advancing our understanding of relationships between weather and climate variations. In the spirit of Professor Sanders' research and teaching, we will then turn our attention to describing key phenomena that bridge the time scales between weather and climate, and our

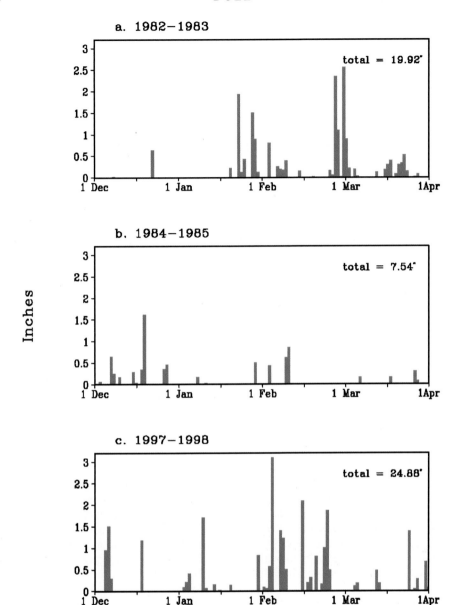

FIG. 1. Daily precipitation at Los Angeles airport for three extended winter seasons (December–April) associated with either El Niño (EN) or La Niña (LN) conditions: (a) 1982/83 EN, (b) 1984/85 LN, and (c) 1997/98 EN.

emerging understanding of the associated dynamical mechanisms. This will be followed by a discussion on the advances in forecast skill and analyses of potential predictability. We conclude with an outlook on prospects and potential directions for further progress in this area.

2. Background

A fundamental problem in meteorology is to advance the lead time of skillful weather forecasts. Sanders (1979), in an update of an earlier study (Sanders 1973), assessed weather forecast skill up to 1978. This assessment was confined to temperature and precipitation fore-

casts for a single station (Logan Airport in Boston) produced by a relatively small sample of forecasters, which consisted of the Massachusetts Institute of Technology (MIT) faculty (including Sanders) and students (including myself for a time). Despite the limited scope of the study, the results appear broadly representative of what was then the state of the art. Put simply, by lead times of 4 days, skill in temperature forecasts, as measured relative to the baseline forecast of climatological-mean values, was marginal, while skill in precipitation forecasts relative to that same baseline was nonexistent. Furthermore, trends in forecast skill evaluated over the period 1966–78 were very small, and in most cases not statistically significant.

Thus, in 1980 forecasts beyond a few days were effectively in the "extended range." While the U.S. National Weather Service did provide outlooks in the form of forecast maps for days 3, 4, and 5, the skill in these products was also limited, in agreement with Sanders' results. Among some MIT forecasters these maps were referred to whimsically as "5-day fantasy charts," in part because in wintertime they had a tendency to show strong cyclogenesis occurring near the East Coast by day 5, with implied major snowstorms for Boston and much of the Northeast corridor that usually failed to materialize.

The status of climate forecasting circa 1980 was, if anything, less advanced. As discussed by Gilman (1985) climate forecasts were at the time largely an empirical art, with no usable, quantitative theory underlying the forecasts, nor any guidance from dynamical model projections. Forecasts were based primarily on statistical relationships, such as linear correlations linking slowly varying oceanic and land surface boundary conditions with time-mean circulation anomalies. El Niño and its atmospheric counterpart, the Southern Oscillation, had attracted the interest of scientists for quite some time (Walker and Bliss 1932; Bjerknes 1969; Wyrtki 1975; Trenberth 1976). However, knowledge was limited and hence information on ENSO had yet to be effectively incorporated into climate forecasts. Indeed, in 1982, the development of one of the largest El Niño events of the century was not fully recognized while it was occurring, much less predicted (National Research Council 1996).

Thus, in the late 1970s, there was no significant forecast skill on time scales longer than a typical synoptic period (approximately 3–5 days) and shorter than a season. There was, however, abundant observational evidence for recurrent flow patterns that persisted beyond the periods associated with synoptic-scale variability. The most commonly cited example was "blocking," which is characterized by a quasi-stationary, persistent, and anomalously strong anticyclone located at mid- to high latitudes, often with a "split" westerly flow defined by separate, well-defined jet maxima located well to the north and south of the anticyclone center (e.g., Namias 1947; Elliott and Smith 1949; Rex 1950a,b; Sumner 1954). Once established, blocking can last for weeks or longer, with the associated persistent flow patterns appearing to divert (block) migratory storm systems far north and south of their normal tracks (Berggren et al. 1949; Petterssen 1956).

Blocking clearly had major implications for extended range forecasting and, indeed, has long been a focus for prediction efforts at the European Centre for Medium-Range Weather Forecasts (ECMWF) and other major operational forecast centers. By 1980, the problem of explaining blocking characteristics, including typical structures, geographical locations, and persistence, had also attracted substantial attention from dynamicists (Green 1977; Charney and DeVore 1979; Tung and Lindzen 1979a,b; McWilliams 1980). Thus, several factors appeared favorable for advancing our understanding of blocking, and thereby potentially improving lead times for extended-range forecasts.

A question remained, however, as to whether blocking was unique, or rather part of a broader class of recurrent phenomena that had time scales longer than typical synoptic periods. If so, could such features be identified and described within a more general and systematic framework? Furthermore, could relationships between the persistent flow anomalies and changes in synoptic-scale storm activity be more rigorously established, and fundamental dynamical mechanisms identified? These were the primary problems considered by myself in my thesis research guided by Professor Sanders. Subsequent progress on these and related questions will be discussed in the remainder of this overview.

3. Major phenomena and mechanisms

To keep the discussion reasonably compact, we will focus on a subset of major, recurrent phenomena that impact extratropical wintertime weather and climate variations over the Pacific–North American region. The characteristic features described here illustrate more general processes that also affect other regions and seasons.

a. Teleconnections and persistent anomalies

We begin by considering the question of whether early studies on blocking could be placed within a more general and systematic framework. Progress in this area has benefited from research that has approached this question from two fronts, one from the "weather side out," and the other from the "climate side in." The former has focused more on initial conditions and details of evolution out to a few weeks, often using synoptic analysis methods, while the latter has placed more emphasis on the effects of boundary conditions and used methods commonly applied in climate and general circulation studies.

The major foundations for identifying systematic behaviors of variability between a week and a season were laid in the late 1970s through pioneering studies of the Northern Hemisphere general circulation led by Blackmon, Lau, and colleagues (Blackmon 1976; Blackmon et al. 1977, 1979; Lau 1978, 1979; Lau and Wallace 1979). Blackmon (1976) described the spatial and temporal characteristics of wintertime variability in the Northern Hemisphere 500-hPa time field. He apportioned the temporal variance into three spectral bands: 1) periods of less than 2 days (high-pass variability); 2) periods between 2.5 and 6 days (bandpass variability); and 3) periods between 10 and 90 days (intraseasonal low-frequency variability), the latter being the time scales of most direct interest here.

Blackmon's analyses of low-frequency variability showed three major regional maxima, with the primary

2a) 500mb low—pass RMS variability

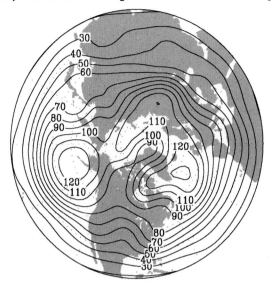

2b) 500mb band—pass RMS variability

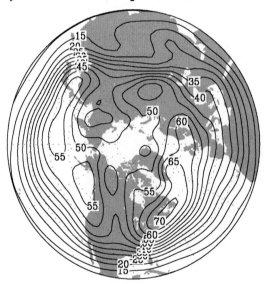

FIG. 2. Winter (DJF) root-mean-square variability of 500-hPa geopotential height fields for (a) low-pass-filtered data and (b) bandpass-filtered data, using the Blackmon (1976) time filters applied to NCEP–NCAR reanalysis data for the years 1950–2004.

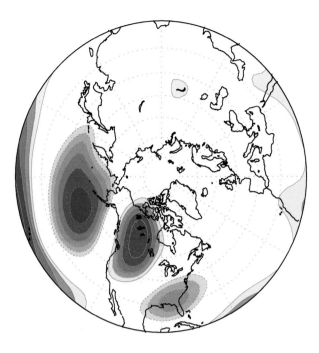

Correlation Coefficient

−0.9 −0.8 −0.7 −0.6 −0.5 −0.4 0.4 0.5 0.6 0.7 0.8 0.9

FIG. 3. Winter (DJF) correlations between the 300-hPa geopotential height field and the PNA index of Barnston and Livezey (1987), derived from the NCEP–NCAR reanalysis data for the same period as in Fig. 2.

centers located over the North Pacific to the south of the Aleutians, over the eastern North Atlantic to the southeast of Greenland, and over northwest Russia (Fig. 2a). The two oceanic low-frequency maxima were centered in the climatological-mean jet exit regions downstream of maxima in bandpass variability (Fig. 2b). Blackmon (1976) and subsequent studies showed that the bandpass variations were associated with migratory synoptic-scale storm systems, with maximum variance occurring near the locations of mean storm tracks de-

scribed in earlier synoptic climatology studies (Petterssen 1956; Palmen and Newton 1969). The bandpass variations exhibited westward phase shifts with height (Blackmon et al. 1979), consistent with the interpretation that they were produced by baroclinic growth processes (Charney 1947; Eady 1949). In contrast, the low-frequency variations had a more equivalent-barotropic structure, with little or no phase tilt with height throughout the troposphere (Blackmon et al. 1979; Schubert 1986).

The early studies by Blackmon, Lau, and collaborators established important characteristics of low-frequency variability, but did not specifically consider associated spatial patterns. Wallace and Gutzler (1981, hereafter WG81) provided a major step forward in this area in their landmark study on "teleconnections," which are defined by significant contemporaneous correlations of a given variable (e.g., 500-hPa heights or sea level pressure) occurring at widely separated geographical locations. While there had been considerable prior research on teleconnections, including monumental efforts on the Southern Oscillation by Walker and Bliss (1932), the WG81 study was particularly significant in providing a simple, objective, and systematic means of identifying major teleconnection patterns, in their study derived from Northern Hemisphere 500-hPa geopotential height and sea level pressure fields. Figure 3 shows one example of a teleconnection pattern dis-

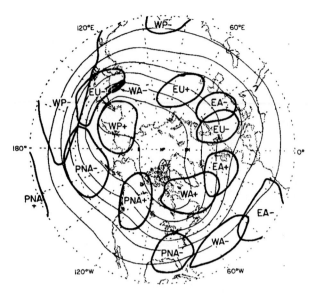

FIG. 4. Schematic of major teleconnection patterns. The map shows ±0.6 isopleths of correlation coefficient for each of the five teleconnection pattern indices and local 500-hPa geopotential heights (heavy lines), superimposed on wintertime mean 500-hPa contours (lighter lines). The contour interval is 120 m. From Wallace and Gutzler (1981).

cussed in detail by WG81, the Pacific–North American (PNA) pattern, as obtained from more recent data. Because of the importance of the PNA pattern for weather and climate variability, as well as its implications for extended-range predictability, we will later discuss this pattern and related dynamical mechanisms in more detail.

WG81 focused on teleconnections in extratropical Northern Hemisphere monthly-average 500-hPa data for three winter months (December–February). They described several strong teleconnection patterns (Fig. 4), many of which had been identified in earlier investigations. These were the western Pacific (WP) and western Atlantic (WA) patterns, which WG81 related to the North Pacific Oscillation (NPO) and North Atlantic Oscillation (NAO) patterns identified previously by Walker and Bliss and analyzed in more detail later by van Loon and Rogers (1978) and Rogers (1981); the PNA pattern, also evident in earlier studies (Klein 1952; Dickson and Namias 1976); an eastern Atlantic (EA) pattern that resembled a pattern described by Sawyer (1970); and a Eurasian pattern (EU). While WG81 did not perform detailed theoretical comparisons, they noted that several of the patterns strongly resembled Rossby wave trains obtained in linearized models on a sphere forced by steady, localized heating or topography (Egger 1977; Hoskins et al. 1977; Opsteegh and Van Den Dool 1980; Hoskins and Karoly 1981; Webster 1981).

Wallace and Gutzler's study presents an important example of what we have termed a climate approach to the problem, with its emphasis on the use of statistical techniques (linear correlations) applied to monthly-av-

erage height anomalies. Such approaches have the capability of identifying major recurrent features, with measures of reproducibility being provided through the estimated statistical significance of the relationships. Furthermore, the use of temporal averaging removes "noise" produced by higher-frequency fluctuations that are not of proximate interest. At the same time, potentially important information may be lost; for example, on how the associated flow patterns develop and evolve in time.

Dole (1982, 1983), influenced by Professor Sanders' synoptic methods, used an alternative approach for identifying persistent features that enabled a more detailed analysis of temporal evolution. Dole defined a "persistent anomaly" event for a given location whenever a 500-hPa height anomaly at that location exceeded a threshold value for longer than a specified duration (e.g., a 500-hPa anomaly of greater than 100 m for more than 10 consecutive days). In common with WG81, an important aspect of this method is that specific patterns are not defined a priori but, rather, are determined through subsequent data analyses. In this aspect, WG81's and Dole's approaches differed substantially from early studies of blocking, in which predefined flow patterns constituted the primary basis for case identification. This enabled WG81 and Dole to address the more general question of whether, and to what extent, blocking patterns were part of a broader class of recurrent low-frequency phenomena.

Dole and Gordon (1983, hereafter DG83) described the geographical distribution and regional persistence characteristics of persistent anomalies. In common with Blackmon (1976), DG83 identified three major regions of frequent occurrence: the North Pacific to the south of the Aleutians, the North Atlantic to the southeast of Greenland, and the region over northern Asia extending northeastward to the Arctic Ocean. While some of the cases identified in DG83 were clearly associated with blocking events, many others were not. For example, DG83's persistent negative anomaly cases over both the Pacific and Atlantic were associated with the persistent, abnormally strong cyclonic circulations in these regions, with intense, eastward-extended jets on their southern flanks. The persistence characteristics of anomalies in each of the regions broadly resembled those seen in a red noise process, with some modest deviations from red noise behavior for large magnitude events, as well as some apparently modest differences between positive and negative anomalies for the strongest events. DG83 also considered temporal characteristics of two primary regional patterns of variability, one essentially identical to the PNA pattern and the other to the EA pattern, but could find no compelling evidence for either strongly preferred durations or the existence of multiple equilibria (Charney and DeVore 1979).

More detailed analyses of persistent anomaly life cycles were provided in the following studies (Dole 1983, 1986a, 1989; Dole and Black 1990, hereafter DB90).

These studies indicated that the events often developed rapidly (a time scale of less than a week), and could break down similarly rapidly, with a typical life cycle from development through decay occurring within a month (for a more recent analysis of temporal characteristics, see Feldstein 2000). Despite the relatively rapid developments, once such events are established, they could project strongly on monthly and, in some cases, even seasonal mean variability, and strongly alter synoptic-scale storm tracks.

Figure 5 illustrates systematic features during development for persistent negative anomaly cases over the central North Pacific (centered near 45°N, 170°W) analyzed by DB90 and Black and Dole (1993). As discussed in DB90, the developments are typically preceded by an intensifying upper-level trough and jet streak over eastern Asia and the far western Pacific 3–5 days before case onset (Figs. 5a,b). At these times, height anomalies over the central Pacific are weak and, if anything, of opposite sign to the subsequent developments. Between days −5 and −1 (Fig. 5c), the upper-level trough and jet maxima propagate eastward and continue to intensify, thereafter becoming quasi-stationary over the central North Pacific. Following this time, the main center continues to intensify, while an alternating series of ridges and troughs develop and amplify in sequence downstream from North America to the central Atlantic (Figs. 5d–f). Subsequent downstream developments can be traced across western Europe and Russia (not shown).

By the early 1980s, several potential mechanisms for low-frequency variability had been proposed. As noted by WG81, linearized dynamical models on a sphere forced by localized heat or topographic sources produced forced Rossby wave responses of approximately the right spatial scales and, for forcing in particular regions, patterns that at least qualitatively resembled the observed teleconnections. Indeed, in a major paper on ENSO, Horel and Wallace (1981, hereafter HW81), obtained highly statistically significant correlations between various ENSO indices and the PNA pattern. HW81's schematic illustration of the hypothesized upper-tropospheric response to tropical diabatic heating associated with ENSO (Fig. 6) shows an arching wave train broadly resembling that of the PNA pattern (cf. Fig. 3).

One clear difficulty in making a direct connection between PNA variability and ENSO was the mismatch between time scales. While strong PNA events could grow and decay within a month and events of opposite sign could occur within a given season (WG81; DG83), ENSO events evolved much more slowly, with typical periods of a few years or longer (Trenberth 1976). A second concern involved the amplitudes of the PNA events, which were much larger than what could be produced in simple models linearized about zonal-mean basic states and forced by realistic values of tropical heating. A third concern involved rather subtle differences in the height patterns over the North Pacific. To better appreciate this last issue and aid in our subsequent discussion of potential mechanisms, it is useful to review a few basic ideas from Rossby wave theory.

b. Potential mechanisms

1) ROSSBY WAVE ENERGY DISPERSION FROM STEADY, LOCALIZED SOURCES

As a simple starting point, consider the dispersion relation for Rossby waves in an unbounded, barotropic β-plane model linearized about a constant zonal-mean zonal flow U (e.g., Holton 2004):

$$\omega = Uk - \frac{\beta k}{K^2}. \tag{1}$$

Here, ω is the frequency; $\beta = df/dy$ is the meridional derivative of the Coriolis parameter f; and $K^2 = k^2 + l^2$ is the total wavenumber squared, where $k\ (=2\pi/L_x)$ and $l\ (=2\pi/L_y)$ are the zonal and meridional wavenumbers, respectively, and L_x and L_y the corresponding zonal and meridional wavelengths, respectively. Because $\omega = kc$, where c is the phase speed, Rossby waves always propagate westward relative to the zonal mean flow. The direction of energy dispersion is related to the group velocity, which is obtained by differentiating (1) with respect to wavenumber. The resulting x and y components of the group velocity are

$$c_{gx} = U + \frac{\beta(k^2 - l^2)}{K^2} \quad \text{and} \tag{2}$$

$$c_{gy} = \frac{\beta kl}{K^2}. \tag{3}$$

From (2), the zonal group speed may be either eastward or westward relative to the zonal mean flow, depending on the structure of the waves. In particular, for waves with $k^2 > l^2$, which appear synoptically as north–south-elongated waves, and is a structure typical of much synoptic-scale variability (Wallace and Lau 1985), the zonal group speed is greater than the mean flow and energy disperses rapidly eastward. Conversely, in zonally elongated waves, more characteristic of low-frequency variability (Wallace and Lau 1985), energy dispersion is westward relative to the zonal mean flow. The direction of meridional energy propagation is determined by the sign of the meridional wavenumber, l, with northward energy propagation when l is positive (northwest–southeast-oriented phase lines) and southward propagation for negative l (southwest–northeast-oriented phase lines).

A particularly interesting case is the wavenumber at which Rossby waves are stationary, as this is a favored response scale for excitation due to geographically fixed forcing (e.g., Holton 2004). The stationary wavenumber is obtained from (1) by setting $\omega = 0$, from which it can be seen that stationary Rossby waves can only exist

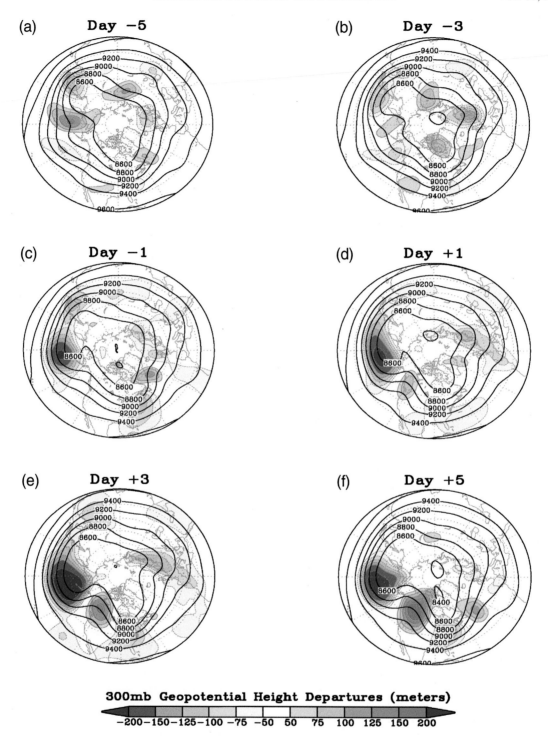

FIG. 5. Time evolution of composite 300-hPa heights (solid lines) and height anomalies (colors) for the 23 persistent negative height anomaly cases over the North Pacific analyzed in Dole and Black (1990), where time is relative to case onset, defined as day 0. For days (a) −5, (b) −3, (c) −1, (d) +1, (e) +3, and (f) +5.

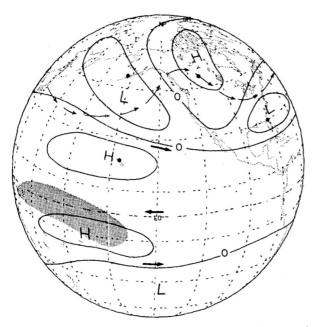

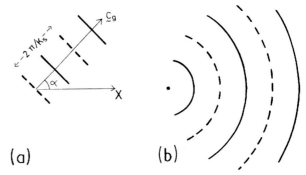

FIG. 7. (a) Stationary Rossby wave in a westerly flow on an infinite β plane with the crests (continuous lines) and troughs (dashed lines) making an angle α with the y axis. The wavelength is $2\pi/K_s$ and the group velocity relative to the ground is $c_g = 2\overline{u}\cos\alpha$. (b) A schematic illustration of the steady vorticity or height field waves forced from local source region on an infinite β plane in an atmosphere with a westerly flow $\overline{u}$ and wave dampening on a time scale T. The wavelength in all directions is $2\pi/K_s$ and the wave train fills a circle with its center at $x = \overline{u}T$ and radius $\overline{u}T$. From Hoskins (1983).

FIG. 6. Schematic illustration of the hypothesized global pattern of mid- and upper-tropospheric geopotential height anomalies (solid lines) during a Northern Hemisphere winter that falls within an episode of warm SSTs in the equatorial Pacific. The arrows in darker type reflect the strengthening of the subtropical jets in both hemispheres along with stronger easterlies near the equator during warm episodes. The arrows in lighter type depict a midtropospheric streamline as distorted by the anomaly pattern, with pronounced troughing over the central Pacific and ridging over western Canada. Shading indicates regions of enhanced cirriform cloudiness and rainfall. From Horel and Wallace (1981).

when the mean flow is westerly. Combining this relation with (2) and (3), for stationary barotropic Rossby waves we obtain the following:

$$c_{gx} = \frac{2Uk^2}{K^2} \quad \text{and} \tag{4}$$

$$c_{gy} = \frac{2Ukl}{K^2}. \tag{5}$$

From (4), energy dispersion in stationary Rossby waves is always eastward and, in the limiting case of stationary waves that are infinitely elongated in the north–south direction, occurs at twice the speed of the mean zonal flow. In addition, (4) and (5) imply that for stationary Rossby waves the group velocity vector is perpendicular to the wave crests; for example, northwest–southeast-oriented stationary waves indicate northeastward energy dispersion. Figure 7 illustrates these basic relationships.

This simple barotropic model can be extended to incorporate additional factors such as meridional shear in the basic state or the effects of spherical geometry in modifying effective β. Hoskins et al. (1977) considered the latter case, and showed that in a constant angular velocity superrotation, which is the spherical analog to constant zonal mean flow, energy dispersion occurs along great circle ray paths. Figure 8 shows one example

of energy dispersion on a sphere away from a fixed, localized source, in which the principal relationships discussed above are clearly evident. Inclusion of more realistic meridional shear results in refraction of the Rossby wave paths away from great circles as well as the formation of waveguides that preferentially organize energy dispersion (Hoskins and Ambrizzi 1993; Newman and Sardeshmukh 1998; Branstator 2002).

Returning to Fig. 6, it can be seen that the schematic ENSO pattern displays phase lines with a northwest–southeast orientation over the subtropical and midlatitude Pacific, consistent with northward energy dispersion by Rossby waves from a source in the tropical mid-Pacific (Hoskins et al. 1977; Plumb 1985). However, examination of the PNA pattern (Fig. 3) shows little evidence of systematic horizontal tilts that would suggest poleward energy propagation from this region. This discrepancy suggested a need for caution in interpreting the PNA pattern as a simple forced Rossby wave response to anomalous heating over the tropical mid-Pacific. However, for many years the terms "ENSO pattern" and "PNA pattern" were often used interchangeably. Mo and Livezey (1986) and Barnston and Livezey (1987) further clarified the observational distinctions between ENSO and PNA patterns, identifying a "tropical Northern Hemisphere" (TNH) pattern as being more directly related to ENSO. Straus and Shukla (2002) also showed that the midlatitude response to El Niño sea surface temperatures differs from the PNA pattern, which they identify as the leading structure of internal variability over the North Pacific in winter.

Despite these issues, it remained possible that ENSO, or other tropical forcing, could alter the likelihood of PNA pattern occurrence, essentially "loading the climate dice" so that events of a given type were more probable in a given year. A second possibility was that WG81's correlation analysis and Dole's use of com-

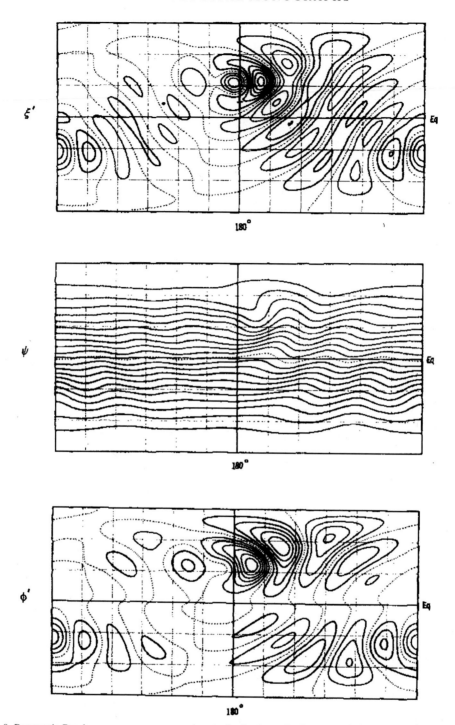

FIG. 8. Barotropic Rossby wave response on a sphere to forcing by a circular mountain located at 30°N in a uniform superrotation flow: (top) perturbation vorticity field, (middle) total streamfunction field, and (bottom) perturbation height field. From Grose and Hoskins (1979).

positing blurred any potential tropical signal, by including a mixture of events arising from both tropical and mid- or high-latitude sources. A third possibility hinged on the wave sources themselves. As noted by Sardeshmukh and Hoskins (1988), Rossby waves could be forced by the advection of vorticity by the divergent

flow as well as by divergence acting on the rotational flow; that is,

$$S_{RW} = -\mathbf{V}_x \cdot \boldsymbol{\nabla}\eta - \eta\boldsymbol{\nabla} \cdot \mathbf{V}, \qquad (6)$$

where S_{RW} is the Rossby wave source, $\mathbf{V}_x$ is the divergent wind component, and η is the absolute vorticity. Therefore, tropical convection could produce Rossby wave sources at latitudes well removed from where the convection itself was occurring; for example, through divergent outflow advecting absolute vorticity in the region of strong vorticity gradients associated with the subtropical jets. Thus, the potential role for tropical forcing remained open, although the simple interpretation of the PNA pattern as a stationary Rossby wave train forced from the tropical mid-Pacific appeared inadequate.

Despite this limitation, work during this period convincingly established the fundamental importance of Rossby wave dynamics for interpreting much low-frequency variability, and arching Rossby wave trains forced by local vorticity sources are now recognized as ubiquitous features in upper-level wind and potential vorticity analyses. As one indication of how our perspective had changed, it is interesting to recall an excellent earlier review on Rossby waves by Platzman (1968), in which he posed the provocative question: "Have Rossby waves ever been observed in the atmosphere?" While Platzman answers affirmatively, perhaps the clearest evidence he had at that time came from spherical harmonic decompositions of 24-h height field tendencies (e.g., Eliasen and Machenhauer 1965), an analysis technique that, while mathematically justified, is not as well-suited for synoptic interpretations of the Rossby wave trains described here, because of the broad band (high number) of individual wave modes required to adequately represent the evolution of observed teleconnection patterns.

2) INSTABILITIES OF ZONALLY VARYING MEAN FLOWS

A second proposed mechanism for producing the low-frequency variations was through the growth of unstable normal modes of the wintertime mean flow. Several candidates were suggested, including baroclinic instability (e.g., Frederiksen 1983) and barotropic instability (Simmons et al. 1983, hereafter SWB) of zonally varying basic states. The study by SWB was particularly noteworthy in showing how barotropic mechanisms could contribute to the development of flow patterns that strongly resembled the PNA and EA patterns. SWB interpreted the development of these patterns as a manifestation of the most rapidly growing mode associated with barotropic instability of the zonally varying climatological basic state (Fig. 9), since termed the SWB mode. To understand the essential development mechanism it is useful to consider the barotropic interactions between the time-mean flow and transient eddies for a mean flow that varies in both zonal and meridional directions.

To an accuracy of approximately 10%, the barotropic energy conversion from the zonally varying time-mean flow to the transient eddies is (Hoskins et al. 1983; SWB):

$$C(\overline{K}, K') \cong (\overline{v'^2} - \overline{u'^2})\frac{\partial \overline{u}}{\partial x} - \overline{u'v'}\frac{\partial \overline{u}}{\partial y}, \qquad (7)$$

where the overbars denote time averages, the primes departures from the time average, and

$$\overline{K} = \frac{\overline{u}^2 + \overline{v}^2}{2}$$

and

$$K' = \frac{\overline{u'^2 + v'^2}}{2}$$

are, respectively, the kinetic energy of the time-mean flow and transient eddies. Following Hoskins et al. 1983 and SWB, this may also be written as

$$C(\overline{K}, K') \cong \mathbf{E} \cdot \boldsymbol{\nabla}\overline{u}, \qquad (8)$$

where

$$\mathbf{E} = [(\overline{v'^2} - \overline{u'^2}), -\overline{u'v'}] \qquad (9)$$

is the barotropic "**E** vector." This latter representation is useful for identifying dynamical relationships between transient eddies and the time-mean flow, as will be discussed in more detail later in this section.

The second term in (7), associated with the meridional shear in the mean zonal flow, is identical in form to the barotropic conversion term for zonally symmetric basic states (e.g., Pedlosky 1979; Holton 2004). This term supports eddy growth in regions where eddy phase lines tilt against the shear. In this case, meridional eddy momentum fluxes transfer zonal momentum away from regions where the zonal flow is initially strong (i.e., the jet) thereby smoothing the jet profile (reducing the meridional shear). The first term is nonzero only when there are zonally varying basic states. This term supports eddy growth when zonally elongated eddies ($\overline{u'^2} > \overline{v'^2}$) enter regions where the climatological mean zonal flow is decreasing eastward ($\partial \overline{u}/\partial x < 0$), as is the case in the jet exit regions over the North Pacific and North Atlantic Oceans. Alternatively, barotropic eddy growth is also supported when meridionally elongated eddies enter regions where the mean zonal flow increases eastward, as in the confluent flow region upstream of the time-mean jet streams.

Figure 10 schematically illustrates the mean flow and perturbation relationships favorable for barotropic eddy growth for basic states characterized by diffluence and strong meridional shear, which are commonly observed states over much of the central North Pacific in wintertime. The barotropic conversion term in (8) can be shown to have the simple geometric interpretation that the eddies will gain energy if the mean flow deformation makes them more "circular," that is, increases eddy isotropy (Farrell 1984; Mak and Cai 1989; Cai 1992; Whitaker and Dole 1995; Black and Dole 2000). Conversely, the sense of the conversions will be from the eddies into the mean flow when mean flow deformation makes the eddies more anisotropic. A familiar synoptic

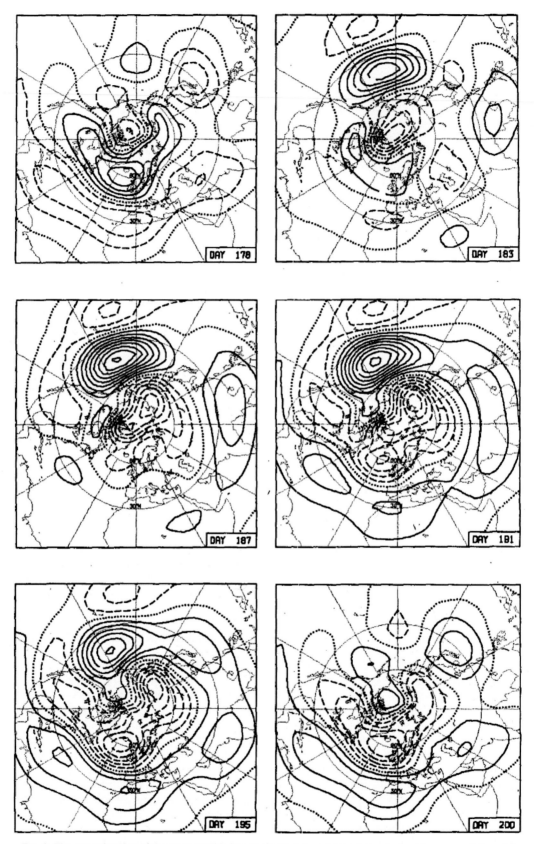

FIG. 9. The streamfunction of the most unstable barotropic mode in a zonal-varying basic state representative of the Northern Hemisphere wintertime time-mean flow, for selected days within its evolution. The contour interval is arbitrary. From Simmons et al. (1983).

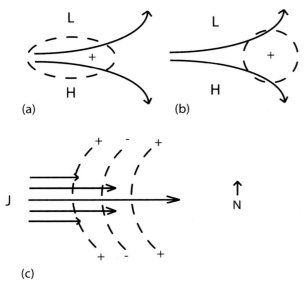

FIG. 10. Schematic illustration of (a) favored structure for barotropic growth for an eddy encountering a diffluent flow downstream of jet maxima, (b) change in eddy structure due to subsequent deformation by large-scale flow, and (c) favored structure for eddy development in a horizontal shear flow. In these figures, the solid lines denote the mean flow streamlines, the dashed lines are the eddy vorticity isopleths, L and H indicate relative minima and maxima in the mean height fields, respectively, and J indicates the mean jet axis.

example is the "shearing out" of upper-level short-wave troughs, with scale collapse occurring along one axis (axis of contraction) while stretching occurs along the orthogonal axis (axis of dilatation).

The SWB theory provided a potential explanation for both the observed locations and favored structures of maximum low-frequency variability over the central North Pacific and eastern North Atlantic Oceans (Blackmon et al. 1984; Wallace and Lau 1985; Dole 1986b; Kushnir and Wallace 1989). However, the SWB theory also had potential inadequacies. One was that observational analyses showed strong thermal advection and westward (upshear) vertical tilts in the growing eddy structures over the central North Pacific during PNA development (Dole 1986a; DB90; Black and Dole 1993), suggesting a potentially significant role for baroclinic growth processes. A second issue was the rate of disturbance growth. When realistic damping is included in the SWB model, the barotropic instability mechanism alone appeared inadequate to account for observed growth rates (Borges and Sardeshmukh 1995; Sardeshmukh et al. 1997). Thus, while barotropic instability on a zonally varying basic state could contribute significantly to growth of the PNA and EA patterns, it seemed unlikely to be the sole source. What had emerged from these studies, however, was the central role of zonal variations in the basic state in accounting for both the development and favored locations of observed low-frequency variations.

3) MULTIPLE-EQUILIBRIA AND FLOW REGIMES

A third theory proposed around 1980 was that persistent anomaly patterns are the manifestation of multiple "quasi-equilibrium" flow states that could exist even with fixed external forcing. Early studies of the wintertime general circulation (e.g., Rossby et al. 1939; Namias 1947, 1950; Willett 1949) suggested that the midlatitude flow tended to vary between two extreme states, one characterized by a relatively weak stationary waves and a strong westerly flow (high-zonal-index state) and the other by highly amplified stationary waves and weak westerlies (low-zonal-index state). In fact, understanding the causes for high- and low-zonal-index states provided a major motivation for Rossby's seminal paper on planetary waves (Rossby et al. 1939). From our previous discussion on Rossby wave dynamics, it is clear that zonal-mean flow variations have significant implications for wave propagation and energy dispersion. There are several potential causes for such variations, and improving our understanding and ability to model zonal-mean wind changes continues to be an important subject for research (e.g., Weickmann and Sardeshmukh 1994; Feldstein and Lee 1998; Feldstein 2001; Weickmann 2003).

Charney and DeVore (1979) showed that, in a simple nonlinear barotropic model forced by topography, two equilibrium states having features similar to high- and low-zonal-index flows could be produced through wave–mean flow interactions, and suggested that blocking might be a metastable equilibrium associated with the low-index state. In this very simple model, the equilibrium states result from a nonlinear balance among zonal flow driving, topographic forcing, and dissipation. This study was followed by extensions to low-order baroclinic models forced by topography (Charney and Straus 1980; Reinhold and Pierrehumbert 1982). Some early observational analyses found support for this theory in explaining blocking events (Charney et al. 1981); however, others found no convincing evidence (DG83). A host of efforts have since searched for multiple quasi-equilibria, or "weather regimes," in observational data (e.g., Hansen and Sutera 1986, 1988; Molteni et al. 1990; Cheng and Wallace 1993), and the debate has been vigorous on the extent to which recurrent nonlinear regimes have been identified (e.g., Nitsche et al. 1994). While questions continue on whether this theory accounts for specific observed states, the possibility that nonlinear quasi-stationary wave regimes can exist and play a significant role in climate dynamics cannot be discounted. Such a possibility has significant implications for climate predictions as well as for projecting future climate changes due to anthropogenic or natural forcing (Palmer 1998, 1999).

4) FORCING BY SYNOPTIC-SCALE TRANSIENT EDDIES

A fourth potential mechanism for low-frequency variations is systematic forcing by synoptic-scale eddies.

In discussing this mechanism, it is helpful to consider two subquestions: first, what are the effects of the low-frequency flow variations on synoptic-scale eddies, and second, what are the effects of synoptic-scale eddies on the low-frequency flow? The first question is directly related to changes in storm tracks and eddy life cycles that accompany persistent flow anomalies, while the second concerns the role of eddy feedbacks on the development, maintenance, and breakdown of persistent flow anomalies. Although it is conceptually useful to consider the two issues separately, it is also important to recognize that interactions between synoptic-scale eddies and the large-scale flow are nonlinear, and that subtle, indirect effects of the interactions can be quite significant. Consequently, interpreting eddy–mean flow relationships from observational data is especially challenging. This section touches on some of the major advances occurring post-1980, focusing on the behavior of synoptic-scale eddies during blocking events.

(i) Effects of large-scale flow anomalies on synoptic-scale activity

Early synoptic investigations indicated that blocking events are accompanied by major changes in synoptic-scale storm activity (e.g., Berggren et al. 1949; Rex 1950a,b). Other studies described cases where intense synoptic-scale cyclogenesis was followed by the downstream development of large-scale flow anomalies, especially blocking (Sanders and Gyakum 1980; Hansen and Chen 1982; Colucci 1985, 1987; Mullen 1987; Tracton 1990). Several characteristic synoptic features are illustrated in Fig. 11 (from Petterssen 1956). This figure shows daily frontal positions over two 10-day periods prior to and following the establishment of blocking over the eastern Atlantic and western Europe. The 10-day period before blocking formation (Fig. 11a) is characterized by a strong, predominantly zonal upper-level flow over the North Atlantic. Frontal systems during this period propagate mainly eastward along a relatively narrow band across the North Atlantic, with a primary genesis region extending from over the southeastern United States across the western Atlantic. In contrast, following the development of blocking (Fig. 11b), the storm track over the central and eastern Atlantic splits into branches well north and south of its initial latitude. Frontal systems approaching the blocking region from the west appear to elongate meridionally before splitting into the northern or southern branches of the flow. During this period, an area of enhanced cyclogenesis also occurs to the east of Greenland. Prior to blocking development, frequent frontal passages led to heavy precipitation over much of western and central Europe; following blocking development, little or no precipitation was observed over the same region (Petterssen 1956).

The early investigations demonstrated that changes in synoptic-scale eddy activity and low-frequency flow variations are related, but the extent and dynamical interpretation of the relationships remained uncertain. Green (1977), Shutts (1986), Dole (1986a,b), Mullen (1986, 1987), Wallace et al. (1988), Lau (1988), and Nakamura and Wallace (1993) among others, documented the relationships more quantitatively by applying temporal filtering techniques to analyze synoptic eddy behavior during blocking events. Figure 12 (from Mullen 1987) shows a good example of this method for evaluating systematic relationships between the time-mean flow and synoptic-scale eddy activity for a composite of Atlantic blocking cases. Note the qualitative similarities of the "bandpass" storm tracks (Fig. 12b) to the earlier synoptic analyses (Fig. 11b), especially the pronounced northward shift over the eastern Atlantic and the indication of a new maximum near southeast Greenland. Nakamura and Wallace (1990) and Neilley (1990) further extended these studies of "mature phase" relationships by analyzing changes in synoptic-scale eddy activity through the life cycles of low-frequency circulation anomalies. These studies showed a tendency for enhanced synoptic-scale variance upstream of developing blocking patterns a few days prior to the onset of blocking, consistent with synoptic descriptions of strong upstream synoptic-scale cyclogenesis preceding blocking development.

To first order, the observed differences in eddy activity associated with persistent flow anomalies conform to synoptic experience and simple theory. In general, the storm tracks approximately coincide with the zones of maximum time-mean baroclinicity. Enhanced variability is also preferentially located near and downstream of major long-wave troughs and, conversely, variability is usually suppressed near and immediately east of major long-wave ridges. However, as shown quite strikingly by Nakamura (1992), baroclinicity alone is not the sole factor modulating the strength of synoptic-scale variability. Nakamura found that maximum baroclinic wave activity (as measured by bandpass variance in heights, meridional heat fluxes, and other variables) occurs in autumn and spring over the western and central North Pacific (Fig. 13), with a midwinter suppression in activity at a time when the climatological-mean baroclinicity is strongest. Nakamura showed that this midwinter suppression occurs when the strength of the upper-tropospheric jet exceeds a threshold of ~45 m s^{-1}. He suggested that this midwinter minimum, which is not observed in the Atlantic storm track, might be related to rapid advection of the disturbances through the highly baroclinic region over the western Pacific as well as increased low-level trapping of the waves. Other possible mechanisms, such as seasonal changes in barotropic shearing deformation and condensational heating (Chang 2001) have also been suggested for this interesting within-season variation, which is manifested in interannual (Nakamura et al. 2002) as well as seasonal time scales, and the problem continues to be actively

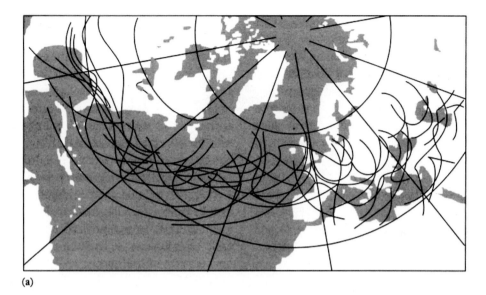

(a)

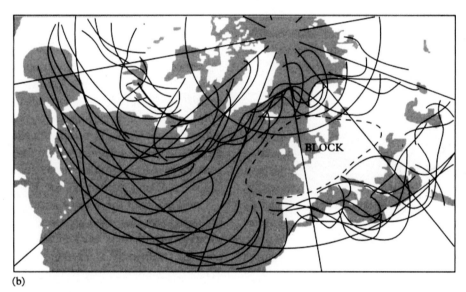

(b)

FIG. 11. Daily frontal positions during two successive 10-day periods (a) preceding the formation and (b) following the development of blocking over the eastern Atlantic and western Europe. From Petterssen (1956).

investigated (Zhang and Held 1999; Chang 2003; Harnik and Chang 2004).

(ii) Advances in dynamical understanding of storm track variations

The progress in describing relationships between low-frequency flow variations and synoptic-scale variability has been accompanied by major advances in our understanding and ability to model storm tracks. For the extratropical wintertime flow, a rough first approximation for preferred storm tracks can be obtained from spatial variations in baroclinicity as shown, for example, by maps of local time-average values of the Eady growth parameter σ (Eady 1949):

$$\sigma = 0.31 \frac{f \frac{\partial \overline{u}}{\partial z}}{N}, \tag{10}$$

where N is the buoyancy frequency (Holton 2004). However, rapid development does not always occur in such regions. Through the 1980s and 1990s, horizontal shear (James 1987) and deformation in the basic-state flow (Mak and Cai 1989; Cai and Mak 1990; Lee 1995a,b; Whitaker and Dole 1995; Branstator 1995) became increasingly recognized as playing important roles in organizing storm tracks and determining differences in synoptic-scale storm life cycles (Thorncroft et al. 1993, and their "type-1" versus "type-2" life cycles).

(a) 500mb geopotential height

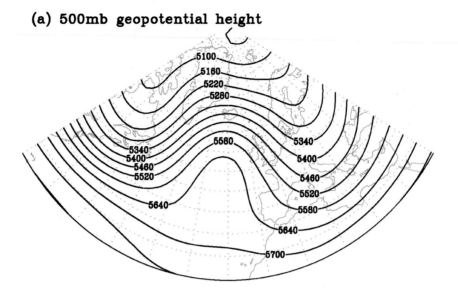

(b) 500mb standard deviation

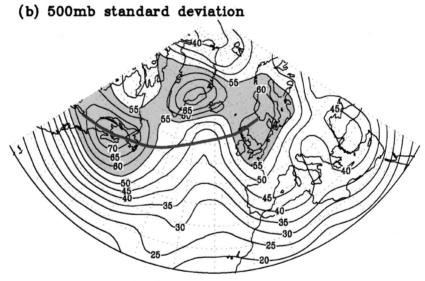

FIG. 12. Distributions of (a) 500-mb composite-mean geopotential height contours and (b) standard deviation of Blackmon's (1976) bandpass-filtered 500-mb geopotential height for observed Atlantic blocking composites. In (b), shading represents values greater than 55 m and the heavy red line represents the axis of the climatological-mean position of the "storm track." From Mullen (1987).

More recent work has reinforced the importance of changes in baroclinicity, horizontal shear and deformation as well as "seeding" by upstream disturbances in accounting for storm-track changes (Chang 2001, 2005; Chang and Fu 2002; Orlanski 2005).

Another crucial theoretical advance was to show that significant transient growth of disturbances could occur even if the flow was formally stable, given the presence of favorably configured perturbations in flows that had either vertical or horizontal shear or deformation (Farrell 1982, 1985, 1984; Mak and Cai 1989). This has led to a linear theory of storm tracks in which many of the observed features of Northern Hemisphere synoptic-scale variability can be deduced simply by considering the eddies as stochastically forced disturbances evolving on a baroclinically *stable* basic-state flow (Whitaker and Sardeshmukh 1998; Zhang and Held 1999).

As discussed by Sanders (1988), among others, the evolving vertical structures obtained in theoretical studies by Farrell (Farrell 1984) and others strongly resemble observed cases of "type-B" cyclogenesis as described by Petterssen (1955). In this form of development, the vertical structure of disturbances evolves systematically, with an initial upshear vertical tilt between the surface low and upper-level trough that decreases as the disturbance intensifies, until the upper-level

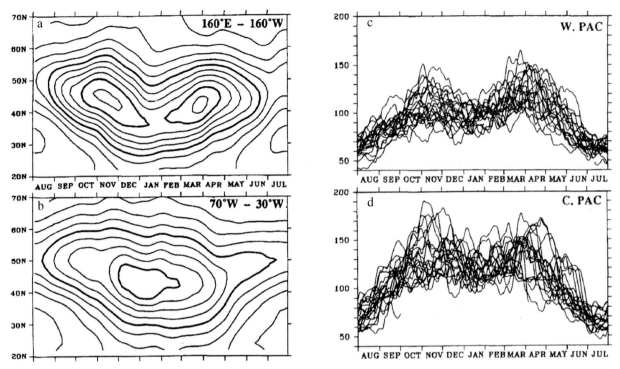

FIG. 13. Latitude–time sections showing the seasonal march of baroclinic wave amplitudes, defined using the temporal filter of Nakamura (1992) in the 250-hPa geopotential heights averaged over the longitude intervals (a) 160°E–160°W and (b) 70°–30°W. Seasonal march of the same fields averaged over (c) the western Pacific (130°–170°E) and (d) central Pacific (160°E–160°W) for individual 12-month periods from August 1965 to July 1984. From Nakamura (1992).

trough becomes approximately in phase with the surface cyclone center. Hoskins et al. (1985, hereafter HMR) provide a simple dynamical interpretation for this type of development in their landmark paper on potential vorticity dynamics. Davis and Emanuel (1991) applied the essential concepts of potential vorticity invertibility discussed in HMR to diagnose the contributions of potential vorticity perturbations at different locations and pressure levels to cyclogenesis. Black and Dole (1993) used a similar method to diagnose sources for the development of cases of strong positive PNA patterns. Their results suggested that nonmodal transient growth involving both barotropic and baroclinic processes likely plays a significant role in these developments (see also Feldstein 2002). Black (1997) subsequently extended these analyses to consider the life cycles of both polarities of the PNA and EA patterns. He found a similar mechanism was responsible for the developments in all types of cases considered. The primary wave source regions were located in the central North Pacific and eastern North Atlantic, respectively, with Rossby wave dispersion from the source regions leading to the establishment of the mature teleconnection patterns.

Beyond these relatively large-scale features, research into storm-track dynamics, most notably by E. Chang and colleagues, shows that synoptic-scale activity is organized into wave packets, which typically disperse energy downstream along baroclinic waveguides at speeds

comparable to the strength of the mean flow. Lee and Held (1993) first noted this behavior in simple models of the Southern Hemisphere general circulation, and Chang (1993) showed that this behavior was also prevalent in the Northern Hemisphere. Figure 14, from Chang et al. (2002) illustrates this behavior over a portion of one winter. Individual trough and ridge systems can be seen propagating within the packets at speeds on the order of 10 m s^{-1}. However, these individual disturbances are imbedded within larger wave packets that propagate more rapidly downstream. This is manifested synoptically by a succession of "downstream developments" (see also Namias and Clapp 1944). The downstream energy fluxes related to this mechanism extend the storm track from regions of strong baroclinicity to regions less favorable for baroclinic development. Chang et al. (2002) provide an excellent review of storm-track dynamics in which this mechanism is discussed in more detail.

(iii) Effects of synoptic-scale eddies on large-scale flow anomalies

We have seen that low-frequency flow anomalies alter synoptic-scale variability and, hence, can contribute to systematic changes in storm behavior like those inferred from Fig. 1. What about the second question, the effect of synoptic-scale eddies on the low-frequency flow?

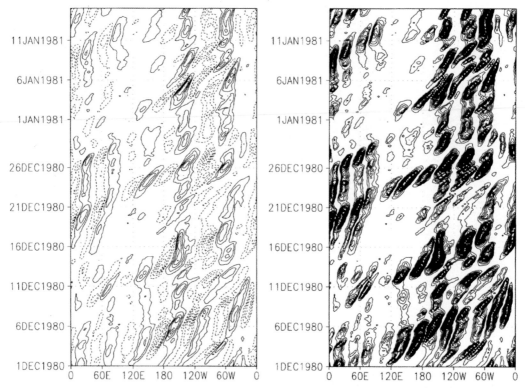

FIG. 14. Hovmöller (longitude–time) diagram of 300-hPa v' and v'^2, unfiltered except for the removal of seasonal mean, for the period 1 Dec 1980 to 14 Jan 1981 (contours 10 m s^{-1} and 100 m s^{-2}). In this figure, v' has been averaged over a 20° latitude band centered on the upper-tropospheric waveguide as defined in Chang and Yu (1999). From Chang et al. (2002).

Here again, there have been substantial advances in understanding. Studies in the late 1970s and early 1980s focused initially on time-mean heat and vorticity budgets during blocking events, in attempts to distinguish relative changes in time-mean and transient eddy forcing (Green 1977; Illari 1984; Shutts 1986; Mullen 1986). To assess the net effects of the eddies, it is necessary to take into account the combined effects of eddy fluxes of vorticity and heat as well as the eddy-induced secondary circulations. The conservation and invertibility principles of potential vorticity dynamics (HMR) play a fundamental role in helping to interpret such relationships.

At the level of quasigeostrophic theory (Holton 2004) the mean flow tendency forced by eddies is due to the convergence of the eddy flux of quasigeostrophic pseudopotential vorticity in the interior of the atmosphere:

$$\frac{\overline{\partial q}}{\partial t} = -\overline{\mathbf{V}} \cdot \nabla \overline{q} - \nabla \cdot (\overline{\mathbf{V}'q'}) + \overline{S},\qquad(11)$$

where the quasigeostrophic pseudopotential vorticity q is

$$q = \frac{1}{f_0}\nabla\Phi + f + \frac{\partial}{\partial p}\left(\frac{f_0}{N_b}\frac{\partial\Phi}{\partial p}\right)\qquad(12)$$

together with the eddy flux convergence of potential temperature, θ, on the lower boundary (Hoskins 1983):

$$\frac{\overline{\partial\theta}}{\partial t} = \overline{\mathbf{V}} \cdot \nabla\overline{\theta} - \nabla \cdot (\overline{\mathbf{V}'\theta'}) + \overline{H}.\qquad(13)$$

In (11) and (13), the overbars denote any simple averaging process, which here we will take to represent a time mean, with the primes denoting deviations from the mean values. In (11) and (13) $\overline{S}$ and $\overline{H}$ represent time-mean frictional and diabatic forcing terms, respectively. It is worth keeping in mind that, while the eddy forcing appears explicitly only in the eddy flux convergence terms, the eddies can also force the mean flow indirectly by altering the time-mean diabatic and frictional forcing terms (Hoskins 1983).

From (11) and (13), a time-mean "tendency equation" for the geopotential height field can be obtained (Lau and Holopainen 1984), which is analogous to the standard quasigeostrophic geopotential tendency equation (Holton 2004), with eddy flux convergence terms for vorticity and temperature replacing the corresponding vorticity advection and thermal advection terms. The eddy contributions to the initial time mean tendency can be written as

$$\left[\frac{1}{f_0}\nabla^2 + f_0\frac{\partial}{\partial p}\left(\frac{1}{N_b}\frac{\partial}{\partial p}\right)\right]\overline{\frac{\partial\Phi}{\partial t}} = D^{\text{HEAT}} + D^{\text{VORT}}, \quad (14)$$

where

$$D^{\text{HEAT}} = f_0\frac{\partial}{\partial p}\left(\frac{\nabla\cdot\overline{V'\theta'}}{N_b}\right) \quad (15)$$

and

$$D^{\text{VORT}} = -\nabla\cdot\overline{V'\zeta'}. \quad (16)$$

Given the eddy forcing terms (15) and (16) together with appropriate boundary conditions for the eddy heat fluxes (Lau and Holopainen 1984), the elliptic equation [(14)] can be solved to give the initial geopotential tendency that would be expected for a specified transient eddy forcing. Lau and Holopainen used this method to determine the initial tendency of the climatological-mean flow due to transient eddies, while Mullen (1987) and Holopainen and Fortelius (1987) applied a similar method to assess the mean-flow tendencies produced by transient eddies during blocking situations. These studies showed that, at least in the vicinity of the blocking patterns, the net upper-level tendency patterns is dominated by the vorticity fluxes [(16)]. Overall, there was net anticyclonic forcing near and upstream of the blocking ridge, tending to reinforce the block and counteracting the tendency for the ridge to be advected eastward by the time-mean flow (Mullen 1987). In a similar analysis of monthly variability, Lau and Nath (1991) found that the initial height tendencies forced by synoptic-scale eddies usually reinforced observed monthly mean upper-level height anomalies. Subsequent studies applying analogous diagnostic approaches indicate that forcing by synoptic-scale eddies can also play a significant role in blocking onset (e.g., Neilley 1990; Nakamura et al. 1997).

An alternative means for assessing wave–mean flow interactions was developed in the 1980s, based on extensions of Eliassen–Palm (EP) flux diagnostics developed for zonal-mean flows to zonally varying time-mean flows (Hoskins et al. 1983; Plumb 1985, 1986; Trenberth 1986). For zonal-mean flows, the EP flux vectors indicate the direction of flux of wave activity, while the EP flux divergence provides a measure of the net eddy forcing of the zonal-mean flow (Andrews and McIntyre 1976; Edmon et al. 1980). Deriving similar relationships for zonally varying time-mean flows requires additional approximations, and hence different formulations have been proposed; see Trenberth (1986) for a discussion of the differences. Here, we will sketch the basic ideas using the formulation of Hoskins et al. (1983).

To illustrate this method, we confine our discussion to the horizontal flux components only, which are then given by the terms in (9). As discussed earlier, the x component of the E vectors is related to horizontal shape asymmetries of the disturbances, with positive (east-ward) values associated with meridionally elongated eddies, and negative values with zonally elongated eddies. The y component is related to horizontal phase tilts, with positive (northward) values being associated with phase lines that slope westward with increasing latitude (i.e., northwest–southeast-tilted ridges and troughs).

As discussed by Hoskins et al., subject to certain approximations, the contribution to the acceleration of the mean westerly flow by transient eddies is given by

$$\left(\frac{\overline{\partial u}}{\partial t} + \overline{u}\frac{\partial\overline{u}}{\partial x} + \overline{v}\frac{\partial\overline{u}}{\partial y}\right)_{\text{TE}} = \nabla\cdot\mathbf{E}. \quad (17)$$

In particular, where $\nabla\cdot\mathbf{E} > 0$ there is a net eddy forcing that tends to increase the westerly component of the mean flow.

Figure 15 (from Shutts 1986) displays E vectors for a case of Atlantic blocking. In this study, E vectors were derived from 300-hPa data that had been filtered to retain periods of less than a week (corresponding to synoptic-scale eddies). Well upstream of the blocked region, the E vectors are mainly directed eastward, indicating eddies that are primarily meridionally elongated. As the eddies approach the region where the large-scale flow becomes strongly diffluent, the magnitudes of the E vectors first increase, and then become very small. The initial increase is the manifestation of baroclinic growth as well as a tendency toward reduction of the eddy zonal wavelengths as disturbances become stretched meridionally in the region of strong large-scale flow deformation. The smaller magnitudes are associated with the very low values of synoptic-scale variability observed in the vicinity of the blocking anticyclone. Shutts interpreted these E-vector patterns as the signature of a barotropic "eddy-straining" mechanism that reduces the east–west scale of the eddies as they encounter the strongly diffluent flow upstream of the block. The variations in the eddies as they approach the block leads to a $\nabla\cdot\mathbf{E} < 0$ near and just to the south of the anticyclonic vorticity center, implying a tendency to decrease the strength of the mean westerly flow in a region where it is already weak. In addition, there is a latitudinal fanning out of E vectors (indicative of "bowed" troughs and ridges) in the westerlies well to the south of the anticyclone giving $\nabla\cdot\mathbf{E} > 0$, implying a tendency for the eddies to increase the strength of the mean westerlies in the region where the westerlies are already anomalously strong. An additional region, where $\nabla\cdot\mathbf{E} > 0$, is located in the westerlies to the north of the blocking anticyclone.

Thus, these analyses suggest that the large-scale flow anomalies associated with blocking alter synoptic-scale eddy activity *and* that these changes result in a net positive eddy feedback that reinforces the large-scale flow anomalies. Subsequent studies have made use of similar techniques to document changes in synoptic-scale eddy structures and diagnose eddy feedbacks onto low-frequency anomalies (Neilley 1990; Higgins and Schubert

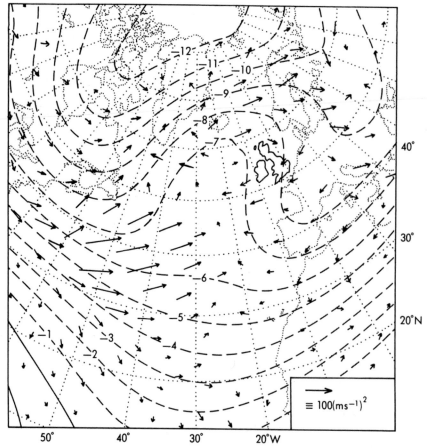

FIG. 15. High-pass-filtered **E** vectors superimposed on the mean streamfunction field for a North Atlantic blocking case for the period 5–22 Feb 1983. From Shutts (1986).

1993, 1994; Nakamura et al. 1997). These studies indicate that positive eddy feedbacks frequently occur in association with both blocking and unusually strong zonal flows. In an interesting diagnostic analysis of an Atlantic blocking case, Hoskins and Sardeshmukh (1987) found that enhanced large-scale diffluence over the western and central Atlantic produced by anomalous tropical heating provided an important catalyst for reorganizing Atlantic synoptic-scale eddy activity, with changes in eddy potential vorticity fluxes then leading to blocking development.

(iv) Additional comments

Aside from the direct effects of the eddy fluxes, synoptic-scale eddies can also exert important indirect effects on the large-scale flow. For example, in midlatitudes, changes in time-mean diabatic heating are related to changes in the storm tracks, primarily through changes in spatial distributions of latent (but also sensible and radiative) heating. The relationships between latent, sensible, and radiative heating distributions and storm tracks are incompletely understood, but certainly are geographically dependent. To the extent that the synoptic-scale eddies influence the mean flow, they also indirectly alter the generation and propagation characteristics of planetary-scale waves.

Interactions of low-frequency flow anomalies with topography can also impact synoptic-scale eddy activity, with variability typically being enhanced on the lee sides of major mountain barriers in regions of stronger-than-normal flow (Hsu and Wallace 1985; Hsu 1987; Dole 1987; Lau 1988). These include secondary maxima to the lee of the U.S. Rockies in the negative PNA cases and to the lee of the Canadian Rockies in the positive PNA cases (Dole 1987), and to the lee of Greenland in the positive EA cases, consistent with Mullen's results (cf. Fig. 12b).

While time-mean budget studies have added significantly to our understanding, their interpretation can be difficult, and in some circumstances they may even obscure important dynamical processes (Held and Hoskins 1985). For example, consider a case in which strong eddy–mean flow interactions occur until the flow approaches a quasi-equilibrium state where the eddy forcing terms becomes very small. Budgets evaluated at this later time may suggest that the role of the eddies is negligible; however, the quasi-equilibrium state may not be obtainable in the absence of the eddies. Reinhold and Pierrehumbert (1982) found such a behavior in a low-order baroclinic model. In establishing causal mechanisms, careful case studies focusing on the time evolution of events are often most useful; in this regard, Professor Sanders' synoptic case studies provide many clear examples. For large-scale flows in which balance conditions are well met, studies focusing on the time evolution of potential vorticity can be particularly illuminating (HMR; Morgan and Nielsen-Gammon 1998).

Despite the differences in methods, synoptic and diagnostic results are consistent in indicating that, in the development and maintenance of blocking and other persistent flow anomalies, synoptic-scale eddies can and often do play a major reinforcing role. The results paint a rich picture of the interplay between synoptic-scale eddies and the large-scale flow on which they evolve. From an operational standpoint, the results suggest that the failure to adequately simulate the large-scale flow will result in errors in predicting the regions of cyclogenesis and subsequent storm tracks. They also suggest that large-scale flow errors can arise from systematic deficiencies in modeling synoptic-scale disturbances.

c. Tropical connections

We have introduced several potential mechanisms for producing extratropical low-frequency variability, including Rossby waves forced by localized diabatic or topographic sources, large-scale flow instabilities (or nonmodal growth), quasi-equilibrium states representing a nonlinear balance between mean flow and forcing terms, and forcing by anomalous synoptic-scale eddy fluxes. While all of the above mechanisms can occur solely through midlatitude processes, the potential impact of tropical diabatic heating on extratropical low-frequency variability has been and continues to be a major focus for research, especially in studies aimed at improving weather and climate forecasts.

There are several reasons for this emphasis, including 1) the existence of coherent low-frequency variations in organized convection that can act as anomalous Rossby wave sources; 2) the relatively strong extratropical response exhibited in both simple and comprehensive general circulation models to variations in tropical forcing (Simmons 1982; Branstator 1985; Newman and Sardeshmukh 1998; Barsugli and Sardeshmukh 2002); and 3) evidence from predictability studies suggesting that much of the potential predictability of weather and climate beyond 1 week is likely related to tropical heating variations (Winkler et al. 2001; Newman et al. 2003; Compo and Sardeshmukh 2004). This section summarizes advances in our understanding of the role of tropical phenomena in forcing extratropical low-frequency variability over the PNA sector, focusing primarily on ENSO and the Madden–Julian oscillation (MJO).

1) ENSO RELATIONSHIPS

As noted previously, the major El Niño event of 1982/83 was not fully recognized, much less predicted, as it was occurring. This situation would soon change dramatically for several reasons, including: 1) rapidly emerging documentation of systematic relationships between ENSO and tropical and extratropical climate variability (e.g., Horel and Wallace 1981; Rasmusson and Wallace 1983); 2) a developing dynamical basis to interpret the relationships between tropical and midlatitude climate variations (Egger 1977; Hoskins et al. 1977; Opsteegh and Van Den Dool 1980; Hoskins and Karoly 1981; Webster 1981); 3) the ability to reproduce observed relationships in general circulation models (Palmer and Mansfield 1984; Lau 1985; Hoerling and Kumar 2000); 4) the successful prediction of El Niño with a simple coupled tropical ocean–atmosphere model (Cane et al. 1986; Zebiak and Cane 1987); and 5) vastly improved observations and expanded research conducted under the Tropical Ocean and Global Atmosphere (TOGA) program. Extensive reviews of ENSO can be found in Philander (1990), National Research Council (1996), articles within a special issue of the *Journal of Geophysical Research—Oceans* (in 1998) and references therein.

(i) Characteristic features

The essence of the ENSO phenomenon arises from coupling of the atmosphere and ocean over the tropical Pacific (Bjerknes 1969). Changes in tropical heating distributions provide anomalous sources for Rossby waves that can propagate into higher latitudes, leading to global climate impacts (Ropelewski and Halpert 1987, 1989; Kiladis and Diaz 1989; Halpert and Ropelewski 1992; Trenberth et al. 1998). Spectral analyses of atmospheric and oceanic variables related to ENSO show significant variability in periods ranging from 2 to 7 yr (Trenberth and Shea 1987; Ropelewski et al. 1992), and on these time scales, ENSO is the dominant mode of global climate variability.

Within this broad time band, the general character of the ENSO variations is that of an irregular oscillation between two extreme states of the tropical Pacific coupled ocean–atmosphere system. One state is characterized by warmer-than-normal sea surface temperatures over the central and eastern tropical Pacific, with below-normal surface pressures in the same region and above-normal surface pressures over the western tropical Pacific, including Indonesia and Australia. This part of the oscillation is generally termed the ENSO warm phase. The other extreme state, the ENSO cold phase, is characterized by below-normal sea surface temperatures over the central and eastern Pacific together with relatively high surface pressure over this region and lower-than-normal surface pressures over the western tropical Pacific. Because of the very strong coupling between

a) El Nino, 31 events

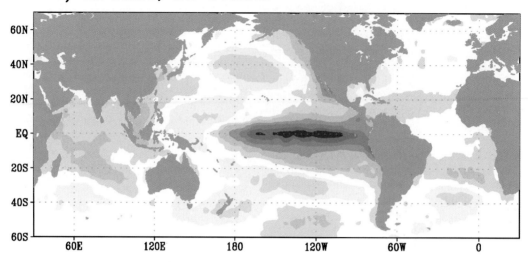

b) La Nina, 24 events

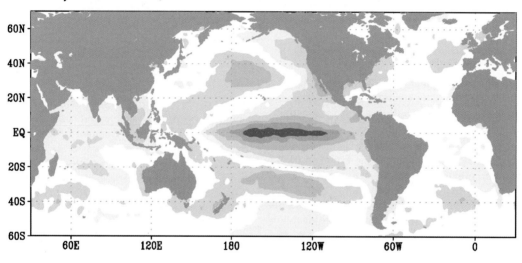

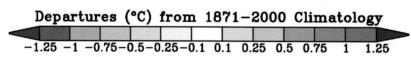

Departures (°C) from 1871–2000 Climatology

−1.25 −1 −0.75 −0.5 −0.25 −0.1 0.1 0.25 0.5 0.75 1 1.25

FIG. 16. Composite November–March SST anomalies for (a) El Niño and (b) La Niña. Cases are from the period 1870–2004, extending the cases of Kiladis and Diaz (1989) using the SST index defined in that study.

sea surface temperatures and sea level pressure over these regions, ENSO warm phase conditions are often simply called El Niño, and cold phase conditions La Niña, although strictly these refer only to the ocean component of the system.

Figure 16 shows composite SST analyses for the two phases. Note that in addition to the strong SST variations in the eastern tropical Pacific, there are systematic variations elsewhere. For example, SST anomalies of the

same sign as the tropical eastern Pacific anomalies extend to higher latitudes along the west coasts of both North America and South America, same-signed SST anomalies also occur over parts of the Indian and tropical Atlantic Oceans, and anomalies of opposite sign are located in the central North Pacific and subtropical South Pacific Oceans. The coastal anomalies are the result of poleward propagation by oceanic Kelvin waves, which are edge waves that propagate parallel to

a) rainfall and tropospheric temperature

b) rainfall, SLP, and surface vector wind

c) SST and surface vector wind

FIG. 17. Composite ENSO surface features. Fields are from NCEP–NCAR reanalyses regressed upon a Pacific cold tongue index defined in Wallace et al. (1998) for the period from January 1985 to December 1993. All regression coefficients are per standard deviation of the cold index. (a) The 1000–200-hPa layer-mean temperature (contour interval 0.1°C; negative contours are dashed) superimposed upon rainfall (anomalies of −1 and −3 cm month^{-1} are shaded blue and greater than 1, 3, 5, and 7 cm month^{-1} are shaded red). (b) SLP (contour interval 0.25 hPa, negative contours are dashed) and surface vector wind superimposed upon rainfall. Only vectors with magnitudes > 0.5 m s^{-1} are plotted. (c) Surface vector wind superimposed upon SST obtained from COADS data. From Wallace et al. (1998, their Fig. 8).

the coastlines with the boundary on the right side in the Northern Hemisphere and on the left side in the Southern Hemisphere (Gill 1982). The anomalies in the North and South Pacific, Atlantic, and Indian Oceans are related to an "atmospheric bridge" mechanism, in which tropical Pacific SSTs lead to changes in the global atmospheric circulation, which can then produce changes in SSTs in the underlying oceans (Alexander et al. 2002).

The "SO" in ENSO reflects the strong tendency for associated sea level pressures to be antiphase between the eastern and western tropical Pacific, as described by Walker (1924). The Southern Oscillation index (SOI), as measured by mean sea level pressure differences between Tahiti and Darwin, tends to be below normal during El Niño and above normal during La Niña (Trenberth 1976; HW81). Figure 17 shows the characteristic surface features associated with ENSO as obtained from

a) El Nino, 7 events

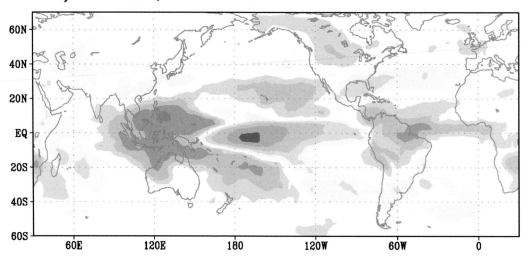

b) La Nina, 5 events

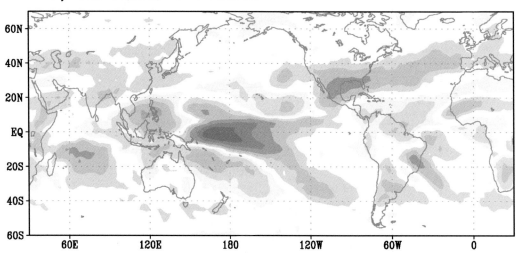

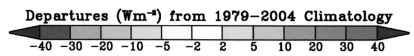

FIG. 18. Composite wintertime OLR anomalies (W m^{-2}) associated with (a) seven El Niño events and (b) five La Niña events over the period from 1979 to 2004.

the National Centers for Environmental Prediction–National Center for Atmospheric Research (NCEP–NCAR) reanalyses. While these features are typical, all events differ in some aspects, and these different "flavors" of ENSO (Trenberth 1997a,b; Trenberth and Stepaniak 2001) can have significant implications, especially for seasonal anomalies and climate forecasts in midlatitudes (Kumar and Hoerling 1997; Hoerling and Kumar 1997; Hoerling et al. 1997).

Related changes in tropical rainfall, as inferred from outgoing longwave radiation (OLR), are quite striking, with a tendency for convection to shift toward the region of warmest SSTs (Fig. 18). As discussed previously, the associated upper-level divergence together with advection of mean vorticity by the anomalous divergent flow provide local sources for Rossby waves. Energy dispersion occurs poleward and eastward from the convective source regions, altering the subtropical jets and extratropical stationary wave patterns. These features can be seen in Fig. 19, which shows respective composite 200-hPa circulation anomalies associated with El Niño and La Niña conditions. During the El Niño events (Fig. 19a) there are flanking anticyclones in the subtropics in both hemispheres located poleward of the regions of maximum anomalous heating, with intensification of the subtropical jets on the poleward sides of the anticyclones. In El Niño winters, the Pacific jet exit region tends to extend well eastward of its normal location, while it remains more confined in the western part of the basin in La Niña years. Figure 20, from Trenberth et al. (1998), provides a schematic illustrating the anticipated upper-level mean flow response to a near-equatorial heat source, as well as expected seasonal-mean storm-track changes, consistent with observed ENSO–storm track relationships (Kok and Opsteegh 1985; Held et al. 1989; Hoerling and Ting 1994).

(ii) Relationships of ENSO to extratropical low-frequency variations

Because of the mismatch in time scales, ENSO is unlikely to be directly responsible for most intraseasonal low-frequency variability. Nevertheless, by modifying the seasonal mean flow and storm tracks, ENSO can effectively load the climate dice by producing conditions that are more (or less) favorable for the development of low-frequency circulation anomalies in a given region. In an early study considering this possibility, Mullen (1989) performed perpetual January experiments with the NCAR general circulation model to assess the responses of Pacific blocking to various tropical and midlatitude sea surface temperature anomalies. His results indicated that the sea surface temperature anomalies did not significantly alter the total blocking frequency over the Pacific, but did affect the preferred locations of blocking formation. He found that ENSO warm phase (El Niño) conditions favored increased blocking along the west coast of North America and suppressed blocking over the mid-Pacific. Conversely, ENSO cold-phase (La Niña) conditions were associated with enhanced mid-Pacific blocking. These basic results were confirmed in later studies (Hoerling and Ting 1994; Renwick and Wallace 1996; Compo et al. 2001). Consistent with the correlations described in HW81, positive PNA patterns are also more likely to occur in the ENSO warm phase, with more negative cases during the ENSO cold phase (Renwick and Wallace 1996; Straus and Shukla 2002).

Studies by Sardeshmukh et al. (2000) and Compo et al. (2001) have examined in more detail the potential influences of El Niño and La Niña on variability across a range of time scales using observational analyses and general circulation model simulations. Sardeshmukh et al. found substantial asymmetries in the remote response to El Niño and La Niña in both time-mean fields and variability. Through analyses of large (180 member) en-

sembles of general circulation model runs, they were able to detect statistically significant changes in probability distributions for two seasons characterized by El Niño [January–March (JFM) 1987] and La Niña (JFM 1989) conditions. They concluded that the changes in distributions were sufficiently large to substantially alter the risks of extreme climate anomalies.

Compo et al. (2001) examined the responses to ENSO across a range of time scales, including variations at synoptic, intraseasonal, monthly, and seasonal time scales. They found that the responses to ENSO differed sharply depending on the time scale of interest; for example, La Niña events were associated with increased variance of the North Pacific near the Aleutians on intraseasonal and monthly time scales, which they attributed to more frequent blocking activity, while at the same time synoptic-scale variability in this region was suppressed. Three mechanisms were suggested as important in accounting for the observed ENSO-induced changes in extratropical variability. First, asymmetries in the tropical heating anomalies related to El Niño versus La Niña conditions could force some of the observed differences. Second, changes in the extratropical base state due to ENSO could alter the stability and potential for growth of some of the low-frequency modes, for example, related to the PNA pattern. Third, changes in storm tracks could alter the synoptic eddy feedbacks onto low-frequency variations. Sardeshmukh et al. (2000) and Compo et al. (2001) both emphasized the need for large ensembles (order of a few hundred members) to reliably estimate systematic changes in extratropical variability due to El Niño or La Niña conditions.

(iii) Relationships of ENSO to extratropical storm variability

Research within the last several years has provided additional insights on the connections between ENSO and synoptic-scale weather phenomena. Barsugli et al. (1999) applied an innovative approach to evaluate the effects of the 1997/98 El Niño on observed synoptic weather events. In their study, they conducted parallel ensembles of medium-range weather forecasts out to 16 days every day throughout the winter of 1997/98, with the sole difference between ensembles being that one ensemble set used as boundary conditions observed SSTs in both the Tropics and midlatitudes, while the other replaced the tropical SSTs with their climatological-mean values. Because tropical SST anomalies during this winter were largely associated with the strong El Niño event, to first order this approach enabled an assessment of the sensitivity of the forecasts to the presence or absence of El Niño conditions. Barsugli et al. termed the ensemble mean differences between the two sets of forecasts the "synoptic El Niño signal," which is a function of initial date and forecast lead time. They found that, on average, an El Niño signal appeared within the first day of the forecasts in the tropical Pacific.

a) El Nino, 14 events

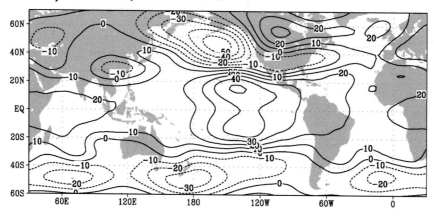

b) La Nina, 10 events

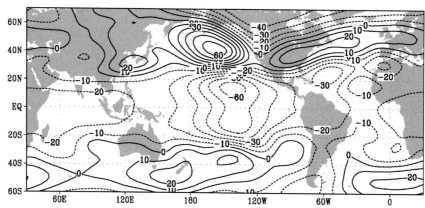

FIG. 19. Composite wintertime 300-hPa height anomalies (m) associated with (a) 14 El Niño and (b) 10 La Niña events over the period from 1950 to 2004, from NCEP–NCAR reanalysis data.

This signal then extended into the extratropics and in some cases substantially altered the midlatitude flow evolution, especially in the second week (days 8–14) of the forecasts.

Interestingly, Barsugli et al. also found that, despite the slow evolution of the SST anomalies, El Niño impacts on the forecasts varied substantially during the course of the season. This suggests that other intraseasonal variations were significantly modulating the effects of the tropical SST anomalies, effectively opening and closing the "window" to strong tropical–extratropical interactions. Figure 21 shows one example of a period of significant extratropical El Niño impacts on 8–14-day forecasts. This period marks the onset of heavy rains in California around 1 February 1998. The precipitation forecasts with observed SSTs that include El Niño conditions (Fig. 21a) show a strong, zonally oriented storm track that extends eastward from the central North Pacific to the western United States, with maximum rainfall near the northern California coast. In contrast, the forecasts where tropical SSTs are replaced with their climatological values (Fig. 21b) have a weak-

er track extending northeastward from the central Pacific to British Columbia, with dry conditions over most of the west coast, including California. The El Niño signal (Fig. 21c) indicates that El Niño forcing contributed substantially to the storm-track changes leading to heavy rainfall in California during this period (Fig. 21d). Barsugli et al. found that for the entire 1997/98 winter season variations in strength of the synoptic El Niño signal broadly coincided with subseasonal rainfall variability over central California, suggesting the importance of tropical–extratropical interactions in modulating rainfall variations for this region and season.

Recent synoptic studies have provided further insights into ENSO impacts on midlatitude storm activity. Shapiro et al. (2001) compared the life cycles of baroclinic storm systems over the North Pacific in two consecutive winters, one characterized by strong El Niño conditions (1997/98) and the other by La Niña conditions (1998/99). The time-mean 250-hPa flow for these two winters is shown in Fig. 22. The contrast in mean flows is clearly evident across the eastern Pacific, with a mean jet axis that extends eastward across Baja California in 1997/

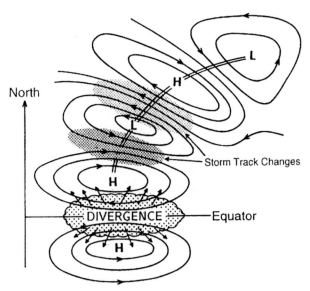

FIG. 20. Schematic of the dominant changes in the upper tropo-sphere, mainly in the Northern Hemisphere, in response to increases in SSTs, enhanced convection, and anomalous upper-tropospheric divergence in the vicinity of the equator (scalloped region). Anomalous outflow into each hemisphere results in subtropical convergence and an anomalous anticyclone pair straddling the equator, as indicated by the streamlines. A wave train of alternating high and low geopotential and streamfunction anomalies results from the quasi-stationary Rossby wave response (linked by the double line). In turn, this typically produces a southward shift in the storm track associated with the subtropical jet stream, leading to enhanced storm-track activity to the south (dark stipple) and diminished activity to the north (light stipple) of the first cyclonic center. Corresponding changes may occur in the Southern Hemisphere. From Trenberth et al. (1998).

98 and is shifted substantially poleward toward the Pacific Northwest in the following year. Shapiro et al. found that the synoptic life cycles and associated eddy fluxes were also quite different in the two winters. Representative examples can be seen in Fig. 23, which shows a pronounced cyclonic roll-up just off the west coast in the region of intense cyclonic shear in the mean zonal flow for an El Niño winter case, while the La Niña winter case shows an example of anticyclonic wave breaking. Shapiro et al.'s results showed that large differences in the mean flows between the El Niño and La Niña winters led to preferential baroclinic life cycles over the North Pacific that closely resembled the life cycle (LC) paradigms LC2 and LC1, respectively, of Thorncroft et al. (1993). Because of the radically differing momentum fluxes between these two life cycles, the results also imply substantial differences in synoptic eddy feedbacks onto the mean flow.

Stepping still further down the scale, recent work shows that El Niño can also impact moisture transports over the eastern Pacific through effects on narrow, concentrated channels termed "atmospheric rivers" (Zhu and Newell 1998; Zhu et al. 2000; see also Newell et al. 1992; Iskenderian 1995). Relatively small but systematic differences in wind direction and moisture transports within these atmospheric rivers have been found

to vary with respect to ENSO phase, while their interaction with the complex local topography helps determine locations of flooding in El Niño versus non–El Niño years (Ralph et al. 2003, 2004, 2005; Andrews et al. 2004). Figure 24 shows an example of an atmospheric river for a case associated with severe flooding in northern California in February 2004. In addition to systematic effects of ENSO on moisture transports, Persson et al. (2005) found that warm coastal SST anomalies associated with strong El Niño events contribute to enhanced upward sensible and latent heat fluxes that can increase convective available potential energy of air parcels reaching the coast. They conclude that when this destabilized air is forced upward by steep coastal terrain deep convection can be enhanced, with an increased risk of coastal flooding. In a recent study, Bao et al. (2006) document the entrainment of tropical water vapor into atmospheric rivers for extreme West Coast flooding events, and find that the direct linkage to the Tropics is most likely in ENSO-neutral years. These and other effects of climate variations on synoptic, mesoscale, and even microscale processes, represent emerging areas for research on the linkages between weather and climate.

2) THE MADDEN–JULIAN OSCILLATION

The previous results indicate that ENSO can alter the probability of occurrence of certain intraseasonal phenomena within a season and region, as well as change the likelihood of extreme weather events. ENSO itself, however, cannot determine the precise timing of such events. Beyond ENSO, a rich diversity of intraseasonal tropical phenomena exist that also modulate large-scale convection and flow fields, and hence can act as potential wave sources for exciting extratropical variability (Kiladis and Weickmann 1997). Of these, the dominant intraseasonal mode has a spectral peak in the range of approximately 40–50 days (Madden and Julian 1971, 1972), and is now generally known as the MJO. This section briefly summarizes the major features of the MJO, with emphasis on potential linkages to extratropical low-frequency variability and storm behavior. More extensive reviews of the MJO can be found in Madden and Julian (1994) and Zhang (2005).

(i) Tropical features

A remarkable aspect of tropical convection is its tendency to organize on scales much larger than that of individual convective elements. This is clearly evident in climatological-mean conditions, as well as on seasonal-to-interannual time scales in association with ENSO. For both climatological-mean and ENSO conditions, this organization is determined largely by ocean boundary conditions, with enhanced convection typically occurring near the region of warmest waters, which shift eastward from the western tropical Pacific toward

(a) NINOSST FORECAST   (b) CLIMOSST FORECAST

(c) SIGNAL   (d) OBSERVED

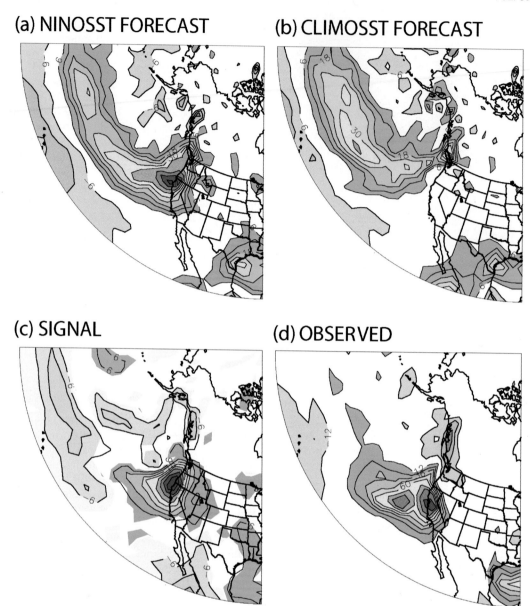

FIG. 21. The 8–14-day mean forecasts of accumulated precipitation anomaly initialized on 24 Jan 1998 and verifying the first week of February 1998 (the "California rain" case): (a) NiñoSST forecast, (b) CLIMOSST forecast, (c) synoptic El Niño signal, and (d) the observed precipitation (estimated from the NCEP–NCAR reanalysis). From Barsugli et al. (1999).

the east-central tropical Pacific during El Niño years. Because of the relatively slow evolution of the SSTs, the associated convective heating anomalies remain nearly stationary on subseasonal time scales.

In contrast, organized convection associated with the MJO does not remain geographically fixed, but rather propagates slowly eastward with time from the tropical Indian Ocean to the west-central Pacific, with an average zonal speed of ~5 m s^{-1} (Weickmann et al. 1985; Knutson and Weickmann 1987; Hendon and Salby 1994). This speed in not constant, but varies for different events and stages of the MJO life cycles. While the direction

of propagation is primarily zonal, meridional propagation is also often apparent in certain regions, for example, across south Asia and the far western Pacific (Lau and Chan 1985), and toward northern Australia. During the Asian summer monsoon, impacts of the MJO often propagate northward into India and modulate rainfall between "active" and "break" periods, with effects that may be predictable 15–30 days in advance (Webster and Hoyos 2004).

Figure 25 illustrates the major modes of large-scale tropical convective organization as inferred from OLR anomalies for two winters, one (1996/97) marked by

a) DJF 1997–98

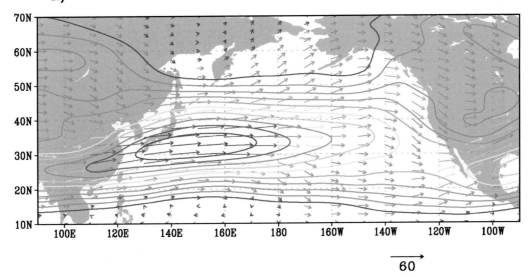

b) DJF 1998–99

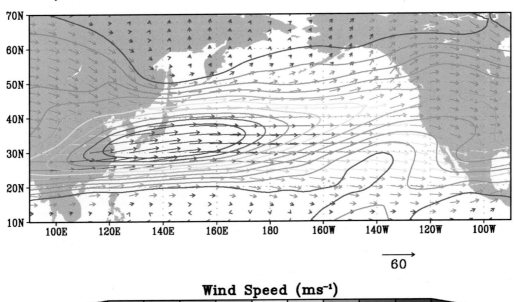

FIG. 22. The 300-hPa time-mean vector winds and wind speeds (m s⁻¹, contours) for winters (a) 1997/98 and (b) 1998/99.

strong MJO activity and relatively weak quasi-stationary (ENSO) forcing, and the other (1997/98) by a strong El Niño event during which MJO activity was nearly absent. In 1996/97, two strong and another two weaker MJO events are evident (Fig. 25a). These events show a tendency for organized convection to develop over the Indian Ocean around 60°E, and propagate slowly eastward until decaying near the date line. While a rough periodicity is evident, this phenomenon occurs over a

relatively broad intraseasonal time band of 30–60 days. The events also exhibit notable differences from case to case, much as found for ENSO. In contrast, during 1997/98 (Fig. 25b), there is much less evidence of the slow eastward propagation of convection that could be related to the MJO. Rather, the predominant convective patterns remain quasi-stationary, with a broad maximum in enhanced convection centered east of the date line near the region of warmest SST anomalies (not shown)

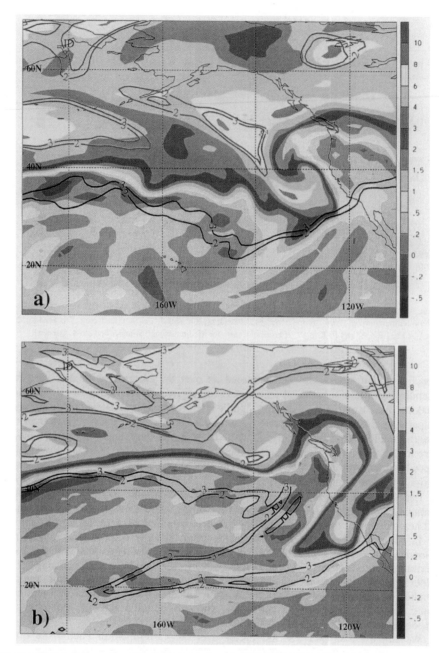

FIG. 23. Potential vorticity at three isentropic levels for (a) 1200 UTC 6 Feb 1998 (El Niño) and (b) 1200 UTC 5 Feb 1999 (La Niña). The 300-K isentropic potential vorticity [IPV; green lines, 2 and 3 potential vorticity units (PVUs, 1 PVU = 10^{-6} K m^2 kg^{-1} s^{-1})]; 320-K IPV (color shading, PVU, as in color bar); and 340-K PV (black lines, 2 and 3 PVUs) (derived from ECMWF analyses). From Shapiro et al. (2001).

and suppressed convection over much of western tropical Pacific.

In addition to ENSO and MJO, other modes of organized tropical convection are also evident in both winters, including westward-propagating features that are associated with equatorial Rossby waves, and very rapid eastward-propagating features that are a manifestation of equatorial Kelvin waves (Wheeler and Kiladis 1999; Wheeler et al. 2000; Wheeler and Weickmann 2001). (These modes of variability are now monitored routinely

for their potential value for medium and extended range predictions, with real-time analyses available online at http://www.cdc.noaa.gov/map/clim/olr_modes/.)

While all the above modes contribute to tropical convective organization, they do not fully account for the overall temporal variations seen in Fig. 25, which are much richer and more complex. Hence composites or schematics should be interpreted as providing only roughly expected relationships. In addition, MJO activity exhibits substantial interannual variability (Hendon

February 16, 2004 Descending Passes
SSM/I Water Vapor (Schluessel algorithm)

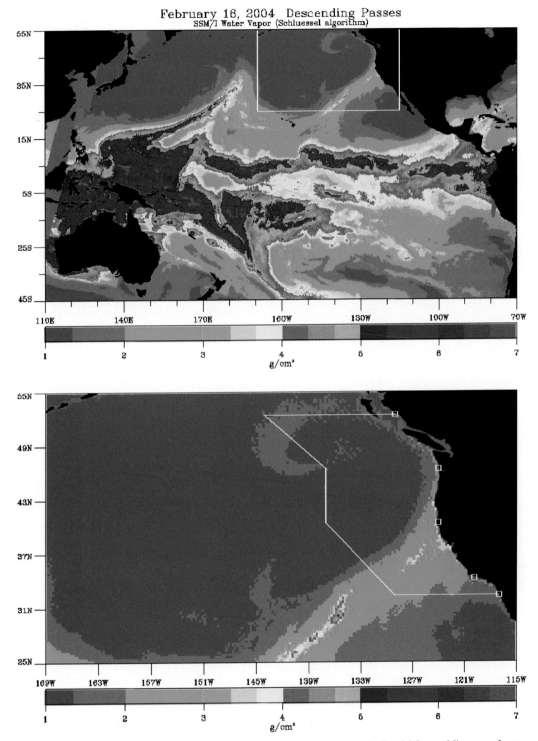

FIG. 24. Composite image of vertically integrated water vapor (cm) taken by the Special Sensor Microwave Imager (SSM/I) on 16 Feb 2004. The top image shows the large-scale connection within the Tropics, with the white box indicating the area of the zoomed-in image for the bottom figure, which shows finer-scale structure within this atmospheric river. This case was associated with severe flooding along the Russian River in northern California. From Bao et al. (2006).

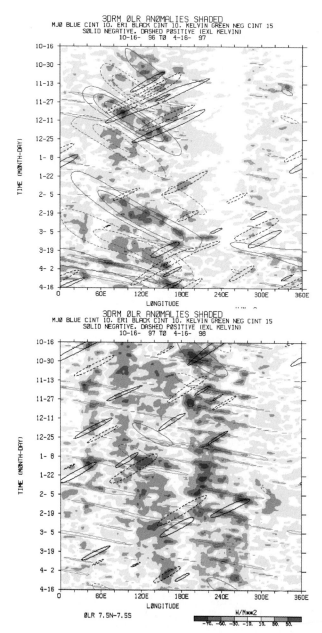

FIG. 25. Hovmöller plot (time–longitude section) of OLR anomalies averaged between 7.5°N and 7.5°S, where the contours illustrate space–time filtered coherent tropical convection modes for (top) 16 Oct 1996–16 Apr 1997 and (bottom) 16 Oct 1997–16 Apr 1998. The blue contours represent the MJO (solid blue for enhanced convective phase, dashed blue for suppressed phase). The green contours are for Kelvin waves and the brown contours for equatorial Rossby wave mode 1. See Wheeler and Kiladis (1999) for additional details. The red shading shows negative OLR anomalies (positive convection anomalies), and the blue shading shows positive OLR anomalies (suppressed convection).

with the idea that an expanded warm pool favors strong MJO activity (Wang and Li 1994). This empirical relationship may provide some guidance on anticipated levels of MJO activity for a given winter.

Figure 26, from Madden and Julian (1972), illustrates some of the principal tropical synoptic features related to the MJO. Typical zonal scales, as determined by the distance between maxima and minima in OLR anomalies, are roughly 10 000–20 000 km. Thus, this phenomenon represents a planetary-scale mode of convective organization. More detailed analyses show that within this large-scale structure organized smaller-scale convective features are embedded that include both eastward- and westward-propagating components (Nakazawa 1988). Convection tends to develop preferentially on the eastern side of the large-scale complex and dissipate on the western side, contributing to the slow eastward propagation.

In the Tropics, large-scale diabatic heating anomalies are approximately balanced by adiabatic cooling (Hoskins and Karoly 1981), with regions of active convection characterized by mean ascending motions, low-level convergence, and upper-level divergence. The descending branches of the circulations associated with organized convection occur far from the convectively active regions as part of the tropical atmospheric response to localized heating (Gill 1980; Salby et al. 1994). Anomalies in velocity potential χ (Holton 2004) whose gradients are directly related to the divergent wind component, provide an alternate, very useful means for monitoring the divergent flows associated with tropical convection. Figure 27 shows one example of such analyses, as obtained for the 200-hPa level, in which the tendency for systematic eastward propagation can be seen with periods of roughly 30–50 days.

(ii) Relationships of the MJO to extratropical circulation patterns

Studies beginning in the early 1980s extended previous research by considering relationships between the MJO and extratropical circulation patterns (Weickmann 1983; Weickmann et al. 1985; Knutson and Weickmann 1987; Kiladis and Weickmann 1992a). A schematic (Fig. 28) from Weickmann et al. (1985) illustrates typical upper-level circulation features observed during the phase of the MJO when the convection is centered over Indonesia. Note the twin flanking anticyclones poleward of the maximum convection, similar to what is usually observed with El Niño events, and broadly resembling solutions that were obtained by Gill (1980) in a two-layer dynamical model forced by localized heating on the equator. To the east of the heating there is strong westerly outflow along the equator, again similar to the Gill solutions. In addition, cyclonic circulations extend into higher latitudes. As the convection propagates eastward, midlatitude flows vary between contracted and extended subtropical jets, which can impact midlatitude

et al. 1999), although in most seasons and years some level of MJO activity is present. Winters with high levels of MJO activity are frequently associated with warmer-than-normal waters over the western Pacific the preceding autumn season (Bergman et al. 2001), consistent

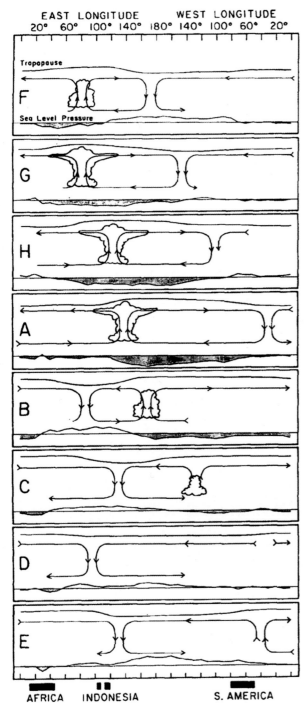

EAST LONGITUDE   WEST LONGITUDE
20° 60° 100° 140° 180° 140° 100° 60° 20°

Tropopause

F

Sea Level Pressure

G

H

A

B

C

D

E

AFRICA   INDONESIA   S. AMERICA

FIG. 26. Schematic of the time–space (zonal plane) variations of the disturbance associated with the 40–50-day oscillation. Dates are indicated symbolically by the letters at the left of each chart and correspond to dates associated with the oscillation in station pressure at Canton, as described in Madden and Julian (1972). Regions of enhanced large-scale convection are indicated schematically by the cumulus and cumulonimbus clouds. The relative tropopause height is indicated at the top of each chart. From Madden and Julian (1972).

weather conditions. Within the Tropics, low-level circulations are approximately antiphase with the upper-level circulations, with low-level inflow into the region of maximum convection. In contrast, at higher latitudes, the low- and upper-level circulations are more nearly in phase, with the upper-level circulation usually substantially stronger.

Subsequent work has refined descriptions of the MJO and evaluated potential connections to extratropical weather systems. Higgins and Mo (1997) found that persistent circulation anomalies over the North Pacific were often preceded by convective anomalies over the tropical western Pacific 1–2 weeks earlier (see also the modeling study by Higgins and Schubert 1996). Several studies have shown that unusually wet conditions in California and the southwest, with drier conditions in the Pacific Northwest are associated with enhanced tropical convection over the central equatorial Pacific (Mo and Higgins 1998a,b; Mo 1999; Jones et al. 2000; Higgins et al. 2000; Whitaker and Weickmann 2001).

Whitaker and Weickmann (2001) developed a statistical prediction model using tropical OLR anomalies as a predictor, first removing the ENSO signal in order to focus on the predictive implications of subseasonal tropical convective variations. They found potential predictability for U.S. west coast precipitation at lead times out to 2 weeks that was connected to subseasonal variations in convection over the central tropical Pacific. Figure 29, adapted from Whitaker and Weickmann, shows a pattern of tropical convective anomalies that is associated 2 weeks later with a doubling of the probability of heavy (upper quintile) rainfall in southern California. The changes in precipitation probabilities on weekly time scales due to the subseasonal variations are roughly comparable to those that would be obtained from an ENSO signal alone (Whitaker and Weickmann 2001). These results suggest that the subseasonal variations in tropical convection can either strongly reinforce or nearly cancel the expected ENSO response.

(iii) Additional comments

While the previous discussion has emphasized the role of the Tropics in forcing extratropical low-frequency variability, it is important to keep in mind that interactions occur in both directions. Wave propagation from higher toward lower latitudes can initiate new convection and produce changes in the tropical wind field (Webster and Holton 1982; Arkin and Webster 1985; Kiladis and Weickmann 1992b; Kiladis 1998). Such changes may initiate new sources for Rossby waves as well as alter the basic state through which the waves propagate.

In addition, surface westerly wind anomalies associated with the MJO can force ocean Kelvin waves that transport warm water eastward from the western Pacific to eastern Pacific. As these waves reach the west coast of South America, they reduce near-coastal upwelling,

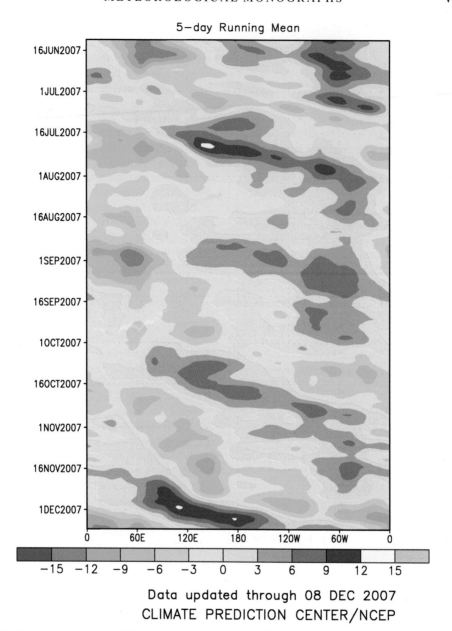

Fig. 27. Time–longitude plot of 200-hPa velocity potential anomalies averaged between 5°N and 5°S. The period is for six months ending 8 Dec 2007. Green colors signify large-scale 200-hPa divergence, while brown colors signify large-scale 200-hPa convergence.

which can result in rapid SST increases in this region. Several studies suggest that through such atmosphere–ocean interactions MJO events can significantly influence the onset and breakdown, and perhaps also the intensity, of ENSO events (Kessler and Kleeman 2000; Zhang and Gottschalck 2002; McPhaden 2004). Perhaps the biggest lesson emerging from studies in this area is the fundamental importance of coupling between the Tropics and high latitudes, and between the atmosphere and ocean, both of which occur on time scales that are sufficiently short to have implications for the predictability of extratropical low-frequency variability.

d. Other phenomena and processes affecting extratropical low-frequency variability

The previous sections have outlined some key phenomena and processes linking weather and climate variability. We describe here a few other systematic behaviors, most of which have been identified within the past decade. Interestingly, while most research has emphasized the effects of tropical ocean–atmosphere interactions on extratropical low-frequency variability, recent research suggests that additional sources of low-frequency variability, and also extended-range predict-

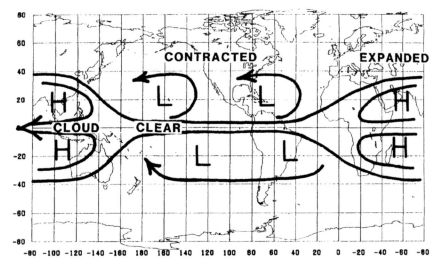

FIG. 28. Schematic of the relationship between OLR and 250-mb circulation for the MJO at a time when convection is at a maximum over Indonesia. From Weickmann et al. (1985).

ability, may be obtained from a very different direction; in particular, circulation variations in the stratosphere and at high latitudes.

Several studies have provided substantial evidence that dynamical coupling between the stratosphere and troposphere can provide an additional source of predictability on time scales from a week to a few months in advance. This is possible in part because the stratospheric circulation varies on relatively long time scales compared with typical tropospheric synoptic-scale disturbances, and thus provides a potential source of memory in the system, just as with the slow tropical atmosphere and ocean variations previously discussed. The difference in characteristic time scales between the stratosphere and troposphere is closely related to the vertical propagation characteristics of Rossby waves (Charney and Drazin 1961). Rapid changes in potential vorticity gradients that occur near the tropopause act as a kind of dynamical low-pass filter that prevents higher-frequency synoptic-scale waves from propagating significantly into the stratosphere (Charney and Drazin 1961; Held 1983). Rossby waves that propagate upward into the stratosphere are characterized by large spatial scales (predominantly zonal wavenumbers 1–3), and are largely forced by topography and quasi-stationary diabatic sources.

The question then becomes to what extent changes in the stratospheric circulation influence tropospheric variability. Recent studies indicate that, at least in some instances, the effects can be substantial. Baldwin and Dunkerton (1999, 2001) found a statistically significant tendency for wind anomalies in the wintertime polar vortex to "propagate" downward from the midstratosphere to the tropopause, with effects extending down to the surface pressure and wind fields. Their results indicated that the midstratospheric polar vortex anomalies lead those in the troposphere by 1–2 weeks. Using

potential vorticity inversion techniques described earlier, Black (2002) showed that the tropospheric circulation anomalies are related to the balanced flow response to potential vorticity anomalies just above the tropopause. At the surface, the resulting circulation anomalies strongly resemble the Arctic Oscillation (AO) described by Thompson and Wallace (1998, 2000). The AO is characterized by variations in the strength of the tropospheric polar vortex and zonal flow along ~55°N, with zonal wind anomalies of the opposite sign near 35°N, which tend to be somewhat more prominent in the Atlantic than in the Pacific sector (Fig. 30). Because of the relatively weaker correlation over the North Pacific, there has been considerable debate on whether the AO and the NAO should be considered as dynamically distinct phenomena (Ambaum et al. 2001; Wallace and Thompson 2002b), and the term "AO–NAO pattern" is sometimes used in the literature.

Thompson et al. (2002) and Baldwin et al. (2003a,b) examined, in more detail, the lag relationships between stratospheric and tropospheric flow variations at high latitudes in the wintertime Northern Hemisphere. Both studies concluded that statistically significant skill exists on time scales of a few weeks to months in predicting persistent anomalies in the tropospheric polar circulation from prior knowledge of the stratospheric conditions. Thompson et al. showed that these variations were related to the phase of the QBO and the upper-level manifestation of the AO, the Northern Annular Mode (NAM), which is a quasi-symmetric circulation pattern characterized by variations in the strength of the polar vortex (Wallace and Thompson 2002a).

One phase of the NAM is characterized by an intense, confined tropospheric polar vortex with strong potential vorticity anomalies along its outer edge, while the other is manifested by a weakened polar vortex with reduced meridional potential vorticity gradients. Because strong

a

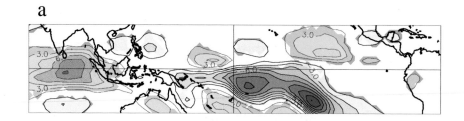

b

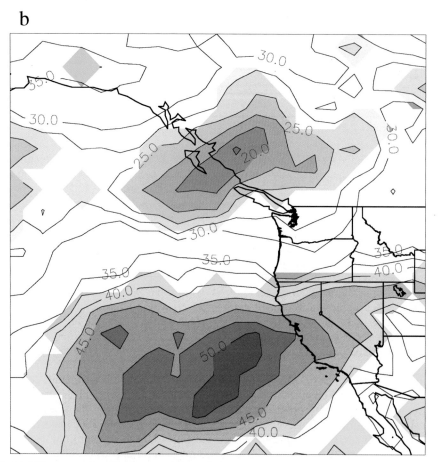

FIG. 29. (a) OLR regressed on leading canonical predictor variable at lag 7 days; map is scaled for one standard deviation of the canonical predictor variable. (b) Probability of precipitation being in the upper tercile given that the leading canonical predictor variable 2 weeks prior is in the upper quintile. Adapted from Whitaker and Weickmann (2001).

positive vorticity gradients increase the horizontal stability of the vortex and inhibit meridional mixing (Charney 1973), surface polar air masses interior to the strong vortex can become extremely cold due to sustained radiative heat losses. Higher vortex stability reduces the likelihood that the cold air will be displaced southward into the midlatitudes. Conversely, a weakened polar vortex allows more mixing and meridional excursions of polar air into midlatitudes. In empirical studies, Thompson et al. (2002) and Thompson and Wallace (2001) found such relationships between changes in the AO

and NAM and the subsequent probability of extreme cold-air outbreaks into midlatitudes, with effects persisting out to approximately 2 months. In addition to altering the frequency and extent of cold-air outbreaks, Thompson and Wallace also identified significant impacts on midlatitude storm behavior and high-latitude blocking. However, current understanding of the physical mechanisms associated with the above statistical relationships remains limited.

With the exception of the MJO, most of the modes of low-frequency variability that we have discussed so

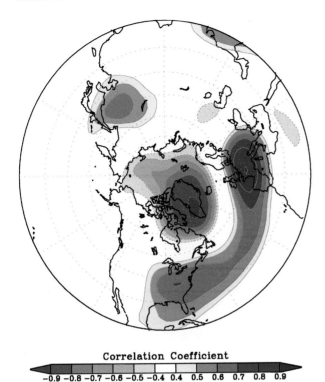

FIG. 30. Correlation of the 500-hPa wintertime (DJF) geopotential heights with the AO index from the NOAA/Climate Prediction Center for the years 1950–2004.

More recently, Branstator (2002) showed that the focusing and trapping effects of the time-mean jet stream act as a waveguide that connects variability around the hemisphere. This behavior is manifested synoptically by chains of perturbations that extend along the jet axis and have maximum amplitudes in the upper troposphere (Fig. 31). Branstator's results suggest that extended wave trains are most efficiently excited by forcing in the vicinity of the jets near Southeast Asia and the east coast of North America. In comparison to the teleconnection patterns identified by WG81 and others, characteristic zonal wavelengths are significantly shorter, and hence rapid downstream dispersion can occur along the jet axis at speeds comparable to the local jet strength [cf. Eqs. (4) and (5)]. This mode of evolution appears well established, but potential predictive implications remain to be determined.

e. Summary

We began this section by considering the problem of whether early studies on blocking could be placed within a more general and systematic framework. Since 1980, our recognition of phenomena and physical processes contributing to extratropical low-frequency variability has advanced considerably. It is vital to recognize that the typical low-frequency patterns that we have described (e.g., teleconnections) are not fixed physical entities, but rather reflect composites of many individual events, each of which will vary in some aspects. The characteristic patterns we described are useful to the extent that they reveal certain dominant or frequently recurrent physical and dynamical processes that operate on these time scales.

Several potential mechanisms were discussed for producing low-frequency flow variations, including Rossby wave dispersion from localized topographic or diabatic sources; large-scale instabilities and initial value growth; quasi-equilibrium states that represent an approximate balance between forcing, advection, and dissipation; and anomalous eddy forcing associated with changes in synoptic-scale storm activity. It is likely that

far are quasi-stationary. Such quasi-stationary features contribute substantially to low-frequency variability at any given geographical location, and frequently produce prolonged anomalies in local and regional weather conditions. However, beyond the MJO there is evidence for other propagating low-frequency phenomena. Branstator (1987) described an example of a long-lived westward-propagating high-latitude mode reminiscent of some early synoptic descriptions of high-latitude blocking anticyclones, which frequently migrate westward (retrogress) when the centers are displaced well north of the main latitudes of the westerlies (Namias 1947; Rex 1950a,b).

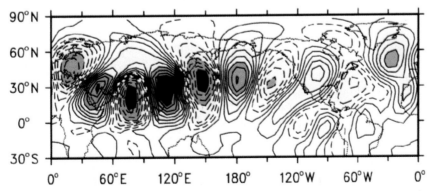

FIG. 31. One-point correlations with a base point at 28.9°N, 112.5°E for internal variability in the winter mean DJF 300-hPa nondivergent meridional winds obtained from the NCAR CCM3. From Branstator (2002).

all of these processes contribute to the total observed spectrum of intraseasonal low-frequency variations, with their relative importance varying from case to case. The specific mechanisms help us understand when and where large low-frequency phenomena will most frequently occur, and what would be the expected flow structures.

For example, consideration of the growth mechanisms suggests why persistent anomalies are more prevalent in the jet exit regions downstream from maxima in synoptic-scale eddy activity, and less favored well upstream of the jet maxima, where barotropic mechanisms would support growth of meridionally elongated eddies [which are associated with rapid eastward energy dispersion; cf. (8)], where the disturbances have yet to enter regions favorable for strong baroclinic growth, and where synoptic-scale feedbacks are generally weaker. To the degree that these processes depend on the mean flow structure and localized sources, they will vary from event to event and year to year as in conjunction with changes in the basic flow and forcing.

Even with steady forcing anomalies, anomalous low-frequency variability will be produced because the climatological-mean flow changes through the year (Newman and Sardeshmukh 1998; Frederisken and Branstator 2005). Changes in responses to a given forcing may be particularly rapid at times of the year when the mean flow itself changes rapidly, as over the central and eastern North Pacific in spring, when a rapid transition occurs from a single- to a double-jet structure. Both the MJO and ENSO also produce systematic changes in the zonal-mean zonal wind fields that tend to propagate slowly poleward from the Tropics into higher latitudes (Black et al. 1996; Kumar and Hoerling 2003). These poleward jet shifts appear to be related to synoptic eddy feedbacks and the response of the meridional circulation to anomalous tropical convection (Weickmann et al. 1997; Feldstein 1998). To the extent that the zonal wind changes are predictable, they may provide another potential source of predictability for extratropical stationary waves.

We then turned our attention to the question of how changes in synoptic-scale variability are related to the low-frequency flow variations. This question is central for two reasons. First, synoptic-scale disturbances are primarily responsible for the midlatitude weather that we seek to predict, especially wintertime precipitation. Hence, understanding the relationships between low-frequency variability, storm tracks, and the evolution of synoptic-scale disturbances is fundamental to the problem of advancing extended range weather predictions. Second, eddy fluxes associated with the synoptic-scale disturbances can force changes in the low-frequency flow, especially in regions of eddy growth and decay. Such interactions appear particularly important in certain classes of events, of which blocking patterns represent one important example.

During blocking events low potential vorticity air is advected far poleward and high potential vorticity air is advected far equatorward of normal. This often leads to breaking Rossby waves, manifested by the formation of closed centers of low potential vorticity (associated with the blocking anticyclone) located far poleward of the subtropical source regions, and closed centers of high potential vorticity (synoptically manifested as cut-off lows) at unusually low latitudes. Such signatures of Rossby wave breaking on upper-level isentropic or constant potential vorticity surfaces have been proposed as a dynamical criterion for defining blocking (Pelly and Hoskins 2003). Observational analyses indicate that blocking occurs preferentially downstream of maxima in synoptic-scale eddy activity in regions where the climatological-mean flow is diffluent. Theoretical analyses of eddy–mean flow interactions (Hoskins 1983; Haines and Marshall 1987) show that regions in which eddy activity decreases eastward will be accompanied by net equatorward fluxes of high potential vorticity (poleward fluxes of low potential vorticity). In regions favorable for blocking, such fluxes tend to weaken the midlatitude westerlies where the mean flow itself is already relatively weak. Thus, theoretical, synoptic and diagnostic analyses are consistent in indicating the likely importance of both the mean flow structure and synoptic-scale eddy feedbacks in the development and maintenance of blocking.

While low-frequency variability can be produced solely through midlatitude processes, tropical heating and flow variations appear to be especially important in determining the potential predictability of the extratropical circulation. Tropical phenomena most likely to enhance predictability are characterized by relatively long time scales, of which we emphasized ENSO and the MJO. However, just as for the teleconnection patterns, it is the dynamical processes that are in fact crucial. Therefore, tropical heating and flow variations other than ENSO and MJO may also excite significant low-frequency variability in the extratropics (Simmons 1982; Barsugli and Sardeshmukh 2002). In addition to tropical–extratropical interactions, there is growing evidence that variability in the stratosphere influences the evolution of the tropospheric circulation, especially at high latitudes.

In summary, over the past 25 years there have been major advances in our ability to describe and understand mechanisms for low-frequency variability. However, compared with synoptic-scale variability, our detailed knowledge of the phenomena and mechanisms connecting short-term weather with longer-term climate variations is still relatively limited, and abundant opportunities remain for future progress in this area.

4. Advances in forecast skill and analyses of potential predictability

Since 1980, skill in short-range model predictions had increased substantially. By measures for which long-term records are available, such as 500-hPa height root-mean-square errors and anomaly correlations (which, as Professor Sanders would be quick to remind us, are not the same as weather) the rate of increase in numerical weather prediction (NWP) model skill over this period has been roughly 1 day per decade (Kalnay et al. 1998; Simmons and Hollingsworth 2002). So, for example, current NWP forecasts of the 500-hPa heights at lead times between 5 and 6 days are roughly as skillful as were 3-day forecasts in 1980.

At first sight, this would still seem to leave prospects for forecast skill beyond a week just out of reach. However, recent research supports the view that skillful forecasts can be made into the second week of the forecast ("week 2"), using a combination of model-based and statistical techniques. Further, some predictability studies suggest that forecast skill of weekly-average conditions beyond week 2 is feasible in many regions. Here, we briefly summarize a few of these recent studies on forecast skill and potential predictability.

a. Advances in forecast skill

Of the many steps that have contributed to forecast improvements over the last few decades, one in particular must be highlighted: the introduction of *ensemble prediction methods*. Ensemble predictions were first implemented operationally at both NCEP and ECMWF in December 1992, and subsequently at other major centers (Tracton and Kalnay 1993; Molteni et al. 1996; Houtekamer et al. 1996). The initial practical motivation for introducing ensemble methods was to advance skill in medium-range weather forecasts. This innovation reflected a fundamental change in forecast strategy for the operational centers from a deterministic approach in which numerical weather predictions were derived from a single run of a "best" model, to a more probabilistic approach in which forecasts were derived from an ensemble of model runs obtained from slightly different (perturbed) initial conditions. In order for the ensemble runs to be completed within operational time constraints, this required a trade-off, generally that the individual ensemble members be run at substantially lower spatial resolution, and perhaps also with less sophisticated physical parameterizations, than the single best model.

Ensemble prediction methods are potentially useful at all time scales, but for medium- and extended-range predictions they become crucial, because any single model run can be quite unrepresentative of the distribution of possible outcomes. Figure 32, from Barsugli et al. 1999, illustrates the basic concepts. Even at the analysis time (Fig. 32a), some uncertainty is present due to inevitable observational (and model) limitations. However, the spread in this initial probability distribution function (PDF) is far smaller than that of either climatological-mean PDFs or those associated with anomalous forcing, such as associated with El Niño, as

Parallel (NINOSST/CLIMOSST) Ensembles

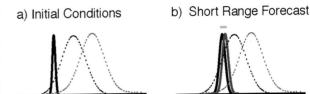

a) Initial Conditions   b) Short Range Forecast

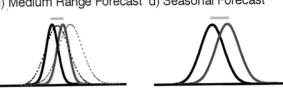

c) Medium Range Forecast   d) Seasonal Forecast

FIG. 32. Schematic probability distribution functions for (a) model initialization, (b) short-term weather forecasts, (c) medium-range weather forecasts, and (d) long-range or seasonal forecasts. The red curve represents the NiñoSST ensemble and the black curves the CLIMOSST ensemble. In (a)–(c) the dotted lines reproduce the long-term or seasonal forecast PDFs from (d). For further details, see text. From Barsugli et al. (1999).

in this illustration. Perturbations to the initial conditions are derived from within the space of possible initial states through a variety of methods (for a recent review, see Buizza et al. 2005). For short-range forecasts (Fig. 32b), the forecast PDF evolves away from the initial state and broadens slightly, but still remains much narrower than, and quite distinct from, either the climatological or El Niño distributions. Thus, these short-range forecasts appear "almost deterministic." At these ranges, the forecasts are fundamentally initial value problems, with proper specification of initial conditions being crucial to forecast skill.

Beyond some lead time, which will depend on flow predictability, forecast variable, and user need, the range of possible outcomes becomes so large that the forecasts must be considered probabilistic. At the time shown in Fig. 32c, the forecast PDF is still strongly affected by initial conditions, but has begun to evolve toward the El Niño PDF, suggesting that within this time range boundary conditions have also begun to substantially influence the forecasts. This is the medium-range forecast problem. Finally, the seasonal forecast (Fig. 32d) shows a PDF whose width is comparable to that of the climatological distribution. Forecast skill in this time range is determined predominantly by the ability to project the response of the atmosphere to the anomalous boundary conditions, in this example associated with El Niño. Note that even though the shift in seasonal means is relatively small compared to the overall spread of the climatological distribution, indicative of a modest signal-to-noise ratio (consistent with most midlatitude responses to ENSO), large changes can occur in relative probabilities at the tails of the distributions (i.e., in the extreme events that can have important practical and forecast implications).

Ensemble prediction methods have contributed substantially to increases in weather and climate forecast skill (Kalnay et al. 1998; Kalnay 2003). Despite the advantages of such methods, systematic model errors and insufficient spread of the model forecasts have limited the utility of distributions derived directly from model output (Hamill et al. 2004a,b). These shortcomings are related to the use of a single model in deriving all ensemble members. In this case, only initial condition errors are being sampled, without accounting for other sources of uncertainties (e.g., in the physical parameterizations of the model).

One proposed approach to addressing this issue is by constructing "superensembles," which are weighted averages of ensemble forecasts derived from multiple models (Krishnamurti et al. 2003; Hagedorn et al. 2005; Doblas Reyes et al. 2005) that each differ in their representations of model physics (and perhaps also resolution). This method has shown the ability to substantially increase some measures of forecast skill, and is likely to see increasing use in operational practice. On a very grand scale, the Climateprediction.net (CPDN) Project (Allen 1999; Allen and Stainforth 2002; Stainforth et al. 2002; Piani et al. 2005) is constructing very large superensembles of climate predictions for the twenty-first century. In the CPDN, the Hadley Centre unified climate model (HadSM3) that includes a simplified, thermodynamic representation of the ocean (i.e., a slab ocean model). This model is integrated over a wide range of physical parameter values and initial conditions using a distributed computing strategy, with the model runs being performed on numerous home and other remote computers for which volunteers have donated computing time for the experiments. This strategy has enabled the CPDN project to produce multithousand member ensembles of climate projections, with nearly 150 000 runs and over 10 000 000 model years completed as of early 2006 (for details, see online at http://www.climateprediction.net/index.php). The results of this fascinating project are beginning to be analyzed (e.g., Piani et al. 2005). It is interesting to contemplate whether such distributed computing efforts could be applied to a broader range of weather and climate prediction problems; for example, in forecasting extreme events such as hurricanes, in which current ensemble sizes may be insufficient for adequately estimating probability distributions.

Beyond multimodel ensembles, another promising method toward incorporating effects of model uncertainties is through the use of stochastic parameterizations. In this method, statistical representations of the uncertainties in modeling physical processes are introduced directly into the parameterization schemes (Buizza et al. 1999; Palmer et al. 2005a,b). More recently, an alternative approach, "ensemble reforecasting," has been introduced that shows considerable promise for improving forecast skill (Hamill et al. 2004a,b, 2006). In essence, the ensemble reforecast (ER) method applies a long (multiyear) set of "reforecasts" together with model output statistics (MOS) methods to statistically adjust forecast distributions obtained directly from the ensemble forecasts. The use of long training periods enables greatly improved estimates of model systematic errors in both the mean and shape of the distributions. To demonstrate this approach, Hamill et al. (2004a,b) generated a 15-member ensemble of 15-day forecast for every day over a 23-yr period from 1979 to 2001, using a 1998 version of the NCEP medium-range forecast model. Forecasts were evaluated by multiple techniques, including comparisons with 6–10- and 8–14-day operational forecasts of temperatures and precipitation over the continental United States for the winters of 2001 and 2002.

Figure 33 shows a comparison of the ER and operational forecast skill for 6–10-day temperature and precipitation forecasts for the two winters. In this reliability diagram (Wilks 2006), points that lie along the diagonal indicate "reliable forecasts"; that is, forecasts whose probabilities accurately reflect the likelihood that an event will actually occur (e.g., for all events when the forecast precipitation probability is 60%, precipitation occurs 60% of the time). The spread in the distributions provides a measure of forecast "resolution" with, for example, forecasts remaining clustered close to climatological probabilities having low resolution. Probabilistic forecast skill is evaluated through a rank probability skill score (RPSS), which provides a measure of the "distance" between the forecast probabilities and the outcome (Epstein 1969; Murphy 1971). Here, the RPSS would have a value of 0 for climatological probability forecasts and 1 for perfect forecasts, the latter requiring that the observed outcome always be predicted with 100% confidence. Comparisons show that the ER forecasts are significantly more skillful than the operational forecasts as measured by the RPSS, due both to increased reliability and resolution. Further analyses indicate that for these years the 8–14-day ER forecasts were also more skillful than the 6–10-day operational forecasts (Hamill et al. 2004a,b), suggesting an improvement in effective forecast lead time of 2–4 days.

Interestingly, Sanders (1979) study also used the RPSS to evaluate skill in 1–4-day probabilistic temperature and precipitation forecasts. While several factors, including differences in temperature and precipitation categories and forecast periods, preclude a precise comparison between studies, a very rough assessment of relative skill is still possible. Such a comparison suggests that probabilistic skill of 6–10-day forecasts of temperature and precipitation in winter using ER is now at roughly the same level as the corresponding cool season (autumn and spring semester at MIT) 3-day temperature and precipitation forecasts evaluated by Sanders for the period 1968–78, while skill in the 8–14-day temperature forecasts from ER is comparable to the av-

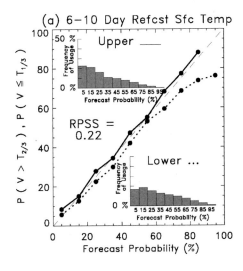

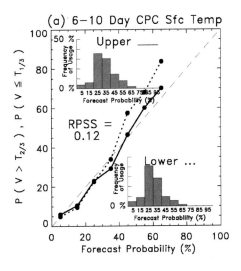

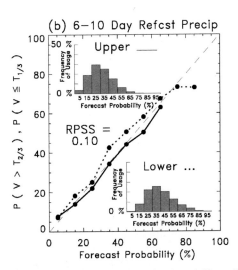

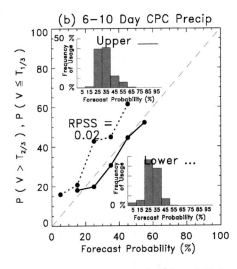

FIG. 33. Reliability diagrams for days 6–10 tercile temperature and precipitation forecasts obtained from (left) the ensemble reforecast method and from (right) NOAA/Climate Prediction Center official forecasts. Forecast skill was evaluated at 484 U.S. stations for a subset of 100 days from the 2001/02 winter. The dotted line denotes lower tercile probability forecasts; the solid line denoted upper tercile forecast probability forecasts. The closeness of the forecast probabilities to the diagonal dashed line provides a measure of forecast reliability; that is, how well forecast and observed probabilities agree. The inset histograms indicate the frequencies with which extreme tercile probability forecasts were made, thus providing a measure of the sharpness of the forecasts. The inset RPSS value indicates the ranked probability skill score. From Hamill et al. (2004b).

erage skill of the 4-day temperature forecasts in Sanders' study. While the skill values at these lead times might be considered marginal, they do suggest that the frontiers of forecast skill have been pushed outward from roughly a 4-day lead time prior to 1980 into the second forecast week today.

b. Potential predictability

If forecast skill has been pushed out into the second week, are we now reaching a point where prospects for additional progress are limited; that is, are we approaching the limits of potential predictability? The seminal studies of Lorenz (1963, 1965, 1969a,b) showed that because of the chaotic nature of atmospheric dynamics, even very small errors in initial conditions inevitably grow until all forecast skill is lost. Estimates by Lorenz, since reinforced in later studies, suggest that on average this *deterministic* predictability limit is reached within a few weeks. Indeed, given the lack of trends in forecast skill in Sanders (1979) study, one might have wondered even at that time whether some sort of predictability limit was already being approached.

However, as Lorenz also clearly recognized, there are different types of prediction problems, and hence also of potential predictability. In addition, as in any theory, the theoretical predictability limits deduced by Lorenz and others are founded on certain assumptions that, when violated, provide opportunities to extend the range of forecasts. One possibility is the existence of phenomena whose intrinsic time scales are much longer than a few weeks. We have previously discussed examples of such phenomena, which include ENSO, QBO, and the MJO. There is clear evidence that such slow variations can lead to *probabilistic* forecast skill at lead times beyond a few weeks. This can occur even if details of the evolution cannot be predicted on these time ranges, as expected from the results of Lorenz and others. Therefore, in considering potential predictability, it is vital to distinguish between deterministic predictability limits and the potential for improved probabilistic forecasts. It is the latter that are at the heart of medium- and extended-range forecasting. Here, we will briefly summarize a few studies suggesting that there remains room for progress in this area.

In a recent study, Newman et al. (2003) examined the potential predictability of weekly-averaged Northern Hemisphere circulation anomalies in winter and summer months. Newman et al. employed the method of linear inverse modeling (LIM), which uses contemporaneous and lagged statistics derived from observations to construct a linear, stable, and stochastically forced model that best approximates the dynamics of the full nonlinear system. This technique has shown great value in understanding and predicting the evolution of tropical sea surface temperatures (Penland and Magorian 1993; Penland and Sardeshmukh 1995; Penland and Matrosova 1998), and despite its great simplicity, is competitive with fully coupled models in forecasting SST anomalies one to four seasons in advance.

In the Newman et al. study, the LIM predictors were derived from 250- and 750-hPa streamfunction fields and vertically integrated diabatic heating fields obtained as iterative solutions of the "chi problem" (Sardeshmukh 1993; Winkler et al. 2001). Newman et al. compared the skill of the 2- and 3-week forecasts obtained from the LIM with the skill of medium-range forecasts obtained directly from the NCEP Medium-Range Forecast (MRF) model. The LIM results showed evidence of skill in upper-level streamfunction fields out to at least the third week (week 3) over much of the Northern Hemisphere in winter. The LIM and MRF forecasts had roughly comparable skill in week 2, with the LIM method perhaps slightly superior by week 3. Diagnostic analyses with the LIM indicated that much of the predictability beyond the first week could be attributed to anomalies in tropical heating, which were poorly forecast in the MRF but more successfully predicted with LIM. Newman et al. suggested that weekly averages of upper-level streamfunction were potentially predictable from 2–3 weeks in advance over large portions of the Northern Hemisphere (Fig. 34), with the greatest source of potential predictability coming from the tropical heating fields. Other factors that may increase potential pre-

dictability, such as stratospheric flow variations, were not considered in this study.

If tropical variations are important, how well do models do, for example, in predicting anomalies associated with the MJO? There is considerable evidence that models are deficient in representing the MJO, as well as other modes related to tropical convection, and that this may be an important factor limiting extended-range predictability (Jones et al. 2000, 2004; Waliser et al. 1999, 2003). In an illuminating study, Hendon et al. (2000) examined medium-range forecast errors associated with the MJO. Figure 35 shows one example of their results, for 850-hPa zonal wind anomalies, where the analyses are conducted relative to phases of the MJO cycle as identified through a principal components analysis. As expected, the observed zonal wind anomalies propagate slowly eastward. However, the forecast wind anomalies instead propagate westward or remain quasi-stationary, and show a pronounced tendency to weaken with increasing forecast lead time. Similar errors are evident in Hendon et al's analyses of the tropical diabatic heating forecasts (not shown here). Such errors in tropical heating and flow fields can propagate into midlatitudes within a few days (Cai et al. 1996) leading to systematic model errors that may compromise any potential predictability that might be gained from the MJO. While more recent models have improved simulations of the MJO, there are still fundamental deficiencies across a large range of models (Lin et al. 2006). This general problem of improving model simulations and predictions of tropical heating and flow fields constitutes a primary target for future progress in advancing model prediction skill.

The above studies have emphasized forecasts and potential predictability extending from weather time scales out. There are also vigorous efforts to determine to what extent climate variations influence the statistics of weather variability (Gershunov and Barnett 1998; Gershunov 1998; Gershunov and Cayan 2003; Cayan et al. 1999). To date, most research in this area has been empirical, but new efforts are emerging to employ ensemble modeling techniques to estimate changes in the probability distributions of weather events. In one recent example, Schubert et al. (2005) applied this approach to estimate the potential impact of ENSO on the probability of extreme winter storms developing along the U.S. Gulf and East Coasts (Schubert et al. 2005), and found significant differences in storm probabilities for El Niño versus La Niña winters. This approach is likely to provide new avenues for forecast progress, and will also further our understanding of the links between weather and climate phenomena.

5. Future directions

The previous sections have highlighted advances in our understanding and capabilities to predict phenomena at the interface between weather and climate. While

Anomaly correlation of forecasts
Based on forecasts made during DJF 1978/79-1999/2000

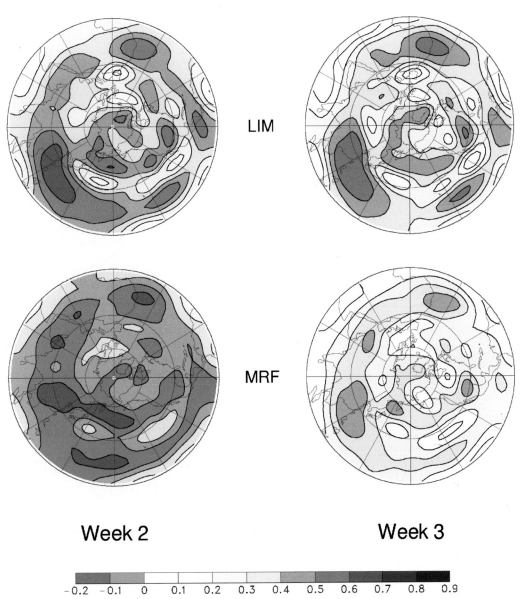

FIG. 34. Comparison of local anomaly correlations of 250-hPa streamfunction wintertime forecasts for the LIM with a 1998 version of the NCEP MRF model: (top) LIM and (bottom) MRF98. The contour interval is 0.1 with negative and zero contours indicated by blue shading and dashed lines. Shading of positive values starts at 0.2; redder shading denotes larger values of correlation, with the reddest shading indicating values above 0.6. From Newman et al. (2003).

progress has been impressive, the research also illuminates areas where significant future gains are possible. Over the next decade, it is likely that improving understanding and capabilities to predict the links between weather and climate will serve as increasingly vital components of an overall research strategy in the earth system science.

In this regard, emerging thrusts in international and national research priorities suggest that over time, artificial distinctions will be removed between weather and climate, as we begin to achieve a more unified understanding of phenomena and processes across time scales. Over the next decade, the World Meteorological Organization (WMO) World Weather Research Program (WWRP) is proposing a major international research and development program called The Observations, Re-

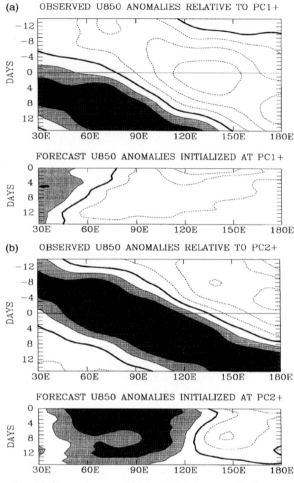

FIG. 35. Composite anomalies (annual cycle removed) of 850-hPa zonal wind relative to (a) maximum PC1+ and (b) maximum PC2+, where PC1+ and PC2+ are principal component patterns representing different phases of the MJO. Anomalies are averaged over latitudes from 5°N to 15°S. (top) The observed anomalies from time −14 to +14 days, where day 0 is the maximum for the principal component. (bottom) The forecast anomalies (mean model error and annual cycle removed) for forecasts initialized at day 0. From Hendon et al. (2000).

search, and Prediction Experiment (THORPEX) to accelerate improvements in forecasts of high-impact weather on time scales from 1 day to 2 weeks (THORPEX 2004). During this same period, the WMO World Climate Research Program (WCRP) is proposing the Coordinated Observation and Prediction of the Earth System (COPES) Program to address the seamless prediction problem on time scales from weeks to centuries in advance (COPES 2005). Within the United States, the Climate Change Science Program (CCSP) 10-yr strategic plan (CCSP 2003) identifies as a key research question the relationships between climate variations and change and extreme events, such as droughts, floods, wildfires, heat waves, and hurricanes. Achieving the objectives of these programs will require a more unified approach than in the past to understanding the

connections between weather and climate. Such an approach is entirely in keeping with the growing demand from the public and decision makers for a seamless suite of weather and climate forecasts that span time scales from very short range forecasts to decadal to centennial climate change projections.

As in the past, it will be fruitful to continue attacking problems on weather–climate linkages from both the weather side out and climate side in. Results from both approaches indicate that a key common issue will be to improve the modeling of tropical convection. Predictability studies described previously suggest that beyond about 10 days, tropical heating variations provide a primary source for potential predictability. By the second week in forecasts, even very simple models that incorporate improved predictions of tropical heating are competitive with operational prediction models, because of major deficiencies in the latter models in predicting the tropical heating fields. Improved observations and simulations of the stratospheric circulation and links to tropospheric weather provide another possible source for extended-range predictability.

Given the chaotic nature of the weather–climate system, probabilistic forecasts will always be at the heart of forecasts at and beyond the medium range. Advances in ensemble prediction techniques as well as data assimilation methods will be central to improving probabilistic estimates. As discussed previously, accurate probabilistic estimates of extreme events will require large ensemble sizes, as well as improved methods for estimating probability distributions from the ensembles. Novel prediction approaches and new applications of ensemble forecasts, including multimodel and reforecast methods, provide significant potential for improving forecast skill, as well for identifying the strengths and limitations of present-generation numerical weather prediction models.

Beyond these steps, there are abundant opportunities for advances through more detailed analyses and improved modeling of the various mechanisms that contribute to low-frequency variability. While several mechanisms were discussed, others not covered in this overview also need thorough consideration. Some examples include influence of land surface properties, including soil moisture, snow, vegetation, and topography; other modes of climate variability beyond those discussed in this review; and the role of tropical ocean variability outside the Pacific. For the latter, emerging evidence suggests the Indian Ocean may hold particular importance (e.g., Hoerling and Kumar 2003).

In considering future research directions, it is useful to recall the three general questions for understanding the links between weather and climate raised at the start of this overview. The first question was how climate variations and change can affect weather phenomena. A central issue in addressing this question will be to determine how probability distribution functions of various weather phenomena, and especially extreme events, are altered by climate variations and change. The study by Schubert et al. (2005) cited earlier provides one recent example of a modeling approach toward addressing this question for the case of wintertime storms along the U.S. Gulf and East Coasts. Advances in ensemble modeling techniques will be fundamental in making further progress in this area. A second important requirement will be to improve methods for modeling or downscaling climate predictions to the scales required for extreme weather events. A third challenge will be to improve our understanding of the mechanisms by which climate variations and change lead to systematic changes in weather events, and the regional and local implications of such connections. To achieve this objective will require improved scientific understanding of the mechanisms by which natural climate variations such as ENSO, AO–NAO, QBO, polar, and stratospheric-tropospheric interactions influence weather phenomena, especially storm-track changes, and determination of the full predictive implications for surface weather conditions.

New lines of research are emerging from questions on how aerosols, trace constituent species, dust transports, and other variables of the climate system influence weather variations. Of vital societal importance is how storm behavior, including changes in storm tracks and intensities, may change in response to anthropogenic forcing. While most current model projections indicate relatively greater warming at higher latitudes, implying a decrease in midlatitude baroclinicity, effects of changes in moisture and planetary wave patterns remain as major sources of uncertainty for projecting future changes in storm tracks and intensities.

Near-term research thrusts related to the second question, how weather phenomena affect climate variations and change, will likely focus on tropical–extratropical interactions, especially the role of tropical convection, boundary layer processes, and ocean–atmosphere–land interactions. Addressing this question will require significant efforts to better understand the hydrological cycle and coupling between short time-scale and climate processes.

The third question, identifying key phenomena and processes that bridge the time scales between weather and climate, will likely also have a major focus on improving the observations, analysis, and modeling of links between tropical and midlatitude phenomena, especially intraseasonal phenomena such as the Madden–Julian oscillation. Variations in the hydrological cycle and ocean–atmosphere interactions at these intermediate time scales are high near-term research priorities. As noted earlier, compared to our knowledge of synoptic-scale storm systems, our detailed understanding of the phenomena and mechanisms connecting short-term weather with longer-term climate variations still remains relatively limited. Attempts to create "synoptic–dynamic models" to aid in interpreting and forecasting features within the medium- to extended-range forecasts are just

being developed (Weickmann and Berry 2007), and there is ample room for more work in this area. In this regard, studies cited previously by Shapiro et al. (2001) and Ralph and colleagues suggest new directions for research linking weather and climate phenomena.

Traditionally, different scientific communities have focused on "weather prediction" and "climate prediction." To make progress over the next decade, we will need to move past this dichotomy and build stronger links between these two communities. We will also need to develop more scientists who are fluent in both weather and climate problems and in the associated methodological approaches. And we will certainly always need scientists like Fred Sanders to make the fundamental connections between observations and analysis of phenomena, dynamical understanding, and the science of weather and climate forecasting.

Acknowledgments. My understanding on this subject has benefited from many years of conversations and interactions with my colleagues at the former NOAA–CIRES Climate Diagnostics Center, now part of the NOAA/Earth System Research Laboratory. I am grateful to Dr. Klaus Weickmann and Professor Lance Bosart for carefully reviewing earlier drafts of this paper and providing many useful suggestions, and to Mr. Jon Eischeid for his extraordinary help in preparing the figures. I also thank Drs. Jeffery Whitaker, Thomas Hamill, and Marty Ralph for their helpful comments and suggestions, and Drs. Robert Black and Steve Mullen for their very thoughtful reviews, which led to substantial improvements in this paper. And, of course, my great thanks to Professor Fred Sanders, whose incisive approaches to problems and enthusiasm for understanding atmospheric phenomena profoundly influenced my thinking, surely the objective as well as the mark of a great teacher.

REFERENCES

Alexander, M. A., I. Blade, M. Newman, J. R. Lanzante, N. C. Lau, and J. D. Scott, 2002: The atmospheric bridge: The influence of ENSO teleconnections on air–sea interaction over the global oceans. *J. Climate,* **15,** 2205–2231.

Allen, M. R., 1999: Do it yourself climate prediction. *Nature,* **401,** 642.

——, and D. A. Stainforth, 2002: Towards objective probabalistic climate forecasting. *Nature,* **419,** 228.

Ambaum, M. H. P., B. J. Hoskins, and D. B. Stephenson, 2001: Arctic Oscillation or North Atlantic Oscillation? *J. Climate,* **14,** 3495–3507.

Andrews, D. G., and M. E. McIntyre, 1976: Planetary waves in horizontal and vertical shear: The generalized Eliassen–Palm relation and the mean zonal acceleration. *J. Atmos. Sci.,* **33,** 2031–2048.

Andrews, E. D., R. C. Antweiler, P. J. Neiman, and F. M. Ralph, 2004: Influence of ENSO on flood frequency along the California coast. *J. Climate,* **17,** 337–348.

Arkin, P. A., and P. J. Webster, 1985: Annual and interannual variability of tropical-extratropical interaction: An empirical study. *Mon. Wea. Rev.,* **113,** 1510–1523.

Baldwin, M. P., and T. J. Dunkerton, 1999: Propagation of the Arctic

Oscillation from the stratosphere to the troposphere. *J. Geophys. Res.,* **104D,** 30 937–30 946.

——, and ——, 2001: Stratospheric harbingers of anomalous weather regimes. *Science,* **294,** 581–584.

——, D. B. Stephenson, D. W. J. Thompson, T. J. Dunkerton, A. J. Charlton, and A. O'Neill, 2003a: Stratospheric memory and skill of extended-range weather forecasts. *Science,* **301,** 636–640.

——, D. W. J. Thompson, E. F. Shuckburgh, W. A. Norton, and N. P. Gillett, 2003b: Weather from the stratosphere? *Science,* **301,** 317.

Bao, J. W., S. A. Michelson, P. J. Neiman, F. M. Ralph, and J. M. Wilczak, 2006: Interpretation of enhanced integrated water vapor bands associated with extratropical cyclones: Their formation and connection to tropical moisture. *Mon. Wea. Rev.,* **134,** 1063–1080.

Barnston, A. G., and R. E. Livezey, 1987: Classification, seasonality, and persistence of low-frequency atmospheric circulation patterns. *Mon. Wea. Rev.,* **115,** 1083–1126.

Barsugli, J. J., and P. D. Sardeshmukh, 2002: Global atmospheric sensitivity to tropical SST anomalies throughout the Indo-Pacific basin. *J. Climate,* **15,** 3427–3442.

——, J. S. Whitaker, A. F. Loughe, P. D. Sardeshmukh, and Z. Toth, 1999: The effect of the 1997/98 El Niño on individual large-scale weather events. *Bull. Amer. Meteor. Soc.,* **80,** 1399–1411.

——, S. I. Shin, and P. D. Sardeshmukh, 2005: Tropical climate regimes and global climate sensitivity in a simple setting. *J. Atmos. Sci.,* **62,** 1226–1240.

Berggren, R., B. Bolin, and C. G. Rossby, 1949: An aerological study of zonal motion, its perturbation and breakdown. *Tellus,* **1,** 14–37.

Bergman, J. W., H. H. Hendon, and K. M. Weickmann, 2001: Intraseasonal air–sea interactions at the onset of El Niño. *J. Climate,* **14,** 1702–1719.

Bjerknes, J., 1969: Atmospheric teleconnections from the equatorial Pacific. *Mon. Wea. Rev.,* **97,** 163–172.

Black, R. X., 1997: Deducing anomalous wave source regions during the life cycles of persistent flow anomalies. *J. Atmos. Sci.,* **54,** 895–907.

——, 2002: Stratospheric forcing of surface climate in the Arctic oscillation. *J. Climate,* **15,** 268–277.

——, and R. M. Dole, 1993: The dynamics of large-scale cyclogenesis over the North Pacific Ocean. *J. Atmos. Sci.,* **50,** 421–442.

——, and ——, 2000: Storm tracks and barotropic deformation in climate models. *J. Climate,* **13,** 2712–2728.

——, and B. A. McDaniel, 2004: Diagnostic case studies of the northern annular mode. *J. Climate,* **17,** 3990–4004.

——, D. A. Salstein, and R. D. Rosen, 1996: Interannual modes of variability in atmospheric angular momentum. *J. Climate,* **9,** 2834–2849.

Blackmon, M. L., 1976: Climatological spectral study of 500 mb geopotential height of the Northern Hemisphere. *J. Atmos. Sci.,* **33,** 1607–1623.

——, J. M. Wallace, N. C. Lau, and S. L. Mullen, 1977: Observational study of Northern Hemisphere wintertime circulation. *J. Atmos. Sci.,* **34,** 1040–1053.

——, R. A. Madden, J. M. Wallace, and D. S. Gutzler, 1979: Geographical variations in the vertical structure of geopotential height fluctuations. *J. Atmos. Sci.,* **36,** 2450–2466.

——, Y. H. Lee, and J. M. Wallace, 1984: Horizontal structure of 500-mb height fluctuations with long, intermediate, and short-time scales. *J. Atmos. Sci.,* **41,** 961–979.

Borges, M. D., and P. D. Sardeshmukh, 1995: Barotropic Rossby wave dynamics of zonally varying upper-level flows during northern Winter. *J. Atmos. Sci.,* **52,** 3779–3796.

Bove, M. C., J. J. O'Brien, J. B. Eisner, C. W. Landsea, and X. Niu, 1998: Effect of El Niño on U.S. landfalling hurricanes, revisited. *Bull. Amer. Meteor. Soc.,* **79,** 2477–2482.

Branstator, G., 1985: Analysis of general circulation model sea surface temperature anomaly simulations using a linear model. Part I: Forced solutions. *J. Atmos. Sci.,* **42,** 2225–2241.

——, 1987: A striking example of the atmospheres leading traveling pattern. *J. Atmos. Sci.,* **44,** 2310–2323.

——, 1995: Organization of storm track anomalies by recurring low-frequency circulation anomalies. *J. Atmos. Sci.,* **52,** 207–226.

——, 2002: Circumglobal teleconnections, the jet stream waveguide, and the North Atlantic oscillation. *J. Climate,* **15,** 1893–1910.

Buizza, R., M. Miller, and T. N. Palmer, 1999: Stochastic representation of model uncertainties in the ECMWF Ensemble Prediction System. *Quart. J. Roy. Meteor. Soc.,* **125,** 2887–2908.

——, P. L. Houtekamer, Z. Toth, G. Pellerin, M. Z. Wei, and Y. J. Zhu, 2005: A comparison of the ECMWF, MSC, and NCEP global ensemble prediction systems. *Mon. Wea. Rev.,* **133,** 1076–1097.

Cai, M., 1992: A physical interpretation for the stability property of a localized disturbance in a deformation flow. *J. Atmos. Sci.,* **49,** 2177–2182.

——, and M. Mak, 1990: On the basic dynamics of regional cyclogenesis. *J. Atmos. Sci.,* **47,** 1417–1442.

——, J. S. Whitaker, R. M. Dole, and K. L. Paine, 1996: Dynamics of systematic errors in the NMC medium range forecast model. *Mon. Wea. Rev.,* **124,** 265–276.

Cane, M. A., S. E. Zebiak, and S. C. Dolan, 1986: Experimental forecasts of El Niño. *Nature,* **321,** 827–832.

Cayan, D. R., K. T. Redmond, and L. G. Riddle, 1999: ENSO and hydrologic extremes in the western United States. *J. Climate,* **12,** 2881–2893.

CCSP, 2003: Strategic plan for the U.S. Climate Change Science Program. Climate Change Science Program and the Subcommittee on Global Change Research Rep., Climate Change Program Office, Washington, DC, 202 pp.

Chang, E. K. M., 1993: Downstream development of baroclinic waves as inferred from regression-analysis. *J. Atmos. Sci.,* **50,** 2038–2053.

——, 2001: The structure of baroclinic wave packets. *J. Atmos. Sci.,* **58,** 1694–1713.

——, 2003: Midwinter suppression of the Pacific storm track activity as seen in aircraft observations. *J. Atmos. Sci.,* **60,** 1345–1358.

——, 2005: The impact of wave packets propagating across Asia on Pacific cyclone development. *Mon. Wea. Rev.,* **133,** 1998–2015.

——, and D. B. Yu, 1999: Characteristics of wave packets in the upper troposphere. Part I: Northern Hemisphere winter. *J. Atmos. Sci.,* **56,** 1708–1728.

——, and Y. F. Fu, 2002: Interdecadal variations in Northern Hemisphere winter storm track intensity. *J. Climate,* **15,** 642–658.

——, S. Y. Lee, and K. L. Swanson, 2002: Storm track dynamics. *J. Climate,* **15,** 2163–2183.

Charney, J. G., 1947: The dynamics of long waves in a baroclinic westerly current. *J. Meteor.,* **4,** 136–162.

——, 1973: Planetary fluid dynamics. *Dynamic Meteorology,* P. Morel, Ed., Reidel, 97–352.

——, and P. G. Drazin, 1961: Propagation of planetary-scale disturbances from the lower to the upper atmosphere. *J. Geophys. Res.,* **66,** 83–109.

——, and J. G. DeVore, 1979: Multiple flow equilibria in the atmosphere and blocking. *J. Atmos. Sci.,* **36,** 1205–1216.

——, and D. M. Straus, 1980: Form-drag instability, multiple equilibria, and propagating planetary waves in baroclinic, orographically forced, planetary wave systems. *J. Atmos. Sci.,* **37,** 1157–1176.

——, J. Shukla, and K. C. Mo, 1981: Comparison of a barotropic blocking theory with observation. *J. Atmos. Sci.,* **38,** 762–779.

Cheng, X., and J. M. Wallace, 1993: Cluster analysis of the Northern Hemisphere wintertime 500-hPa height field: Spatial patterns. *J. Atmos. Sci.,* **50,** 2674–2696.

Colucci, S. J., 1985: Explosive cyclogenesis and large-scale circulation changes: Implications for atmospheric blocking. *J. Atmos. Sci.,* **42,** 2701–2717.

——, 1987: Comparative diagnosis of blocking versus nonblocking planetary-scale circulation changes during synoptic-scale cyclogenesis. *J. Atmos. Sci.,* **44,** 124–139.

Compo, G. P., and P. D. Sardeshmukh, 2004: Storm track predictability on seasonal and decadal scales. *J. Climate,* **17,** 3701–3720.

——,——, and C. Penland, 2001: Changes of subseasonal variability associated with El Niño. *J. Climate,* **14,** 3356–3374.

COPES, 2005: The World Climate Research Programme Strategic Framework 2005–2015: Coordinated Observation and Prediction of the Earth System (COPES). WCRP-123, WMO/TDF-1291, World Meteorological Organization, Geneva, Switzerland, 65 pp.

Davis, C. A., and K. A. Emanuel, 1991: Potential vorticity diagnostics of cyclogenesis. *Mon. Wea. Rev.,* **119,** 1929–1953.

Dickson, R. R., and J. Namias, 1976: North American influences on circulation and climate of the North Atlantic sector. *Mon. Wea. Rev.,* **104,** 1255–1265.

Doblas-Reyes, F. J., R. Hagedorn, and T. N. Palmer, 2005: The rationale behind the success of multi-model ensembles in seasonal forecasting. II. Calibration and combination. *Tellus,* **57A,** 234–252.

Dole, R. M., 1982: Persistent anomalies of the extratropical Northern Hemisphere wintertime circulation. Ph.D. thesis, Massachusetts Institute of Technology, 226 pp.

——, 1983: Persistent anomalies of the extratropical Northern Hemisphere wintertime circulation. *Large-Scale Dynamical Processes in the Atmosphere,* B. J. Hoskins and R. P. Pearce, Eds., Academic Press, 95–110.

——, 1986a: The life cycles of persistent anomalies and blocking over the North Pacific. *Advances in Geophysics,* Vol. 29, Academic Press, 31–69.

——, 1986b: Persistent anomalies of the extratropical Northern Hemisphere wintertime circulation: Structure. *Mon. Wea. Rev.,* **114,** 178–207.

——, 1987: Persistent large-scale flow anomalies. Part II: Relationships to variations in synoptic-scale eddy activity and cyclogenesis. *The Nature and Prediction of Extratropical Weather Systems,* Vol. II, European Centre for Medium-Range Weather Forecasts, 73–122.

——, 1989: Life cycles of persistent anomalies. Part I: Evolution of 500 mb height fields. *Mon. Wea. Rev.,* **117,** 177–211.

——, and N. D. Gordon, 1983: Persistent anomalies of the extratropical Northern Hemisphere wintertime circulation: Geographical distribution and regional persistence characteristics. *Mon. Wea. Rev.,* **111,** 1567–1586.

——, and R. X. Black, 1990: Life cycles of persistent anomalies. Part II: The development of persistent negative height anomalies over the North Pacific Ocean. *Mon. Wea. Rev.,* **118,** 824–846.

Eady, E. T., 1949: Long waves and cyclone waves. *Tellus,* **1,** 33–52.

Edmon, H. J., B. J. Hoskins, and M. E. McIntyre, 1980: Eliassen–Palm cross-sections for the troposphere. *J. Atmos. Sci.,* **37,** 2600–2616.

Egger, J., 1977: On the linear theory of the atmospheric response to sea surface temperature anomalies. *J. Atmos. Sci.,* **34,** 603–614.

Eliasen, E., and B. Machenhauer, 1965: A study of the fluctuations of the atmospheric planetary flow patterns represented by spherical harmonics. *Tellus,* **17,** 220–238.

Elliott, R. D., and T. B. Smith, 1949: A study of the effects of large blocking highs on the general circulation of the Northern Hemisphere westerlies. *J. Meteor.,* **6,** 67–85.

Emanuel, K. A., 2001: The contribution of tropical cyclones to the oceans' meridional heat transport. *J. Geophys. Res.,* **106,** 14 777–14 781.

——, 2002: A simple model of multiple climate regimes. *J. Geophys. Res.,* **107,** 4077, doi: 10.1029/2001JD001002.

Epstein, E. S., 1969: A scoring system for probability forecasts of ranked categories. *J. Appl. Meteor.,* **8,** 985–987.

Farrell, B., 1982: The initial growth of disturbances in a baroclinic flow. *J. Atmos. Sci.,* **39,** 1663–1686.

——, 1984: Modal and non-modal baroclinic waves. *J. Atmos. Sci.,* **41,** 668–673.

——, 1985: Transient growth of damped baroclinic waves. *J. Atmos. Sci.,* **42,** 2718–2727.

Feldstein, S. B., 1998: An observational study of the intraseasonal poleward propagation of zonal mean flow anomalies. *J. Atmos. Sci.,* **55,** 2516–2529.

——, 2000: The timescale, power spectra, and climate noise properties of teleconnection patterns. *J. Climate,* **13,** 4430–4440.

——, 2001: Friction torque dynamics associated with intraseasonal length-of-day variability. *J. Atmos. Sci.,* **58,** 2942–2953.

——, 2002: Fundamental mechanisms of the growth and decay of the PNA teleconnection pattern. *Quart. J. Roy. Meteor. Soc.,* **128,** 775–796.

——, and S. Lee, 1998: Is the atmospheric zonal index driven by an eddy feedback? *J. Atmos. Sci.,* **55,** 3077–3086.

Frederiksen, J. S., 1983: A unified three-dimensional instability theory of the onset of blocking and cyclogenesis. Part II: Teleconnection patterns. *J. Atmos. Sci.,* **40,** 2593–2609.

——, and G. Branstator, 2005: Seasonal variability of teleconnection patterns. *J. Atmos. Sci.,* **62,** 1346–1365.

Gershunov, A., 1998: ENSO influence on intraseasonal extreme rainfall and temperature frequencies in the contiguous United States: Implications for long-range predictability. *J. Climate,* **11,** 3192–3203.

——, and T. P. Barnett, 1998: ENSO influence on intraseasonal extreme rainfall and temperature frequencies in the contiguous United States: Observations and model results. *J. Climate,* **11,** 1575–1586.

——, and D. R. Cayan, 2003: Heavy daily precipitation frequency over the contiguous United States: Sources of climatic variability and seasonal predictability. *J. Climate,* **16,** 2752–2765.

Gill, A. E., 1980: Some simple solutions for heat-induced tropical circulation. *Quart. J. Roy. Meteor. Soc.,* **106,** 447–462.

——, 1982: *Atmosphere–Ocean Dynamics.* Academic Press, 662 pp.

Gilman, D. L., 1985: Long-range forecasting—The present and the future. *Bull. Amer. Meteor. Soc.,* **66,** 159–164.

Gray, W. M., 1984a: Atlantic seasonal hurricane frequency. Part I: El Niño and 30 mb quasi-biennial oscillation influences. *Mon. Wea. Rev.,* **112,** 1649–1668.

——, 1984b: Atlantic seasonal hurricane frequency. Part II: Forecasting its variability. *Mon. Wea. Rev.,* **112,** 1669–1683.

Green, J. S. A., 1977: The weather during July 1976: Some dynamical considerations of the drought. *Weather,* **32,** 120–128.

Grose, W. L., and B. J. Hoskins, 1979: Influence of orography on large-scale atmospheric flow. *J. Atmos. Sci.,* **36,** 223–234.

Hagedorn, R., F. J. Doblas-Reyes, and T. N. Palmer, 2005: The rationale behind the success of multi-model ensembles in seasonal forecasting. I. Basic concept. *Tellus,* **57A,** 219–233.

Haines, K., and J. Marshall, 1987: Eddy-forced coherent structures as a prototype of atmospheric blocking. *Quart. J. Roy. Meteor. Soc.,* **113,** 681–704.

Halpert, M. S., and C. F. Ropelewski, 1992: Surface temperature patterns associated with the Southern Oscillation. *J. Climate,* **5,** 577–593.

Hamill, T. M., J. S. Whitaker, and X. Wei, 2004a: Medium-range ensemble "re-forecasting." *Bull. Amer. Meteor. Soc.,* **85,** 507–508.

——, ——, and ——, 2004b: Ensemble reforecasting: Improving medium-range forecast skill using retrospective forecasts. *Mon. Wea. Rev.,* **132,** 1434–1447.

——, ——, and S. L. Mullen, 2006: Reforecasts: An important dataset for improving weather predictions. *Bull. Amer. Meteor. Soc.,* **87,** 33–46.

Hansen, A. R., and T.-C. Chen, 1982: A spectral energetics analysis of atmospheric blocking. *Mon. Wea. Rev.,* **110,** 1146–1165.

——, and A. Sutera, 1986: On the probability density distribution of planetary-scale atmospheric wave amplitude. *J. Atmos. Sci.,* **43,** 3250–3265.

——, and ——, 1988: Planetary wave amplitude bimodality in the Southern Hemisphere. *J. Atmos. Sci.,* **45,** 3771–3783.

Harnik, N., and E. K. M. Chang, 2004: The effects of variations in jet width on the growth of baroclinic waves: Implications for midwinter Pacific storm track variability. *J. Atmos. Sci.,* **61,** 23–40.

Held, I. M., 1983: Stationary and quasi-stationary eddies in the extratropical troposphere: Theory. *Large-scale Dynamical Processes in the Atmosphere,* B. J. Hoskins and R. P. Pearce, Eds., Academic Press, 127–168.

——, and B. J. Hoskins, 1985: Large-scale eddies and the general-circulation of the troposphere. *Advances in Geophysics,* Vol. 28, Academic Press, 3–31.

——, S. W. Lyons, and S. Nigam, 1989: Transients and the extratropical response to El Niño. *J. Atmos. Sci.,* **46,** 163–174.

Hendon, H. H., and M. L. Salby, 1994: The life-cycle of the Madden–Julian Oscillation. *J. Atmos. Sci.,* **51,** 2225–2237.

——, C. Zhang, and J. D. Glick, 1999: Interannual variation of the Madden–Julian Oscillation during austral summer. *J. Climate,* **12,** 2538–2550.

——, B. Liebmann, M. Newman, J. D. Glick, and J. E. Schemm, 2000: Medium-range forecast errors associated with active episodes of the Madden–Julian Oscillation. *Mon. Wea. Rev.,* **128,** 69–86.

Higgins, R. W., and S. D. Schubert, 1993: Low-frequency synoptic-eddy activity in the Pacific storm track. *J. Atmos. Sci.,* **50,** 1672–1690.

——, and ——, 1994: Simulated life-cycles of persistent anticyclonic anomalies over the North Pacific—Role of synoptic-scale eddies. *J. Atmos. Sci.,* **51,** 3238–3260.

——, and ——, 1996: Simulations of persistent North Pacific circulation anomalies and interhemispheric teleconnections. *J. Atmos. Sci.,* **53,** 188–207.

——, and K. C. Mo, 1997: Persistent North Pacific circulation anomalies and the tropical intraseasonal oscillation. *J. Climate,* **10,** 223–244.

——, J. K. E. Schemm, W. Shi, and A. Leetmaa, 2000: Extreme precipitation events in the western United States related to tropical forcing. *J. Climate,* **13,** 793–820.

Hoerling, M. P., and M. Ting, 1994: Organization of extratropical transients during El Niño. *J. Climate,* **7,** 745–766.

——, and A. Kumar, 1997: Origins of extreme climate states during the 1982–83 ENSO winter. *J. Climate,* **10,** 2859–2870.

——, and ——, 2000: Understanding and predicting extratropical teleconnections related to ENSO. *El Niño and the Southern Oscillation: Multiscacle Variability and Global and Regional Impacts,* H. F. Diaz and V. Markgraf, Eds., Cambridge University Press, 57–88.

——, and ——, 2003: The perfect ocean for drought. *Science,* **299,** 691–694.

——, ——, and M. Zhong, 1997: El Niño, La Niña, and the nonlinearity of their teleconnections. *J. Climate,* **10,** 1769–1786.

Holopainen, E., and C. Fortelius, 1987: High-frequency transient eddies and blocking. *J. Atmos. Sci.,* **44,** 1632–1645.

Holton, J. R., 2004. *An Introduction to Dynamic Meteorology.* Elsevier Academic Press, 535 pp.

Horel, J. D., and J. M. Wallace, 1981: Planetary-scale atmospheric phenomena associated with the Southern Oscillation. *Mon. Wea. Rev.,* **109,** 813–829.

Hoskins, B. J., 1983: Dynamical processes in the atmosphere and the use of models. *Quart. J. Roy. Meteor. Soc.,* **109,** 1–21.

——, and D. J. Karoly, 1981: The steady linear response of a spherical atmosphere to thermal and orographic forcing. *J. Atmos. Sci.,* **38,** 1179–1196.

——, and P. D. Sardeshmukh, 1987: A diagnostic study of the dynamics of the Northern-Hemisphere winter of 1985–86. *Quart. J. Roy. Meteor. Soc.,* **113,** 759–778.

——, and T. Ambrizzi, 1993: Rossby-wave propagation on a realistic longitudinally varying flow. *J. Atmos. Sci.,* **50,** 1661–1671.

——, A. J. Simmons, and D. G. Andrews, 1977: Energy dispersion in a barotropic atmosphere. *Quart. J. Roy. Meteor. Soc.,* **103,** 553–567.

——, I. N. James, and G. H. White, 1983: The shape, propagation and mean-flow interaction of large-scale weather systems. *J. Atmos. Sci.,* **40,** 1595–1612.

——, M. E. McIntyre, and A. W. Robertson, 1985: On the use and significance of isentropic potential vorticity maps. *Quart. J. Roy. Meteor. Soc.,* **111,** 877–946.

Houtekamer, P. L., L. Lefaivre, J. Derome, H. Ritchie, and H. L. Mitchell, 1996: A system simulation approach to ensemble prediction. *Mon. Wea. Rev.,* **124,** 1225–1242.

Hsu, H. H., 1987: Propagation of low-level circulation features in the vicinity of mountain-ranges. *Mon. Wea. Rev.,* **115,** 1864–1892.

——, and J. M. Wallace, 1985: Vertical structure of wintertime teleconnection patterns. *J. Atmos. Sci.,* **42,** 1693–1710.

Illari, L., 1984: A diagnostic study of the potential vorticity in a warm blocking anticyclone. *J. Atmos. Sci.,* **41,** 3518–3526.

Iskenderian, H., 1995: A 10-year climatology of Northern Hemisphere tropical cloud plumes and their composite flow patterns. *J. Climate,* **8,** 1630–1637.

James, I., 1987: Suppression of baroclinic instability in horizontally sheared flows. *J. Atmos. Sci.,* **44,** 3710–3720.

Jones, C., D. E. Waliser, J. K. E. Schemm, and W. K. M. Lau, 2000: Prediction skill of the Madden and Julian Oscillation in dynamical extended range forecasts. *Climate Dyn.,* **16,** 273–289.

——, ——, K. M. Lau, and W. Stern, 2004: The Madden–Julian oscillation and its impact on Northern Hemisphere weather predictability. *Mon. Wea. Rev.,* **132,** 1462–1471.

Kalnay, E., 2003: *Atmospheric Modeling, Data Assimilation, and Predictability.* Cambridge University Press, 341 pp.

——, S. J. Lord, and R. D. McPherson, 1998: Maturity of operational numerical weather prediction: Medium range. *Bull. Amer. Meteor. Soc.,* **79,** 2753–2769.

Kessler, W. S., and R. Kleeman, 2000: Rectification of the Madden–Julian oscillation into the ENSO cycle. *J. Climate,* **13,** 3560–3575.

Kiladis, G. N., 1998: Observations of Rossby waves linked to convection over the eastern tropical Pacific. *J. Atmos. Sci.,* **55,** 321–339.

——, and H. F. Diaz, 1989: Global climatic anomalies associated with extremes in the Southern Oscillation. *J. Climate,* **2,** 1069–1090.

——, and K. M. Weickmann, 1992a: Circulation anomalies associated with tropical convection during northern winter. *Mon. Wea. Rev.,* **120,** 1900–1923.

——, and ——, 1992b: Extratropical forcing of tropical Pacific convection during northern winter. *Mon. Wea. Rev.,* **120,** 1924–1938.

——, and ——, 1997: Horizontal structure and seasonality of large-scale circulations associated with submonthly tropical convection. *Mon. Wea. Rev.,* **125,** 1997–2013.

Klein, W. H., 1952: Some empirical characteristics of long waves on monthly mean charts. *Mon. Wea. Rev.,* **80,** 203–219.

Klotzbach, P. J., and W. M. Gray, 2003: Forecasting September Atlantic basin tropical cyclone activity. *Wea. Forecasting,* **18,** 1109–1128.

——, and ——, 2004: Updated 6–11-month prediction of Atlantic basin seasonal hurricane activity. *Wea. Forecasting,* **19,** 917–934.

Knutson, T. R., and K. M. Weickmann, 1987: 30–60 day atmospheric oscillations—Composite life-cycles of convection and circulation anomalies. *Mon. Wea. Rev.,* **115,** 1407–1436.

Kok, C. J., and J. D. Opsteegh, 1985: Possible causes of anomalies in seasonal mean circulation patterns during the 1982–83 El Niño event. *J. Atmos. Sci.,* **42,** 677–694.

Krishnamurti, T. N., and Coauthors, 2003: Improved skill for the anomaly correlation of geopotential heights at 500 hPa. *Mon. Wea. Rev.,* **131,** 1082–1102.

Kumar, A., and M. P. Hoerling, 1997: Interpretation and implications of the observed inter-El Niño variability. *J. Climate,* **10,** 83–91.

——, and ——, 2003: The nature and causes for the delayed atmospheric response to El Niño. *J. Climate,* **16,** 1391–1403.

Kushnir, Y., and J. M. Wallace, 1989: Low-frequency variability in the Northern Hemisphere winter: Geographical distribution, structure and time-scale dependence. *J. Atmos. Sci., 46,* 3122–3142.

Lau, K. M., and P. H. Chan, 1985: Aspects of the 40–50 day oscillation during the Northern winter as inferred from outgoing longwave radiation. *Mon. Wea. Rev., 113,* 1889–1909.

Lau, N. C., 1978: Three-dimensional structure of observed transient eddy statistics of Northern Hemisphere wintertime circulation. *J. Atmos. Sci., 35,* 1900–1923.

——, 1979: Structure and energetics of transient disturbances in the Northern Hemisphere wintertime circulation. *J. Atmos. Sci., 36,* 982–995.

——, 1985: Modeling the seasonal dependence of the atmospheric response to observed El Niños in 1972–76. *Mon. Wea. Rev., 113,* 1970–1996.

——, 1988: Variability of the observed midlatitude storm tracks in relation to low-frequency changes in the circulation pattern. *J. Atmos. Sci., 45,* 2718–2743.

——, and J. M. Wallace, 1979: Distribution of horizontal transports by transient eddies in the Northern Hemisphere wintertime circulation. *J. Atmos. Sci., 36,* 1844–1861.

——, and E. O. Holopainen, 1984: Transient eddy forcing of the time-mean flow as identified by geopotential tendencies. *J. Atmos. Sci., 41,* 313–328.

——, and M. J. Nath, 1991: Variability of the baroclinic and barotropic transient eddy forcing associated with monthly changes in the midlatitude storm tracks. *J. Atmos. Sci., 48,* 2589–2613.

Lee, S., 1995a: Linear modes and storm tracks in a two-level primitive equation model. *J. Atmos. Sci., 52,* 1841–1862.

——, 1995b: Localized storm tracks in the absence of local instability. *J. Atmos. Sci., 52,* 977–989.

——, and I. M. Held, 1993: Baroclinic wave-packets in models and observations. *J. Atmos. Sci., 50,* 1413–1428.

Lin, J. L., and Coauthors, 2006: Tropical intraseasonal variability in 14 IPCC AR4 climate models. Part I: Convective signals. *J. Climate, 19,* 2665–2690.

Lorenz, E. N., 1963: Deterministic nonperiodic flow. *J. Atmos. Sci., 20,* 130–141.

——, 1965: A study of the predictability of a 28-variable atmospheric model. *Tellus, 17,* 321–333.

——, 1968a: The predictability of a flow which possesses many scales of motion. *Tellus, 21,* 289–307.

——, 1969b: Atmospheric predictability as revealed by naturally occurring analogues. *J. Atmos. Sci., 26,* 636–646.

Madden, R. A., and P. R. Julian, 1971: Detection of a 40–50 day oscillation in the zonal wind in the tropical Pacific. *J. Atmos. Sci., 28,* 702–708.

——, and ——, 1972: Description of global-scale circulation cells in the Tropics with a 40–50 day period. *J. Atmos. Sci., 29,* 1109–1123.

——, and ——, 1994: Observations of the 40–50-day tropical oscillation—A review. *Mon. Wea. Rev., 122,* 814–837.

Mak, M., and M. Cai, 1989: Local barotropic instability. *J. Atmos. Sci., 46,* 3289–3311.

McPhaden, M. J., 2004: Evolution of the 2002/03 El Niño. *Bull. Amer. Meteor. Soc., 85,* 677–695.

McWilliams, J. C., 1980: An application of equivalent modons to atmospheric blocking. *Dyn. Atmos. Oceans, 5,* 43–66.

Mo, K. C., 1999: Alternating wet and dry episodes over California and intraseasonal oscillations. *Mon. Wea. Rev., 127,* 2759–2776.

——, and R. E. Livezey, 1986: Tropical–extratropical geopotential height teleconnections during the Northern Hemisphere winter. *Mon. Wea. Rev., 114,* 2488–2515.

——, and R. W. Higgins, 1998a: Tropical convection and precipitation regimes in the western United States. *J. Climate, 11,* 2404–2423.

——, and ——, 1998b: Tropical influences on California precipitation. *J. Climate, 11,* 412–430.

Molteni, F., S. Tibaldi, and T. N. Palmer, 1990: Regimes in the wintertime circulation over northern extratropics. 1. Observational evidence. *Quart. J. Roy. Meteor. Soc., 116,* 31–67.

——, R. Buizza, T. N. Palmer, and T. Petroliagis, 1996: The ECMWF ensemble prediction system: Methodology and validation. *Quart. J. Roy. Meteor. Soc., 122,* 73–119.

Morgan, M. C., and J. W. Nielsen-Gammon, 1998: Using tropopause maps to diagnose midlatitude weather systems. *Mon. Wea. Rev., 126,* 2555–2579.

Mullen, S. L., 1986: The local balances of vorticity and heat for blocking anticyclones in a spectral general-circulation model. *J. Atmos. Sci., 43,* 1406–1441.

——, 1987: Transient eddy forcing of blocking flows. *J. Atmos. Sci., 44,* 3–22.

——, 1989: Model experiments on the impact of Pacific sea surface temperature anomalies on blocking frequency. *J. Climate, 2,* 997–1013.

Murphy, A. H., 1971: A note on the ranked probability score. *J. Appl. Meteor., 10,* 155–156.

Nakamura, H., 1992: Midwinter suppression of baroclinic wave activity in the Pacific. *J. Atmos. Sci., 49,* 1629–1642.

——, and J. M. Wallace, 1990: Observed changes in baroclinic wave activity during the life-cycles of low-frequency circulation anomalies. *J. Atmos. Sci., 47,* 1100–1116.

——, and ——, 1993: Synoptic behavior of baroclinic eddies during the blocking onset. *Mon. Wea. Rev., 121,* 1892–1903.

——, M. Nakamura, and J. L. Anderson, 1997: The role of high- and low-frequency dynamics in blocking formation. *Mon. Wea. Rev., 125,* 2074–2093.

——, T. Izumi, and T. Sampe, 2002: Interanual and decadal modulations recently observed in the Pacific storm track activity and East Asian winter monsoon. *J. Climate, 15,* 1855–1874.

Nakazawa, T., 1988: Tropical super clusters within intraseasonal variations over the western Pacific. *J. Meteor. Soc. Japan, 66,* 823–839.

Namias, J., 1947: Physical nature of some fluctuations in the speed of the zonal circulation. *J. Meteor., 4,* 125–133.

——, 1950: The index cycle and its role in the general circulation. *J. Meteor., 7,* 130–139.

——, and P. F. Clapp, 1944: Studies of the motion and development of long waves in the westerlies. *J. Meteor., 1,* 57–77.

National Research Council, 1996: *Learning to Predict Climate Variations Associated with El Niño and the Southern Oscillation: Accomplishments and Legacies of the TOGA Program.* National Academy Press, 171 pp.

Neilley, P. P., 1990: Interaction between synoptic-scale eddies and the large-scale flow during the life cycles of persistent flow anomalies. Ph.D. disseration, Massachusetts Institute of Technology, 272 pp. [Available from Department of Earth Atmospheric and Planetary Sciences, Massachusetts Institue of Technology, Cambridge, MA 02139.]

Newell, R. E., N. E. Newell, Y. Zhu, and C. Scott, 1992: Tropospheric rivers—A pilot study. *Geophys. Res. Lett., 19,* 2401–2404.

Newman, M., and P. D. Sardeshmukh, 1998: The impact of the annual cycle on the North Pacific–North American response to remote low-frequency forcing. *J. Atmos. Sci., 55,* 1336–1353.

——, ——, C. R. Winkler, and J. S. Whitaker, 2003: A study of subseasonal predictability. *Mon. Wea. Rev., 131,* 1715–1732.

Nitsche, G., J. M. Wallace, and C. Kooperberg, 1994: Is there evidence of multiple equilibria in planetary wave amplitude statistics. *J. Atmos. Sci., 51,* 314–322.

Opsteegh, J. D., and H. M. Van Den Dool, 1980: Seasonal differences in the stationary response of a linearized primitive equation model: Prospects for long-range weather forecasting? *J. Atmos. Sci., 37,* 2169–2185.

Orlanski, I., 2005: A new look at the Pacific storm track variability: Sensitivity to tropical SSTs and to upstream seeding. *J. Atmos. Sci., 62,* 1367–1390.

Owens, B. F., and C. W. Landsea, 2003: Assessing the skill of operational Atlantic seasonal tropical cyclone forecasts. *Wea. Forecasting, 18,* 45–54.

Palmen, E., and C. W. Newton, 1969: *Atmospheric Circulation Systems.* Academic Press, 603 pp.

Palmer, T. N., 1998: Nonlinear dynamics and climate change: Rossby's legacy. *Bull. Amer. Meteor. Soc., 79,* 1411–1423.

——, 1999: A nonlinear dynamical perspective on climate prediction. *J. Climate, 12,* 575–591.

——, and D. A. Mansfield, 1984: Response of two atmospheric general circulation models to sea-surface temperatures in the tropical East and West Pacific. *Nature, 310,* 483–485.

——, F. J. Doblas-Reyes, R. Hagedorn, and A. Weisheimer, 2005a: Probabilistic prediction of climate using multi-model ensembles: From basics to applications. *Philos. Trans. Roy. Soc., 360,* 1991–1998.

——, G. J. Shutts, R. Hagedorn, F. J. Doblas-Reyes, T. Jung, and M. Leutbecher, 2005b: Representing model uncertainty in weather and climate prediction. *Annu. Rev. Earth Planet. Sci., 33,* 163–193.

Pedlosky, J., 1979: *Geophysical Fluid Dynamics.* Springer-Verlag, 624 pp.

Pelly, J. L., and B. J. Hoskins, 2003: A new perspective on blocking. *J. Atmos. Sci., 60,* 743–755.

Penland, C., and T. Magorian, 1993: Prediction of Niño-3 sea surface temperatures using linear inverse modeling. *J. Climate, 6,* 1067–1076.

——, and P. D. Sardeshmukh, 1995: The optimal growth of tropical sea surface temperature anomalies. *J. Climate, 8,* 1999–2024.

——, and L. Matrosova, 1998: Prediction of tropical Atlantic sea surface temperatures using linear inverse modeling. *J. Climate,* 11, 483–496.

Persson, P. O. G., P. J. Neiman, B. Walter, J. W. Bao, and F. M. Ralph, 2005: Contributions from California coastal-zone surface fluxes to heavy coastal precipitation: A CALJET case study during the strong El Niño of 1998. *Mon. Wea. Rev., 133,* 1175–1198.

Petterssen, S., 1955: A general survey of some factors influencing development at sea level. *J. Meteor., 12,* 36–42.

——, 1956: *Weather Forecasting and Analysis.* Vol. 1, McGraw-Hill, 428 pp.

Philander, S. G., 1990: *El Niño, La Niña, and the Southern Oscillation.* Academic Press, 293 pp.

Piani, C., D. J. Frame, D. A. Stainforth, and M. R. Allen, 2005: Constraints on climate change from a multi-thousand member ensemble of simulations. *Geophys. Res. Lett., 32,* L23825, doi: 10.1029/2005GL024452.

Platzman, G. W., 1968: The Rossby wave. *Quart. J. Roy. Meteor. Soc., 94,* 225–248.

Plumb, R. A., 1985: On the three-dimensional propagation of stationary waves. *J. Atmos. Sci., 42,* 217–229.

——, 1986: Three-dimensional propagation of transient quasi-geostrophic eddies and its relationship with the eddy forcing of the time-mean flow. *J. Atmos. Sci., 43,* 1657–1678.

Ralph, F. M., P. J. Neiman, D. E. Kingsmill, P. O. G. Persson, A. B. White, E. T. Strem, E. D. Andrews, and R. C. Antweiler, 2003: The impact of a prominent rain shadow on flooding in California's Santa Cruz Mountains: A CALJET case study and sensitivity to the ENSO cycle. *J. Hydrometeor., 4,* 1243–1264.

——,——, and G. A. Wick, 2004: Satellite and CALJET aircraft observations of atmospheric rivers over the eastern North Pacific Ocean during the winter of 1997/98. *Mon. Wea. Rev., 132,* 1721–1745.

——, ——, and R. Rotunno, 2005: Dropsonde observations in low-level jets over the northeastern Pacific Ocean from CALJET-1998 and PACJET-2001: Mean vertical profile and atmospheric river characteristics. *Mon. Wea. Rev., 133,* 889–910.

Rasmusson, E. M., and J. M. Wallace, 1983: Meteorological aspects of the El Niño–Southern Oscillation. *Science, 222,* 1195–1202.

Reinhold, B. B., and R. T. Pierrehumbert, 1982: Dynamics of weather regimes: Quasi-stationary waves and blocking. *Mon. Wea. Rev.,* 110, 1105–1145.

Renwick, J. A., and J. M. Wallace, 1996: Relationships between North Pacific wintertime blocking, El Niño, and the PNA pattern. *Mon. Wea. Rev., 124,* 2071–2076.

Rex, D., 1950a: Blocking action in the middle troposphere and its effects on regional climate. I. An aerological study of blocking. *Tellus, 2,* 196–211.

——, 1950b: Blocking action in the middle troposphere and its effects on regional climate. II. The climatology of blocking action. *Tellus, 2,* 275–301.

Rogers, J. C., 1981: Spatial variability of seasonal sea-level pressure and 500 mb height anomalies. *Mon. Wea. Rev., 109,* 2093–2106.

Ropelewski, C. F., and M. S. Halpert, 1987: Global and regional scale precipitation patterns associated with the El Niño–Southern Oscillation. *Mon. Wea. Rev., 115,* 1606–1626.

——, and ——, 1989: Precipitation patterns associated with the high index phase of the Southern Oscillation. *J. Climate, 2,* 268–284.

——, ——, and X. Wang, 1992: Observed tropospheric biennial variability and its relationship to the Southern Oscillation. *J. Climate, 5,* 594–614.

Rossby, C. G., and Coauthors, 1939: Relations between variations in the intensity of the zonal circulation of the atmosphere and the displacements of the semi-permanent centers of actions. *Tellus,* 2, 275–301.

Salby, M. L., R. R. Garcia, and H. H. Hendon, 1994: Planetary-scale circulations in the presence of climatological and wave-induced heating. *J. Atmos. Sci., 51,* 2344–2367.

Sanders, F., 1973: Skill in forecasting daily temperature and precipitation—Some experimental results. *Bull. Amer. Meteor. Soc.,* 54, 1171–1179.

——, 1979: Trends in skill of daily forecasts of temperature and precipitation, 1966–78. *Bull. Amer. Meteor. Soc., 60,* 763–769.

——, 1988: Life history of mobile troughs in the upper westerlies. *Mon. Wea. Rev., 116,* 2629–2648.

——, and J. R. Gyakum, 1980: Synoptic-dynamic climatology of the "bomb." *Mon. Wea. Rev., 108,* 1589–1606.

Sardeshmukh, P. D., 1993: The baroclinic chi problem and its application to the diagnosis of atmospheric heating rates. *J. Atmos. Sci., 50,* 1099–1112.

——, and B. J. Hoskins, 1988: The generation of global rotational flow by steady idealized tropical divergence. *J. Atmos. Sci., 45,* 1228–1251.

——, M. Newman, and M. D. Borges, 1997: Free barotropic Rossby wave dynamics of the wintertime low-frequency flow. *J. Atmos. Sci., 54,* 5–23.

——, G. P. Compo, and C. Penland, 2000: Changes of probability associated with El Niño. *J. Climate, 13,* 4268–4286.

Sawyer, J. S., 1970: Observational characteristics of atmospheric fluctuations with a time scale of a month. *Quart. J. Roy. Meteor. Soc., 96,* 610–625.

Schonher, T., and S. E. Nicholson, 1989: The relationship between California rainfall and ENSO events. *J. Climate, 2,* 1258–1269.

Schubert, S. D., 1986: The structure, energetics, and evolution of the dominant frequency-dependent three-dimensional atmospheric modes. *J. Atmos. Sci., 43,* 1210–1237.

——, Y. Chang, M. Suarez, and P. Pegion, 2005: On the relationship between ENSO and extreme weather over the contiguous U. S. *U.S. CLIVAR Variations, 3,* 1–4.

Shapiro, M. A., H. Wernli, N. A. Bond, and R. Langland, 2001: The influence of the 1997–99 El Niño–Southern Oscillation on extratropical baroclinic life cycles over the eastern North Pacific. *Quart. J. Roy. Meteor. Soc., 127,* 331–342.

Shutts, G. J., 1986: A case study of eddy forcing during an Atlantic blocking episode. *Advances in Geophysics,* Vol. 29, Academic Press, 135–162.

Simmons, A. J., 1982: The forcing of stationary wave motion by tropical diabatic heating. *Quart. J. Roy. Meteor. Soc., 108,* 503–534.

——, and A. Hollingsworth, 2002: Some aspects of the improvement in skill of numerical weather prediction. *Quart. J. Roy. Meteor. Soc., 128,* 647–677.

——, J. M. Wallace, and G. W. Branstator, 1983: Barotropic wave-propagation and instability, and atmospheric teleconnection patterns. *J. Atmos. Sci.,* **40,** 1363–1392.

Stainforth, D., J. Kettleborough, M. Allen, M. Collins, A. Heaps, and J. Murphy, 2002: Distributed computing for public-interest climate modeling research. *Comput. Sci. Eng.,* **4,** 82–89.

Straus, D. M., and J. Shukla, 2002: Does ENSO force the PNA? *J. Climate,* **15,** 2340–2358.

Sumner, E. J., 1954: A study of blocking in the Atlantic-European sector of the Northern Hemisphere. *Quart. J. Roy. Meteor. Soc.,* **80,** 402–416.

Thompson, D. W. J., and J. M. Wallace, 1998: The Arctic Oscillation signature in the wintertime geopotential height and temperature fields. *Geophys. Res. Lett.,* **25,** 1297–1300.

——, and ——, 2000: Annular modes in the extratropical circulation. Part I: Month-to-month variability. *J. Climate,* **13,** 1000–1016.

——, and ——, 2001: Regional climate impacts of the Northern Hemisphere annular mode. *Science,* **293,** 85–89.

——, M. P. Baldwin, and J. M. Wallace, 2002: Stratospheric connection to Northern Hemisphere wintertime weather: Implications for prediction. *J. Climate,* **15,** 1421–1428.

Thorncroft, C. D., B. J. Hoskins, and M. F. McIntyre, 1993: Two paradigms of baroclinic-wave life-cycle behavior. *Quart. J. Roy. Meteor. Soc.,* **119,** 17–55.

THORPEX, 2004: A global atmospheric research programme for the beginning of the 21st century. *WMO Bulletin,* Vol. 54, No. 3.

Tracton, M. S., 1990: Predictability and its relationship to scale interaction processes in blocking. *Mon. Wea. Rev.,* **118,** 1666–1695.

——, and E. Kalnay, 1993: Operational ensemble prediction at the National Meteorological Center—Practical aspects. *Wea. Forecasting,* **8,** 379–398.

Trenberth, K. E., 1976: Fluctuations and trends in indexes of Southern Hemispheric circulation. *Quart. J. Roy. Meteor. Soc.,* **102,** 65–75.

——, 1986: An assessment of the impact of transient eddies on the zonal flow during a blocking episode using localized Eliassen–Palm flux diagnostics. *J. Atmos. Sci.,* **43,** 2070–2087.

——, 1997a: Short-term climate variations: Recent accomplishments and issues for future progress. *Bull. Amer. Meteor. Soc.,* **78,** 1081–1096.

——, 1997b: The definition of El Niño. *Bull. Amer. Meteor. Soc.,* **78,** 2771–2777.

——, and D. J. Shea, 1987: On the evolution of the Southern Oscillation. *Mon. Wea. Rev.,* **115,** 3078–3096.

——, and D. P. Stepaniak, 2001: Indices of El Niño evolution. *J. Climate,* **14,** 1697–1701.

——, G. W. Branstator, D. Karoly, A. Kumar, N. C. Lau, and C. Ropelewski, 1998: Progress during TOGA in understanding and modeling global teleconnections associated with tropical sea surface temperatures. *J. Geophys. Res.,* **103C,** 14 291–14 324.

Tung, K. K., and R. S. Lindzen, 1979a: A theory of stationary long waves. Part I: A simple theory of blocking. *Mon. Wea. Rev.,* **107,** 714–734.

——, and ——, 1979b: A theory of stationary long waves. Part II: Resonant Rossby waves in the presence of realistic vertical shears. *Mon. Wea. Rev.,* **107,** 735–750.

van Loon, H., and J. C. Rogers, 1978: the seesaw in winter temperatures between Greenland and Northern Europe. Part I: General description. *Mon. Wea. Rev.,* **106,** 296–310.

Vitart, F., and T. N. Stockdale, 2001: Seasonal forecasting of tropical storms using coupled GCM integrations. *Mon. Wea. Rev.,* **129,** 2521–2537.

Walker, G. T., 1924: Correlation in seasonal variations of weather. IX. A further study of world weather. *Memo. Indian Meteor. Dept.,* **24,** 275–332.

——, and E. W. Bliss, 1932: World Weather V. *Memo. Roy. Meteor. Soc.,* **4,** 53–84.

Waliser, D. E., C. Jones, J. K. E. Schemm, and N. E. Graham, 1999: A statistical extended-range tropical forecast model based on the slow evolution of the Madden–Julian oscillation. *J. Climate,* **12,** 1918–1939.

——, K. M. Lau, W. Stern, and C. Jones, 2003: Potential predictability of the Madden–Julian oscillation. *Bull. Amer. Meteor. Soc.,* **84,** 33–50.

Wallace, J. M., and D. S. Gutzler, 1981: Teleconnections in the geopotential height field during the Northern Hemisphere winter. *Mon. Wea. Rev.,* **109,** 784–812.

——, and N. C. Lau, 1985: On the role of barotropic energy conversions in the general circulation. *Advances in Geophysics,* Vol. 28, Academic Press, 33–74.

——, and D. W. J. Thompson, 2002a: Annular modes and climate prediction. *Phys. Today,* **55,** 28–33.

——, and ——, 2002b: The Pacific center of action of the Northern Hemisphere annular mode: Real or artifact? *J. Climate,* **15,** 1987–1991.

——, G. H. Lim, and M. L. Blackmon, 1988: Relationship between cyclone tracks, anticyclone tracks, and baroclinic wave-guides. *J. Atmos. Sci.,* **45,** 439–462.

——, E. M. Rasmusson, T. P. Mitchell, V. E. Kousky, E. S. Sarachik, and H. von Storch, 1998: The structure and evolution of ENSO-related climate variability in the tropical Pacific: Lessons from TOGA. *J. Geophys. Res.,* **103** (C7), 14 241–14 259.

Wang, B., and T. Li, 1994: Convective interaction with boundary-layer dynamics in the development of a tropical intraseasonal system. *J. Atmos. Sci.,* **51,** 1386–1400.

Webster, P. J., 1981: Mechanisms determining the atmospheric response to sea surface temperature anomalies. *J. Atmos. Sci.,* **38,** 554–571.

——, and J. R. Holton, 1982: Cross-equatorial response to middle-latitude forcing in a zonally varying basic state. *J. Atmos. Sci.,* **39,** 722–733.

——, and C. Hoyos, 2004: Prediction of monsoon rainfall and river discharge on 15–30-day time scales. *Bull. Amer. Meteor. Soc.,* **85,** 1745–1765.

Weickmann, K., 1983: Intraseasonal circulation and outgoing longwave radiation modes during Northern Hemisphere winter. *Mon. Wea. Rev.,* **111,** 1838–1858.

——, 2003: Mountains, the global frictional torque, and the circulation over the Pacific–North American region. *Mon. Wea. Rev.,* **131,** 2608–2622.

——, and P. D. Sardeshmukh, 1994: The atmospheric angular momentum cycle associated with a Madden–Julian oscillation. *J. Atmos. Sci.,* **51,** 3194–3208.

——, and E. Berry, 2007: A synoptic–dynamic model of subseasonal atmospheric variability. *Mon. Wea. Rev.,* **135,** 449–474.

——, G. R. Lussky, and J. E. Kutzbach, 1985: Intraseasonal (30–60 day) fluctuations of outgoing longwave radiation and 250 mb streamfunction during northern winter. *Mon. Wea. Rev.,* **113,** 941–961.

——, G. N. Kiladis, and P. D. Sardeshmukh, 1997: The dynamics of intraseasonal atmospheric angular momentum oscillations. *J. Atmos. Sci.,* **54,** 1445–1461.

Wheeler, M., and G. N. Kiladis, 1999: Convectively coupled equatorial waves: Analysis of clouds and temperature in the wavenumber-frequency domain. *J. Atmos. Sci.,* **56,** 374–399.

——, and K. M. Weickmann, 2001: Real-time monitoring and prediction of modes of coherent synoptic to intraseasonal tropical variability. *Mon. Wea. Rev.,* **129,** 2677–2694.

——, G. N. Kiladis, and P. J. Webster, 2000: Large-scale dynamical fields associated with convectively coupled equatorial waves. *J. Atmos. Sci.,* **57,** 613–640.

Whitaker, J. S., and R. M. Dole, 1995: Organization of storm tracks in zonally varying flows. *J. Atmos. Sci.,* **52,** 1178–1191.

——, and P. D. Sardeshmukh, 1998: A linear theory of extratropical synoptic eddy statistics. *J. Atmos. Sci.,* **55,** 237–258.

——, and K. M. Weickmann, 2001: Subseasonal variations of tropical convection and week-2 prediction of wintertime western North American rainfall. *J. Climate,* **14,** 3279–3288.

Wilks, D. S., 2006: *Statistical Methods in the Atmospheric Sciences.* 2d ed. Academic Press, 627 pp.

Willett, H. C., 1949: Long-period fluctuations of the general circulation of the atmosphere. *J. Meteor.,* **6,** 34–50.

Winkler, C. R., M. Newman, and P. D. Sardeshmukh, 2001: A linear model of wintertime low-frequency variability. Part I: Formulation and forecast skill. *J. Climate,* **14,** 4474–4494.

Wyrtki, K., 1975: El Niño—The dynamic response of the equatorial Pacific Ocean to atmospheric forcing. *J. Phys. Oceanogr.,* **5,** 572–584.

Zebiak, S. E., and M. A. Cane, 1987: A model El Niño–Southern Oscillation. *Mon. Wea. Rev.,* **115,** 2262–2278.

Zhang, C., 2005: Madden–Julian oscillation. *Rev. Geophys.,* **43,** RG2003, doi: 10.1029/2004RG000158.

——, and J. Gottschalck, 2002: SST anomalies of ENSO and the Madden–Julian Oscillation in the Equatorial Pacific. *J. Climate,* **15,** 2429–2445.

Zhang, Y., and I. M. Held, 1999: A linear stochastic model of a GCM's midlatitude storm tracks. *J. Atmos. Sci.,* **56,** 3416–3435.

Zhu, Y., and R. E. Newell, 1998: A proposed algorithm for moisture fluxes from atmospheric rivers. *Mon. Wea. Rev.,* **126,** 725–735.

——, ——, and W. G. Read, 2000: Factors controlling upper-troposphere water vapor. *J. Climate,* **13,** 836–848.

Chapter 16

Closed Anticyclones of the Subtropics and Midlatitudes: A 54-Yr Climatology (1950–2003) and Three Case Studies

THOMAS J. GALARNEAU JR., LANCE F. BOSART, AND ANANTHA R. AIYYER*

Department of Earth and Atmospheric Sciences, University at Albany, State University of New York, Albany, New York

(Manuscript received 1 December 2005, in final form 15 June 2006)

Galarneau   **Bosart**   **Aiyyer**

ABSTRACT

The pioneering large-scale studies of cyclone frequency, location, and intensity conducted by Fred Sanders prompt similar questions about lesser-studied anticyclone development. The results of a climatology of closed anticyclones (CAs) at 200, 500, and 850 hPa, with an emphasis on the subtropics and midlatitudes, is presented to assess the seasonally varying distribution and hemispheric differences of these features. To construct the CA climatology, a counting program was applied to twice-daily 2.5° NCEP–NCAR reanalysis 200-, 500-, and 850-hPa geopotential height fields for the period 1950–2003. Stationary CAs, defined as those CAs that were located at a particular location for consecutive time periods, were counted only once.

The climatology results show that 200-hPa CAs occur preferentially during summer over subtropical continental regions, while 500-hPa CAs occur preferentially over subtropical oceans in all seasons and over subtropical continents in summer. Conversely, 850-hPa CAs occur preferentially over oceanic regions beneath upper-level midocean troughs, and are most prominent in the Northern Hemisphere, and over midlatitude continents in winter.

Three case studies of objectively identified CAs that produced heat waves over the United States, Europe, and Australia in 1995, 2003, and 2004, respectively, are presented to supplement the climatological results. The case studies, examining the subset of CAs than can produce heat waves, illustrate how climatologically hot continental tropical air masses produced over arid and semiarid regions of the subtropics and lower midlatitudes can become abnormally hot in conjunction with dynamically driven upper-level ridge amplification. Subsequently, these abnormally hot air masses are advected downstream away from their source regions in conjunction with transient disturbances embedded in anomalously strong westerly jets.

* Current affiliation: Department of Marine, Earth, and Atmospheric Sciences, North Carolina State University, Raleigh, North Carolina.

Corresponding author address: Thomas J. Galarneau, Department of Earth and Atmospheric Sciences, University at Albany, State University of New York, ES-234, 1400 Washington Ave., Albany, NY 12222.
E-mail: tomjr@atmos.albany.edu

1. Introduction

a. Motivation from Fred's work

Fred Sanders recognized the importance of understanding how large-scale flow patterns and orographic barriers impacted the structure, location, and frequency of cyclones. Sanders and Gyakum (1980) mapped for the first time the distribution of rapidly intensifying cyclones ("bombs") in the Northern Hemisphere (NH) and discovered that they occurred preferentially over the western margins of ocean basins adjacent to the east coasts of continents in winter. These regions are favorable for explosive cyclogenesis because of the presence of baroclinicity augmented by differential diabatic heating between the cold continents and the warm Gulf Stream and Kuroshio. Sanders (1988) used 9 yr of once-daily 0000 UTC NH 500-hPa maps to deduce that trough births occurred preferentially in midlatitudes over and downstream of major orographic barriers such as the Rockies and Greenland. Conversely, trough deaths occurred preferentially over the eastern sides of ocean basins upstream of major orographic barriers. Sanders and Davis (1988) examined a 16-yr sample of western Atlantic explosive cyclogenesis and showed that the precursor disturbance originated over western Canada 5 days earlier. The most robust explosive cyclogenesis cases occurred with anomalously warm (cold) conditions in polar regions (midlatitudes). Weaker cases were associated with the opposite large-scale thermal anomaly. This behavior is consistent with the negative and positive phase of the Arctic Oscillation (AO), also called the NH Annular Mode (NAM; e.g., Thompson and Wallace 2000; Thompson et al. 2000; Ambaum et al. 2001; Benedict et al. 2004). A climatology of explosive cyclogenesis in two general circulation models by Sanders and Mullen (1996) confirmed many of the details originally presented in Sanders and Gyakum (1980) and Roebber (1984).

These pioneering large-scale studies of cyclone frequency, location, and intensity prompt questions about lesser-studied ridge developments and have motivated us to construct a global climatology of subtropical and midlatitude 200-, 500-, and 850-hPa closed anticyclones (CAs). A subset of these CAs are important because they can be associated with significant heat waves, especially when continental tropical (cT) air masses produced beneath CAs "escape" from their normally hot and arid source regions in response to midlatitude dynamical forcing. Three case studies from the United States, Europe, and Australia in 1995, 2003, and 2004, respectively, are presented in the appendix to illustrate the impact of displaced CA-produced cT air masses.

b. Selected heat wave and drought studies

The United States has experienced several extreme heat waves and droughts during the twentieth century. Namias (1955) investigated the cause of summertime drought over the central United States in conjunction with 700-hPa anticyclones. He found a persistent "three-cell" pattern during periods of extreme drought with a 700-hPa closed anticyclone over the Gulf of Alaska and a subsequent downstream trough (ridge) over the west coast of the United States (central United States). Namias (1955) suggested that factors contributing to the three-cell pattern likely included differing ground sur-

face characteristics, Pacific sea surface temperature anomalies, and varying intensity of solar radiation.

Examples of intense heat waves and drought over the United States include the infamous Dust Bowl of the 1930s and extended periods of heat in 1980 and 1988 over the Great Plains (e.g., Namias 1982, 1991; Schubert et al. 2004). Several studies suggested that the 1980 and 1988 events were associated with persistent summer continental CAs that acted as blocks (e.g., Livezey 1980; Namias 1982, 1991; Lyon and Dole 1995; Nkemdirim and Weber 1999). A positive feedback process between heat-related soil moisture depletion and convective inhibition leading to further soil moisture depletion was documented as playing a role in maintaining the 1980 and 1988 summer droughts over the plains (e.g., Hao and Bosart 1987; Chang and Smith 2001; Hong and Kalnay 2002). Studies by Trenberth et al. (1988) and Trenberth and Branstator (1992) suggested that anomalous sea surface temperatures (SSTs) in the equatorial central Pacific Ocean played a role in the 1988 heat wave event. The configuration of these SST anomalies in the tropical Pacific resulted in anomalous atmospheric heating in the vicinity of the intertropical convergence zone (ITCZ) near 180° longitude. This heating forced a wave train of high and low pressure anomalies across North America, placing a trough over the west coast of North America and a subsequent ridge over the Great Plains.

In July 1995, an extreme but short-lived heat wave occurred over the north-central United States, killing 525 people in the city of Chicago, Illinois, among a total of 830 heat-related deaths across the Great Plains (e.g., Changnon et al. 1996). A distinguishing feature of this heat wave was the extremely high surface dewpoint temperatures (>28°C) that developed through evapotranspiration processes [National Oceanic and Atmospheric Administration (NOAA) 1995; Changnon et al. 1996; Kunkel et al. 1996; Livezey and Tinker 1996]. Practical lessons learned from the 1995 Chicago heat wave were successfully applied to the 1999 Great Plains heat wave event, resulting in a death toll 75% smaller than the 1995 event (Palecki et al. 2001).

Prominent heat waves occurred over parts of Europe and Australia in 2003 and 2004, respectively. The extreme nature of the 2003 European heat wave illustrated how synergistic interactions between 1) large-scale dynamical forcing associated with CAs that allowed hot Saharan air to reach southwestern Europe, 2) regional influences associated with airmass sources, 3) land use and soil moisture availability, and 4) local impacts related to the surface energy budget all combined to create very favorable conditions for persistent heat (e.g., Black et al. 2004; Burt 2004; Burt and Eden 2004; Fink et al. 2004; Baldi et al. 2005). Cassou et al. (2005) examined outgoing longwave radiation (OLR) anomalies to show that the ITCZ was shifted anomalously poleward across the eastern Atlantic and western Africa in the summer of 2003. They used these observations and the results of a climate simulation to suggest that the poleward shift in the position of the ITCZ over western Africa might possibly be linked to an enhanced monsoon circulation with increased compensating subsidence to the south (north) near the equator (over southern Europe). They also noted that other factors may have played an important role in the European summer 2003 heat wave (e.g., extratropical circulation influences on the Tropics, soil moisture anomalies, and sea surface temperature anomalies). Finally, a preliminary analysis of the February 2004 Australian heat wave by Galarneau and Bosart (2006) showed that a strong subtropical continental CA positioned equatorward of a strong jet (similar to the July 1995 U.S. heat wave) set the stage for fierce heat in southern Australia with northerly low-level flow over the continental interior.

c. Previous anticyclone studies

Extreme weather events like the examples discussed above have provided motivation for many past studies that have sought to understand the structure and evolution of both cyclones and anticyclones. Several climatological studies have documented preferential regions of NH surface and upper-level anticyclones (e.g., Petterssen 1956; Klein 1957, 1958; Klein and Winston 1958; Bell and Bosart 1989; Parker et al. 1989). Petterssen (1956) constructed a winter and summer climatology of surface anticyclone and anticyclogenesis frequency during the period 1899–1939. He showed that subtropical anticyclones occur preferentially in the western Pacific and Atlantic Ocean basins and often merge with cold continental anticyclones moving equatorward over the continents in winter. In summer, subtropical anticyclones occur preferentially in the western Atlantic and Pacific Ocean basins, as in winter, but are more frequent. Cold continental anticyclones occur preferentially in the polar latitudes during summer, mostly eliminating the interaction between subtropical anticyclones and cold continental anticyclones seen in winter.

Petterssen (1956) also noted the influence of inland water bodies on anticyclones, showing how the Great Lakes, the Mediterranean Sea, the Black Sea, and the Caspian Sea have cold continental anticyclone maxima in summer and minima in winter. He suggested that this observed distribution is due to the water being colder (warmer) than the surrounding land in summer (winter) in recognition that relatively cooler (warmer) surfaces favor anticyclonic (cyclonic) vorticity generation. Similarly, upper-level anticyclone climatological studies have shown that the frequency of anticyclone centers is maximized over the subtropical oceans in all seasons and over subtropical continents during summer (e.g., Bell and Bosart 1989; Parker et al. 1989). These centers are displaced substantially northward along western coastal regions of North America and Europe (Bell and Bosart 1989), and maxima are also noted north of the main belt of westerlies in the northern Pacific and north-

ern Atlantic Ocean in a band from northwest Europe to central Asia (e.g., Zishka and Smith 1980; Bell and Bosart 1989; Parker et al. 1989; Alberta et al. 1991).

Many climatological studies of cyclones and anticyclones involved tracking individual weather systems either subjectively or objectively (e.g., Klein 1957, 1958; Sanders 1988). In an alternative approach, Blackmon (1976) calculated the variance of geopotential height from gridded datasets in a frequency band considered to be associated with synoptic time scales. He found that regions of low 500-hPa height variance were consistent with persistent large-scale features such as blocking anticyclones, while regions of large height variance denoted areas with frequent trough and ridge passages and their attendant surface cyclones and anticyclones. He identified regions of large height variance associated with storm tracks, a practice that has continued to the present day (e.g., Blackmon et al. 1977; Wallace et al. 1988; Chang and Yu 1999; Chang 1999; Hoskins and Hodges 2002, 2005; Anderson et al. 2003).

Hoskins and Hodges (2002, hereafter HH02) used both of the above tracking schemes in their study of storm tracks during the NH winter for the period 1979–2000 using the 15-yr European Centre for Medium-Range Weather Forecasts (ECMWF) Re-Analysis (ERA-15) supplemented by the National Centers for Environmental Prediction–National Center for Atmospheric Research (NCEP–NCAR; Kalnay et al. 1996) and Goddard Earth Observing System-11 (GEOS-1; Schubert et al. 1995) reanalyses. They used a standard 2–6-day bandpass filter for the variance analysis and compared it with the objective feature-tracking scheme. Their results showed the North Atlantic and Pacific upper-tropospheric storm tracks occurred in a band spiraling eastward around the NH beginning in midlatitudes over the western side of the respective ocean basin and ending in polar latitudes over the eastern side of the same ocean basin. The lower-tropospheric storm tracks occupied a more limited longitudinal sector, remaining closer to regions of baroclinicity along the western sides of ocean basins, and were likely triggered by upper-tropospheric disturbances moving within the aforementioned band of activity. The objective feature-based tracking, which allows cyclones and anticyclones to be considered separately, revealed that cyclones are generally stronger and more coherent then their anticyclone counterparts.

Simultaneous examination of surface and upper-level climatologies reveals variations in vertical structure. For example, Bell and Bosart (1989) show a 500-hPa anticyclone maximum over the central United States in summer while Petterssen (1956) shows a surface anticyclone minimum. This characteristic, also seen over India in summer, demonstrates how a warm-core anticyclone intensifies upward and may be associated with a "heat low" at the surface. The opposite case, with an anticyclone at the surface and cyclonic flow aloft, can be seen in winter with the cold core Siberian and Yukon anticyclones. This structural variation with height is not usually seen in warm subtropical oceanic anticyclones, however, because of their quasi-equivalent barotropic structure.

Climatological studies of Southern Hemisphere (SH) anticyclones show similar characteristics as their NH counterparts, with maxima over the subtropical oceans in all seasons and subtropical continents in summer (e.g., Taljaard 1967; LeMarshall et al. 1985; Leighton and Deslandes 1991; Leighton 1994; Leighton and Nowak 1995; Pezza and Ambrizzi 2003). Pezza and Ambrizzi (2003), who documented that subtropical anticyclones are more prevalent in the South Atlantic Ocean (South Pacific Ocean) during El Niño (La Niña) years, provided evidence of intraseasonal variability.

Several studies have addressed cyclones, anticyclones, and their respective storm tracks in the SH using the techniques described above for the NH (e.g., Trenberth 1991; Sinclair 1994, 1995, 1996; Sinclair et al. 1997; Hoskins and Hodges 2005, hereafter HH05). HH05 applied the methodology from HH02 to the 40-yr ECMWF Re-Analysis (ERA-40) in order to provide a new perspective on Southern Hemisphere (SH) storm tracks. They found that the general picture remains the same as compared with previous storm-track climatologies (e.g., Trenberth 1991; Sinclair 1994) with the differences lying in the details. In summer, they found a rather circular high-latitude storm track, while in winter the storm track was more asymmetric with a spiral from the Atlantic and Indian Oceans eastward and poleward to Antarctica. A subtropical jet–related lower-latitude storm track was found over the Pacific, again tending to spiral poleward.

Subtropical anticyclones are quasi-permanent features of the subtropical Pacific and Atlantic Oceans and are usually related to radiative cooling and the descending arm of the Hadley cell (e.g., Hoskins 1996; Davis et al. 1997). Hoskins (1996) showed that summer subtropical anticyclones are stronger and more longitudinally dependent than their winter counterparts, and that monsoonal latent heat release over neighboring continents to the east is the likely cause of summer subtropical anticyclones. As this diabatic heating over the continents moves poleward, deep descent is induced poleward and westward of it where the flow on isentropic surfaces is directed toward higher pressure in the vicinity of the core diabatic heating region (Hoskins 1996). Orographic effects, radiative cooling, suppressed convection, oceanic upwelling, and cold sea surface temperatures are viewed as amplifiers of this descent (e.g., Hoskins 1996; Chen et al. 2001; Rodwell and Hoskins 2001; Shaffrey et al. 2002; Liu et al. 2004). Alternatively, Miyasaka and Nakamura (2005) suggested that cold advection and advection of the earth's vorticity in the northerlies east of the subtropical anticyclone must be counterbalanced by subsidence warming and associated vortex-tube shrinking to produce anticyclonic vorticity to maintain the high. They further suggest that monsoonal latent heating may not be the direct instigator for summer subtropical anticyclones, much less the primary cause. This is supported by evidence that the summer subtropical anticyclone can be well established prior to the onset of the summer monsoon.

d. Blocking anticyclone studies

Warm season subtropical and midlatitude continental CAs can also be associated with atmospheric blocking (e.g., Rex 1950a,b). The climatology of blocking (e.g., Lejenäs and Økland 1983; Trenberth and Mo 1985; Blackmon et al. 1986; Sinclair 1996; Wiedenmann et al. 2002; Pelly and Hoskins 2003a), the evolution and maintenance of blocking (e.g., Colucci 1985; Mullen 1986, 1987, 1989; Lupo and Smith 1998; Lupo and Bosart 1999), and the prediction of blocking (e.g., Ceppa and Colucci 1989; Colucci and Baumhefner 1998; Colucci 2001; Pelly and Hoskins 2003b) are all relevant to the distribution of subtropical and midlatitude CAs.

Mullen (1986, 1987, 1989) studied how a blocking flow is maintained, using a spectral general circulation model (GCM), and found that transient eddy vorticity fluxes act to shift the block upstream while the time-mean flow acts to shift the block downstream, thus counteracting each other. He also found that eddy heat fluxes in the periphery of the anticyclone are balanced by adiabatic warming in the troposphere and diabatic heating at the surface (Mullen 1986, 1987). It is likely that anomalous synoptic-scale eddy heat and potential vorticity fluxes associated with blocking anticyclones and heat and moisture fluxes associated with regional- and local-scale circulations may interact constructively to help maintain continental blocking CAs.

e. Outline of paper

This paper will focus on a global climatology of CAs and three representative case studies of a subset of CAs that produce heat waves from different parts of the globe. Section 2 describes the datasets and programs used to construct the climatology and three case studies. Section 3 presents the climatological results, while section 4 provides an overview discussion of the findings from the climatology. Section 5 will provide the conclusions from the climatology. The appendix will show the results, and provide discussion and critical conclusions, of the findings of three case studies of CAs that were associated with significant heat waves.

2. Data and methods

a. Climatology

The CA climatology was constructed using the twice-daily (0000 and 1200 UTC) geopotential height fields from the NCEP–NCAR reanalysis (Kalnay et al. 1996;

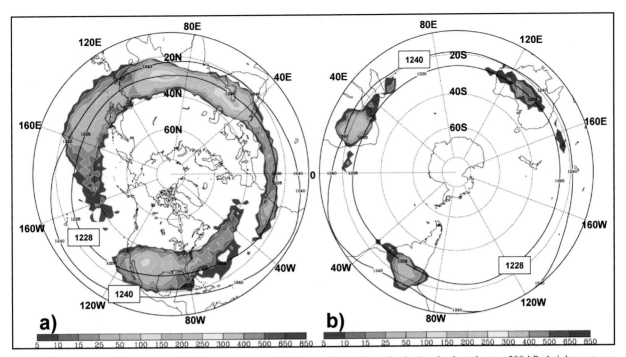

FIG. 1. Total number of 2CA events 1236 dam or greater (shaded according to the color bar) and selected mean 200-hPa height contours (solid; 1228 and 1240 dam) during the period 1950–2003 for the (a) NH and (b) SH.

Kistler et al. 2001). The data are archived on a 2.5° latitude–longitude grid. The climatology spans the period 1950–2003. The objective method used to identify the closed circulation associated with anticyclones is adapted from Bell and Bosart (1989) and proceeds as follows. At the 850-hPa level, any grid point was deemed a potential anticyclone center if its geopotential height satisfied two criteria: (i) it exceeded a threshold of 162 dam and (ii) it equaled or exceeded the height at the 8 surrounding grid points. All potential anticyclone centers were subjected to a further test to determine whether they were associated with closed circulations. This was accomplished by interpolating the heights to a cylindrical grid around each potential center, and testing whether, compared to the center value, the height decreased by at least 15 m along each radial arm of the new grid. This ensured that at least one closed contour, 15 m less than the maximum value, was present for each anticyclone. The procedure for the 500 (200)-hPa level was identical except for the center threshold value of 588 (1236) dam and the requirement of a closed contour with a difference of 30 (60) m from the anticyclone center. The above thresholds were chosen to focus the climatology on closed anticyclones in the subtropics and midlatitudes. Stationary anticyclones, defined as those anticyclones that were located at a particular grid point for consecutive time periods, were counted only once. The NH (SH) winter season is defined as December–February (June–August), the spring season as March–May (September–November), the summer season as June–August (December–February),

and the autumn season as September–November (March–May).

b. Case studies

Analyses and diagnostic calculations prepared in this manuscript for the case studies were derived from the 2.5° NCEP–NCAR reanalysis (Kalnay et al. 1996; Kistler et al. 2001). For the European and Australian cases, the 500-hPa height and absolute vorticity diagnostics, air parcel trajectory analyses, and 850-hPa isotherm continuity analyses are derived from the NCEP 1.0° Global Forecast System (GFS) analyses. Mortality and surface temperature statistics for the U.S. case were obtained from Donoghue et al. (1995) and Nashold et al. (1996). European mortality statistics were obtained from Johnson et al. (2005) and Pirard et al. (2005). Surface temperature statistics for France (United Kingdom) were obtained from Météo-France (Met Office). Surface temperature statistics for the Australian case were obtained from the Australian Bureau of Meteorology.

3. Climatology

a. 200-hPa closed anticyclones

An overview of 200-hPa CA (2CA) frequency shows clearly that the maximum in anticyclone frequency is greatest over continental regions and western ocean basins between 20° and 40°N (Fig. 1a). Within these regions, frequency maxima are located over the southwest

United States and northern Mexico (100–150 events), the Middle East (150–200 events), the Tibetan Plateau (200–250 events), and the western Pacific Ocean (50–100 events). Also apparent is the absence of 2CA activity in the persistent midocean troughs located over the central and eastern Pacific and Atlantic Oceans. This lack of 2CA activity is supported by the mean position of the 200-hPa height contours, which show broad mean troughs in these regions. In the SH, 2CAs are seen almost exclusively over continental regions between 10° and 30°S (Fig. 1b). Frequency maxima are located over central South America just east of the Andes Mountains (25–50 events), over southern Africa eastward to Madagascar (50–100 events), and over Australia (15–25 events). As in the NH, 2CA activity is minimized in the midocean troughs over the southern Atlantic and southeast Pacific Oceans.

In the NH, most 2CAs occur in summer (Fig. 2c), and to a lesser extent in autumn (Fig. 2d), with little to no activity occurring in winter and spring (Figs. 2a,b) based on our 1236-dam threshold. In winter, the mean 200-hPa height contours are farthest equatorward compared with other seasons and show the strong zonal jet over the western Pacific Ocean near 20°N (Fig. 2a). In spring, other than a small maximum over Southeast Asia (15–25 events), there continues to be minimal 2CA activity as the mean 200-hPa height contours shift poleward and become more zonally oriented (Fig. 2b). In summer, there is an eruption of 2CA activity over continental regions and the western sides of the Atlantic and Pacific Oceans with concentrated maxima located over the southwest United States and northern Mexico (100–150 events), the Middle East (150–200 events), the Tibetan Plateau (150–200 events), and the western Pacific Ocean (50–100 events; Fig. 2c). In association with the dramatic increase in 2CA activity, the mean 200-hPa height pattern becomes strongly amplified over the Pacific and Atlantic Oceans as the midocean troughs are now flanked on either side by strong anticyclones. In autumn, the areal extent of 2CA activity is lessened, especially over the eastern Atlantic Ocean where the axis of the midocean trough moves eastward to 30°W (Fig. 2d). As in summer, 2CAs occur preferentially over the continental regions and western Pacific Ocean. Primary maxima are located over northern Mexico (15–25 events), the Middle East (25–50 events), the Tibetan Plateau (25–50 events), and the western Pacific Ocean (25–50 events).

The NH intermonthly variability shows that 2CAs occur preferentially during the period May–November with a peak in total, continental (land), and oceanic (water) activity in July (5.0 events day^{-1}), July (3.5 events day^{-1}), and August (1.5 events day^{-1}), respectively (Fig. 3a). As was seen in Fig. 2, continental 2CA events compose the bulk of the total number of events. The delay in peak activity over oceans versus over continents is likely related to the oceanic thermal lag relative to the land. The result is that water bodies take a

longer time to warm relative to landmasses; hence, 2CA activity is delayed over water bodies relative to landmasses. Subsequently, there is a significant decrease in 2CAs over continental regions by September because of the cooling of the continents relative to the oceans, showing that water bodies take a longer time to cool relative to landmasses as well. A NH interannual variability of ~0.3 events day^{-1} with a minimum of 0.9 events day^{-1} in 1968 and a maximum of 1.7 events day^{-1} in 1999 is observed (Fig. 3b). Continental anticyclones compose 60% of the total number of 2CA events throughout the 54-yr period on average.

The SH interseasonal variability shows that almost all 2CAs occur during summer (Fig. 4c) with the exception of a small maximum (5–10 events) over southern Africa during autumn (Fig. 4d). In winter, the mean 200-hPa heights show a zonal large-scale pattern over the SH with the exception of the midocean troughs seen over the eastern Pacific and Atlantic Oceans (Fig. 4a). Spring is similar to winter except that the midocean troughs become more amplified in response to ridging over the Andes Mountains and the Namib and Kalahari Deserts of southern Africa (Fig. 4b). In summer, an eruption of 2CAs near 30°S over the continental regions of South America, southern Africa, and central Australia occurs in conjunction with more amplified 200-hPa heights. Primary 2CA maxima are located over interior South America east of the Andes Mountains (25–50 events), over southern Africa stretching from just inland of the Skeleton Coast eastward to Madagascar with peak activity over interior southern Africa (50–100 events), and over central Australia (15–25 events; Fig. 4c). In autumn, the large-scale hemispheric pattern transitions back to a more zonal configuration and 2CAs are limited to a small maximum in southern Africa (5–10 events; Fig. 4d).

The 2CAs in the SH occur preferentially from December to March with a peak in total, continental, and oceanic activity all in February (0.9, 0.6, and 0.3 events day^{-1}, respectively; Fig. 5a). When compared with their NH counterparts, SH 2CAs occur less frequently (peak total frequency of 0.9 SH versus 5.0 NH events day^{-1}) and over a shorter period of a given year on average (4 months in SH versus 6 months in NH; cf. Figs. 3a and 5a). Also noteworthy is the shorter intermonthly lag between continental and oceanic 2CAs when compared with the NH (cf. Figs. 3a and 5a). This difference is probably related to the relatively smaller continental extent in the SH. In the SH, a year-to-year variation of ~0.2 events day^{-1} is found with a maximum total frequency of 0.5 events day^{-1} in 1974 and minimum of 0.02 events day^{-1} in 1955 (Fig. 5b). Continental anticyclones compose 65% of the total number of 2CA events throughout the 54-yr period on average. There is also the suggestion of more events per day beginning in the late 1970s, coinciding with the start of the incorporation of global satellite data in modern data assimilation and initialization systems.

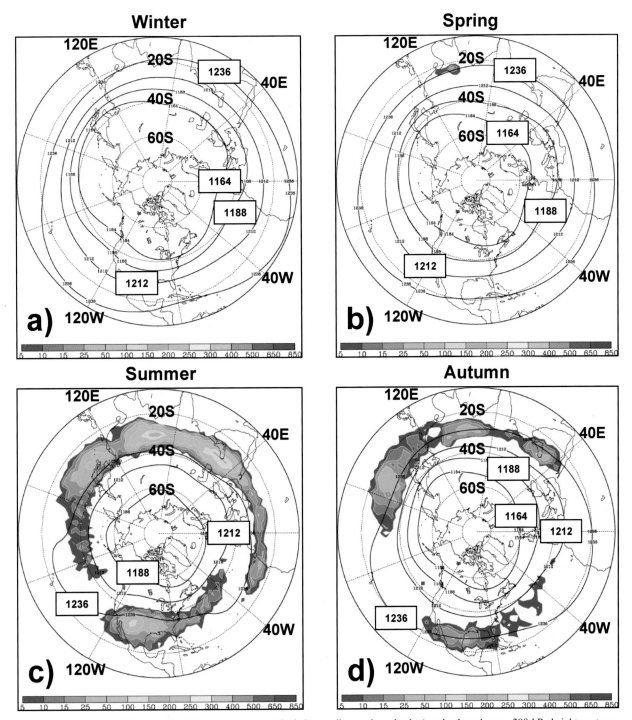

Fig. 2. Total number of 2CA events 1236 dam or greater (shaded according to the color bar) and selected mean 200-hPa height contours (solid; 1164, 1188, 1212, and 1236 dam) in the NH during the period 1950–2003 for (a) winter, (b) spring, (c) summer, and (d) autumn.

b. 500-hPa closed anticyclones

An overview of 500-hPa CA frequency demonstrates that unlike 2CAs, 500-hPa anticyclones (5CAs) are more uniformly distributed over the subtropical and midlatitude regions of both hemispheres (Fig. 6). In the NH, 5CAs occur preferentially over continental regions

and the western sides of ocean basins (Fig. 6a). Specifically, maxima are found over the southeast United States (250–300 events), northwest Africa (300–400 events), the Middle East (300–400 events), the western Pacific (250–300 events), the Atlantic Ocean (150–200 events), and the Caribbean (100–150 events). In the SH,

NH 200 hPa Closed Anticyclone Events 1950–2003

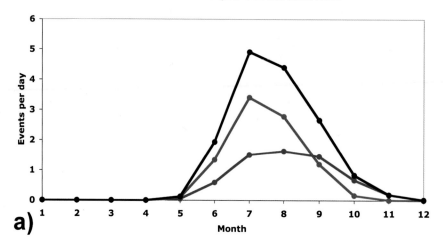

a)

NH 200 hPa Closed Anticyclone Events 1950–2003

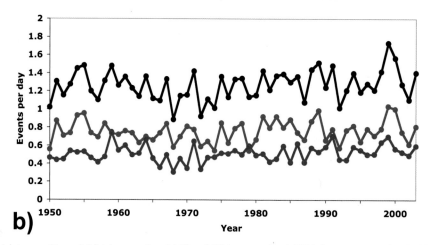

b)

FIG. 3. The (a) intermonthly and (b) interannual variability of 2CA events day⁻¹ 1236 dam or greater for the NH. Total (black), continental (red), and oceanic (blue) are plotted.

5CAs occur preferentially over oceanic regions and the western sides of continents (Fig. 6b). Maxima are found over southwest Africa (300–400 events), east of Madagascar (50–100 events), over western Australia (200–250 events), over the east-central Pacific Ocean near 120°W (100–150 events), over the Andes Mountains (100–150 events) and the eastern coast of South America (50–100 events), and the central Atlantic Ocean (100–150 events). Also noteworthy is the greater meridional extent and oceanic coverage of 5CAs versus 2CAs.

The NH interseasonal variability shows that 5CAs occur preferentially over oceanic regions in winter and over both oceanic and continental regions in summer while being absent over south-central and eastern Asia (Fig. 7). In winter, 5CA maxima occur on the equatorward side of a strong jet, as evident from the strong height gradient seen in the mean 500-hPa height con-

tours (Fig. 7a). Specifically, they occur over the western Pacific Ocean near 15°–20°N (100–150 events) and from the Caribbean eastward into the western Atlantic Ocean near 20°–30°N (50–100 events). In spring, the areal extent of 5CAs expands eastward across both ocean basins and onto continental regions (Fig. 7b). 5CA maxima are found south of a strong zonal jet across the western and central Pacific Ocean (50–100 events), over central America (50–100 events), and over North Africa (50–100 events), and the Middle East (50–100 events). The summer maximum in 5CA activity is shifted poleward in response to increased insolation (Fig. 7c). Primary 5CA maxima are located over the southwest United States (200–250 events), the eastern Atlantic Ocean (100–150 events), northwest Africa, especially over the Moroccan Highlands (300–400 events), the Middle East over the Saudi Arabian desert (50–100 events), and the higher terrain over Iran (250–

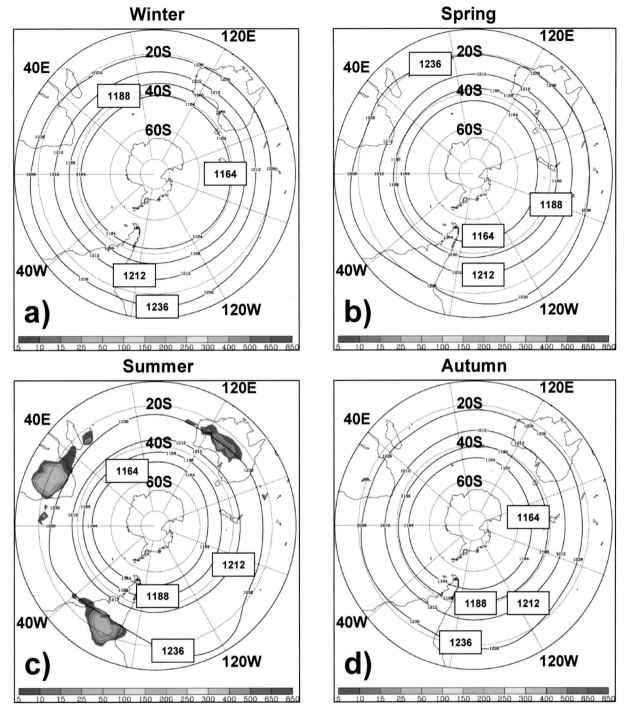

FIG. 4. Same as in Fig. 2, but for the SH.

300 events). In autumn, the areal coverage of 5CAs remains comparable to summer except that the magnitude of the continental maxima is 50% less in response to decreasing insolation (Fig. 7d). Oceanic 5CA maxima, however, are comparable in autumn and summer with a greater meridional extent in summer and autumn as compared with winter and spring. Similarly, the mid-

ocean troughs near 40° and 120°W are less apparent at 500 hPa than at 200 hPa.

Other subtle differences include 1) the northwest Africa maximum apparent at 500 hPa in summer is virtually absent at 200 hPa, 2) the southwest U.S. maximum has a larger magnitude at 500 hPa than at 200 hPa in summer, and 3) the 200-hPa maximum over the Ti-

SH 200 hPa Closed Anticyclone Events 1950-2003

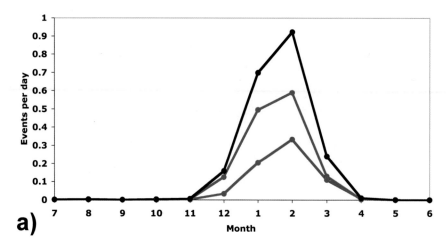

a)

SH 200 hPa Closed Anticyclone Events 1950-2003

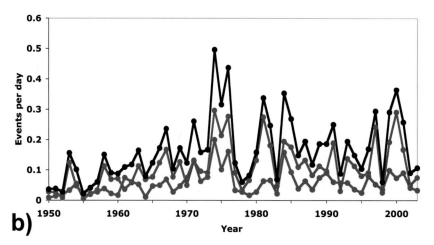

b)

Fig. 5. Same as in Fig. 3, but for the SH.

betan Plateau in summer and winter is virtually absent at 500 hPa (Figs. 2a,c and 7a,c). Comparison of Fig. 7c with the 500-hPa anticyclone frequency climatology given by Bell and Bosart (1989; their Fig. 5c) reveals a much weaker maximum over the Tibetan Plateau in our results. This discrepancy likely reflects our choice of a 588-dam height contour threshold to define closed subtropical anticyclones. Over India in summer, the mean 500-hPa height pattern is cyclonic with heights < 585 dam, while over the Tibetan Plateau the mean 500-hPa heights are closer to 582 dam along the anticyclonic shear side of the westerly flow (not shown). Given the summer 500-hPa height climatology in this part of the world, it is probable that our analysis will capture fewer events than Bell and Bosart (1989) who did not use a minimum 500-hPa height contour threshold. The 5CA events captured by our analysis over India likely represent breaks in the summer monsoon period.

The intermonthly variability of 5CAs shows a peak

from June to September (Fig. 8a). Peak 5CA frequencies for total, continental, and oceanic regions occur in July with values near 9.0, 5.5, and 3.5 events day^{-1}, respectively. The number of 5CAs comprise mostly oceanic events throughout the year, in particular in winter. The oceanic events also have a slower decrease in frequency after July when compared with continental events.

Interannual 5CA variability in the NH averages 0.5–1.0 events day^{-1} with a maximum total frequency of 6.0 events day^{-1} in 1999 and a minimum of 3.5 events day^{-1} in 1968 (Fig. 8b). The oceanic frequency is typically double the continental frequency with a weak increasing trend since the late 1970s. This is the reverse of 2CA behavior where the intermonthly and interannual variability of continental 2C As are more frequent than oceanic 2CAs overall (Fig. 3). Oceanic anticyclones compose 69% of the total number of 5CAs, while continental anticyclones compose 65% of the total number of 2CAs on average for the 54-yr period 1950–2003.

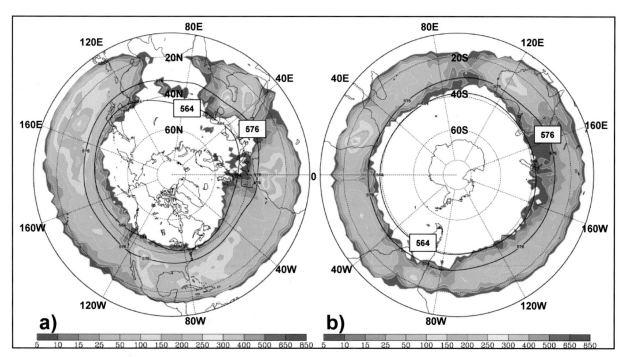

FIG. 6. Total number of 5CA events 588 dam or greater (shaded according to the color bar) and selected mean 500-hPa height contours (solid; 564 and 576 dam) during the period 1950–2003 for the (a) NH and (b) SH.

The interseasonal variability of 5CAs in the SH is generally confined to between 15° and 35°S (Fig. 9). In winter, 5CAs occur in a narrow band approximately 10° latitude in width centered on 20°S in two regions (Fig. 9a). The first region has several maxima in the 15–25 event range and stretches from western Australia eastward to 80°W (just west of the Andes Mountains). The second region has several maxima in the 15–25 event range as well and stretches from eastern South America eastward to 70°E (just east of Madagascar). In spring, anticyclone activity over the Pacific Ocean diminishes greatly and becomes more prominent over continental regions and the Atlantic Ocean (Fig. 9b). Primary maxima are found over the Andes Mountains (25–50 events) and the eastern coast of South America (25–50 events), across the Atlantic Ocean (15–25 events), over southern Africa (150–200 events), over and east of Madagascar (15–25 events), and over western Australia (25–50 events). In summer, 5CAs are found throughout the SH in a band between 20° and 40°S (Fig. 9c). Primary maxima are located over the Andes Mountains (50–100 events) and the eastern coast of South America (25–50 events), the Atlantic Ocean (50–100 events), southern Africa (150–200 events), the Indian Ocean just east of Madagascar (50–100 events), western Australia (100–150 events), and the Pacific Ocean near 100°W (50–100 events). In autumn, 5CAs are found throughout the SH in a band between 20° and 35°S (Fig. 9d). The largest difference between 2CAs and 5CAs are that 5CAs are abundant over continental and oceanic regions while 2CAs are primarily continental features (Figs. 4 and 9).

In the SH, 5CAs occur throughout the year but their frequency is maximized from December to April with a peak in total, oceanic, and continental frequency in February (5.0, 3.8, and 1.2 events day⁻¹, respectively; Fig. 10a). Oceanic 5CAs occur more frequently than continental 5CAs throughout the year except during October with the largest difference occurring the period December–April. On interannual time scales there are on average 0.5 events day⁻¹ with a maximum total frequency of 2.8 events day⁻¹ in 2001 and a minimum of 0.9 events day⁻¹ in 1950. The oceanic frequency is typically 1.5–2.0 times the continental frequency on average (Fig. 10b). The signature of the satellite era may be apparent beginning in the mid-1970s. As in the NH, continental 2CAs (oceanic 5CAs) are more frequent than their oceanic (continental) counterparts on the overall for intermonthly and interannual variability (Figs. 5 and 10).

c. 850-hPa closed anticyclones

An overview of 850-hPa CA (8CA) frequency shows that unlike 2CAs and 5CAs, 8CAs occur preferentially over oceanic regions in midlatitudes (Fig. 11). In the NH, 8CA maxima are prevalent over the Atlantic (300–400 events) and eastern Pacific (250–300 events) Oceans between 30° and 50°N, over extreme northern Africa (25–50 events), and over northern Europe and Siberia (50–100 events; Fig. 11a). The maximum indicated over the Himalayas–Tibetan Plateau (>850 events) is likely to be unrealistic since the 850-hPa sur-

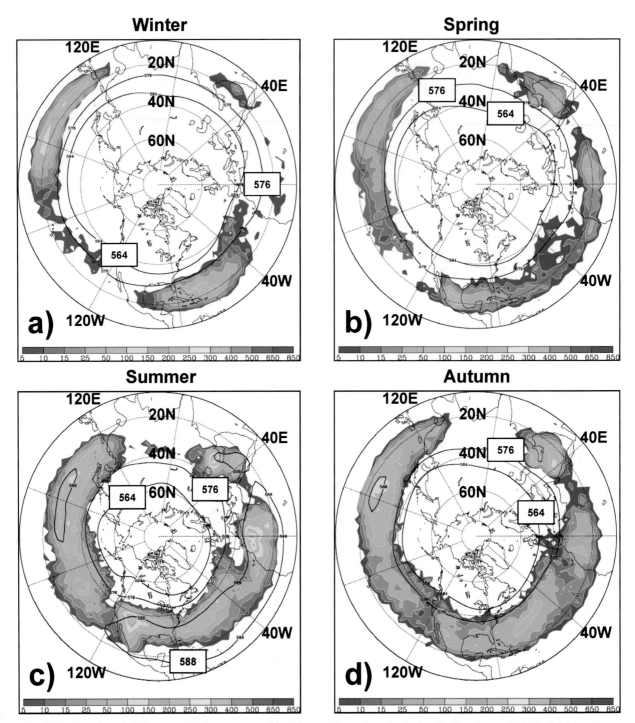

Fig. 7. Total number of 5CA events 588 dam or greater (shaded according to the color bar) and selected mean 500-hPa height contours (solid; 564, 576, and 588 dam) in the NH during the period 1950–2003 for (a) winter, (b) spring, (c) summer, and (d) autumn.

face is below ground. In the SH, 8CA maxima are prevalent throughout the SH between 30° and 50°S except over South America (Fig. 11b). Within this band, primary maxima are located over the Pacific Ocean near 100°W (100–150 events), over the Atlantic Ocean near 15°W (50–100 events), over southern Africa (50–100 events), over the Indian Ocean near 60°E, and over

southern Australia near 140°E (25–50 events). The maxima over the Andes Mountains and Antarctica are also likely to be unrealistic.

The NH interseasonal variability shows that 8CAs are prevalent over the Atlantic and eastern Pacific Oceans throughout the year and over northern Europe and Asia during winter, spring, and autumn (Fig. 12). In winter,

NH 500 hPa Closed Anticyclone Events 1950-2003

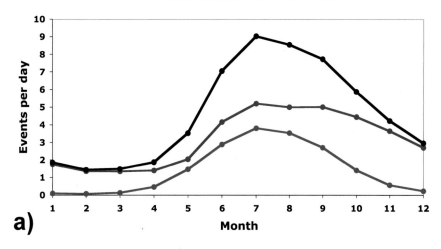

a)

NH 500 hPa Closed Anticyclone Events 1950-2003

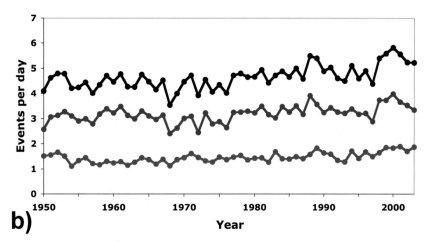

b)

FIG. 8. The (a) intermonthly and (b) interannual variability of 5CA events day^{-1} 588 dam or greater for the NH. Total (black), continental (red), and oceanic (blue) are plotted.

8CAs occur preferentially over the eastern Pacific Ocean near 40°N and 140°W (25–50 events), over the Atlantic Ocean from near the east coast of the United States to western Europe (50–100 events), and over northern Asia near 60°E (15–25 events; Fig. 12a). In spring, 8CAs occur preferentially in the same regions as in winter, except that the Pacific Ocean maximum expands westward to near 180° longitude, the western edge of Atlantic Ocean maximum contracts eastward to near 60°W, and the northern Asia maximum decreases in magnitude (10–15 events; Fig. 12b). In summer, the eastern Pacific and Atlantic Ocean maxima occupy the same region and increase in magnitude to 150–200 and 200–250 events, respectively (Fig. 12c). These maxima occur in a region of climatologically high 850-hPa heights, as denoted by the closed 156-dam 850-hPa height contour south of a corridor of zonal westerly flow located near 40°N. The northern Asia maximum, seen in winter and spring, has

disappeared. In autumn, the eastern Pacific and Atlantic maxima are again in the same location as in summer, but have decreased in magnitude (25–50 and 50–100 events, respectively; Fig. 12d). The northern Asia maximum (25–50 events) has returned in response to decreased insolation. The location of 8CA minima below regions of 2CA maxima is consistent with deep warm core CAs with 850-hPa troughs (not shown) situated beneath 200-hPa ridges (Figs. 2 and 12). In particular, this signature is found over the southwest United States and from the western Pacific eastward to northwest Africa between 20° and 40°N.

In the NH, 8CAs occur preferentially from May to September with a peak in total and oceanic frequency in July (3.9 and 3.8 events day^{-1}, respectively; Fig. 13a). The peak in continental frequency occurs in November; however, it is likely that the frequencies for 8CAs are skewed because of the Himalayan and Tibetan Plateau

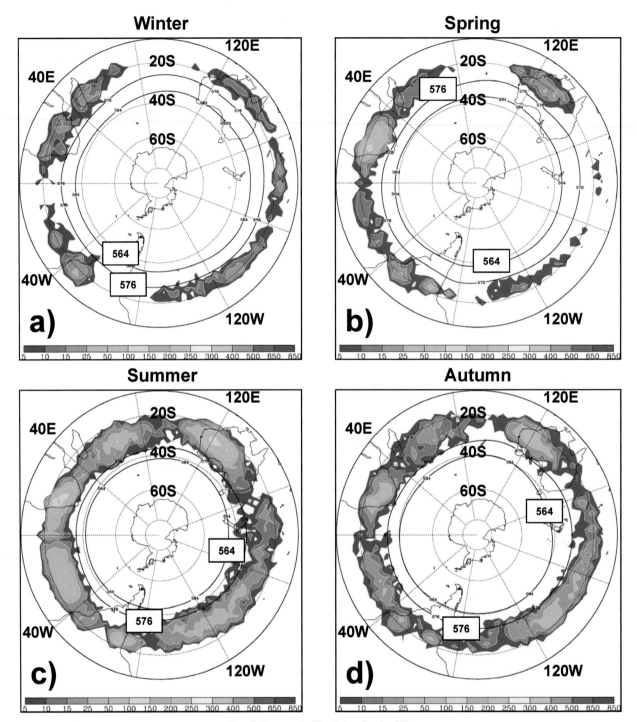

Fig. 9. Same as in Fig. 7, but for the SH.

maximum. Overall, oceanic 8CAs dominate the total frequency distribution. The interannual 8CA variability ranges from 0.5 to 1.0 events day^{-1} with a maximum total frequency of 3.1 events day^{-1} in 1998 and a minimum of 1.5 events day^{-1} in 1968 (Fig. 13b). The oceanic frequency likely dominates the total frequency by more than is suggested in Fig. 13b since the Himalayan

and Tibetan Plateau maxima skew the continental frequencies.

The SH interseasonal variability shows that 8CAs occur preferentially over midlatitude oceans and continents in winter and midlatitude oceans in summer (Fig. 14). In winter, 8CAs flank both sides of the Andes Mountains between 30°–40°S near 100°W (15–25 events)

SH 500 hPa Closed Anticyclone Events 1950-2003

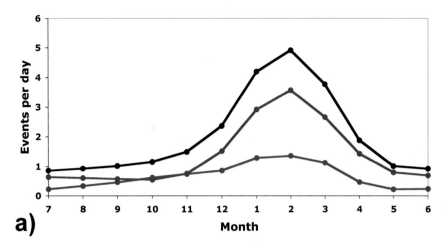

a)

SH 500 hPa Closed Anticyclone Events 1950-2003

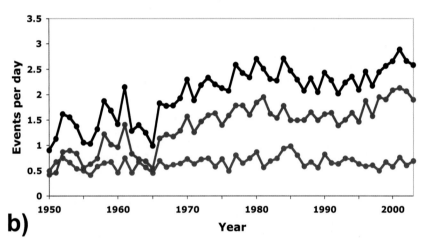

b)

FIG. 10. Same as in Fig. 8, but for the SH.

and 40°W eastward to 15°E (25–50 events), and lie from southern Africa eastward to 90°E (50–100 events), and occur across southern Australia eastward to near 150°W (10–15 events; Fig. 14a). In spring, the areal coverage of 8CAs diminishes throughout the SH, with maxima occurring preferentially over oceanic regions as 850-hPa troughs develop over continental regions and are most pronounced in the lee of the Andes Mountains over central South America (Fig. 14b). Primary maxima are located along 40°S over the Pacific Ocean near 100°W (15–25 events), over the Atlantic Ocean from 40°W eastward to 10°E (15–25 events), and over the Indian Ocean from 40° to 90°E (15–25 events). In summer, 8CA maxima continue over oceanic regions as continental troughs become more pronounced as insolation is at a maximum (Fig. 14c). The eastern Pacific Ocean 8CA maximum near 100°W grows in size and magnitude (25–50 events), while the Atlantic and Indian Ocean maxima diminish in size while maintaining sim-

ilar magnitude (15–25 events for both regions) as compared with spring. As in the NH, 8CA minima (maxima) are found below 2CA maxima (minima), pointing to the likely warm core structure of these features (Figs. 4c and 14c). In autumn, 8CA maxima return to southern Australia eastward across the Pacific Ocean to 100°W with peak values of 15–25 events (Fig. 14d). The Atlantic Ocean maximum grows zonally beginning at 40°W and extends eastward to 10°E (10–15 events), while the Indian Ocean maximum also grows zonally beginning near 40°E and stretches eastward to 120°E (25–50 events) where it meets the southern Australia maximum.

In the SH, 8CAs occur preferentially from February to September with a peak in total, oceanic, and continental frequencies in August (1.6 events day^{-1}), August (1.3 events day^{-1}), and June (0.6 events day^{-1}), respectively (Fig. 15a). It is likely, however, that the monthly continental 8CA frequencies are skewed by

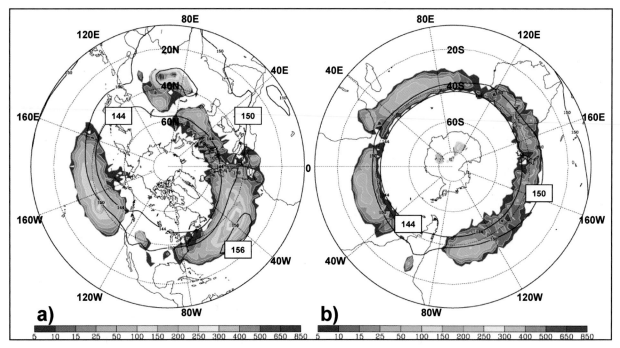

FIG. 11. Total number of 8CA events 162 dam or greater (shaded according to the color bar) and selected mean 850 hPa height contours (solid; 144, 150, and 156 dam) during the period 1950–2003 for the (a) NH and (b) SH.

unrealistic numbers where the 850-hPa surface is below ground over Antarctica and the Andes Mountains in autumn and winter. As in the NH summer, the total frequency from November to February is entirely an oceanic signature as continental regions typically have 850-hPa troughs during peak insolation. In the SH 8CAs exhibit a year-to-year variation of 0.3 events day⁻¹ on average with an increasing trend in total and oceanic events throughout the climatological period (Fig. 15b). This is likely because of the advent of satellite measurements, especially over oceanic regions. The continental 8CAs do not show a similar increasing trend. A maximum in total frequency of 1.4 events day⁻¹ occurred in 1997 while a minimum of 0.5 events day⁻¹ occurred in 1957.

4. Discussion

A total of 54 yr of 2.5° NCEP–NCAR reanalysis–derived geopotential height data were used to construct an objective global climatology of closed 200-, 500-, and 850-hPa subtropical and midlatitude anticyclones. The frequency of 5CA centers is maximized over the subtropical oceans in all seasons and over the subtropical continents in summer in agreement with Bell and Bosart (1989) and Parker et al. (1989). This agreement lends support to the different methodologies, data sources, and record lengths used in the three papers. The geopotential height threshold used for 2CAs, 5CAs, and 8CAs (1236, 588, and 162 dam, respectively) in this study eliminated cold anticyclones in higher latitudes

found in previous studies (e.g., Bell and Bosart 1989; Parker et al. 1989) in favor of CAs in subtropical and lower midlatitudes. For example, the distribution of summer 5CAs shown in Bell and Bosart (1989; their Fig. 5c) is very similar to the distribution of 5CAs in Fig. 7c except north of the main belt of westerlies. Bell and Bosart (1989) indicate that secondary 5CA maxima are situated across northern Siberia, Alaska, and northern Canada and Greenland whereas no such maxima appear in Fig. 7c. Additionally, Bell and Bosart (1989; their Fig. 5c) found a prominent summer 5CA maximum over northern India and the Tibetan Plateau where only a very weak maximum appears in Fig. 7c. Our use of a 588-dam geopotential height threshold for 5CAs likely explains this frequency difference (over northern India and the Tibetan Plateau the mean summer 500-hPa heights are <588 dam). Weak cyclonic flow prevails over India while stronger westerlies are found on the anticyclonic shear side of the jet over the Tibetan Plateau (Fig. 7c). Given this circulation climatology, the frequency of 5CAs is reduced over this region in our summer climatology.

a. Interseasonal geographical variability

Our findings indicate that 2CAs are found predominantly in the subtropical latitudes over continental regions in summer and autumn in the NH and in summer in the SH. They are also found over the western Pacific and Atlantic Oceans in the NH. The 2CA minima are located over oceanic regions, especially over the eastern

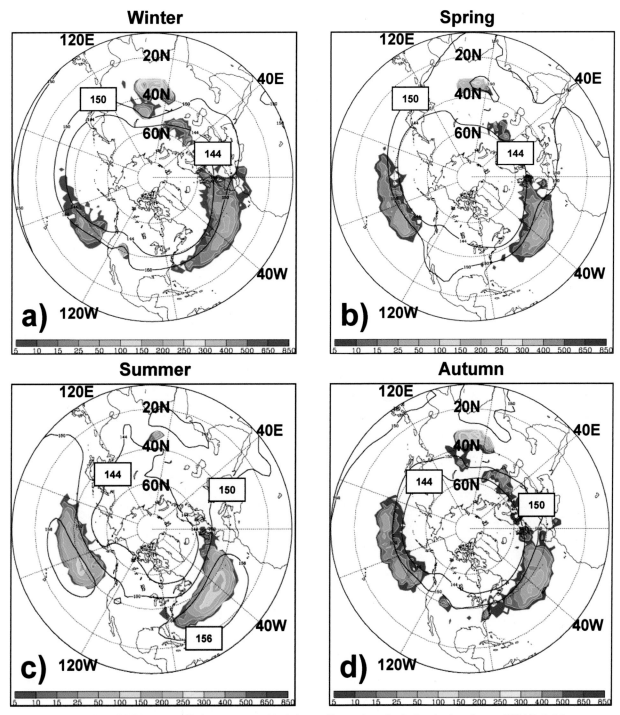

FIG. 12. Total number of 8CA events 162 dam or greater (shaded according to the color bar) and selected mean 850-hPa height contours (solid; 144, 150, and 156 dam) in the NH during the period 1950–2003 for (a) winter, (b) spring, (c) summer, and (d) autumn.

sides of ocean basins where the persistent midocean troughs are present.

Midocean troughs, the modern-day version of tropical upper-tropospheric troughs (TUTTs), were originally documented by Sadler (1967) and have been further discussed in their role in tropical cyclone development (Sadler 1976). These midocean troughs are most likely manifest by potential vorticity (PV) streamers that extend from northeast to southwest into low latitudes. Tropical cloud plumes can also provide clues to the persistence and behavior of midocean troughs and are often observed to occur ahead of these features where synoptic-scale transient features reach low latitudes and excite a tropical cloud plume (e.g., McGuirk et al. 1987,

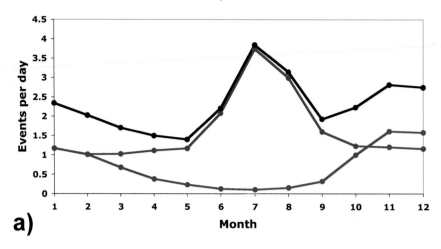

a)

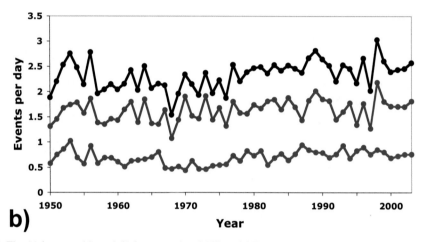

b)

FIG. 13. The (a) intermonthly and (b) interannual variability of 8CA events day^{-1} 162 dam or greater for the NH. Total (black), continental (red), and oceanic (blue) are plotted.

1988; Iskendarian 1995). Iskendarian (1995) constructed a 10-yr climatology of NH tropical cloud plumes and their composite flow patterns and found that they are most common near midocean troughs between 20° and 60°W over the Atlantic and 120°W–180° over the Pacific. There is a relative minimum in plume activity during February and March, which is consistent with the mean 200-hPa height field that shows weaker midocean troughs in the eastern Pacific and Atlantic Oceans in winter and spring (Figs. 2a,b).

The 2CAs are more prevalent over the NH than the SH because of the larger areal coverage of continental regions in the NH and prominent monsoonal circulations over North America and, in particular, southern Asia. Over the eastern Atlantic Ocean and Africa the 2CA distribution may be influenced by 1) the westward extension of the Asiatic summer monsoon anticyclone (Rodwell and Hoskins 1996), 2) convective systems

moving westward off the African coast (Cassou et al. 2005), and 3) the split-flow vorticity dipole associated with midocean troughs (Miyasaka and Nakamura 2005). In the SH, 2CAs in summer are concentrated over continental regions in 10°–15° latitude bands centered near 20°S (Fig. 4c). The 2CA maxima are found over and to the east of the Andes Mountains, over the Kalahari Desert of southern Africa, and over the Great Victoria Desert of west-central Australia (Fig. 4c). The summer concentration of 2CAs over central South America and southern Africa lies well poleward of the 250-hPa bandpass-filtered geopotential height variance maxima (storm track) and 250-hPa relative vorticity track density maxima shown in Figs. 2a and 10a of HH05, respectively. Over Australia, the 2CA maximum is found along the equatorward side of the 250-hPa geopotential height variance maximum and 250-hPa relative vorticity track density maximum associated with the northern branch

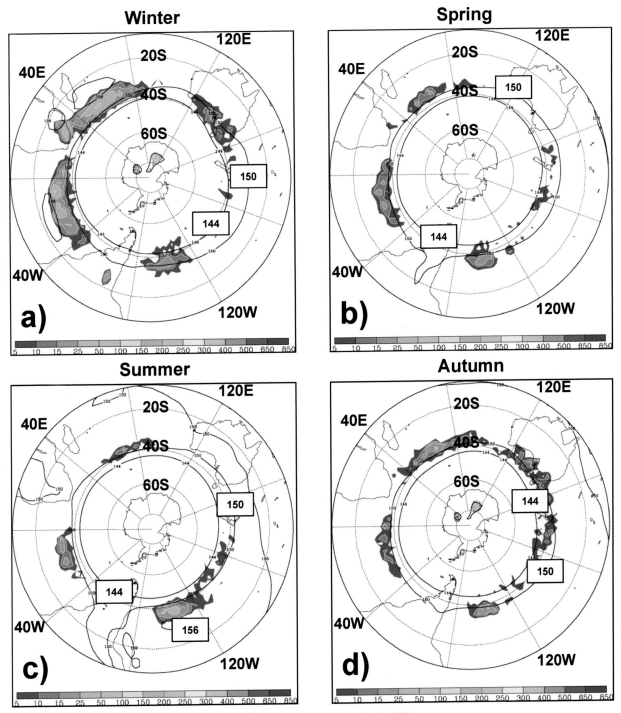

Fig. 14. Same as in Fig. 12, but for the SH.

of the split storm track in that region, suggestive of a possible relationship with the subtropical jet in that region (HH05, their Figs. 2a and 10a). No 2CAs reached our five-event threshold for plotting purposes in the other seasons with the exception of a tiny region over southern Africa in autumn (Fig. 4d).

The 5CAs are predominantly found in subtropical lat-

itudes over oceanic regions in all seasons and continental regions in summer (Fig. 7c). Significant maxima are located over the western sides of continents, in particular over the intermountain region of North America, northwest Africa, the higher terrain of the Middle East, and the western sides of the Atlantic and Pacific Oceans in the NH. Although the midocean trough signature seen

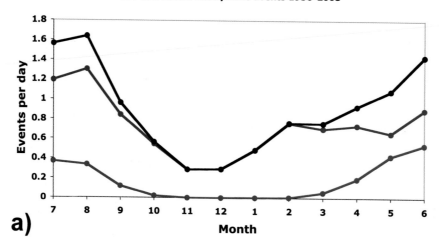

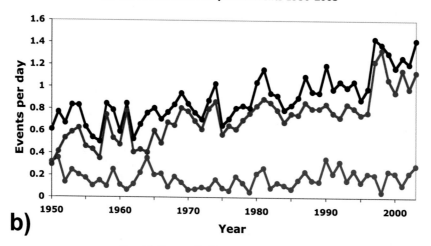

FIG. 15. Same as in Fig. 13, but for the SH.

over the eastern Pacific and Atlantic Oceans at 200 hPa is less apparent at 500 hPa, it is still associated with a 5CA minimum. Comparison of Figs. 2c and 7c with Figs. 1–3 from Chen (2005) suggests that the Saharan 5CAs may be linked to the African easterly jet and differential zonal heating across the west coast of Africa while the 200-hPa flow in the same region is dominated by the westward extension of the Indian summer monsoon anticyclone. Although the distribution of 5CAs over Africa in summer features a distinct maximum near the Atlas Mountains that is very "monsoonlike" (e.g., Rodwell and Hoskins 1996), there appears to be a weaker secondary maximum within the broad maximum between 15° and 20°N over much of Africa (Fig. 7c). This secondary 5CA maximum overlaps (in part) the east–west-oriented 600-hPa velocity potential minimum seen in Fig. 3 from Chen (2005) and may be associated with the midlevel Saharan anticyclone described by Chen (2005, his Fig. 2). Given the observed distribution of

5CAs and 2CAs from near the midocean trough over the central Atlantic eastward to Africa, the question arises as to whether the African easterly jet is enhanced during periods when 5CAs are especially robust, whether the midlevel Saharan anticyclone is more intense in these situations, and whether the configuration of the split-flow vorticity dipole associated with the Atlantic midocean trough (e.g., Miyasaka and Nakamura 2005) can be related to the evolution of the observed 2CA and 5CA circulations over Africa.

The distribution of 5CAs in the SH is much more uniform as compared with the NH (Fig. 6b). These features maximize in subtropical latitudes, especially in summer and autumn. Oceanic 5CAs occur in all seasons while continental 5CAs are evident from spring to autumn with relative maxima observed over southern Africa, western Australia, and western and eastern South America (Fig. 9). The greater percentage of oceanic coverage in the SH and the relatively limited east–west

extent of continents in subtropical latitudes as compared
to the NH likely accounts for the relatively more uniform distribution of 5CAs in the SH.

The 8CAs are found mostly over the ocean basins in
all seasons in both hemispheres (Fig. 11). Over the NH,
8CAs maximize over the central and eastern Atlantic
and eastern Pacific Oceans in all seasons with a peak
in summer (Fig. 12). The NH summer 8CA peak is
consistent with the observed mean distribution of oceanic subtropical anticyclones in both ocean basins while
the cool season 8CA oceanic maximum in both ocean
basins may also reflect where dynamically driven subsidence would be favored on the anticyclonic shear side
of the mean upper-level jet. The greater latitudinal extent
of the eastern Atlantic 8CA maximum as compared to
its eastern Pacific counterpart may also reflect the commonly observed occurrences of higher-latitude blocking
in this region.

In the NH the 8CA maximum in the central and eastern Atlantic has an inland extension of individual maxima from autumn to spring with maxima north of the
Caspian Sea and from the Taklimakan Desert eastward
across the northern Tibetan Plateau to the Gobi Desert
(Figs. 12a,b,d). These individual maxima likely reflect
a combination of dynamical forcing (associated with
ridge building and blocking episodes) and thermodynamical forcing (low-level anticyclogenesis associated
with occasional snow cover) with the caveat that the
maximum near the northern edge of the Tibetan Plateau
may be artificial. Elsewhere, a small 8CA maximum is
situated over the intermountain region of western North
America in autumn and winter where the 850-hPa level
is close to the surface (Figs. 12a,d). This maximum
likely reflects the tendency of surface anticyclones to
form over snow- or ice-covered surfaces (e.g., Hobbs
1945). The absence of a cool season 5CA/2CA signature
over the intermountain region is consistent with the relatively shallow nature of near-surface-based cold anticyclones. Similarly, the absence of an autumn and winter 8CA maximum over northeast Asia, a usually snow-
covered region that is home to frequent occurrences of
cold ≥1040-hPa sea level anticyclones, likely indicates
that cool season CAs in this part of the world are usually
cold core, occur on the cyclonic shear side of the jet,
and are generally below our 162-dam threshold for
8CAs. The assorted autumn–spring continental 8CA
maxima disappear in summer in response to strong surface heating and trough development at 850-hPa beneath
the common upper-level anticyclones seen at 500 and
200 hPa.

A comparison of NH and SH 8CAs reveals that SH
8CAs are less numerous overall than their NH counterparts (Figs. 12 and 14). Unlike in the NH, however,
SH oceanic 8CAs maximize in winter instead of summer
with prominent maxima found over the South Atlantic
Ocean and from extreme southeastern Africa eastward
to the central South Indian Ocean (Fig. 14). Besides the
southeastern Africa 8CA maximum, continental 8CA

maxima are also observed across southern Australia.
The axis of maximum 8CA frequency in the SH is zonally oriented and latitudinally confined between 30° and
40°S (Fig. 14). This 8CA distribution indicates that most
SH 8CAs are situated on the equatorward side of the
storm tracks (HH05, their Figs. 4a and 11). The absence
of 8CA maxima poleward of 40°S is likely an indication
that, unlike in the NH, there is a relative sparsity of
land at higher latitudes of the SH that can support the
formation of surface-based cold anticyclones in addition
to our choice of a 162-dam geopotential height contour
threshold for selecting 8CAs.

A comparison of the locations of 8CAs and 2CAs in
the NH reveals that in summer the 2CA maxima end
just to the west of the northeast–southwest-oriented midocean troughs over the Atlantic and Pacific Oceans,
while 8CA maxima are located farther east closer to the
200-hPa trough axis. The westward displacement with
height of the 2CA maxima relative to the 8CA maxima
suggests that these features have a baroclinic structure
consistent with the existence of westerly shear (Figs. 2c
and 12c). The same signature is seen in autumn, but it
is weaker, while in winter and spring there are too few
occurrences of 2CAs to permit a comparison (Figs. 2a–c
and 12a–c).

The near collocation of 8CA maxima with the upper-
level midocean troughs over the eastern Atlantic and
Pacific Oceans in the NH is consistent with the findings
of Miyasaka and Nakamura (2005, their Fig. 1). They
show that surface subtropical anticyclones over the
North Pacific and Atlantic Oceans are concurrent with
upper-level vorticity dipoles associated with a double-
jet structure. The northern region of these subtropical
anticyclones exhibits a quasi-equivalent barotropic
structure due to the weaker flow and associated weaker
vertical wind shear while the southern portion is characterized by a more baroclinic structure where the low-
level mean flow is easterly and the vertical wind shears
are subsequently greater. The associated ridge over
trough configuration is favorable for inducing descent
over the surface anticyclone. Miyasaka and Nakamura
(2005) also suggest that Rossby wave train activity that
originates from the North Pacific subtropical anticyclone and is strengthened over Hudson Bay may reinforce the North Atlantic subtropical anticyclone by producing enhanced subsidence over the surface anticyclone center (cf. their Figs. 1b and 3b with our Figs. 2c
and 12c). Also noteworthy is the similarity in location
of surface subtropical anticyclones in Miyasaka and
Nakamura (2005) and 8CAs seen in Fig. 12c. This geographical similarity suggests that subtropical 8CAs may
be a reasonable proxy for surface subtropical anticyclones.

A comparison of the NH winter distribution of 8CAs
(Fig. 12a) with the 850-hPa negative relative vorticity
track density map from HH02 (their Fig. 8a) suggests
that the 8CA maxima are found along and to the east
of the corridor of 850-hPa negative relative vorticity

track density frequency maximum (see HH02 for track density definition). Similarly, the HH02 negative relative vorticity track density maximum north of the Caspian Sea corresponds reasonably well with a similar 8CA maximum seen in Fig. 12a. In the SH, 8CA maxima in all seasons lie equatorward of the storm tracks as defined by the 250-hPa bandpass-filtered geopotential height variance maxima and 850-hPa relative vorticity track density maxima as shown in Figs. 2 and 11 from HH05, respectively. The anticyclones in HH05 are progressive features embedded in the storm tracks, while the 8CAs here are quasi-stationary features that lie equatorward of the storm tracks.

b. Intermonthly and interannual variability

The intermonthly frequency of 2CAs exhibits a broad summer maximum in both hemispheres over continental and oceanic regions (Figs. 3a and 5a). In the NH, 2CAs are most numerous over the oceans from July to September with a weaker August maximum while over the land a more distinct maximum is found in July when the continental landmasses are warmest (Fig. 3a). In the SH, the 2CA intermonthly frequency maximum is sharper and peaks in February over both continental and oceanic regions, consistent with the greater areal extent of the oceans, which tend to be warmest in February (Fig. 5a). The broader peak in the frequency of 2CA events in the NH may also be a reflection of the relatively larger areal distribution of landmasses in subtropical regions and the large east–west extent of the western Pacific oceanic warm pool. Although the frequency of 5CA events peaks in summer in both hemispheres, 5CAs, unlike 2CAs, are more numerous over oceanic than continental regions, and the difference in numbers is consistent with larger 500–200-hPa thickness values (and reduced 500-hPa heights) in continental 2CAs.

In contrast, the intermonthly frequency of 8CAs exhibits some significant differences from 2CAs. Over the NH a prominent summer (July) maximum is almost exclusively associated with oceanic 8CAs, while oceanic and continental regions contribute almost equally to the secondary winter maximum in 8CAs (Fig. 13a). The NH summer 8CA maximum occurs over the central and eastern Pacific and Atlantic Oceans in conjunction with the midocean troughs (Figs. 12c and 13a). The tendency for few warm season 8CA events to be observed over continental regions where 2CA events are maximized is consistent with the occurrence of 1000–200-hPa thickness maxima over continental regions in summer in conjunction with monsoon circulations and strong sensible heating over elevated and semiarid terrain. These large 1000–200-hPa thickness values disappear in winter when low-level heat lows are replaced by cold surface anticyclones and upper-level anticyclones are replaced by troughs. The intermonthly frequency of 8CAs in the SH is similar to the NH with the exception that the maximum activity is delayed until August in

reflection of the greater thermal lag over the oceans (Fig. 15a).

Petterssen (1956) computed the percentage frequency of anticyclones in winter and summer in squares 100 000 km^2 in size based upon once-daily sea level pressure analyses for the period 1899–1939. A comparison of his Fig. 13.7.2 (winter) with our winter 8CA results in Fig. 12a shows good agreement, with anticyclone maxima in both representations appearing in the eastern portions of the Pacific and Atlantic Oceans. The maximum over the intermountain region of the United States is also replicated in both analyses. Discrepancies are found over northeastern Alaska, northwestern Canada, and northeastern Asia where the cold surface-based anticyclone maximum in Fig. 13.7.2 of Petterssen (1956) is not seen in our analysis, which emphasizes warmer 8CAs of subtropical and lower midlatitudes (Fig. 12a). Over northeastern Asia, a surface anticyclone maximum that is located near 50°N in Fig. 13.7.2 from Petterssen (1956) is a manifestation of the cold Siberian anticyclone of winter. The higher-frequency anticyclone "freeway" that extends southeastward from this maximum to near 30°N over eastern China marks the preferred pathway of cold surface anticyclones that break off the main anticyclone reservoir over Siberia and move equatorward to the east of the Tibetan Plateau. As expected, neither feature is seen in our winter 8CA analysis, which emphasizes warmer systems (Fig. 12a). Instead, our winter 8CA maximum over eastern Asia is located farther south along the northern flank of the Tibetan Plateau and likely represents lower-latitude (warm) ridging well to the north and east of which an upward increase in anticyclonic vorticity advection likely contributes to cold anticyclogenesis over snow-covered terrain. In summer when cold anticyclones are mostly absent, excellent agreement is noted between the surface anticyclone maxima over the eastern ocean basins of the NH shown in Fig. 13.7.4 of Petterssen (1956) and our Fig. 12a.

In the NH, 2CAs have an interannual variability of approximately ±0.2 events day^{-1} during the period 1950–2003 with a slight increasing trend in the annual number of 2CA events since the late 1960s (Fig. 3b). A slightly larger interannual variability (±0.5 events day^{-1}) is seen for NH 5CAs (Fig. 8b). Most of the interannual variability and the slight increasing trend in the annual number of 5CA events since the mid-1970s is a reflection of an increasing number of oceanic 5CAs. 8CAs have an interannual variability of ±0.5 events day^{-1} with the larger fluctuations driven by oceanic 8CAs (Fig. 13b). The same increasing trend in the number of CA events, seen at 200 and 500 hPa, is also apparent at 850 hPa (Fig. 13b). The interannual variability of the number of CA events in the SH is more erratic than in the NH prior to the incorporation of systematic satellite measurements into model data analysis and initialization schemes, rendering the credibility of interannual trends in the number of CA events in the

SH before 1978 small. Since the beginning of the satellite era there is little evidence for a significant trend in the number of 2CA and 5CA events in either hemisphere while the interannual variability remains near ±0.3 and ±0.5 events day⁻¹, respectively (Figs. 8b and 10b). The number of 8CA events in the SH, however, show an increasing trend since 1978 with an interannual variability of ±0.2 events day⁻¹ (Fig. 15b). This increasing 8CA trend is mostly a result of increasing numbers of oceanic 8CAs while the number of continental 8CA events has remained almost constant.

Finally, when addressing the credibility of the intermonthly and interannual CA trends, the effects of including satellite observations in the analyses, particularly over oceanic regions, have to be considered. Another issue that needs to be considered is the accuracy of the NCEP–NCAR reanalysis fields used to construct the present climatology. For example, Trenberth et al. (2002) suggest that the standard 17-level reanalysis pressure level archive used in the NCEP–NCAR and ECMWF reanalysis does not adequately represent the actual three-dimensional structure of the atmosphere. Accordingly, caution is required when attempting to interpret CA trends, especially in the SH before the modern satellite era. It should be noted, however, that although the interannual variability is large especially when comparing the pre- and postsatellite era, the basic mean structures are consistent. Making inferences on CA trends other than what was discussed above is beyond the scope of this study.

5. Conclusions

The results of a 54-yr objectively prepared climatology of CAs, with an emphasis on the subtropics and midlatitudes, on the 850-, 500-, and 200-hPa pressure surfaces are presented. The CA climatology is supplemented by three case studies of CAs from different parts of the world that were associated with significant heat waves. The CA climatology was constructed from geopotential height fields derived from the 2.5° NCEP–NCAR reanalysis. A limiting geopotential height contour threshold was placed on the CAs at each pressure level to emphasize the distribution of CAs in the subtropical and midlatitudes.

The 2CAs are found predominantly in subtropical latitudes over continental regions in summer and autumn in the NH, over continental regions of the SH in summer only, and over the central and western North Pacific and North Atlantic Oceans in summer and autumn. A relative minimum in 2CA frequency is observed over the eastern Pacific and Atlantic Oceans in summer and autumn over and to the east of the locations of the persistent midocean troughs. An exception is a small 2CA summer maximum west of subtropical Africa that marks the westward extension of the continental maximum. The 2 CAs are more numerous in the NH than the SH and the difference is attributed to the greater areal extent of landmasses in the subtropics and midlatitudes and the more prominent monsoon circulations in the NH.

The distribution of 5CAs differs from 2CAs in that maxima occur over both continents and oceans in both hemispheres with an overall summer maximum. Continental 5CA maxima are best defined in summer and are concentrated over the western portions of continents in regions of elevated and semiarid terrain. The distribution of 5CAs in the SH is much more uniform as compared to the NH. Oceanic 5CAs occur in all seasons while continental 5CAs are evident from spring to autumn with relative maxima observed over southern Africa, western Australia, and western and eastern South America. The greater percentage of water coverage in the SH, and the relatively limited east–west extent of continents in subtropical latitudes as compared to the NH, likely accounts for the more uniform distribution of 5CAs in the SH.

The 8CAs maximize mostly over ocean basins in all seasons in both hemispheres. Over the NH, 8CA maxima occur over the central and eastern Atlantic and eastern Pacific Oceans in all seasons with a peak in summer, and are consistent with the observed distribution of mean oceanic subtropical anticyclones. In the NH, the 8CA maximum in the central and eastern Atlantic has a broken inland extension from autumn to spring with maxima north of the Caspian Sea, and from the Taklimakan Desert eastward across the northern Tibetan Plateau to the Gobi Desert where dynamical anticyclogenesis and low-level anticyclogenesis associated with snow cover can combine to permit CA formation. A small 8CA maximum is situated over the often snow-covered intermountain region of western North America in autumn and winter where the 850-hPa level is close to the ground and CAs are favored.

The absence of a cool season 5CA/2CA signature over the intermountain region of the United States is consistent with the relatively shallow nature of near-surface-based cold anticyclones beneath a warm dynamical ridge aloft. The corresponding absence of a cool season 8CA maximum over northeast Asia, a usually cold, snow-covered region, reflects both our emphasis on mapping warmer CAs of lower latitudes and that cold 8CAs in this part of the world weaken upward on the cyclonic shear side of the jet. Continental 8CA maxima disappear in summer in response to strong surface heating and trough development at 850 hPa beneath upper-level anticyclones.

A comparison of NH and SH 8CAs reveals that SH 8CAs are less numerous overall than their NH counterparts. Unlike in the NH, however, SH oceanic 8CAs maximize in winter instead of summer with prominent maxima found over the South Atlantic Ocean and from extreme southeastern Africa eastward to the central South Indian Ocean. Continental 8CA maxima are observed over southern Africa and across southern Australia and along with their oceanic counterparts are situated on the equatorward side of the SH storm tracks.

Summer 2CAs in both hemispheres are situated just to the west of the northeast–southwest-oriented midocean troughs over the Atlantic and Pacific Oceans while 8CA maxima are located farther east closer to the 200-hPa trough axis. The near collocation of 8CA maxima with the upper-level midocean troughs over the eastern Atlantic and Pacific Oceans in the NH is consistent with findings elsewhere and suggests the importance of dynamically driven baroclinic circulations to the maintenance of these features.

On intermonthly time scales, 2CA events in both hemispheres are most numerous in summer and more common over land, with the oceanic maximum lagging the continental maximum by upward of 1 month in response to the thermal lag over the oceans. Although the frequency of 5CA events peaks in summer in both hemispheres, 5CAs, unlike 2CAs, are more numerous over oceanic than continental regions, and the difference in numbers is consistent with larger 500–200-hPa thickness values (and reduced 500-hPa heights) in continental 2CAs. On intermonthly time scales, 8CA events peak strongly in summer in both hemispheres and are almost exclusively oceanic events. An exception is over the NH where a secondary 8CA maxima with almost equal contributions from continental and oceanic regions is observed. The tendency for few warm season 8CA events to be observed over continental regions where 2CA events are maximized is consistent with the occurrence of 1000–200-hPa thickness maxima over continental regions in summer in conjunction with monsoon circulations and strong sensible heating over the elevated and semiarid terrain.

Acknowledgments. This work was supported by the National Science Foundation Grants ATM-0233172 and ATM-0646907. We thank Dan Keyser and two anonymous reviewers whose suggestions and comments greatly helped to improve this manuscript. We thank Harald Richter for providing the Australian surface temperature data and satellite imagery (not shown) for the February 2004 heat wave case study. Discussions with Tim Hewson, Eyad Atallah, Ron McTaggart-Cowan, Harold Richter, and Mike Montgomery also contributed to this work. Celeste Iovinella is thanked for her help in preparing this manuscript.

APPENDIX

Three Case Studies of Closed Anticyclones Associated with Heat Waves

The purpose of this appendix is to provide a case study analysis of a subset of CAs that can produce extreme heat waves over continental regions. Each heat wave was associated with a CA that met the selection criteria used in the climatological portion of this paper. These case studies are used to reinforce the point that significant heat waves can occur when climatologically

warm cT air masses, produced over arid and semiarid regions in the subtropics and lower midlatitudes, are further warmed in conjunction with enhanced dynamically driven subsidence during upper-level ridge amplification. When these extremely warm cT air masses are subsequently advected downstream by transient disturbances embedded in anomalously strong westerlies they may become associated with significant heat waves as will be illustrated.

a. Results

1) 10–15 JULY 1995 U.S. HEAT WAVE

An intense short-lived heat wave devastated portions of the northern Great Plains and Great Lakes region of the United States during the period 10–15 July 1995. Nearly 800 heat-related deaths were reported in Chicago (CHI), Illinois, and Milwaukee, Wisconsin, both during and following the days of most intense heat (Donoghue et al. 1995; Nashold et al. 1996). A noteworthy aspect of this heat wave was the combination of very high temperatures ($>40°C$) and very high dewpoint temperatures ($>25°C$) that resulted in near-surface $\theta_e > 400$ K in places (e.g., NOAA 1995; Changnon et al. 1996; Kunkel et al. 1996). At 0000 UTC 13 July 1995, the northern Great Plains and Great Lakes region was covered by an extremely warm, moist, and shallow air mass as evident by the Davenport (DVN), Iowa, sounding (Fig. A2a) and nearby soundings from Green Bay, Wisconsin, and Chanhassen, Minnesota (not shown). Within this hot and humid air mass, the mixed-layer depths were ~100 hPa, the base of the subsidence inversion capping this air was located near 900 hPa, and the strength of the capping inversion was 7°–8°C. Evapotranspiration likely resulted in further moistening of the air in the shallow boundary layer below the subsidence inversion, given that the observed high dewpoint air (~27°C) was *not* advected into the upper Midwest from the Gulf of Mexico (e.g., Changnon et al. 1996; Kunkel et al. 1996). Here our analysis is restricted to the role of the evolution of the large-scale flow in contributing to the severity of the heat wave.

During 6–10 July 1995, a deep trough over the eastern Pacific led to downstream ridging and the formation of a 5CA over the Intermountain West (Fig. A1a). This "amplification phase" cultivated the hot cT air that was advected eastward during the "eastward progression phase" of 11–15 July 1995 (Fig. A1b). The combination of an anomalously strong CA (cyclone) over the western United States (Hudson Bay) resulted in an anomalously strong zonally oriented jet near the Canadian–U.S. border (Fig. A1b). The equatorward jet exit region that was positioned over the northern Great Plains and the Great Lakes region likely favored enhanced subsidence. This subsidence probably contributed to further warming of the already hot cT air mass that was moving eastward

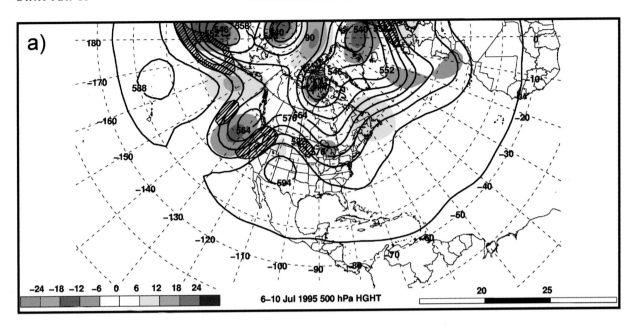

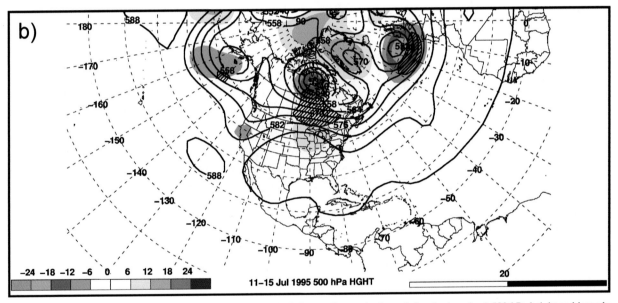

FIG. A1. Mean (solid contours every 6 dam) and anomaly (shaded according to the lower-left color bar; dam) 500-hPa height and isotachs (stippled shading according to the lower-right color bar; m s⁻¹) for the period (a) 6–10 and (b) 11–15 Jul 1995. Anomaly based upon 6–10 and 11–15 Jul long-term mean for 1968–1996. Data source: 2.5° NCEP–NCAR reanalysis.

from the intermountain western United States as seen in the air parcel trajectories ending at 0000 UTC 13 July 1995 near CHI (Figs. A2b–d).

At 0000 UTC 5 July 1995, a 500-hPa trough was positioned just west of British Columbia over the eastern Pacific with a downstream "miniridge" centered over the California coast (Fig. A3a). This trough began to dig southward by 0000 UTC 7 July 1995, resulting in downstream amplification of the ridge over the Intermountain West and the formation of a 5CA (Fig. A3b). This amplification continued at 0000 UTC 9 July 1995,

allowing hot cT air over the southwest United States and Mexican Plateau to surge northward (Fig. A3c). By 0000 UTC 11 July 1995, the 5CA attenuated slightly and moved eastward onto the Great Plains in response to upstream trough deamplification (Fig. A3d). By 0000 UTC 13 July 1995, the 5CA became more zonally oriented and progressive, which allowed the hot cT air over the elevated intermountain western United States to be advected eastward. Concurrently, the deep cyclone over northern Canada strengthened and began to move southward, resulting in a strong jet across southern Canada.

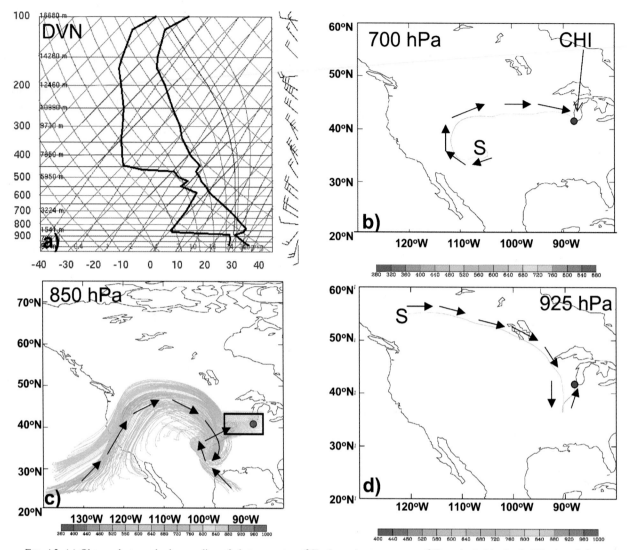

FIG. A2. (a) Observed atmospheric sounding of air temperature (°C), dewpoint temperature (°C), and wind barbs (half barb = 2.5 m s⁻¹, full barb = 5.0 m s⁻¹) for DVN at 0000 UTC 13 Jul 1995. Backward trajectory analysis for 168 h beginning 0000 UTC 13 Jul 1995 for CHI (blue dot) at (b) 700, (c) 850, and (d) 925 hPa. Arrows denote direction of air parcel movement. The 850-hPa analysis shows 231 air parcel backward trajectories beginning in the black box. Air parcel pressure level shaded according to the color bar. Data source: University of Wyoming sounding archive and 2.5° NCEP–NCAR reanalysis.

Subsidence in the equatorward jet exit region likely contributed to further warming of the already hot cT air (Figs. A2b–d and A3e). By 0000 UTC 15 July 1995, the 5CA and associated hot air reached the U.S. east coast as the now-weakened eastern Pacific trough crossed the Rockies onto the northern Great Plains, ending the extreme heat (Fig. A3f).

The strong amplification of the eastern Pacific trough and central U.S. 5CA occurred in response to a downstream development-producing wave packet beginning near 160°E on 4 July (Fig. A4a). The initial amplification of the heat wave–producing 5CA began near 115°W on 7 July, which allowed hot air to moved northward and further "cook" over the elevated heat source that is the Intermountain West (Fig. A4b). The hot air was then advected eastward away from its source region

as the anticyclone deamplified and became progressive in response to the wave packet passing to the east (Figs. A4a,b). The initial amplification and subsequent eastward movement of the hot cT air is summarized in the 850-hPa 21°C isotherm continuity map (Fig. A5).

2) 1–15 AUGUST 2003 EUROPEAN HEAT WAVE

Over much of western Europe the summer of 2003 was noteworthy for well-above-normal air temperatures. In particular, the period 1–15 August 2003 was characterized by extreme warmth when temperatures frequently reached 35°–40°C and contributed to approximately 40 000 heat-related deaths over western Europe (e.g., Hémon et al. 2003; Burt 2004; Burt and Eden 2004; Pirard et al. 2005). Comprehensive meteorolog-

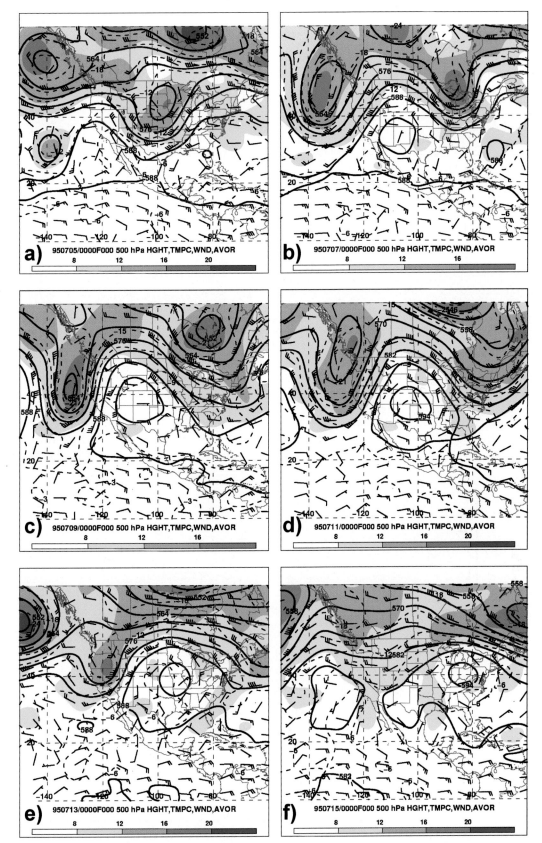

FIG. A3. 500-hPa height (solid contours every 6 dam), temperature (dashed contours every 3°C), absolute vorticity (shaded according to the color bar; ×10⁻⁵ s⁻¹), and wind barbs (half barb = 2.5 m s⁻¹, full barb = 5 m s⁻¹, pennant = 25 m s⁻¹) at 0000 UTC (a) 5, (b) 7, (c) 9, (d) 11, (e) 13, and (f) 15 Jul 1995. Data source: 2.5° NCEP–NCAR reanalysis.

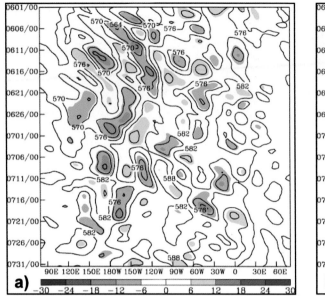
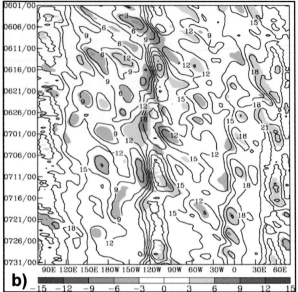

FIG. A4. Time–longitude analysis of (a) 500-hPa height (solid contours every 6 dam) and anomaly (shaded according to the color bar; dam) and (b) 850-hPa temperature (solid contours every 3°C) and anomaly (shaded according to the color bar; °C) for the period 1 Jun–31 Jul 1995 averaged over 30°–50°N. Anomaly based upon 5-day running mean for 1950–2003. Data source: 2.5° NCEP–NCAR reanalysis.

ical and hydrological analyses of this European heat wave can be found in Black et al. (2004), Burt (2004), and Fink et al. (2004). In this paper, the 1–15 August 2003 heat wave will be examined from a large-scale perspective with an emphasis on the contributory role of a CA to the extreme heat.

The mean 500-hPa height from the period 1 June–31 August illustrates the persistence of the heat wave–producing ridge over Europe (Fig. A6). The anomalously strong ridge (+6 dam) is positioned downstream of an anomalously deep trough (−6 dam) over the eastern Atlantic and upstream of a persistent trough over eastern

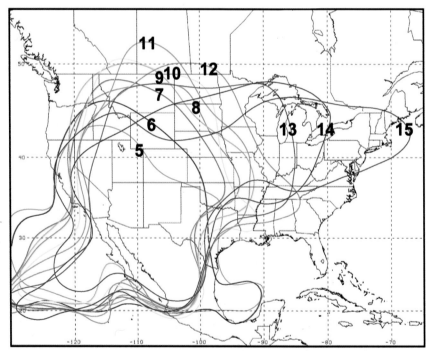

FIG. A5. 850-hPa 21°C isotherm continuity map for 0000 UTC 5–15 Jul 1995. Contours plotted in order of color spectrum beginning with blue on 5 Jul and ending with red on 15 Jul 1995. Data source: 2.5° NCEP–NCAR reanalysis.

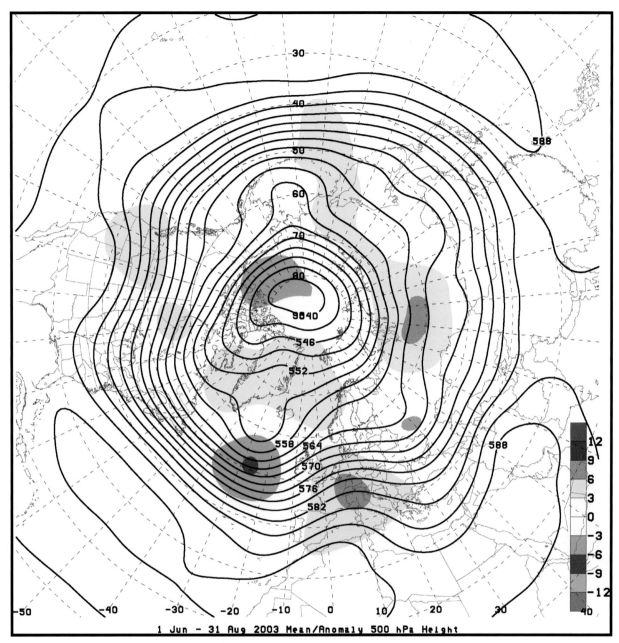

FIG. A6. Mean (solid contours every 3 dam) and anomaly (shaded according to the color bar; dam) 500-hPa height for the period 1 Jun–31 Aug 2003. Anomaly based upon 1 Jun–31 Aug long-term mean for 1968–96. Data source: 2.5° NCEP–NCAR reanalysis.

Europe and extreme western Asia. Concurrently, the mean 850-hPa temperatures for the same period show anomalous warmth (+3°C) centered on France in association with the persistent ridge (Fig. A7a). For perspective purposes, the mean 850-hPa temperatures for the period 1–15 August show anomalous warmth approaching +9°C as the mean 21°C isotherm approaches central France (Fig. A7b). The persistence of these above-normal 500-hPa heights and 850-hPa temperatures near 0° longitude is demonstrated in Hovmöller diagrams (Figs. A8a,b). Brief interruptions of the per-

sistent 500-hPa ridge and 850-hPa warmth occur during 25 June–6 July and again from 23 to 26 July in association with transient disturbances that temporarily break down the ridge (Figs. A8a,b). As these transient disturbances continued downstream, the ridge was allowed to rebuild poleward. This quasi-blocking pattern continued until 15 August when the persistent trough over the eastern Atlantic made landfall on the European coast, subsequently transitioning the amplified flow regime seen for much of the summer to be a more zonal pattern.

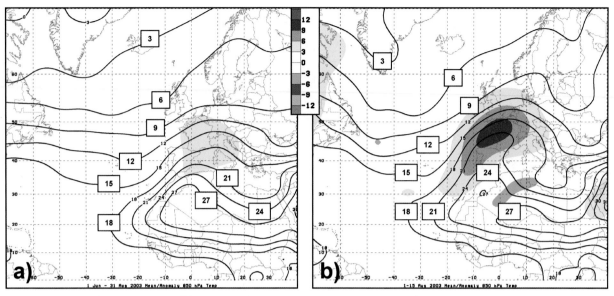

FIG. A7. Mean (solid contours every 3°C) and anomaly (shaded according to the color bar; °C) 850-hPa temperature for the period (a) 1 Jun–31 Aug and (b) 1–15 Aug 2003. Anomaly based upon (a) 1 Jun–31 Aug and (b) 1–15 Aug long-term mean for 1968–96. Data source: 2.5° NCEP–NCAR reanalysis.

The 500-hPa height and vorticity charts show ridge building under way over Europe by 1800 UTC 4 August 2003 in response to an amplifying trough over the central Atlantic (Fig. A9a). The cyclone over Scandanavia was impeding the poleward amplification of the ridge, but by 1800 UTC 6 August 2003 this cyclone was moving southward into the downstream trough over extreme eastern Europe (Fig. A9b). In response, the western European ridge amplified poleward due to continued trough deepening over the eastern Atlantic by 1800 UTC

8 August 2003 (Fig. A9c). While the eastern Atlantic trough slowly moved east-northeastward the eastern European trough remained stationary, resulting in a continued narrowing of the European ridge (Fig. A9d). This ridge narrowing, occurring in the presence of the eastern European trough, which stretched southwestward over the Mediterranean into northwest Africa, enabled a 5CA circulation to form. This inference is supported by trajectory analysis for 168 h ending at 1800 UTC 10 August 2003 over northern France and the southern United

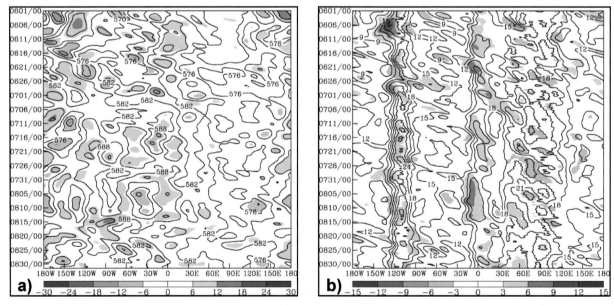

FIG. A8. Time–longitude analysis of (a) 500-hPa height (solid contours every 6 dam) and anomaly (shaded according to the color bar; dam) and (b) 850-hPa temperature (solid contours every 3°C) and anomaly (shaded according to the color bar; °C) for the period 1 Jun–31 Aug 2003 averaged over 30°–50°N. Anomaly based upon 5-day running mean for 1950–2003. Data source: 2.5° NCEP–NCAR reanalysis.

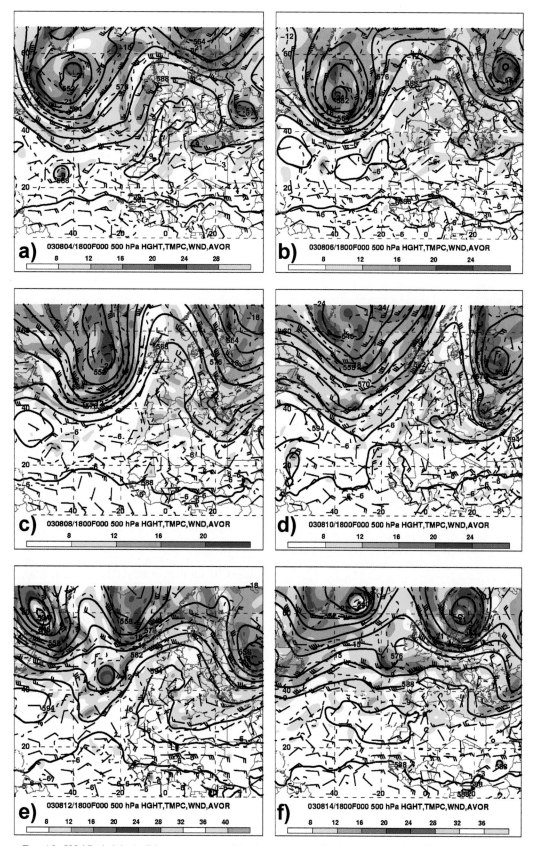

FIG. A9. 500-hPa height (solid contours every 6 dam), temperature (dashed contours every 3°C), absolute vorticity (shaded according to the color bar; ×10⁻⁵ s⁻¹), and wind barbs (half barb = 2.5 m s⁻¹, full barb = 5 m s⁻¹, pennant = 25 m s⁻¹) at 1800 UTC (a) 4, (b) 6, (c) 8, (d) 10, (e) 12, and (f) 14 Aug 2003. Data source: NCEP 1.0° GFS analyses.

Kingdom (Figs. A10b–d). The initial source regions of the hot air at 850 hPa are located over northwest Africa, southern Spain, and the upstream westerlies in the 600–700-hPa layer. Subsequently, the air from these source regions was drawn into the anticyclone circulation where it was trapped as it slowly descended, suggesting that subsidence warming was a factor in the intensity of the heat wave (Fig. A10) as was found independently by Black et al. (2004). By 1800 UTC 12 August 2003 the eastern Atlantic trough was beginning to move on-shore in response to the pattern becoming more progressive, subsequently ending the heat wave by 1800 UTC 14 August 2003 (Figs. A9e,f).

The 21°C 850-hPa isotherm continuity map shows the movement of the hot air from 1 to 15 August (Fig. A11). The initial poleward push of the hot air on 1–4 August was associated with ridge building. From 4 to 12 August the 500-hPa ridge remained anchored over Europe (Fig. A9), while the hot air at 850 hPa slowly recirculated and subsided as evidenced by the trajectory analysis (Fig. A10). No farther poleward push of the hot air occurred while a slight erosion of the hot air mass was observed on the upstream side of the ridge because of the eastern Atlantic trough coming slowly onshore. Also noteworthy is the trough over eastern Europe that stretched southwestward over the Mediterranean from 4 to 10 August (Figs. A9b–d and A11). This trough likely prevented the hot air from reaching central and southern Italy until it lifted out after 10 August. Finally, from 13 to 15 August the hot air mass eroded quickly from the northwest as the eastern Atlantic trough came onshore (Fig. A11).

3) 1–22 FEBRUARY 2004 AUSTRALIAN HEAT WAVE

The period 1–22 February 2004 was marked by an intense heat wave over the southern half of Australia, in particular over northern Victoria where new records for the number of days with maximum daily temperatures above 40°C (9) were set at Mildura Aerodome (MA), Ouyen Post Office (OPO), and Walpeup Research Centre (WRC). Figure A12 shows the maximum temperature and anomaly for 7-day periods ending 10 and 17 February, respectively. The hottest air on the Australian continent (>40°C; +6°C anomaly) from 4 to 10 February was found just inland from western Australian eastward into northern Victoria and New South Wales (Figs. A12a,b). During 11–17 February, the hottest air (>45°C; >+6°C anomaly) shifted eastward and was located primarily over the southeastern portion of the Australian continent (Figs. A12c,d). On 14 February, 11 stations within Victoria reported maximum temperatures over 45°C, with MA, OPO, and WRC reporting 45.6°, 46.7°, and 46.3°C, respectively (see online at http://www.bom.gov.au/announcements/media_releases/vic/20040301.shtml).

The large-scale pattern over Australia changed dramatically between January and February 2004. The Jan-uary mean 500-hPa height pattern shows an anomalously deep trough (−6 dam) over Victoria and Tasmania, with anomalously high heights (+12 dam) over the southern Indian and Pacific Oceans (Fig. A13a). A 588-dam CA was anchored over western Australia during the same period. A major large-scale pattern change occurred at the very end of January and persisted through 22 February. During 1–22 February an anticyclone over Australia replaced the previous month's trough while anomalously deep troughs (−12 dam) developed over the southern Indian and Pacific Oceans (Fig. A13b). The mean 200-hPa height pattern for 1–22 February 2004 shows anomalous ridging (troughing) over the Australian continent (the Southern Ocean) resulting in an anomalously strong jet immediately poleward of Australia (Fig. A13c). This jet, analogous to the July 1995 U.S. case, likely resulted in further warming of the hot cT air over the continent via enhanced subsidence on its anticyclonic shear side. In response to development of the Australia anticyclone, warm 850-hPa temperatures (>24°C) covered the southern half of Australia with anomalous warmth (+6°C) over northern Victoria and New South Wales from 7 to 22 February (Fig. A14).

Inspection of time–longitude Hovmöller diagrams suggests that several 5CAs develop in succession over Australia (120°–150°E) from 9 to 21 February (Figs. A15a and A16). Prior to the onset of the heat wave in eastern Australia, a zonal pattern with embedded short-wave troughs is in place poleward of Australia with a persistent 588-dam 5CA positioned over the continent (Figs. A16a,b). Several episodes of downstream development occur from 5 to 22 February, spanning the globe, with the first wave packet building the initial heat wave–producing an anticyclone near 150°E on 9–10 February (Fig. A15a). By 1200 UTC 14 February 2004, this initial anticyclone, now over eastern Australia, continues to amplify in response to continued downstream trough amplification (Figs. A16c,d). Subsequently, hot cT air builds eastward as evidenced by the >21°C air at 850 hPa positioned over southeast Australia (Figs. A17c,d). A second wave packet triggers ridge building slightly upstream, near 135°E, of the initial anticyclone (Figs. A15 and A16e) by 1200 UTC 17 February 2004. This second anticyclone was the strongest seen (+12 dam anomaly) over the Australian continent from 1 to 22 February (Fig. A15a). The largest positive 850-hPa temperature anomaly (+9°C) was located over Australia with the initial anticyclone growth from 10 to 13 February (Fig. A15b). By 1200 UTC 20 February 2004, the end of the heat wave is signaled as an amplifying trough and associated cold front off the southwest coast of Australia prepares to come onshore while the broad anticyclone over the Australian continent begins to move eastward (Figs. A16f and A17f). Enhanced northerly flow at 850 hPa ahead of this trough provides a temporary resurgence of hotter air (near 33°C at 850 hPa)

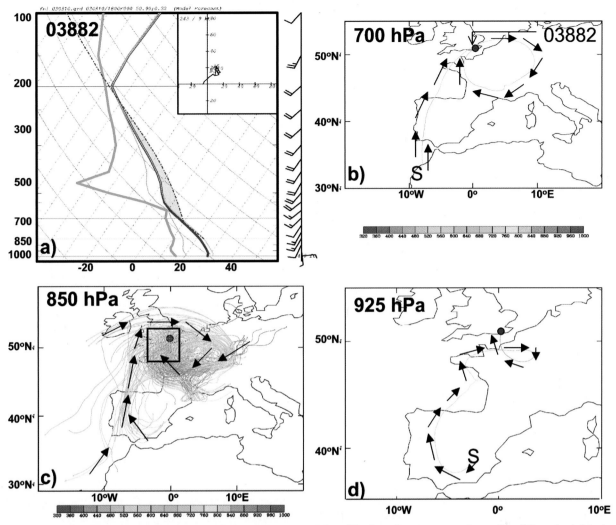

FIG. A10. (a) Model atmospheric sounding of air temperature (red line; °C), dewpoint temperature (green line; °C), and wind barbs (m s⁻¹; half barb = 2.5 m s⁻¹, full barb = 5.0 m s⁻¹) and hodograph (insert) for Herstmonceux, United Kingdom (03882) at 1800 UTC 10 Aug 2003. Backward trajectory analysis for 168 h beginning 1800 UTC 10 Aug 2003 for Herstmonceux (blue dot) at (b) 700, (c) 850, and (d) 925 hPa. Arrows denote direction of air parcel movement. The 850-hPa analysis shows 121 air parcel backward trajectories beginning in the black box. Air parcel pressure level shaded according to the color bar. Data source: NCEP 1.0° GFS analyses.

into southeast Australia (Figs. A17e,f) just prior to the end of the intense heat.

A backward trajectory analysis beginning at 850 hPa on 1200 UTC 14 February 2004 over southeast Australia, during the first phase of the intense heat, shows that the air parcels originate in the quasi-persistent downstream trough over and north of New Zealand (Figs. A16d and A18a). Most of these air parcels begin near 20°S, 165°E between 680 and 640 hPa, while other air parcels originate near 35°–40°S, 155°–160°E at 360 hPa. Still other air parcels begin in the upstream westerlies near 600 hPa, while the rest of the parcels originate farther eastward (near 170°–175°E) and at lower levels (near 900 hPa) than all other parcels discussed above. These air parcel source regions suggest that the intense Australian heat wave has a dynamical lineage. Air parcels subside and warm behind the downstream

trough axis, whereupon they move westward in the tropical easterlies and reach northeastern Australia. Once these air parcels reach the continent they turn southward and warm further in the deep mixed layer over the arid continental interior. The backward trajectory analysis for parcels beginning at 925 hPa shows source regions in the same area as for 850 hPa except that most of the air parcels originate near 900 hPa (Fig. A18b). These air parcels also move westward and then southward over the arid continental interior where they are heated further.

An 850-hPa 21°C isotherm continuity map summarizes the four stages of the southeast Australia heat wave during February 2004 (Fig. A19). The first stage, 1–6 February, shows the hot air initially over the west coast of Australia on 1–2 February (Figs. A19a and A17a). The hot air then expands eastward from 3 to 6 February

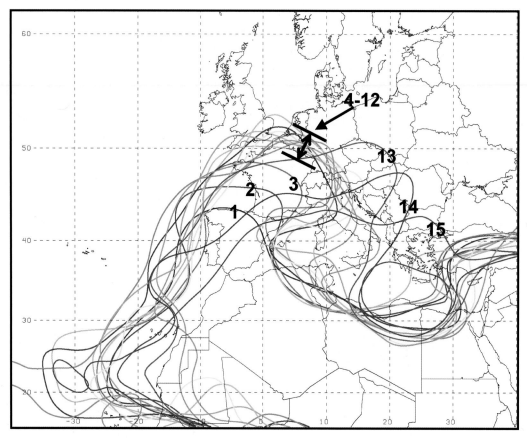

Fig. A11. 850-hPa 21°C isotherm continuity map for 1800 UTC 1–15 Aug 2003. Contours plotted in order of color spectrum beginning with violet on 1 Aug and ending with red on 15 Aug 2003. Data source: NCEP 1.0° GFS analyses.

as the initial anticyclone builds over central and eastern Australia. In the second stage, 7–12 February, the hot air remains anchored over the Australian continent with the continental 5CA locked in place (Figs. A19b, A16b,c, and A17b,c). In the third stage, 13–18 February, the hot air mass erodes slightly on its west and south side as the initial 5CA deamplifies and moves eastward. Subsequently, the hot air mass reintensifies over western Australia and, to a lesser extent, over northern Australia as the next 5CA strengthens (Figs. A19c, A16d,e, and A17d,e). In the final stage, 19–22 February, the heat erodes steadily from the southwest as a trough and associated cold front moves onshore concurrent with the second anticyclone deamplifying and moving eastward (Figs. A19d, A16f, and A17f).

b. Discussion

Bjerknes and Solberg (1922) proposed that airmass characteristics (e.g., temperature) could be determined almost exclusively by the nature of the underlying surface in the airmass source regions. Simple airmass classification schemes were based on whether the underlying surface was continental or maritime and whether the surface was warm or cold. Emanuel (2008) has re-

viewed and revisited the airmass concept in an essay elsewhere in this monograph volume. His essay is designed to motivate fresh studies on the role of thermodynamical and dynamical processes in the formation of air masses, specifically arctic air masses and their associated continental anticyclones. Motivated by Emanuel (2008), case studies of heat wave events that were associated with CAs that met our objective definition were presented in section 4. These heat wave–producing CA case studies included the following: 1) July 1995 over the midwestern United States, 2) August 2003 over much of western Europe, and 3) February 2004 over Australia. These cases were examined to determine the evolution of the upper-level circulations and to establish the source of the associated hot air masses. At issue is whether the hot air masses associated with the heat waves result from surface thermodynamical processes (e.g., sensible heating over arid soil) and/or free atmosphere dynamical processes (e.g., subsidence warming associated with strong jet-induced vertical circulations). To help address these issues a common denominator of the three case studies was an assessment of the airmass source regions and airmass warming mechanisms.

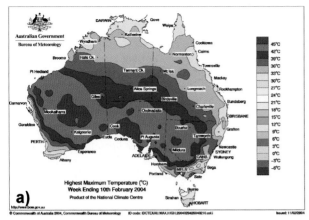
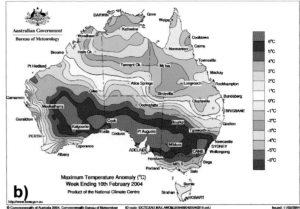
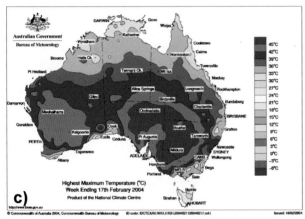
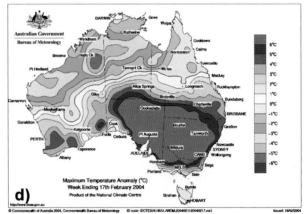

FIG. A12. (a) Maximum temperature (shaded with solid contours; °C) and (b) anomaly (shaded with solid contours; °C) during the period 4–10 Feb 2004. Anomaly based upon 4–10 Feb long-term mean for 1961–90. (c) Maximum temperature (shaded with solid contours; °C) and (d) anomaly (shaded with solid contours; °C) during the period 11–17 Feb 2004. Anomaly based upon 11–17 Feb long-term mean for 1961–90. Figure obtained from the climate summary report at the Australian Bureau of Meteorology.

1) 10–15 JULY 1995 U.S. HEAT WAVE

The structure of the 6–15 July 1995 heat wave and associated CA is very similar to the heat wave scenario as described in Namias (1955), and discussed earlier in this paper. This heat wave scenario, dubbed the "three-cell pattern," has an anticyclone positioned over the Gulf of Alaska and a trough (ridge) positioned over the extreme western United States (central United States). A similar pattern is noted in the mean 500-hPa height during the ridge amplification period of 6–10 July 1995 (Fig. A1a). This ridge amplification period was followed by the eastward progression of the CA during 11–15 July 1995 (Fig. A1b). The eastward progression of this CA enabled the hot cT air to move eastward away from its climatological elevated, semiarid source region, ultimately producing a heat wave over the northern Great Plains and Great Lakes region.

A time–longitude analysis of 850-hPa temperature suggests that hot cT air is produced in situ near 120°W on 4–5 July in association with the 500-hPa ridge building (Fig. A4). Physical mechanisms to produce the initial cT air mass over the Mexican Plateau appear to be both dynamic through 500-hPa ridge building and thermodynamic through sensible heating over elevated, semiarid terrain that continues as the cT air mass expands poleward in response to pattern amplification (Figs. A3–A5). Formation of a strong 500-hPa jet over the U.S.–Canadian border occurs in response to hot cT air expanding poleward and cool continental polar air expanding equatorward across Canada behind a strengthening trough over the Hudson Bay (Figs. A3 and A5). The result of jet formation is that the hot cT air mass is diverted eastward along the equatorward side of the jet axis where dynamically driven subsidence likely maintains and reinforces the hot cT air mass over the northern Great Plains and Great Lakes region (Figs. A2 and A5). The air parcel trajectory analysis shows that the air over CHI at 0000 UTC 13 July 1995 originated from the upstream side of the CA over the eastern Pacific Ocean. This air moved around the poleward side of the CA, subsequently descending in the aforementioned equatorward jet exit region (Figs. A2b–d).

A thorough analysis of the July 1995 heat wave, including the critical role of surface processes in enhanc-

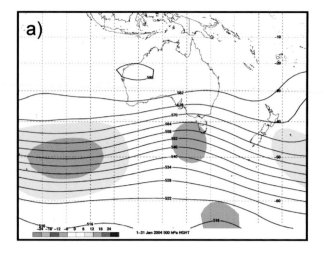

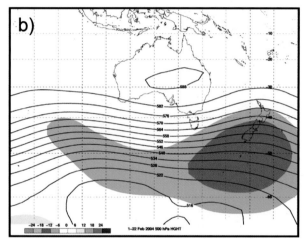

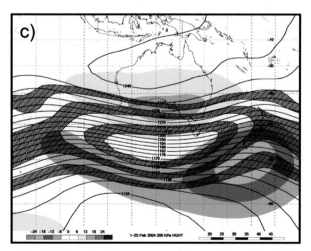

FIG. A13. Mean (solid contours every 6 dam) and anomaly (shaded according to the color bar; dam) 500-hPa height for the period (a) 1–31 Jan and (b) 1–22 Feb 2004. Mean (solid contours every 6 dam) and anomaly (shaded according to the lower left color bar; dam) 200-hPa height and isotachs (stippled shading according to the lower-right color bar; m s⁻¹) for (c) 1–22 Feb 2004. Anomaly based upon 1–31 Jan and 1–22 Feb long-term mean for 1968–96. Data source: 2.5° NCEP–NCAR reanalysis.

ing heat stress on people and livestock, can be found in NOAA (1995). Our analysis here is restricted to the role of large-scale processes responsible for 5CA formation and amplification over the western United States, with the associated development of a very hot cT air mass, followed by the subsequent eastward expansion of the 5CA and the associated cT air mass along the equatorward periphery of an anomalously strong westerly jet across southern Canada. Sinking motion along the equatorward periphery of the aforementioned strong jet likely resulted in a widespread and relatively low subsidence inversion over the Midwest, and set the table for the generation of very high surface θ_e values as moisture evaporated from the surface became concentrated near the ground below the relatively shallow and strong subsidence inversion. The observed sounding for DVN shown in Fig. A2a is considered representative for CHI. It shows the presence of a relatively low subsidence inversion in the 900–800-hPa layer that likely contributed to the development of extreme surface θ_e values through surface evapotranspiration processes over previously moist ground in the shallow boundary layer below the subsidence inversion. The heat stresses associated with these high θ_e values contributed to the excessive loss of life in the CHI area (Fig. A2a; NOAA 1995; Changnon et al. 1996; Kunkel et al. 1996).

2) 1–15 AUGUST 2003 EUROPEAN HEAT WAVE

Jones and Moberg (2003) and Klein Tank and Können (2003) have argued that Europe has warmed significantly over the last 25 yr, approaching +0.5°C decade⁻¹, and that extreme weather events have become more common. According to Cassou et al. (2005), two characteristic European "warm patterns," whose frequency has been increasing, have been identified based on the nonlinear cluster analysis approach of Cheng and Wallace (1993). The two warm patterns identified by Cassou et al. (2005) include a blocking pattern identified by Liu (1994) and an Atlantic low pattern. The former is characterized by an anomalous 500-hPa anticyclone centered near Scandinavia and the latter is marked by an anomalous 500-hPa cyclone centered to the south of Iceland. The Atlantic low pattern also features a pronounced anticyclonic anomaly from northwest Africa and the Iberian Peninsula northeastward to northwest Russia.

Our view from looking at daily weather maps is that the two warm patterns identified by Cassou et al. (2005) may not be mutually exclusive in the sense that slight rearrangements in the large-scale flow pattern (e.g., the ridge elongates meridionally and contracts zonally) can result in a pattern change. Cassou et al. (2005) indicated (see their Table 1) that both warm patterns were evident in the summer of 2003 with the blocking pattern occurring 20% of the time (mostly in August) and the Atlantic low pattern occurring 40% of the time (mostly in June). Their conclusion that a blocking pattern prevailed in August 2003 is supported by our results that

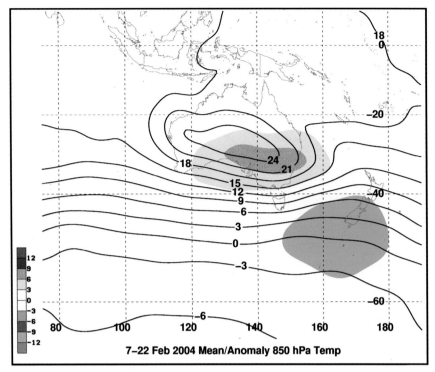

FIG. A14. Mean (solid contours every 3°C) and anomaly (shaded according to the color bar; °C) 850-hPa temperature for the period 7–22 Feb 2004. Anomaly based upon 7–22 Feb long-term mean for 1968–96. Data source: 2.5° NCEP–NCAR reanalysis.

much of the period from 1 to 15 August 2003 met our definition of a CA, and by the winds and trajectories that show anticyclonic circulations with easterly wind components over western Europe (Figs. A9 and A10).

Our Hovmöller diagram of 500-hPa height anomalies for the summer of 2003 (Fig. A8a) shows that negative (positive) height anomalies were present near 30°W (0°) in the first half of June and again in the first half of

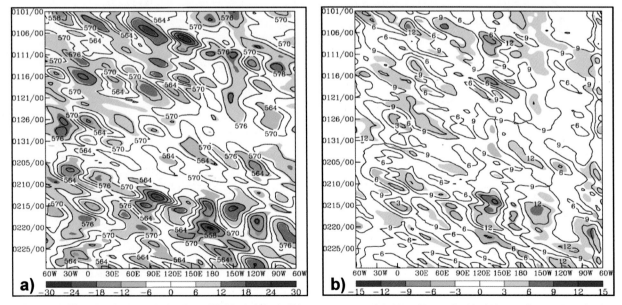

FIG. A15. Time–longitude analysis of (a) 500-hPa height (solid contours every 6 dam) and anomaly (shaded according to the color bar; dam) and (b) 850-hPa temperature (solid contours every 3°C) and anomaly (shaded according to the color bar; °C) for the period 1 Jan–29 Feb 2004 averaged over 30°–50°S. Anomaly based upon 5-day running mean for 1950–2003. Data source: 2.5° NCEP–NCAR reanalysis.

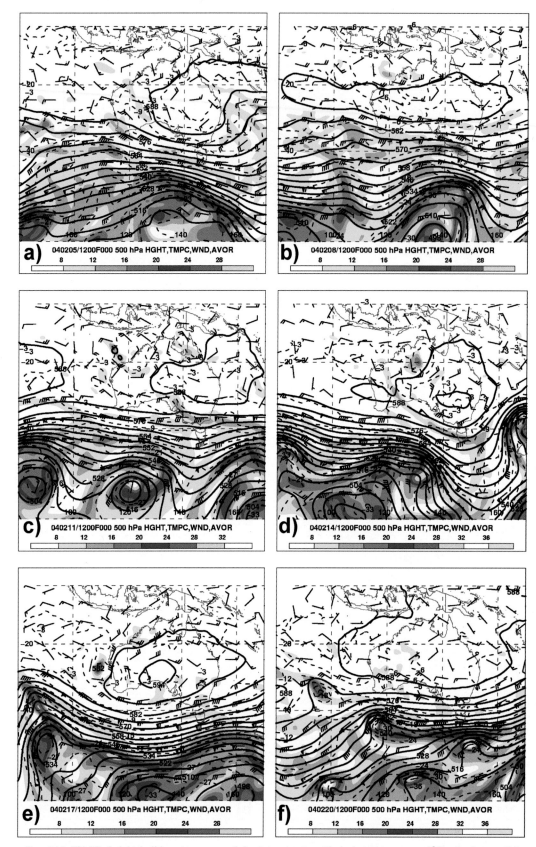

FIG. A16. 500-hPa height (solid contours every 6 dam), temperature (dashed contours every 3°C), absolute vorticity (shaded according to the color bar; × 10⁻⁵ s⁻¹), and wind barbs (half barb = 2.5 m s⁻¹, full barb = 5 m s⁻¹, pennant = 25 m s⁻¹) at 1200 UTC (a) 5, (b) 8, (c) 11, (d) 14, (e) 17, and (f) 20 Feb 2004. Data source: NCEP 1.0° GFS analyses.

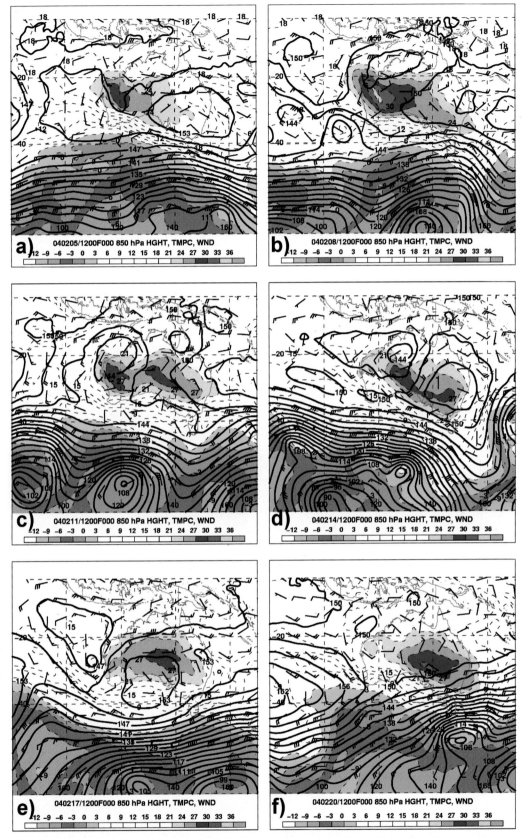

FIG. A17. 850-hPa height (solid contours every 3 dam), temperature (shaded according to the color bar with dashed contours every 3°C), and wind barbs (half barb = 2.5 m s⁻¹, full barb = 5 m s⁻¹, pennant = 25 m s⁻¹) at 1200 UTC (a) 5, (b) 8, (c) 11, (d) 14, (e) 17, and (f) 20 Feb 2004. Data source: NCEP 1.0° GFS analyses.

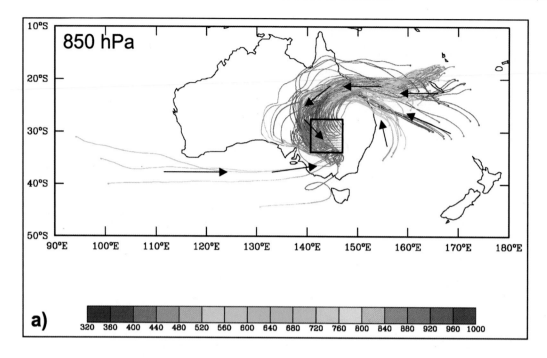

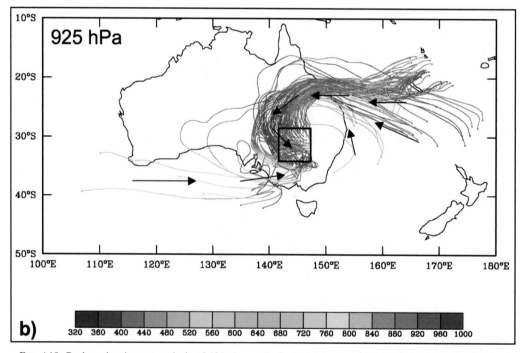

FIG. A18. Backward trajectory analysis of 121 air parcels for 168 h beginning at 1200 UTC 12 Feb 2004 in the black box at (a) 850 and (b) 925 hPa. Arrows denote direction of air parcel movement. Air parcel pressure level shaded according to the color bar. Data source: NCEP 1.0° GFS analyses.

August. Although the 500-hPa height anomaly pattern in the first half of June was somewhat retrogressive, indicative of a possible block, there is otherwise little to distinguish between the June and August height/anomaly patterns on the basis of the Hovmöller plots alone.

Consistent with the results discussed in the preceding paragraphs, Figs. A6–A11 establish the extreme nature of the large-scale circulation patterns and associated heat over western Europe and vicinity in the first half of August 2003. For example, it is quite apparent that 5CAs over western Europe are climatologically rare

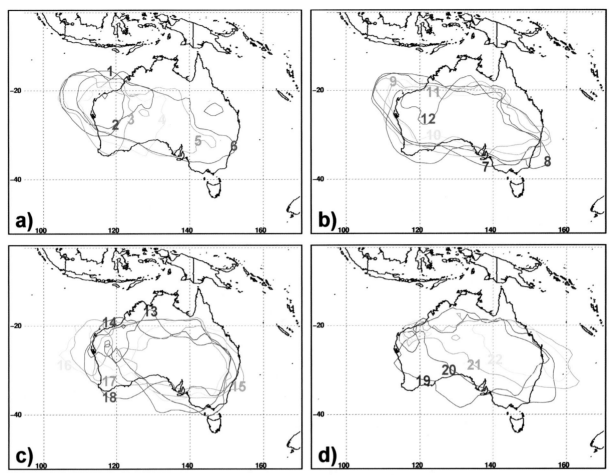

FIG. A19. 850-hPa 21°C isotherm continuity map for 1200 UTC (a) 1–6, (b) 7–12, (c) 13–18, and (d) 19–22 Feb 2004. Contours plotted in order of color spectrum. Data source: NCEP 1.0° GFS analyses.

events based on the thresholds used in this paper when comparing the 5CA seen in Fig. A9a with the summer 5CA climatology seen in Fig. 7c. An analysis of air parcel trajectories and corresponding wind fields shows anticyclonic circulations, with easterly wind components over southern Europe, which supports the counting program that flagged this case as a CA. Hot cT air was generated in situ near 0° longitude on 25–26 July within the persistent 500-hPa anticyclone over western Europe (Fig. A8). The persistence of this 500-hPa anticyclone through the first part of August 2003 is indicated by a lack of progressive features at 500 hPa as shown in Fig. A8. Poleward of 50°N, however, the flow is more progressive. The heat wave ends toward mid-August as the ridge deamplifies, the flow becomes more zonal as the westerlies shift equatorward, and Atlantic air is finally able to move inland across western Europe. Prior to this time, repeated shortwave disturbances on the upstream side of the deep trough near 40°W help to anchor and deepen the trough, which allows the downstream ridge over western Europe to amplify and lock

in place, analogous to block maintenance processes described by Mullen (1986, 1987, 1989; Fig. A9).

The hot cT air mass has a primary source over the Iberian Peninsula where additional warming likely occurs through sensible heating over elevated terrain (Figs. A10b, d). Subsequently, the hot cT air then expands northward in a narrow corridor into France as the 500-hPa ridge amplifies (Figs. A9 and A11). Recirculation of hot cT air then occurs over France with likely further dynamically driven warming due to subsidence (Fig. A10c), an assertion that is also in agreement with the comprehensive analysis of Black et al. (2004) and Fink et al. (2004). The air parcel trajectories also suggest a secondary cT air source from northwest Africa (Figs. A7 and A11). Without air parcel trajectories showing a primary source from northwest Africa, the observed poleward expansion of the 21°C 850-hPa isotherm across Spain and France likely occurs through in situ dynamically driven subsidence and/or enhanced in situ sensible heating over unusually arid soils that may have been partially dried by the earlier heat in June and July in these same regions as discussed by Black et al. (2004) and Baldi et al. (2005).

3) 1–22 FEBRUARY 2004 AUSTRALIAN HEAT WAVE

The mean 500-hPa height for the period 1–22 February 2004 shows a 588-dam 5CA anchored over southeast Australia. In a climatological sense, it is more common to find 5CAs along the western coast of Australia during summer, decreasing in frequency eastward across the Australian continent. The mean 5CA over eastern Australia evolves from two occasions of ridge building over the western coast, and is followed by eastward progression of the anticyclone, resulting in a regime transition from the mean trough observed during January (Figs. A13a,b). The ridge is more anomalous at 200 hPa than 500 hPa, which is reflected in the 45 m s^{-1} jet evident along the southern coast of Australia at 200 hPa (Fig. A13c). This flow configuration is similar to the July 1995 U.S. case in which an anomalously deep trough lays poleward of a strong anticyclone, resulting in a strong zonal jet (cf. Figs. A1b and A13c).

The hottest temperatures over eastern Australia occur during the period 10–17 February. The onset of these high temperatures is heralded by downstream amplification under way over western Australia on 10 February. The hottest temperatures over eastern Australia develop after 12–13 February when a 500-hPa ridge builds in from the western coast through the aforementioned downstream amplification and eastward progression (Fig. A15). This situation suggests that subsidence warming on the anticyclonic shear side of the strong jet reinforces and further warms an already hot air mass produced from sensible heating over the arid continental interior (Figs. A15 and A19). In this context, the long-lived Australian heat wave event of February 2004 bears some similarity to the short-lived U.S. heat wave event of July 1995.

An air parcel trajectory analysis shows that the initial source of the hot air resides over the warm tropical Coral Sea (Fig. A18). Warm moist maritime tropical (mT) air over the Coral Sea moves westward in the tropical easterlies, then turns southward over the Australian continent. The mT air is allowed to move southward because of the north-northwest flow ahead of weak cold fronts that are associated with progressive disturbances embedded in the westerly flow well south of Australia (Fig. A16). This trajectory path allows the warm moist mT air to undergo additional warming through sensible heating and to dry out over the arid continental interior as it transitions to a cT air mass (Fig. A17). As this new cT air mass moves southward, it is warmed further by subsidence on the equatorward side of the strong jet over coastal southern Australia (Figs. A16–A19).

c. Conclusions

Three case studies of objectively identified CAs, that were associated with severe continental heat waves over the United States (July 1995), Europe (August 2003), and Australia (February 2004), are presented to illustrate the structure and evolution of CAs and their influences on producing extreme summer warmth. Common to these cases was the buildup of hot cT air masses in response to dynamic ridge building aloft over elevated and semiarid terrain. In the U.S. and Australian case, subsequent to ridge building, the hot cT air was advected eastward away from its source region along the equatorward flank of a strong jet stream. Additional subsidence warming along the equatorward flank of the jet likely further heated an already very warm air mass. In the U.S. case, a capping inversion associated with the subsidence likely contributed to the buildup of an extremely hot and humid air mass by restricting surface evapotranspiration to a relative shallow layer with an unfortunate effect on the population. In the European case, the 2CA was more persistent and more confined to the longitudes of the Iberian Peninsula and northwest Africa, setting up a situation that favored lighter winds aloft and slow subsidence warming of recirculating air parcels within a persistent anticyclonic flow.

REFERENCES

Alberta, T. L., S. J. Colucci, and J. C. Davenport, 1991: Rapid 500-mb cyclogenesis and anticyclogenesis. *Mon. Wea. Rev.,* **119,** 1186–1204.

Ambaum, M. H. P., B. J. Hoskins, and D. B. Stephenson, 2001: Arctic Oscillation or North Atlantic Oscillation? *J. Climate,* **14,** 3495–3507.

Anderson, D., K. I. Hodges, and B. J. Hoskins, 2003: Sensitivity of feature-based analysis methods of storm tracks to the form of background field removal. *Mon. Wea. Rev.,* **131,** 565–573.

Baldi, M., M. Pasqui, F. Cesarone, and G. DeChiara, 2005: Heat waves in the Mediterranean region: Analysis and model results. Preprints, *16th Conf. on Climate Variability and Change,* San Diego, CA, Amer. Meteor. Soc., P10.5.

Bell, G. D., and L. F. Bosart, 1989: A 15-year climatology of Northern Hemisphere 500 mb closed cyclone and anticyclone centers. *Mon. Wea. Rev.,* **117,** 2142–2163.

Benedict, J. J., S. Lee, and S. B. Feldstein, 2004: Synoptic view of the North Atlantic Oscillation. *J. Atmos. Sci.,* **61,** 121–144.

Bjerknes, J., and H. Solberg, 1922: Life cycles of cyclones and the polar front theory of atmospheric circulations. *Geophys. Publ.,* **3,** 1–18.

Black, E., M. Blackburn, G. Harrison, B. Hoskins, and J. Methven, 2004: Factors contributing to the summer 2003 European heatwave. *Weather,* **59,** 217–223.

Blackmon, M. L., 1976: A climatological spectral study of the 500 mb geopotential height of the Northern Hemisphere. *J. Atmos. Sci.,* **33,** 1607–1623.

——, J. M. Wallace, N.-C. Lau, and S. L. Mullen, 1977: An observational study of the Northern Hemisphere wintertime circulation. *J. Atmos. Sci.,* **34,** 1040–1053.

——, S. L. Mullen, and G. T. Bates, 1986: The climatology of blocking events in a perpetual January simulation of a spectral general circulation model. *J. Atmos. Sci.,* **43,** 1379–1405.

Burt, S., 2004: The August 2003 heatwave in the United Kingdom. Part I: Maximum temperatures and historical precedents. *Weather,* **59,** 199–208.

——, and P. Eden, 2004: The August 2003 heatwave in the United Kingdom. Part II: The hottest sites. *Weather,* **59,** 239–246.

Cassou, C., L. Terray, and A. S. Phillips, 2005: Tropical Atlantic influence on European heat waves. *J. Climate,* **18,** 2805–2811.

Ceppa, T. K., and S. J. Colucci, 1989: Predictability of 500 mb cyclones and anticyclones as a function of their persistence. *Mon. Wea. Rev.*, **117**, 887–900.

Chang, E. K. M., 1999: Characteristics of wave packets in the upper troposphere. Part II: Seasonal and hemispheric variations. *J. Atmos. Sci.*, **56**, 1729–1747.

——, and D. B. Yu, 1999: Characteristics of wave packets in the upper troposphere. Part I: Northern Hemisphere winter. *J. Atmos. Sci.*, **56**, 1708–1728.

Chang, F., and E. A. Smith, 2001: Hydrological and dynamical characteristics of summertime droughts over the U.S. Great Plains. *J. Climate*, **14**, 2296–2316.

Changnon, S. A., K. E. Kunkel, and B. C. Reinke, 1996: Impacts and responses to the 1995 heat wave: A call to action. *Bull. Amer. Meteor. Soc.*, **77**, 1497–1506.

Chen, P., M. P. Hoerling, and R. M. Dole, 2001: The origin of the subtropical anticyclones. *J. Atmos. Sci.*, **58**, 1827–1835.

Chen, T.-C., 2005: Maintenance of the midtropospheric North African summer circulation: Saharan high and African easterly jet. *J. Climate*, **18**, 2943–2962.

Cheng, X., and J. Wallace, 1993: Cluster analysis of the Northern Hemisphere winter 500-hPa height field: Spatial patterns. *J. Atmos. Sci.*, **50**, 2674–2696.

Colucci, S. J., 1985: Explosive cyclogenesis and large-scale circulation changes: Implications for atmospheric blocking. *J. Atmos. Sci.*, **42**, 2701–2717.

——, 2001: Planetary scale preconditioning for the onset of blocking. *J. Atmos. Sci.*, **58**, 933–942.

——, and D. P. Baumhefner, 1998: Numerical prediction of the onset of blocking: A case study with forecast ensembles. *Mon. Wea. Rev.*, **126**, 773–784.

Davis, R. E., B. P. Hayden, D. A. Gay, W. L. Phillips, and G. V. Jones, 1997: The North Atlantic subtropical anticyclone. *J. Climate*, **10**, 728–744.

Donoghue, E. R., and Coauthors, 1995: Heat-related mortality—Chicago, July 1995. *Morb. Mort. Week. Rep.*, **44**, 577–579.

Emanuel, K., 2008: Back to Norway. *Synoptic–Dynamic Meteorology and Weather Analysis and Forecasting: A Tribute to Fred Sanders, Meteor. Monogr.*, No. 55, Amer. Meteor. Soc.

Fink, A. H., T. Brücher, A. Krüger, G. C. Leckebusch, J. G. Pinto, and U. Ulbrich, 2004: The 2003 European summer heatwaves and drought–synoptic diagnosis and impacts. *Weather*, **59**, 209–216.

Galarneau, T. J., Jr., and L. F. Bosart, 2006: Ridge rollers: Mesoscale disturbances on the periphery of cutoff anticyclones. Preprints, *Symp. on Challenges of Severe Convective Storms*, Atlanta, GA, Amer. Meteor. Soc., P1.11.

Hao, W., and L. F. Bosart, 1987: A moisture budget analysis of the protracted heat wave in the southern Plains during the summer of 1980. *Wea. Forecasting*, **2**, 269–288.

Hémon, D., E. Jougla, J. Clavel, F. Laurent, S. Bellec, and G. Pavillon, 2003: Surmortalité liée á la canicule d'août 2003 en France (High mortality during the heat wave in August 2003 in France). *Bull. Epidemiol. Hebdomadaire*, **45–46**, 221–225.

Hobbs, W. H., 1945: The Greenland glacial anticyclone. *J. Meteor.*, **2**, 143–153.

Hong, S., and E. Kalnay, 2002: The 1998 Oklahoma–Texas drought: Mechanistic experiments with NCEP global and regional models. *J. Climate*, **15**, 945–963.

Hoskins, B., 1996: On the existence and strength of the summer subtropical anticyclones. *Bull. Amer. Meteor. Soc.*, **77**, 1287–1292.

——, and K. I. Hodges, 2002: New perspectives of the Northern Hemisphere winter storm tracks. *J. Atmos. Sci.*, **59**, 1041–1061.

——, and K. I. Hodges, 2005: A new perspective on Southern Hemisphere storm tracks. *J. Climate*, **18**, 4108–4129.

Iskendarian, H., 1995: A 10-year climatology of Northern Hemisphere tropical cloud plumes and their composite flow patterns. *J. Climate*, **8**, 1630–1637.

Johnson, H., R. S. Kovats, G. McGregor, J. Stedman, M. Gibbs, H.

Walton, L. Cook, and E. Black, 2005: The impact of the 2003 heat wave on mortality and hospital admissions in England. *Health Stat. Quart.*, **25**, 6–11.

Jones, P. D., and A. Moberg, 2003: Hemispheric and large-scale surface air temperature variations: An extensive revision and an update to 2001. *J. Climate*, **16**, 206–223.

Kalnay, E., and Coauthors, 1996: The NCEP/NCAR 40-Year Reanalysis Project. *Bull. Amer. Meteor. Soc.*, **77**, 437–472.

Kistler, R., and Coauthors, 2001: The NCEP–NCAR 50-Year Reanalysis: Monthly means CD-ROM and documentation. *Bull. Amer. Meteor. Soc.*, **82**, 247–267.

Klein, W. H., 1957: Principle tracks and mean frequencies of cyclones and anticyclones in the Northern Hemisphere. Research Paper 40, U.S. Weather Bureau, 60 pp.

——, 1958: The frequency of cyclones and anticyclones in relation to the mean circulation. *J. Meteor.*, **15**, 98–102.

——, and J. Winston, 1958: Geographical frequency of troughs and ridges on mean 700-mb charts. *Mon. Wea. Rev.*, **86**, 60–70.

Klein Tank, A. M. G., and G. P. Können, 2003: Trends in indices of daily temperature and precipitation extremes in Europe, 1946–99. *J. Climate*, **16**, 3665–3680.

Kunkel, K. E., S. A. Changnon, B. C. Reinke, and R. W. Arritt, 1996: The July 1995 heat wave in the Midwest: A climatic perspective and critical weather factors. *Bull. Amer. Meteor. Soc.*, **77**, 1507–1518.

Leighton, R. M., 1994: Relationship of anomalies of (anti)cyclonicity to some significant weather events over the Australian region. *Aust. Meteor. Mag.*, **43**, 255–261.

——, and R. Deslandes, 1991: Monthly anticyclonicity and cyclonicity in the Australian region: Averages for January, April, July and October. *Aust. Meteor. Mag.*, **39**, 149–154.

——, and H. Nowak, 1995: Variations in seasonal and annual anticyclonicity across the eastern Australian region during the 29-year period 1965–1993. *Aust. Meteor. Mag.*, **44**, 299–308.

Lejenäs, H., and H. Økland, 1983: Characteristics of Northern Hemisphere blocking as determined from a long time series of observational data. *Tellus*, **35A**, 350–362.

Le Marshall, J. F., G. A. M. Kelly, and D. J. Karoly, 1985: An atmospheric climatology of the southern hemisphere based on ten years of daily numerical analyses (1972–82). I: Overview. *Aust. Meteor. Mag.*, **33**, 65–85.

Liu, Q., 1994: On the definition and persistence of blocking. *Tellus*, **46A**, 286–298.

Liu, Y., G. Wu, and R. Ren, 2004: Relationship between the subtropical anticyclone and diabatic heating. *J. Climate*, **17**, 682–698.

Livezey, R. E., 1980: Weather and circulation of 1980—Climax of historical heat wave and drought over the United States. *Mon. Wea. Rev.*, **108**, 1708–1716.

——, and R. Tinker, 1996: Some meteorological, climatological, and microclimatological considerations of the severe U.S. heat wave of mid-July 1995. *Bull. Amer. Meteor. Soc.*, **77**, 2043–2054.

Lupo, A. R., and P. J. Smith, 1998: The interactions between a midlatitude blocking anticyclone and synoptic-scale cyclones that occurred during the summer season. *Mon. Wea. Rev.*, **126**, 502–515.

——, and L. F. Bosart, 1999: An analysis of a relatively rare case of continental blocking. *Quart. J. Roy. Meteor. Soc.*, **125**, 107–138.

Lyon, B., and R. M. Dole, 1995: A diagnostic comparison of the 1980 and 1988 U.S. summer heat wave–droughts. *J. Climate*, **8**, 1658–1675.

McGuirk, J. P., A. H. Thompson, and N. R. Smith, 1987: Moisture bursts over the tropical Pacific Ocean. *Mon. Wea. Rev.*, **115**, 787–798.

——, ——, and J. R. Schaeffer, 1988: An eastern Pacific tropical plume. *Mon. Wea. Rev.*, **116**, 2505–2521.

Miyasaka, T., and H. Nakamura, 2005: Structure and formation mechanisms of the Northern Hemisphere summertime subtropical highs. *J. Climate*, **18**, 5046–5065.

Mullen, S. L., 1986: The local balances of vorticity and heat for blocking anticyclones in a spectral general circulation model. *J. Atmos. Sci.,* **43,** 1406–1441.

——, 1987: Transient eddy forcing of blocking flows. *J. Atmos. Sci.,* **44,** 3–22.

——, 1989: The impact of orography on blocking frequency in a general circulation model. *J. Climate,* **2,** 1554–1560.

Namias, J., 1955: Some meteorological aspects of drought: With special reference to the summers of 1952–54 over the United States. *Mon. Wea. Rev.,* **83,** 199–205.

——, 1982: Anatomy of Great Plains protracted heat waves (especially the 1980 U.S. summer drought). *Mon. Wea. Rev.,* **110,** 824–838.

——, 1991: Spring and summer 1988 drought over the contiguous United States—Causes and prediction. *J. Climate,* **4,** 54–65.

Nashold, R., P. Remington, P. Peterson, J. Jentzen, and R. Kapella, 1996: Heat-wave-related mortality—Milwaukee, Wisconsin, July 1995. *Morb. Mort. Week. Rep.,* **45,** 505–507.

Nkemdirim, L., and L. Weber, 1999: Comparison between the droughts of the 1930s and the 1980s in the southern prairies of Canada. *J. Climate,* **12,** 2434–2450.

NOAA, 1955: Natural Disaster Survey Report: July 1995 heat wave. National Oceanic and Atmospheric Administration, 52 pp. [Available from the U. S. Dept. of Commerce, National Weather Service, 1325 East–West Highway, Silver Spring, MD, 20910.]

Palecki, M. A., S. A. Changnon, and K. E. Kunkel, 2001: The nature and impacts of the July 1999 heat wave in the Midwestern United States: Learning from the lessons of 1995. *Bull. Amer. Meteor. Soc.,* **82,** 1353–1367.

Parker, S. S., J. T. Hawes, S. J. Colucci, and B. P. Hayden, 1989: Climatology of 500 mb cyclones and anticyclones, 1950–1985. *Mon. Wea. Rev.,* **117,** 558–570.

Pelly, J. L., and B. J. Hoskins, 2003a: A new perspective on blocking. *J. Atmos. Sci.,* **60,** 743–755.

——, and ——, 2003b: How well does the ECMWF Ensemble Prediction System predict blocking? *Quart. J. Roy. Meteor. Soc.,* **129,** 1683–1702.

Petterssen, S., 1956: *Motion and Motion Systems.* Vol. 2., *Weather Analysis and Forecasting,* McGraw-Hill, 428 pp.

Pezza, A. B., and T. Ambrizzi, 2003: Variability of Southern Hemisphere cyclone and anticyclone behavior: Further analysis. *J. Climate,* **16,** 1075–1083.

Pirard, P., S. Vandentorren, M. Pascal, K. Laaidi, A. Le Tertre, S. Cassadou, and M. Ledrans, 2005: Summary of the mortality impact assessment of the 2003 heat wave in France. *Euro Surveill.,* **10** (7), 153–156. [Available online at http://www.eurosurveillance.org/em/v10n07/1007-224.asp.]

Rex, D. F., 1950a: Blocking action in the middle troposphere and its effects on regional climate. I: An aerological study of blocking action. *Tellus,* **2,** 196–211.

——, 1950b: Blocking action in the middle troposphere westerlies and its effects on regional climate. II: The climatology of blocking action. *Tellus,* **2,** 1577–1589.

Rodwell, M. J., and B. J. Hoskins, 1996: Monsoons and the dynamics of deserts. *Quart. J. Roy. Meteor. Soc.,* **122,** 1385–1404.

——, and ——, 2001: Subtropical anticyclones and summer monsoons. *J. Climate,* **14,** 3192–3211.

Roebber, P. J., 1984: Statistical analysis and updated climatology of explosive cyclones. *Mon. Wea. Rev.,* **112,** 1577–1589.

Sadler, J. C., 1967: The tropical upper tropospheric trough as a secondary source of typhoons and a primary source of trade wind disturbances. Hawaii Institute of Geophysics Rep. 67-12, 44 pp.

[Available from Dept. of Meteorology, University of Hawaii at Manoa, Honolulu, HI 96822.]

——, 1976: A role of the tropical upper tropospheric trough in early season typhoon development. *Mon. Wea. Rev.,* **104,** 1266–1278.

Sanders, F., 1988: Life history of mobile troughs in the upper westerlies. *Mon. Wea. Rev.,* **116,** 2629–2648.

——, and J. R. Gyakum, 1980: Synoptic-dynamic climatology of the "bomb." *Mon. Wea. Rev.,* **108,** 1589–1606.

——, and C. A. Davis, 1988: Patterns of thickness anomaly for explosive cyclogenesis over the west-central North Atlantic Ocean. *Mon. Wea. Rev.,* **116,** 2725–2730.

——, and S. L. Mullen, 1996: The climatology of explosive cyclogenesis in two general circulation models. *Mon. Wea. Rev.,* **124,** 1948–1954.

Schubert, S. D., and Coauthors, 1995: A multiyear assimilation with the *GEOS-1* system: Overview and results. NASA Tech. Memo. 104606, Vol. 6, 183 pp.

——, M. J. Suarez, P. J. Pegion, R. D. Koster, and J. T. Bacmeister, 2004: On the cause of the 1930s dust bowl. *Science,* **303,** 1855–1859.

Shaffrey, L. C., B. J. Hoskins, and R. Lu, 2002: The relationship between the North American summer monsoon, the Rocky Mountains and the North Pacific subtropical anticyclone in HadAM3. *Quart. J. Roy. Meteor. Soc.,* **128,** 2607–2622.

Sinclair, M. R., 1994: An objective cyclone climatology for the Southern Hemisphere. *Mon. Wea. Rev.,* **122,** 2239–2256.

——, 1995: Climatology of cyclogenesis for the Southern Hemisphere. *Mon. Wea. Rev.,* **123,** 1601–1619.

——, 1996: A climatology of anticyclones and blocking for the Southern Hemisphere. *Mon. Wea. Rev.,* **124,** 245–263.

——, J. A. Renwick, and J. W. Kidson, 1997: Low-frequency variability of Southern Hemisphere sea level pressure and weather system activity. *Mon. Wea. Rev.,* **125,** 2531–2543.

Taljaard, J. J., 1967: Development, distribution, and movement of cyclones and anticyclones in the Southern Hemisphere during the IGY. *J. Appl. Meteor.,* **6,** 973–987.

Thompson, D. W. J., and J. M. Wallace, 2000: Annular modes in the extratropical circulation. Part I: Month-to-month variability. *J. Climate,* **13,** 1000–1016.

——, ——, and G. C. Hegerl, 2000: Annular modes in the extratropical circulation. Part II: Trends. *J. Climate,* **13,** 1018–1036.

Trenberth, K. E., 1991: Storm tracks in the Southern Hemisphere. *J. Atmos. Sci.,* **48,** 2159–2178.

——, and K. C. Mo, 1985: Blocking in the Southern Hemisphere. *Mon. Wea. Rev.,* **113,** 3–21.

——, and G. W. Branstator, 1992: Issues in establishing causes of the 1988 drought over North America. *J. Climate,* **5,** 159–172.

——, ——, and P. A. Arkin, 1988: Origins of the 1988 North American drought. *Science,* **242,** 1640–1645.

——, D. P. Stepaniak, and J. M. Caron, 2002: Accuracy of atmospheric energy budgets from analyses. *J. Climate,* **15,** 3343–3360.

Wallace, J. M., G.-H. Lim, and M. L. Blackmon, 1988: Relationship between cyclones tracks, anticyclone tracks, and baroclinic waveguides. *J. Atmos. Sci.,* **45,** 439–462.

Wiedenmann, J. M., A. R. Lupo, I. I. Mokhov, and E. A. Tikhonova, 2002: The climatology of blocking anticyclones for the Northern and Southern Hemispheres: Block intensity as a diagnostic. *J. Climate,* **15,** 3459–3473.

Zishka, K. M., and P. J. Smith, 1980: The climatology of cyclones and anticyclones over North America and surrounding ocean environs for January and July, 1950–77. *Mon. Wea. Rev.,* **108,** 387–401.

Fred Sanders holding one of his famous pipes while holding court in front of the Green Building's 16th-floor map board. This photo includes a young Kerry Emanuel and was taken by Eddie Nelson, who worked for years with Fred taking care of the MIT map room, among other tasks; Eddie formally retired in 1999.

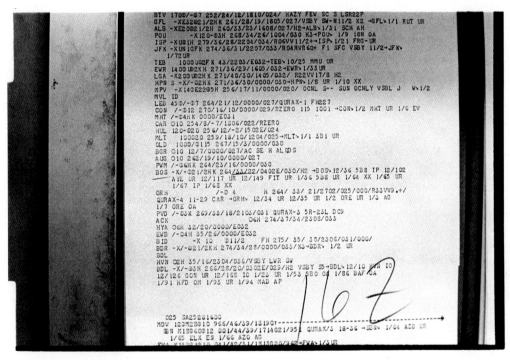

During the 1960s, 1970s, and 1980s, before the Internet, weather data were disseminated via teletype circuits and facsimile machines. This photograph (1970) shows the hourly surface observations printed on paper rolls, which were posted on a clipboard in the hallway of the 16th floor of the Green Building. Facsimile maps were hung up on clipboards to the right of the teletype rolls. Students and faculty would eagerly await the latest "hourlies," which were hung up by Eddie Nelson. Fred Sanders was frequently seen peering over the hourlies and making comments to the assembly of nearby students. Courtesy of Howie Bluestein.

The "bet board" in the map room on the 16th floor in the Green Building at MIT (1973). Fred Sanders, Eddie Nelson (who was in charge of the maps, teletype machine, and facsimile machines), and students would frequently make bets on forecast meteorological events. Some of the names on the bet board include "Olde Dad," Fred Sanders; "EN", Eddie Nelson; "DOC," Doc Willett; "H Cb B," Howie Bluestein; "MUD," Steve Mudrick; "DAN S," Dan Schwarzkopf; "Lutz," Mark Lutz; "Stern/BS," Bill Stern; "Katz/DIK," Dave Katz; "JB," John M. Brown; and "NM," Norm MacDonald. "Doc" (Hurd C.) Willett was a longtime professor at MIT, who joined its staff in 1929. He, along with Jerome Namias, pioneered the five-day forecasting at the Weather Bureau in Washington, D.C., and was coauthor, with Fred Sanders, of the textbook *Descriptive Meteorology* (Academic Press, 2nd ed., 1959). Doc, who passed away in 1992, frequently had lunch with Fred Sanders in the Walker Memorial cafeteria, across from the Green Building. Doc was an active participant in the forecasting contest and a frequent visitor to the maps on the 16th floor, as were MIT Professors Ed Lorenz and Victor Starr. For more about Doc Willett, see MIT's press release, written on the occasion of his passing, at http://web.mit.edu/newsoffice/1992/willett-0401.html. Courtesy of Howie Bluestein.

(left to right, standing) Len Keshishian, John Gyakum, Fred Sanders (cum pipe), Neil Gordon, Frank Marks, and (seated) Howie Bluestein (cum beard) in front of the Green Building at MIT, spring 1976. Courtesy of Howie Bluestein.

Eddie Nelson (who was in charge of the maps, teletype machine, and facsimile machines) in the MIT map room, 1999. Not pictured are long-time MIT staff members Annie Corrigan and Isabel Kole (both deceased) who worked tirelessly with Fred Sanders to help maintain the MIT map room (Annie) and to draft endless numbers of figures and maps for publication (Isabel). Courtesy of Howie Bluestein.

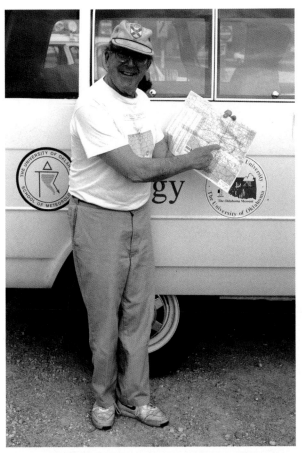

Fred Sanders showing our location on a map, before embarking on a storm chase in Norman, OK, 10 April 1994. He never saw a tornado during all the years he went storm chasing. Courtesy of Howie Bluestein.

Fred Sanders displaying Lou Wicker's (right) voodoo doll (an unsuccessful appeasement to the weather gods), before a storm chase in Norman, OK, 3 April 1981. Fred was not just a passive member of the chase crew looking out the window at the storm—he spent much time plotting and analyzing weather maps inside the car during the chase. Courtesy of Howie Bluestein.

(left to right) Howie Bluestein, Fred Sanders, and Lance Bosart at the 10th Cyclone Workshop in Val Morin, Quebec, Canada, September 1997. Courtesy of Dave Schultz.

(left to right) Dick Reed, Dave Schultz, and Fred Sanders at the 10th Cyclone Workshop in Val Morin, Quebec, Canada, September 1997. Courtesy of Dave Schultz.

front, center (left to right) Chester Newton, Dick Reed, Fred Sanders, and other participants at the 10th Cyclone Workshop in Val Morin, Quebec, Canada, September 1997. Courtesy of Dave Schultz.

Participants listening to a presentation. (left to right) John Nielsen-Gammon, Fred Sanders, Lance Bosart, Michael Morgan, Chris Davis, and Jon Martin at the 12th Cyclone Workshop in Val Morin, Quebec, Canada, September 2003. Courtesy of Fuqing Zhang.

(left to right) Fred Sanders, Lance Bosart, and Fuqing Zhang at the 12th Cyclone Workshop in Val Morin, Quebec, Canada, September 2003. Courtesy of Fuqing Zhang.

(left to right) Dick Reed, Steve Mullen, and Fred Sanders at the 10th Cyclone Workshop in Val Morin, Quebec, Canada, September 1997. Courtesy of Chuck Doswell.

Generations of Fred Sanders' students gathered at the 10th Cyclone Workshop in Val Morin, Quebec, Canada, September 1997. (left to right) back row: Ed Bracken, Steve Tracton, Dave Schultz, Michael Morgan; middle row: Joshua Watson, Steve Colucci, Howie Bluestein, Fred Sanders, Paul Roebber, Lance Bosart, John Nielsen-Gammon, Kerry Emanuel, and John Gyakum; front row: Eyad Atallah and Eric Hoffman. Courtesy of Chuck Doswell.

The University of Washington group. (left to right) Dave Schultz, Jim Steenburgh, Steve Mullen, Dick Reed, Cliff Mass, and Warren Blier at the 10th Cyclone Workshop in Val Morin, Quebec, Canada, September 1997. Courtesy of Chuck Doswell.

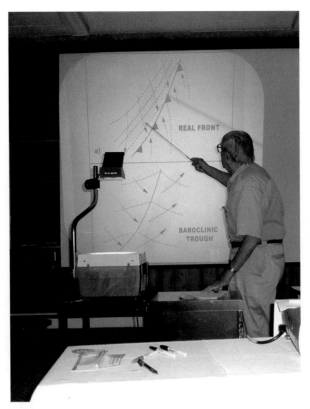

Fred Sanders, explaining the difference between a real front and a baroclinic trough (it's what's up "front" that counts) at the 10th Cyclone Workshop in Val Morin, Quebec, Canada, September 1997. Courtesy of Howie Bluestein.

(left to right) Lance Bosart, Fred Sanders, and Howie Bluestein at the 12th Cyclone Workshop in Val Morin, Quebec, Canada, September 2003. Courtesy of Howie Bluestein.

(left to right) Fred Sanders, Steve Tracton, and Howie Bluestein, along the coast in Monterey, CA, during the 11th Cyclone Workshop, August 2000. Courtesy of Steve Tracton.

(left to right) John M. Brown, Fred Sanders, Lance Bosart, and Howie Bluestein at the Eighth AMS Conference on Mesoscale Processes in Boulder, CO, June 1999. Courtesy of Howie Bluestein.

(left to right) John M. Brown, Fred Sanders, and Howie Bluestein at the 16th AMS Severe Storms Conference in Kananaskis Park, Alberta, Canada, 25 October 1990. Courtesy of Howie Bluestein.

Howie Bluestein, speaking at the Sanders Symposium banquet in Seattle, WA, January 2004. Courtesy of Dave Schultz.

Lance Bosart and Howie Bluestein (flower children) in "typical" attire at the Sanders Symposium in Seattle, WA, January 2004. Courtesy of Howie Bluestein.

Lance Bosart, speaking at the Sanders Symposium in Seattle, WA, January 2004. Courtesy of Dave Schultz.

John M. Brown, speaking at the Sanders Symposium in Seattle, WA, January 2004. Courtesy of Dave Schultz.

Bob Burpee, speaking at the Sanders Symposium in Seattle, WA, January 2004. Courtesy of Dave Schultz.

Kerry Emanuel, speaking at the Sanders Symposium in Seattle, WA, January 2004. Courtesy of Dave Schultz.

Todd ("Glickperson" as Fred Sanders used to call him) Glickman, speaking at the Sanders Symposium banquet in Seattle, WA, January 2004. Courtesy of Dave Schultz.

Eric Nyberg, Fred Sanders' longtime sailing partner, at the Sanders Symposium in Seattle, WA, January 2004. Ed regaled the audience at the Sanders Symposium banquet with tales of what it was like sailing with Fred. Courtesy of Howie Bluestein.

Neil Gordon at the Sanders Symposium in Seattle, WA, January 2004. Courtesy of Howie Bluestein.

Jim Holton, listening to a speaker, at the Sanders Symposium in Seattle, WA, January 2004. Courtesy of Howie Bluestein.

Steve Mudrick, speaking at the Sanders Symposium banquet in Seattle, WA, January 2004. Courtesy of Dave Schultz.

Mike Wallace, speaking at the Sanders Symposium in Seattle, WA, January 2004. Courtesy of Dave Schultz.

Rich Pasarelli at the Sanders Symposium banquet in Seattle, WA, January 2004. Courtesy of Dave Schultz.

Norm Phillips and John M. Brown in deep conversation (but not deep convection) at the Sanders Symposium in Seattle, WA, January 2004. Courtesy of Howie Bluestein.

Fred Sanders with Dick Reed at the Sanders Symposium banquet in Seattle, WA, January 2004. Bill Kuo looks on as Dick inspects the AMS monograph in his honor. Courtesy of Howie Bluestein.

Paul Roebber at the Sanders Symposium in Seattle, WA, January 2004. Courtesy of Dave Schultz.

Fred Sanders, attentively listening to a speaker, at the Sanders Symposium in Seattle, WA, January 2004. Courtesy of Dave Schultz.

Fred Sanders (side view), speaking "transparently" at the Sanders Symposium in Seattle, WA, January 2004. Courtesy of Howie Bluestein.

(left to right) Mel Shapiro, Howie Bluestein, and Lance Bosart at the Sanders Symposium in Seattle, WA, January 2004. Courtesy of Howie Bluestein.

Joanne and Bob Simpson, enrapt by one of the speakers at the Sanders Symposium in Seattle, WA, January 2004. Courtesy of Dave Schultz.

Louis Uccellini and Lance Bosart at the Sanders Symposium in Seattle, WA, January 2004. Courtesy of Dave Schultz.

Chris Weiss in front of his poster, at the Sanders Symposium in Seattle, WA, January 2004. Courtesy of Howie Bluestein.

Ed Zipser, Fred Sanders' dear colleague, at the Sanders Symposium in Seattle, WA, January 2004. Courtesy of Howie Bluestein.

Nancy and Fred Sanders in Tucson, Arizona, March 1999. In the late 1990s, Fred was a frequent visitor at the University of Arizona during the winter, as a guest of Steve Mullen. Courtesy of Howie Bluestein.

Fred and Nancy Sanders in Brittany, France, in the early 1990s. Courtesy of Kerry Emanuel.

Fred Sanders' children and grandchildren; rear view of the Sanders' T-shirt. Courtesy of the Sanders family.

Fred Sanders' children and grandchildren; front view of the Sanders' T-shirt on a "dog-day" afternoon. Courtesy of the Sanders family.

Group of Fred Sanders' students and colleagues at the Sanders Symposium in Seattle, WA, January 2004: (left to right) Dave Schultz, Ed Kessler, John M. Brown, Lance Bosart, Bob Burpee, Todd Glickman, Steve Tracton, John Gyakum, Fred Sanders, Eric Hoffman, Frank Colby, Louis Uccellini, Randy Dole (mostly hidden), Kerry Emanuel, and Howie Bluestein. Courtesy of Howie Bluestein.

(left) Lance Bosart and (right) Tom Galarneau in Norman, OK, in Howie Bluestein's office in May 2007, "hooked" on a radar image. Courtesy of Howie Bluestein.

Howie Bluestein and Lance Bosart, 2008, proudly holding a copy of Howie's *The Elements of Style* (Strunk and White), which Fred had given to his students in preparation for writing their theses. Courtesy of Howie Bluestein.

APPENDIX A

A Career with Fronts: Real Ones and Bogus Ones

—Frederick Sanders
Marblehead, Massachusetts
September 2004

My introduction to frontal concepts occurred during my initial training in meteorology in the U.S. Army Air Corps at the Massachusetts Institute of Technology in 1944. I had never been a boyhood weather nut. At MIT I was taught that the features on the surface weather maps, and their evolution, were to be understood in light of the Polar Front Theory. This was a comprehensive view of the general circulation of the atmosphere and of cyclones in particular as the struggle between less dense tropical air and more dense polar air meeting at a discontinuity, put forward in a series of remarkable papers by J. Bjerknes and his collaborators (e.g., Bjerknes 1918; Bjerknes and Solberg 1921, 1922) in Bergen, Norway. It was explained to us that the discontinuity of density (actually temperature) between the two air masses might appear as only a strong gradient when viewed from close up but could be regarded as a discontinuity when seen from a distance.

Besides the intellectual appeal of this comprehensive theory, the associated map analyses were aesthetically pleasing, with great sweeps of the frontal systems—blue for advancing cold air, red for advancing warm air, and purple for occlusions—the result of a cold front overtaking a warm front.

Mischief quickly arose, however, after Bjerknes and Solberg (1922) made the astute observation that whereas the thermal boundary marking the polar front is often very distinct near the center of a young cyclone, it may be indistinct in other places where the cold and warm currents have flowed nearly parallel for some time. In these cases it is suggested that indirect methods can be used to locate the front. Among these are differences in trajectory, differences in humidity, and differences in temperature between air and sea. This abandonment of the thermal definition of a front raises the question of what a definition might be and validates differences of opinion depending on which criterion is considered most relevant. Worse still, note that we are invited to consider that the polar front is present even when the observations conspire to hide it. This is simply not good

science, and the present chaotic state of frontal analysis is thus no surprise.

After separation from the U.S. Army Air Corps, I spent two years as a forecaster at the International Aviation Forecast Unit at La Guardia Field, New York. I worked with a superb group of colleagues, many fresh from the Service as I was. We faithfully adhered to the Norwegian frontal concepts, easily done over the data-sparse Atlantic Ocean. Cold and warm fronts and occluded fronts in our forecast vertical cross sections showed the ideal cloud patterns given by Bjerknes (1918), in the absence of observations to the contrary. I recall one spectacular episode in which three occlusions were closely parallel on the periphery of a deep cyclone south of Iceland. The frontal and cloud structures in a vertical section strained the imagination.

We had reservations about the Polar Front Theory. It was immediately obvious that cyclones did not develop spontaneously from infinitesimal perturbations of a preexisting front, and we found that numerous oceanic cyclones continued to deepen after occlusion, contrary to the Norwegian writ. But we kept our reservations to ourselves.

I returned to MIT in 1949 as a graduate student and found a kindred spirit in Dick Reed, who had just completed his doctoral degree. We were especially interested in conditions at upper levels, which were routinely mapped only after the early 1940s. Bjerknes and his collaborators had only scant data from upper levels: mountaintop observations and scattered soundings by kite or aircraft. It was immediately clear that cyclogenesis at the surface was a result of interaction with a preexisting upper-level trough, and that surface fronts were as much a consequence of cyclogenesis as a cause of it.

My own doctoral research was centered on examples of intense surface fronts in the central United States. In a striking case (Sanders 1955), it was clear that a surface front had by far the most intense temperature gradient in the surface boundary layer, being much weaker only about 1000 m above the surface. The reason for this weakness lay in the intense updraft at the edge of the surface front. The warm air cooled adiabatically in this ascent and the cold air warmed in adjacent descent. Thus the gradient of temperature in the rising air was weakened, except very near the ground, below the level of significant vertical motion. The practical result was that if a front was not present in the surface observations it was not present at all. This characteristic was found theoretically by Hoskins and Bretherton (1972), who further showed the crucial importance of advection of temperature near the surface by the usually neglected horizontal limb of the ageostrophic circulation.

As said, Reed and I were also interested in conditions at upper levels. Temperature inversions, potentially considered evidence for a frontal zone, were discounted as "subsidence inversions" since the relative humidity at the top of the stable layer was characteristically very low, whereas lifted warm air would be moist. On occasion, however, there was strong vertical wind shear through this layer, indicating strong horizontal temperature gradient, provided the layer sloped. I considered mainly surface phenomena while Reed concentrated on upper levels, describing typical examples and behavior (e.g., Reed 1955). He showed that the frontal temperature gradient at upper levels was a consequence of strong descent and adiabatic warming of the warm air, rather than a reversal of the Norwegian picture of warm air rising. He further found that some of the air within the upper-level zone originated in the vicinity of the polar tropopause or within the polar stratosphere. This led to the concept of a folded tropopause, which has since been much studied and discussed.

The question arises as to whether an upper-level front is continuous with a surface front, yielding a single frontal structure through the entire troposphere. I think not, because the latter is present in an environment of ascent on the east side of an upper trough, while the former occurs in a region of descent on the west side of the trough. The frontogenetical mechanisms are entirely different, so I see no reason why the zones should be continuous. Proof, of course, could be found most convincingly if both zones could be found in a single sounding. But such observations have not been presented.

My most recent research on fronts and frontal analysis was prompted by a 1991 workshop at the National Meteorological Center as summarized by Uccellini et al. 1992. This meeting was prompted by complaints from the field about the poor quality of surface map analyses, especially the frontal analyses.

My return to research on fronts was inspired by the automated analyses of surface potential temperature.

The method was developed by Eric Hoffman as a graduate student at the University at Albany. The maps were available at 3-h intervals on the Internet and made possible a serious consideration of these fields, which I did not find possible on the basis of a limited amount of manual calculation. The first result of this consideration was a study of the occurrence of intense surface baroclinic zones in the United States (Sanders and Hoffman 2002). An intense zone was arbitrarily defined as one in which the gradient of potential temperature was greater than 7°C 100 km^{-1}. Three areas of high frequency of such zones, one along the Pacific coast (most prominent during the daytime in summer), one off the Atlantic and Gulf coasts (most prominent during the night in winter), and one along the eastern slopes of the North American Cordillera (with little diurnal or seasonal variation) were seen. The zones obviously reflected horizontal variation of diabatic heating due to surface heat flux, an effect usually neglected in studies of frontogenesis. Most importantly, they corresponded only moderately to surface fronts analyzed by the National Centers for Environmental Prediction (NCEP). Analyzed fronts lay along the warm edge of an intense baroclinic zone only about half the time. Some examples in Oklahoma of abrupt temperature *increases* analyzed as cold-front passages were presented by Sanders and Kessler (1999), and the degree of correspondence between analyzed fronts and baroclinic zones was confirmed by Sanders (2005). It appears that in operational analysis, a distinction should be made between features that correspond closely to the Bjerknes model and those that do not. Research to this end is under way.

REFERENCES

Bjerknes, J., 1918: On the structure of moving cyclones. *Geof. Publ.,* **1** (2), 1–8.
——, and H. Solberg, 1921: Meteorological conditions for the formation of rain. *Geof. Publ.,* **2** (3), 60 pp.
——, and ——, 1922: Life cycle of cyclones and the Polar Front theory of atmospheric circulation. *Geof. Publ.,* **3,** 18 pp.
Reed, R. J., 1955: A study of a characteristic type of upper-level frontogenesis. *J. Meteor.,* **12,** 226–237.
Sanders, F., 1955: An investigation of the structure and dynamics of an intense surface frontal zone. *J. Meteor.,* **12,** 542–552.
——, 2005: Real front or baroclinic trough? *Wea. Forecasting,* **20,** 647–651.
——, and E. Kessler, 1999: Frontal analysis in the light of abrupt temperature changes in a shallow valley. *Mon. Wea. Rev.,* **127,** 1125–1133.
——, and E. G. Hoffman, 2002: A climatology of surface baroclinic zones. *Wea. Forecasting,* **17,** 774–782.
Uccellini, L. W., S. F. Corfidi, N. W. Junker, P. J. Kocin, and D. A. Olson, 1992: Report on the surface analysis workshop at the National Meteorological Center 25–28 March 1991. *Bull. Amer. Meteor. Soc.,* **73,** 459–471.

APPENDIX B

Fred Sanders' Students

Ph.D. Recipients

L. F. Bosart 1969
D. G. Baker 1971
R. W. Burpee 1971
M S. Tracton 1972
J. M. Brown 1975
H. B. Bluestein 1976
N. D. Gordon 1978
J. R. Gyakum 1981
R. M. Dole 1982
F. P. Colby Jr. 1983
G. Huffman 1983
F. Marks 1983
B. R. Colman 1984
J. Du 1996 (with Steve Mullen)

M.S. Recipients

H. S. Muench 1956
R. C. Copeland 1957
A. J. Wagner 1958
G. B. Brown 1959
M. T. Mulkern 1959
T. N. Carlson 1960
R. W. Barnes 1961
C. W. C. Rogers 1961
H. M. Woolf 1961
M. O. Bunde 1962
L. L. Leblanc 1962
R. C. Gammill 1963
S. Barr 1965
M. B. Lawrence 1964

H. M. Poppe 1964
K. A. Campana 1965
H. E. Headlee 1965
J. A. Neilon 1965
D. A. Olson 1965
J. Plotkin 1965
L. F. Bosart 1966
R. W. Burpee 1966
R. S. Donaldson 1966
G. F. King 1966
J. J. Owens 1966
C. S. Ahn 1967
D. G. Baker 1967
R. T. Bergh 1967
W. T. Sommers 1967
M. S. Tracton 1969
R. A. Anawalt 1971
H. B. Bluestein 1972
J. M. McNeely 1972
F. R. Williams 1972
J. F. Gaertner 1973
C. Leary 1973
M. P. Lutz 1973
H. S. Rosenblum 1973
A. J. Garrett 1974
J. E. Kester 1974
R. W. Wilcox 1974
R. J. Paine 1975
A. L. Adams 1976
J. Stokes 1976
N. J. B. Gordon 1977
J. R. Gyakum 1977

W. Jensen 1978
D. A. Miller 1978
J. P. Sheldon 1978
F. P. Colby Jr. 1979
G. S. Domm 1980
M. J. Rocha 1981
D. A. Clark 1983
R. P. Callahan 1983
P. J. Roebber 1983

B.S. Recipients[1]

P. R. Leavitt 1956
M. T. Mulkern 1957
C. W. C. Rogers 1958
D. Kennard 1961
G. Perry III 1961
J. C. Dodge 1963
C. Leary 1970
T. J. Matejka 1972
R. Edson 1975
D. Katz 1975
T. Glickman 1977

---

[1] The list of B.S. students whom Fred advised is likely incomplete. MIT students interested in meteorology and weather had the option of obtaining a B.S. degree in Geology and Geophysics, where they were advised formally by a faculty member, but informally by Fred if they worked with him on undergraduate research projects or otherwise wanted advice from him about careers and graduate school in meteorology.